Data Science and Risk Analytics in Finance and Insurance

This book presents statistics and data science methods for risk analytics in quantitative finance and insurance. Part I covers the background, financial models, and data analytical methods for market risk, credit risk, and operational risk in financial instruments, as well as models of risk premium and insolvency in insurance contracts. Part II provides an overview of machine learning (including supervised, unsupervised, and reinforcement learning), Monte Carlo simulation, and sequential analysis techniques for risk analytics. In Part III, the book offers a non-technical introduction to four key areas in financial technology: artificial intelligence, blockchain, cloud computing, and big data analytics.

Key Features:

- Provides a comprehensive and in-depth overview of data science methods for financial and insurance risks.
- Unravels bandits, Markov decision processes, reinforcement learning, and their interconnections.
- Promotes sequential surveillance and predictive analytics for abrupt changes in risk factors.
- Introduces the ABCDs of FinTech: Artificial intelligence, blockchain, cloud computing, and big data analytics.
- Includes supplements and exercises to facilitate deeper comprehension.

Tze Leung Lai is the Ray Lyman Wilbur Professor and Professor of Statistics at Stanford University. He received the COPSS Presidents' Award in 1983. He has published extensively on sequential statistical analysis and a wide range of applications in the biomedical sciences, engineering, and finance.

Haipeng Xing is a Professor of Applied Mathematics and Statistics at State University of New York, Stony Brook. His research interests include sequential statistical methods and its applications, econometrics, quantitative finance, and recursive methods in macroeconomics.

Chapman & Hall/CRC Financial Mathematics Series

Series Editors

M.A.H. Dempster
Centre for Financial Research
Department of Pure Mathematics and
Statistics
University of Cambridge, UK

Rama Cont
Mathematical Institute
University of Oxford, UK

Dilip B. Madan
Robert H. Smith School of Business
University of Maryland, USA

Robert A. Jarrow
Ronald P. & Susan E. Lynch Professor of
Investment Management aSamuel Curtis
Johnson Graduate School of Management
Cornell University

Recently Published Titles

Geometry of Derivation with Applications
Norman L. Johnson

Foundations of Quantitative Finance
Book I: Measure Spaces and Measurable Functions
Robert R. Reitano

Foundations of Quantitative Finance
Book II: Probability Spaces and Random Variables
Robert R. Reitano

Foundations of Quantitative Finance
Book III: The Integrais of Riemann, Lebesgue and (Riemann-)Stieltjes
Robert R. Reitano

Foundations of Quantitative Finance
Book IV: Distribution Functions and Expectations
Robert R. Reitano

Foundations of Quantitative Finance
Book V: General Measure and Integration Theory
Robert R. Reitano

Computational Methods in Finance, Second Edition
Ali Hirsa

Interest Rate Modeling
Theory and Practice, Third Edition
Lixin Wu

Data Science and Risk Analytics in Finance and Insurance
Tze Leung Lai and Haipeng Xing

For more information about this series please visit: https://www.routledge.com/Chapman-and-HallCRC-Financial-Mathematics-Series/book-series/CHFINANCMTH

Data Science and Risk Analytics in Finance and Insurance

Tze Leung Lai and Haipeng Xing

CRC Press

Taylor & Francis Group

Boca Raton London New York

CRC Press is an imprint of the
Taylor & Francis Group, an **informa** business

A CHAPMAN & HALL BOOK

Designed cover image: © Shutterstock, Stock Photo ID: 1640575732, Photo Contributor: NicoElNino

First edition published 2025
by CRC Press
2385 NW Executive Center Drive, Suite 320, Boca Raton FL 33431

and by CRC Press
4 Park Square, Milton Park, Abingdon, Oxon, OX14 4RN

CRC Press is an imprint of Taylor & Francis Group, LLC

ISBN: 978-1-439-83948-5 (hbk)
ISBN: 978-1-032-85049-8 (pbk)
ISBN: 978-1-315-11704-1 (ebk)

DOI: 10.1201/9781315117041

Typeset in CMR10 font
by KnowledgeWorks Global Ltd.

Publisher's note: This book has been prepared from camera-ready copy provided by the authors.

To Lai's wife and children who have supported and encouraged us throughout this journey.

To Xing's parents, wife, and children, whose love and understanding have been our greatest support.

Contents

III   Data and Risk Analytics in FinTech   401

Preface

Quantitative and data science methods have been widely used in finance and insurance to evaluate the price and risk of financial instruments and insurance contracts during the past decades. In recent years, advancements in financial technology have transformed the traditional way of financial services, transactions, and interactions. New types of data, such as social media, smartphone, and satellite data, can be collected and analyzed to improve the efficiency of financial services and enhance risk management and compliance process. Moreover, due to explosive growth of financial data, the need of data-driven decision process is also increasing considerably. These developments motivate financial researchers and practitioners to explore and exploit more data analytical methods for financial modeling and risk analysis, in addition to traditional mathematical modeling and statistical methods in finance and insurance.

The goal of this book is to provide a textbook treatment of basic and advanced data and risk analytics in finance and insurance. The content of the book is derived from parts of courses taught by the authors at their respective universities, as well as from talks and mini-courses delivered by the authors at seminars, workshops, and conferences held across the United States, Europe, and Asia over the past decade. We have designed this book to be accessible to both graduate-level students and practitioners interested in gaining a comprehensive understanding of data and risk analytics within the realms of finance and insurance. To achieve this, we offer a balanced mix of theoretical principles and practical applications. Additionally, we provide real-data examples and case studies to illustrate key concepts and demonstrate their applications in finance and insurance.

The materials of the book consists of three parts. Part I of the book, consisting of Chapters 1–4, presents the background, financial models, and data analytical methods for market risk, credit risk, and operational risk in financial instruments, as well as probabilitic and statistical models of risk premium and insolvency in insurance contracts. Part II of the book, comprising Chapters 5–8, provides an overview of machine learning (supervised learning, unsupervised learning, and reinforcement learning), Monte Carlo simulation, and sequential analysis techniques for risk analytics. In contrast to the conventional quantitative methods for risk management presented in Part I, the data analytics methods in Part II are less commonly addressed in traditional quantitative risk management but have gained significance in big data analytics and financial technology in recent years. One example is the increasing attention that reinforcement learning, one of the three key branches of machine learning, has garnered in the financial industry in recent years. However, there is a lack of concise overviews exploring its connections to bandits and Markov decision processes within the context of data analytics. To address this issue, Chapter 6 introduces multi-armed bandit problems, Markov decision processes, reinforcement learning algorithms, and their interconnections. Another example is the growing significance of sequential surveillance and predictive analytics in the financial and insurance industries. However, there appears to be a shortage of systematic introductions to this topic with a focus on financial applications. To bridge this gap, Chapter 8 provides a detailed description of statistical methods for sequential hypothesis testing and change-point detection, along with their application to financial surveillance and risk factor monitoring. Part III of the book, comprising only Chapter 9, provides a non-technical overview of four key areas in financial technology: artificial intelligence, blockchain, cloud computing, and big data analytics, which are collectively referred to as the ABCDs of FinTech.

Although this book tries to be comprehensive on widely used methods for data and risk analysis in finance and insurance, it is acknowledged that its coverage cannot be exhaustive. Consequently, to supplement the content of each chapter, additional references are provided at the end. It's worth noting that these supplementary references are selected with a certain bias, and there may be oversights in the coverage of certain subjects. We extend our apologies for any potential omissions. Additionally, each chapter is supplemented with exercises and problems to help readers reinforce their understanding of the concepts presented. As some examples in the book involve data analysis using R or Python, interested readers can access the data and R or Rython scripts for the examples used in the book via the following website.

`://www.ams.sunysb.edu/~xing/DSrisk/index.html`.

Acknowledgment

We are grateful to the audience and students who have attended our talks and mini-courses at seminars, workshops, and conferences over the past several years. We would like to extend our gratitude to the following students for helping us improve the course material and develop exercises and projects. In particular, we want to thank Ling Chen, Shaojie Deng, Pei He, Ming Jing, Yuming Kuang, Jacob Longuil, Chung Kwan Pong, Milan Shen, Yong Su, Kevin Sun, Ka Wai Tsang, Zhiyu Wang, Hongsong Yuan, Feng Zhang, Shilin Zhu. Despite our best efforts to ensure the accuracy and reliability of the information presented, we acknowledge the possibility of errors. Readers are encouraged to provide feedback and suggestions by emailing them to `haipeng.xing@stonybrook.edu`.

Department of Statistics, Stanford University   *Tze Leung Lai*
Department of Applied Mathematics and Statistics,
State University of New York, Stony Brook   *Haipeng Xing*
                                                                 March 2024

Part I

Background and Basic Analytics

Chapter 1

Risk management and regulation

Risk management in finance and insurance is the process of identifying, assessing, and controlling the various types of financial and insurance risk that financial institutions face. The aim of risk management for financial institutions is to protect their assets, maintain financial stability, and make the informed decisions about their risk exposure. Risk management is an internal process within the institution. In contrast to this, regulation refers to external rules, laws, and policies issued by the regulatory authorities to control the behaviors and operations of financial institutions and financial markets. It is designed to protect consumers' interest and maintain the stability and integrity of the financial system.

This chapter provides background on financial risk management and related regulatory frameworks in finance and insurance. Section 1.1 explains the main types of financial risk and their Basel regulatory requirements. Section 1.2 describes regulations on insurance companies. An overview of the book is provided in Section 1.3.

1.1 Financial risk and Basel regulatory framework

Financial risks can be broadly classified into several categories, namely market risk, credit risk, operational risk, liquidity risk, and legal risk. *Market risk* is the risk of loss due to changes in the value of tradable or traded assets. *Credit risk* is the risk of loss due to the failure of the counterparty to pay the promised obligation. *Operational risk* is the risk of loss caused by inadequate or failed internal processes, people and systems, or external events. *Liquidity risk* is the risk of loss arising from the inability either to meet payment obligations or to liquidate positions with little price impact. *Legal risk* is the risk of loss arising from failure to comply with statutory or regulatory obligations.

The Basel Accords are a series of three sequential banking regulation agreements that were developed by the *Basel Committee on Bank Supervision* (BCBS). They are designed to ensure that financial institutions maintain enough capital on account to meet their financial obligations and absorb unexpected losses. Basel I Accord, which was developed in 1988 by the BCBS and

DOI: 10.1201/9781315117041-1

later adopted by the central banks of the G10 countries in 1992, represents the first step toward an international regulatory framework by setting minimum capital requirements for commercial banks to manage their risk. Basel II Accord, proposed in 2001 and finalized in 2005, was a more comprehensive and risk-sensitive framework. Basel II establishes a three-pillar system of regulation: minimum capital requirement (Pillar 1), supervisory review (Pillar 2), and market discipline (Pillar 3). In response to the global financial crisis of 2007–2008, Basel III was introduced to impose more stringent capital and liquidity requirements. The Basel Accords are not binding laws for banks or national authorities, but they are generally adopted by individual countries through their national banking regulations and are key references for banking supervision.

1.1.1 Market risk

Market risk is the risk of a change in the value of a financial position or portfolio caused by price changes of underlying components. Investors who hold assets such as stocks, mutual funds, bonds, or exchange-traded funds are exposed to market risk. To mitigate market risk, strategies such as diversification and hedging can be used. However, market risk cannot be completely eliminated by these strategies; hence, market risk management is essential for banks and financial institutions.

Market risk didn't receive enough attention in the original 1988 Basel Accord, which only concerned credit risk. In April 1993, the Basel Committee published a consultative paper to incorporate market risk in the Cooke ratio and agreed to compute the capital charge for market risks with an internal model. In January 1996, the Basel Committee published the 1996 amendment to incorporate market risk into the capital accord. Due to the big impact of the 2008 Global Financial Crisis on market risk, the Basel Committee developed and published a comprehensive suite of capital rules, the *Fundamental Review of the Trading Book* (FRTB) in January 2019, which is Basel III framework for computing the minimum capital requirements for market risk and comes into effect on January 2022.

According to the BCBS, the risks subject to market risk capital requirements include but are not limited to (i) default risk, interest rate risk, credit spread risk, equity risk, foreign exchange (FX) risk, and commodities risk for trading book instruments; and (ii) FX risk and commodities risk for banking book instruments. Moreover, the Basel Committee makes the distinction between the trading book and the banking book. Instruments to be included in the trading book are subject to market risk capital requirements, while instruments to be included in the banking book are subject to credit risk capital requirements. Here, the trading book refers to positions in assets held with trading intent or for hedging other elements of the trading book, and the banking book refers to positions in assets that are expected to be held until maturity. Therefore, the first task of market risk management for banks is to define trading book assets and banking book assets.

The Basel regulatory framework

In Basel I/II/III framework, banks can choose two ways to compute the capital charge: the *standardized measurement* method and the *internal model-based* approach. The standardized measurement method has been implemented by banks at the end of the 1990s. In this method, five main risk categories are identified: interest rate risk, equity risk, currency risk, commodity risk, and price risk on options and derivatives. For each category, a capital charge is calculated to cover the general market risk and the specific risk.

In contrast to the standardized measurement method, the internal model-based approach can significantly reduce banks' capital requirements. The use of an internal model usually needs the approval of the supervisory authority. Specifically, the bank must meet certain criteria on different topics, including the risk management system, the specification of market risk factors, the properties of the internal model, the stress testing framework, the treatment of the specific risk, and the backtesting procedure. Banks can choose their own internal models, which should satisfy some quantitative criteria. For example, the value-at-risk (VaR) should be computed on a daily basis with a 99% confidence level, and the minimum holding period of the VaR is 10 trading days. Moreover, the sample period for calculating the value-at-risk is at least one year and the bank must update the data set frequently (every month at least).

Stress testing and backtesting

The goal of *stress testing* is to identify and manage situations that can cause extraordinary losses. Stress testing is required by the Basel Committee as one of the conditions to be satisfied when using internal models. However, unlike those for backtesting, the guidelines for stress testing are vague. Stress testing uses extreme scenarios that might occur, given the current economic environment and uncertainties. A simple method to create scenarios is to move key market variables sequentially by a large amount, ignoring correlations. While this method has often been used, regulators and financial institutions have recently realized that identification of extreme scenarios should be based on particular portfolios held by the institution. For instance, a highly leveraged portfolio with a long position in corporate bonds offset by a short position in treasury bonds may have sharp losses if the correlations between the rates of corporate and treasury bonds decrease significantly.

Backtesting is a statistical procedure for testing whether the projected losses under a VaR model differ significantly from the actual losses. It played a basic role in the Basel Committee's decision to allow internal VaR models for capital requirements in the 1996 amendment, which outlines the framework of backtests. Suppose a financial institution provides $100(1 - \alpha)\%$ 1-day VaR values for a total of n days. The backtest counts the number of times, denoted by τ, and the actual loss exceeds the previous day's VaR. If the internal VaR model is accurate, the probability of such exceedances is α, and therefore the expected number of exceedances is $n\alpha$. If the number of exceedances is significantly larger than $n\alpha$, the VaR model is considered to be underestimating risk and the regulator may impose penalties for allocating insufficient capital to cover the risk. Statistical significance of the number of exceedances is measured by the p-value of a binomial test that assumes the occurrences of exceedance to be independent Bernoulli trials so that the number of occurrences follows a binomial(n, α) distribution. Using the normal approximation to the binomial distribution, the test statistic is

$$Z = (\tau - n\alpha)/\sqrt{n\alpha(1 - \alpha)}.$$

The backtesting procedure adopted by the Basel Committee consists of recording the number of exceedances of the 99% daily VaR over the last year with $n = 250$ trading days. Up to four exceedances are considered acceptable (p-value > 0.83, binomial test), putting the bank in a "green" zone that has no penalty. For $5 \leq \tau \leq 9$, the bank falls into a "yellow" zone in which the penalty is up to the regulator, depending on the circumstances that cause the exceedances. For $\tau \geq 10$, the bank falls into a "red" zone and receives an automatic penalty. The penalty is in the form of an increase in the *multiplicative factor* in setting the bank's market capital charge. The general market capital charge is set at the larger of the previous day's VaR or the average VaR of the last 60 business days times the multiplicative factor determined by local regulators, subject to a minimum value of 3. The multiplicative factor is increased progressively from the yellow zone to the red zone; see Jorion (2006, pp. 148–151).

1.1.2   Credit risk

Credit risk, also known as default risk or counterparty risk, refers to the risk that an obligor fails to meet his/her financial obligations during a determined time period. Entities and individuals that are involved in lending or borrowing are exposed to credit risk. For example, investors who purchase corporate bonds or government bonds from countries with higher credit risk are exposed to credit risk, and banks that lend money to individuals and financial institutions are also exposed to credit risk. Credit risk management is primarily concerned with assessing and managing the risk of obligors not fulfilling their financial obligations due to various factors. Credit risk is the first type of risk that is required by the Basel regulatory framework.

The credit market

The credit market can be broadly categorized as the traditional debt market of loans, the bond market, and the market of credit transfer. The *traditional debt market of loans* is based on banking intermediation, in contrast to the financial market of debt securities. This loan market can be distinguished into two kinds. The first kind is related to counterparties, which can be further divided into four types: sovereign, financial, corporate, and retail. The second kind of the loan market deals with loan products, including mortgage and housing debt, consumer credit, and student loans. Contrary to loan instruments, *bonds* are debt securities that are traded in financial markets. Bonds are issued in the primary market and traded via the secondary market. The bond issuance market is dominated by central and local governments, public entities, and corporates. Government and public entities use the bond market to fund their budget deficit and projects, and large firms also use the bond market for investment and expansion.

Besides the above two markets, credit transfer instruments have been developed since the 1990s. There are two kinds of credit transfer instruments. The first is *credit securitization*, which is a process of transforming illiquid and non-tradable assets into tradable securities. The credit securities resulting from the securitization process are referred to as *asset-backed securities* (ABS), which is a bond whose coupons are derived from a collateral pool of assets. ABS can be further divided into three categories: *mortgage-backed securities* (MBS), including residential and commercial MBS; *collateralized debt obligations* (CDO), including collateralized loan obligations and collateralized bond obligations; and ABS, including auto loans, student loans, credit cards, and revolving credit. Different from credit securitization, the second kind of credit transfer instrument refers to credit derivatives that are financial instruments whose payoff explicitly depends on credit events. Typical products include *credit default swap* (CDS), *basket default swap* (BDS), and CDO. A CDS is an insurance derivative that transfers the credit risk from one party to another. A BDS is similar to a CDS, but the underlying asset is a basket of reference entities instead of one single reference entity. A CDO has the form of a multi-name CDS and has the ABS structure; see more discussion on CDS and CDO in Section 3.7.

Capital requirement

There are generally two types of credit risk. One is the *default risk* that arises when the borrower is unable to pay the principal or interests, and the other is the *downgrading risk* that concerns debt securities. Default of an obligor is usually triggered when the bank believes that the obligor is unlikely to pay its credit obligations to the banking group in full or when the obligor has past due more than 90 days on any material credit obligation to the banking group. Downgrading risk is a little tricky. If the counterparty is rated by a credit rating agency, downgrading risk can be measured by a single or multi-notch downgrade. However, one needs to realize that the credit quality of the counterparty may have already deteriorated before the downgrade announcement.

The Basel Accords set minimum capital requirements for banks to manage their risk. Basel I introduces a framework to calculate the Cooke ratio, which is a set of ratios of Tier 1 and Tier 2 capitals. The Cooke ratio was criticized for the lack of economic rationale with respect to risk weights in the calculation. Hence, Basel II develops standardized and internal rating-based approaches for the credit risk capital requirement. These approaches take into account the default probability of the counterparty and avoid regulatory arbitrage. These methods are also used in Basel III for computing the capital requirement for credit risk.

Credit risk modeling

The primary goal of credit risk modeling deals with the valuation of exposure at default, the loss given default, the probability of default, correlated default, and other quantities involving changes of the obligor' credit quality. There are, in principle, two approaches to model credit risk: structural approaches and reduced-form approaches. Chapter 3 provides a brief overview on these methods to model obligors' default risk and rating transition risk.

1.1.3   Operational risk

Operational risk is defined as the risk of loss resulting from inadequate or failed internal processes, people, and systems or from external events. However, losses resulting from strategic decisions or reputational risk are not included in the definition of operational risk. The operational risk was integrated into Basel II Accord in 2001 and is today considered as one of the major risks for the banking industry. Basel II framework describes three methods to calculate the operational risk capital requirement: the basic indicator approach, the standardized approach, and the advanced measurement approach. Since January 2022, the three approaches has been replaced by the *standardized measurement approach* (SMA) in Basel III Accord. The SMA can be considered as a mix of the three approaches of Basel II framework. The standard models for calculating the capital requirement use the loss distribution approach, which is also one of the important methods of assessing insurance risk as described in Chapter 4.

1.1.4   Liquidity risk

There are two notions of *liquidity risk*: the *market liquidity* and the *funding liquidity*. The market liquidity refers to the ease with which assets can be traded in a market, but their prices are not significantly affected. Market liquidity can be assessed by different measurements, such as volume-based measures, transaction cost measures, equilibrium price-based measures, and market-impact measures. The funding liquidity concerns the ability of banks or corporates to meet their liabilities.

In Basel III, the market liquidity is included in the market risk framework by considering five liquidity horizons: 10, 20, 40, 60, and 120 days. For funding liquidity, Basel III proposes two minimum standards: the liquidity coverage ratio and the net stable funding ratio. The funding liquidity in banks or financial institutions is usually managed by effective *asset and liability management*, which is the process of dealing with mismatch risk between assets and liabilities in a balance sheet.

The statistical models and methods introduced in this book can be applied to measure market liquidity risk. However, market liquidity models involving bid-ask spread, price and/or volume impact, and market microstructure are not covered in the book. We refer the interested reader to detailed surveys by O'Hara (1998) and Hasbrouck (2007).

1.2   Insurance, reinsurance, and regulation

Insurance and reinsurance are two important risk management concepts in finance. Insurance is a contract according to which the insured pays a premium to an insurer in return for an obligation to compensate some possible losses of the policy-holder. Insurance transfers risk from the insured to the insurer and provides the insured with some financial protection against certain risks. Insurance companies usually design and provide a wide range of insurance policies such as life insurance, health insurance, property insurance (e.g., home and auto insurance), liability insurance, and so on. These policies are usually purchased by individuals, businesses, and other entities to protect themselves or their assets from unexpected events such as accidents, illnesses, property damage, or liability claims. When a covered loss occurs, the policyholder can file a claim with the insurance company. The insurance company then assess the claim and pays out compensation according to the insurance policy.

Reinsurance is also a financial mechanism that transfers risks from one party to another. However, reinsurance is a separate and distinct industry from insurance. Reinsurance companies provide financial protection to primary insurers (i.e., insurance companies) and are not directly involved with individual policyholders. Reinsurance allows primary insurers to transfer a portion of their risk to reinsurers and shares some risk with primary insurers. This helps primary insurers manage their exposure to large and unexpected losses. There are different types of reinsurance arrangements. For example, proportional reinsurance allows the reinsurer to share premiums and losses with the ceding insurer in proportion to their agreed-upon percentage, and excess of loss reinsurance protects the ceding insurer from catastrophic losses by covering losses that exceed a predetermined amount. Primary insurance companies purchase reinsurance to protect their financial stability and ensure they have the resources to meet their policyholders' claims in catastrophe events.

Compared to insurance through which individuals and businesses protect themselves against specific risks by purchasing coverage directly from an insurance company, reinsurance is a secondary market where primary insurers transfer a portion of their risks to other specialized companies. Reinsurance helps primary insurers manage their risk exposure and maintain their financial stability. It plays a critical role in maintaining the overall stability and resilience of the insurance industry as a whole.

In contrast to the way that Basel III provides a regulatory framework for banks, there isn't a single global regulatory framework for insurance companies. Instead, there are several international organizations and agreements that aim to harmonize insurance regulation and promote international cooperation, e.g., the *International Association of Insurance Supervisors* (IAIS). While international organizations like the IAIS work to develop common principles and standards for insurance regulation, countries or regions have their own national or regional authorities that set rules and regulations to implement and enforce regulatory standards.

1.2.1 State vs. federal insurance regulation

The regulatory framework for the insurance industry in the United States is a combination of state and federal regulations. Insurance regulation in the United States is primarily conducted at the state level. Each state has its own insurance departments that are responsible for supervising insurance companies and their products within its jurisdiction. State insurance commissioners are responsible for licensing insurance companies, setting premium rates, reviewing policy forms, and ensuring the financial solvency of insurers. State insurance laws and regulations may vary from state to state, but they generally follow the model regulations and guidelines developed by the *National Association of Insurance Commissioners* (NAIC), which is a non-profit organization that facilitates cooperation among state insurance regulators.

As the insurance industry is primarily regulated at the state level, federal laws and agencies also play a critical role in insurance regulation, particularly for certain types of insurance and issues. Some federal agencies with insurance oversight responsibilities include the *Federal Insurance Office* (FIO), the Federal Reserve, and the *National Flood Insurance Program* (NFIP). As a part of the U.S. Department of the Treasury, the FIO monitors the insurance industry and provides recommendations on insurance-related matters to the federal government. But it does not have the authority to directly regulate the insurance industry or insurance companies. The Federal Reserve supervises insurance companies that are designated as *systemically important financial institutions* (SIFIs) and falls under its jurisdiction. It conducts consolidated supervision for SIFI insurers to ensure their financial stability. The NFIP is operated by the Federal Emergency Management Agency and provides flood insurance in the United States. It is subject to federal regulation.

Besides the regulatory activities at the state and the federal levels, some insurance regulatory activities are conducted through interstate compacts. These agreements allow several states to collaborate on certain regulatory functions, such as product approvals and licensing. The most well-known example is the Interstate Insurance Product Regulation Compact, which streamlines the approval process for certain insurance products.

In summary, the regulatory framework for the insurance industry in the United States is a complex network of state and federal regulations, in which state insurance departments are the primary regulators. Federal laws and agencies play a critical role in some specific areas, e.g., the oversight of systemically important insurers. State regulations, though in general are consistent across states due to the influence of the NAIC, can still have variations and nuances based on state-specific laws and regulations.

1.2.2 Solvency II

Solvency II is a regulatory framework for insurance and reinsurance companies operating in the *European Union* (EU). It was introduced to create a consistent and comprehensive set of rules and requirements for the insurance industry within the EU and its primary goal is to enhance policyholder protection and ensure the financial stability of insurance companies. Solvency II became effective on January 1, 2016, replacing the earlier Solvency I framework.

Key components in Solvency II are three pillars. Pillar I sets the quantitative requirements. In these requirements, the solvency capital requirement represents the amount of capital needed to ensure that an insurer is able to meet its obligations over a one-year period with a 99.5% confidence level, and the minimum capital requirement is a lower threshold designed to trigger regulatory action if capital falls below this level. Pillar II sets the qualitative requirements and focuses on the supervisory review process. It allows regulators to assess whether insurers' own risk assessments are in line with regulatory expectations. Pillar III aims to enhance transparency and disclosure. It requires insurers to disclose relevant information about their financial condition, risk profile, and governance practices to market participants, regulators, and other stakeholders.

Besides Pillars I, II, and III, Solvency II also requires insurers to perform a comprehensive assessment themselves to assess their overall solvency position by taking into account their specific risk profile, business strategy, and economic environment. Solvency II also introduces a risk-based supervision approach, through which supervisory authorities can assess insurers' solvency and risk management practices based on the nature, scale, and complexity of their operations.

Solvency II is designed to create a uniform regulatory framework for insurers across member states and to harmonize insurance regulation in the EU. It enhances consumer protection, improves market stability, reduces regulatory arbitrage, and promotes prudent risk management and capital adequacy within the EU insurance industry. Insurance and reinsurance companies subject to Solvency II must comply with its requirements, which include reporting regularly to regulatory authorities and maintaining adequate levels of capital to cover their risks. Non-compliance with Solvency II regulations can result in various consequences, which include fines, restrictions on business operations, or even the revocation of an insurer's license.

1.3   Overview of chapters

The materials of the book consist of three parts. Part I of the book, consisting of Chapters 2–4, presents the background and basic analytics for market risk, credit risk, operational risk in financial instruments and risk premium in insurance contracts.

Chapter 2 describes the basic concepts and methods in risk management, mainly for market risk. It starts with the representation of the loss function of a financial portfolio as a function of risk factors, and discusses its linear and linear-quadratic approximation and several risk measures such as value-at-risk (VaR) and expected shortfall (ES) for the loss. The valuation of VaR and ES is presented when changes of risk factors are assumed cross-sectionally and temporally independent and normally distributed. The rest of Chapter 2 presents statistical methods that characterize distributions of risk factor changes when the above assumptions on risk factor changes are violated. In particular, Section 2.2 assumes temporal independence and presents several statistical models and methods for multivariate risk factor changes, which include dimension reduction methods that reduce the dimensionality of the data, copula methods that map univariate marginal distributions to a joint distribution, and rank-based dependence measures that describe the concordance of random variables or their tails. Section 2.3 relaxes the temporal independence constraint on risk factor changes and describes univariate and multivariate linear time series models and univariate volatility and multivariate covariance models for risk factor changes and their volatilities or covariance matrices. Section 2.4 drops the normal distribution assumption and introduces extreme value theory and quantile regression to model the non-Gaussian tail behavior of risk factor changes.

Chapter 3 provides an overview of financial derivatives and their pricing theory and highlights analytical methods for credit risk. It starts with a review of the classic Black-Scholes-Merton option pricing theory in Section 3.1 and several stochastic volatility models in Section 3.2. Then a review of stochastic interest models such as models of short rate and forward rate dynamics is followed in Section 3.3. To better introduce the reduced-form or the intensity models for stochastic hazard rates, Section 3.4 presents basic concepts in survival analysis and introduces several types of regression models that link individuals' hazard rates with individuals' covariates. These regression models include proportional hazard rate models, additive hazard models, and accelerated failure time models. Following these survival models, Sections 3.5 and 3.6 present commonly used finance models for credit default risk and credit rating migrations. Section 3.5 introduces two fundamental modeling approaches, namely structural models and intensity models, designed to model credit default risk dynamics. Section 3.6 describes discrete- and continuous-time Markovian models, providing insights into the intricate dynamics of credit rating migrations. In addition to this, Section 3.6 also explains piecewise time-homogenenous Markov chain models that accommodate structural breaks within the dynamics of credit rating migrations, concepts of recurrent events and competing risk in survival analysis, and modulated Markov and semi-Markov models that incorporate firm-specific covariates into the dynamics of firms' rating transitions. Sections 3.7 and 3.8 extend models and methods on credit risk from corporates' bonds to credit derivatives. In particular, Section 3.7 introduces credit default swaps and credit default obligations, which are typical single-name and multi-name credit derivatives, respectively. To characterize the joint distribution and dependence of a set of credit events, Section 3.8 introduces concepts of static and dynamic frailty variables and presents mixture models and copula approaches to model credit risk in which credit events are assumed to be conditionally independent given a set of unobserved common factors or frailties.

As many of the concepts and techniques introduced in Chapters 2 and 3 originate from actuarial science and are related to actuarial methodology, Chapter 4 introduces fundamental concepts relevant to insurance contracts and quantitative models for insurance risk. It specifically focuses on the valuation of the risk premium and the calculation of ruin probability in insurance contracts. The methods explained in this chapter can also be applied to operational risk management in financial institutions. Specifically, Section 4.1 provides an overview of utility and premium principles in insurance. Section 4.2 introduces individual and collective risk models for the distribution of aggregate claims. Compound distributions, De Pril's recursions, and Panjer recursions are introduced to calculate aggregate claim distributions. In contrast to the collective risk models that consider the aggregate amount of claims paid out in a single time period, Section 4.3 introduces the ruin theory that studies the evolution of an insurance fund over time. Defining a stochastic surplus process in continuous time, the ruin probability, its Lungberg's exponential bound, and the distribution of maximal aggregate loss are presented. As models in Sections 4.1–4.3 use probabilistic approaches, Sections 4.4 and 4.5 present the credibility theory and its extensions, which are statistical models of estimating premiums for a group of insurance contracts based on historical data. Since the credibility theory is a Bayesian inference method, Section 4.4 first explains basic concepts on Bayes inference and empirical Bayes, and then presents several models for credibility theory, including the Bühlmann-Straub model, linear mixed models, generalized linear models, and generalized linear mixed models. Section 4.5 further extends the credibility theory to an evolutionary credibility theory which can evaluate the individual premium recursively.

Part II of the book, consisting of Chapters 5–8, provides an overview of machine learning, Monte Carlo simulation, and sequential analysis methods for risk analytics. Compared to conventional quantitative methods for risk management presented in Part I, data analytics methods in Part II are not often covered in traditional quantitative risk management, but have become increasingly important in big data analytics and financial technology during recent years. Specifically, machine learning algorithms provide data-driven decisions based on vast amount of finance data with complex structures, Monte Carlo methods can estimate financial risk efficiently via simulations, and sequential analysis methods provide on-line assessment on the stability of interested global or specific risk factors. Specific focus of Chapters 5–8 is outlined as follows.

Chapter 5 introduces supervised and unsupervised learning methods in machine learning. Specifically, Section 5.1 starts with various types of linear regression and regularization and describes spline, basis expansion, and the cross-validation method. Section 5.2 presents commonly used linear and nonlinear classification methods, which include logistic and ordered probit models, discriminant analysis, classification and regression trees, and nearest neighbor methods. Section 5.3 extends linear models for classification to produce nonlinear boundaries for classication in an enlarged and transformed feature space. Concepts of optimal separating hyperplane, kernel function and support vector expansion, soft margin and introduced, and the dual problem, the risk minimization problem, and support vector regression are also explained. The supervised learning methods in Sections 5.1–5.3 rely on individual learners, to achieve better performance, Section 5.4 introduces ensemble learning which aims for better learning performance by combining multiple individual learners and presents three types of ensemble learning methods: generalized additive models, boosting, and bagging, and random forest algorithms. In addition to the learning methods in Sections 5.1–5.4, Section 5.5 presents neural networks that are popular machine learning techniques mimicking the mechanism of learning in biological organisms. The section first introduces feed-forward neural networks and the back-propagation algorithm, and then presents the architecture of several commonly used deep learning neural networks, such as radial basis function networks, convolutional neural networks, recurrent neural networks, generative adversarial networks, and Boltzmann machines. In contrast to supervised learning methods, unsupervised learning methods aim to explore and discover hidden and interesting patterns in unlabeled datasets. Section 5.6 introduces one such method, clustering, which partitions a data set into disjoint subsets based on user-specified performance and distance measures.

Besides supervised and unsupervised learning, the third type of machine learning is the reinforcement learning, which enables an intelligent agent to learn the optimal behavior by balancing exploration and exploitation in environment so as to maximize the cumulative reward. Chapter 6 introduces multi-armed bandit problems, Markov decision processes, and reinforcement learning algorithms. Section 6.1 presents two kinds of approaches to formulate the multi-armed bandit problems: the Bayesian and the frequentist. The optimality of the Gittins index policy for the Bayesian multi-armed bandit is explained, and for the frequentist multi-armed bandit, lower bounds on regret and Lai-Robbins policy that achieves asymptotic optimality are described. As theoretical underpinnings of reinforcement learning, Sections 6.2–6.4 introduce finite- and infinite-horizon *Markov decision processes* (MDP) and partially observed MDP (POMDP). In particular, Section 6.2 introduces finite-horizon MDP and the Bellman equation for the value function of the optimization problem. The backward induction for the value function and the forward induction for the maximal expected discounted reward are also introduced. To illustrate the theory, Section 6.2 also presents a finite-horizon linear-quadratic regulator problem and its solution. Section 6.3 extends the MDP from finite-horizon to infinite-horizon with discounting. In such a case, transition kernels and rewards don't depend on time, and hence the state- and action-value functions don't depend on time either, Bellman equation becomes a fixed point equation and the optimal policy becomes stationary. The MDP in Sections 6.2 and 6.3 assume that the state process is completely observed, whereas in many applications, the decision maker can only have partial information about the state process. Section 6.4 considers this scenario and introduces the partially observed MDP. It derives a filter equation that accounts for the conditional distribution of the unobserved state given the up-to-date information of observable states and actions. For illustration purpose, Section 6.4 also introduces a partially observable linear-quadratic regulator problem and derives the Kalman filter and the certainty-equivalence principle for the problem. The connection between POMDP and MDP is also established by formulating the POMDP as a standard MDP. Provided the overview on the MDP in Sections 6.2–6.4, Sections 6.5–6.6 present two types of reinforcement learning methods that deal with how agents learn to make decisions in an environment based on exploration-exploitation trade-offs. The reinforcement learning methods can deal with the case that the agent doesn't have a complete knowledge of the environment or the case that the agent has access to a model of the environment but exact solutions of the problem are challenging to obtain. Specifically, Section 6.5 presents value function based reinforcement learning, such as temporal difference learning, Q-learning, and SARSA algorithms, and Section 6.6 describes policy-based reinforcement learning approaches, such as policy gradient methods and actor-critic algorithms.

In addition to machine learning methods, another important method of determining financial risk is Monte Carlo simulation. Chapter 7 provides an overview of Monte Carlo methods with emphasis on rare event simulation. In particular, Section 7.1 describes methods of simulating uni- and multi-variate random variables, and Section 7.2 explains how to simulate commonly used stochastic processes in finance and insurance, including discrete- and continuous-time Markov chains, compound Poisson processes, stochastic differential equations, jump-diffusion processes, and Lévy processes. To reduce the standard error of the estimate and improve the efficiency of the estimate that are based on Monte Carlo methods, Section 7.3 presents variance reduction methods. In many applications, generating independent and identically distributed samples from the target distribution is infeasible, in such cases, dependent samples can be drawn as long as the sample mean converges to the true mean. Such sample methods are called *Markov chain Monte Carlo* (MCMC) simulations. Section 7.4 introduces three main MCMC samplers: the Gibbs sampler, the Metropolis-Hastings sampler, and the reversible jump sampler. As many problems of evaluating financial risk involves rare-event simulation, Sections 7.5 and 7.6 introduce importance sampling and sequential Monte Carlo methods that simulate rare events much more efficiently than direct Monte Carlo. Specifically, Section 7.5 describes an importance sampler for light and heavy-tailed distributions, which may or may not depend on the state, and Section 7.6 presents a generic form of sequential importance sampling with resampling methods and their applications to simulating dynamic portfolio risks.

Instead of evaluating financial risk, an important concern in risk management and regulation is how to online monitor the instability of global or interested risk factors for compliance and regulation. Chapter 8 introduces sequential analysis methods which provide a systematic treatment on online surveillance, estimation, and prediction for financial portfolios and systems. Section 8.1 explains two kinds of risk management processes that involve surveillance analytics, one is the active risk management of financial portfolios and the other is the surveillance of instability of financial systems. Sequential analysis methods are described afterwards. In particular, Section 8.2 presents sequential hypothesis testing methods in which the sample size is not fixed in advance. The sequential probability ratio test, multiple simple hypothesis test, and some simple composite hypothesis testing problems are introduced. To deal with the problem of online surveillance of risk factors, frequentist and Bayesian methods for quickest change-point detection are presented. Section 8.3 describes the frequestist methods, which consist of commonly used control charts, the minimax formulation and optimality of the CUSUM algorithm, and generalized likelihood ratio rules. The Bayeisan methods for change-point detection are introduced in Section 8.4. It first explains the Shiryaev's procedure on change-point detection and its optimal property, and then detection procedures for composite hypothesis. Section 8.4 also presents a hidden Markov filtering approach for multiple change-point detection and estimation, which allows explicit and recursive calculation of filters and smoothers and hence greatly reduces the computational complexity for multiple change-point problems. As applications of the presented sequential analysis methods, Section

8.5 discusses two applications of multiple change-point problems in financial modeling, one is the surveillance of the instability in credit rating transition dynamics and the other is to introduce a class of change-point time series models that help us understand components of asset return dynamics. Section 8.6 considers the prediction problem for the instability of risk factors and introduces a generic model for change-point prediction based on the hidden Markov filtering approach in Section 8.4.

Part III of the book, consisting only of Chapter 9, links the data and risk analytical methods to financial technology. There are four key areas in financial technology, artificial intelligence, blockchain, cloud computing, big data and data analytics, which are called the *ABCDs of FinTech*. These four key areas are briefly described with non-technical words in Sections 9.1–9.4, respectively. For each subject, we present a brief history of the technology and explain its application to financial services and its impact on the usage of data and risk analytics.

Supplements

Hull (2023) gives a clear and succinct introduction on various aspects of financial risk, including market risk, credit risk, operational risk, liquidity risk and model risk, and financial institutions regulations. Roncalli (2020) provides a comprehensive review on financial regulation and various regulation standards concerning market risk, credit risk, operational risk, and liquidity risk. It explains various practical applications of risk management in the financial sector and provides a discussion of the mathematical and statistical tools used in risk management.

Vaughan and Vaughan (2014) gives a thorough introduction to the field of insurance. It emphasizes on the insurance product and the use of insurance within the risk management framework. Rejda and McNamara (2020) provides an in-depth overview of various risk concepts and themes. It covers the basics of risk and insurance, and traditional and enterprise risk management.

Chapter 2

Basic concepts and methods in risk management

This chapter presents several risk measures and some basic statistical methods for them such as time series models, multivariate approaches, and extreme value theory. In particular, Section 2.1 first explains the idea of the loss distribution and its first- and second-order approximations, and introduces several risk measures that are commonly used in risk management, including value-at-risk, tail conditional median, expected shortfall, and general coherent risk measures. From the statistical point of view, value-at-risk is the quantile of the loss, and expected shortfall is the tail conditional mean. Hence, in order to evaluate the risk measures, one needs to understand the distribution of the loss function for assets or portfolios.

When the loss distribution can be represented as independent univariate normal random variables which depend on a single risk factor, the valuation of the aforementioned risk measures is usually not difficult. Some complication arises when independence or normality assumptions are violated. Sections 2.2–2.4 present methods to address these concerns. In particular, Section 2.2 assumes the loss distribution is dependent upon several risk factors and introduces multivariate approaches to model the joint behavior of risk factors. The section first relaxes the univariate normality assumption and introduces multivariate normal and normal mixture distributions. Then two dimension reductions methods, principal component analysis and factor analysis, are presented. To characterize the joint distribution of several risk factors via marginal distributions, the section also introduces copula methods. Moreover, since normality assumptions for risk factors are usually violated, the linear correlation concept becomes inappropriate to account for the dependence of risk factors. To account for this, dependence measures are introduced to describe the concordance of random variables and/or their tail behavior.

The joint distributions of risk factors in Section 2.2 are assumed independent over time, this is usually not true in practice. To address this issue, Section 2.3 describes a set of time series models to account for the temporal dependence of risk factors. Specifically, Section 2.3 first presents the concept of weakly stationary series and univariate ARMA models for temporal behavior of conditional mean of individual risk factors. To account for the conditional heteroskedasticity of individual risk factors, two types of volatility models, ARCH/GARCH models and stochastic volatility models, are also introduced.

DOI: 10.1201/9781315117041-2

These univariate time series are then extended to multivariate ARMA and multivariate volatility models so that the temporal dependence of several risk factors can be explained.

The models on distributions of risk factors in Sections 2.2 and 2.3 do not deal with distributions of extreme values of risk factors or characterize the dependence of quantiles of random variables on explanatory variables. To fill in this gap, extreme value theory and quantile regression are introduced in Section 2.4. Two types of commonly used extreme value distributions are introduced, one is based on the block maxima and the other is based on threshold exceedance. Besides distributions of extreme values, Section 2.4 also introduces quantile regression to estimate the conditional quantiles of response variables (or risk factors) across values of explanatory variables.

2.1 Loss distributions and risk measures

This section first defines the loss distribution for a portfolio of assets and liabilities, its first- and second-order approximations, and then describes several commonly used risk measures.

2.1.1 Loss distribution

Consider a portfolio of assets and liabilities which could be a collection of stocks, bonds, or derivatives. Denote by V_t the value of the portfolio at time t, which can be determined from the information up to time t. Consider the value change of the portfolio in the time period $[t, t + \Delta t]$. For simplicity, assume that the portfolio composition remains invariant and there are no intermediate payments of income during the period $[t, t + \Delta t]$. Since the value of the portfolio at $t + \Delta t$ is $V_{t+\Delta t}$, the change in value of the portfolio is $\Delta V_t := V_{t+\Delta t} - V_t$ and the loss of holding the portfolio during the period $[t, t + \Delta t]$ is $L_t^{\Delta t} := -\Delta V_t$. Then $L_t^{\Delta t}$ can be viewed as a random variable at time t, its distribution is called the *loss distribution*.

Suppose that the value of the portfolio at time t can be written as a function of time t and a d-dimensional state variable $\mathbf{Z}_t = (Z_{t,1}, \ldots, Z_{t,d})'$ of *risk factors*,

$$V_t = f(t, \mathbf{Z}_t) \tag{2.1}$$

for some measurable function $f : \mathbb{R}^+ \times \mathbb{R}^d \to \mathbb{R}$. Let $\mathbf{X}_{t+\Delta t} = \mathbf{Z}_{t+\Delta t} - \mathbf{Z}_t = (X_{t+\Delta t,1}, \ldots, X_{t+\Delta t,d})'$ be the risk-factor changes over the time period $[t, t + \Delta t]$. Then the portfolio loss is expressed as

$$L_t^{\Delta t} = -f(t + \Delta t, \mathbf{Z}_t + \mathbf{X}_{t+\Delta t}) + f(t, \mathbf{Z}_t). \tag{2.2}$$

Linearized loss

Under the assumption that f is differentiable, the first-order approximation of (2.2) represents the loss as a linear function of the risk-factor changes and is expressed as

$$L_t^{\Delta t} \approx L_{t,\text{linear}}^{\Delta t} := -\frac{\partial f}{\partial t}(t, \mathbf{Z}_t)\Delta t + \sum_{i=1}^{d} \frac{\partial f}{\partial Z_{t,i}}(t, \mathbf{Z}_t)X_{t+\Delta t,i}. \qquad (2.3)$$

The following two examples show the linearized loss for a stock portfolio and a European call option.

Example 2.1 (Stock portfolio) *Consider a portfolio of d stocks. At each time t, denote by $S_{t,i}$ the price of the ith stock ($i = 1, \ldots, d$) and n_i the number of shares of stock i in the portfolio. Then the value of the portfolio is*

$$V_t = \sum_{i=1}^{d} n_i S_{t,i} = \sum_{i=1}^{d} n_i e^{Z_{t,i}},$$

where $Z_{t,i} = \ln S_{t,i}$ is the logarithmic price of the ith stock. Assume that the number of stocks remains same and there are no intermediate payments of income during the period $[t, t + \Delta t]$. The change in value of the portfolio is

$$\Delta V_t = \sum_{i=1}^{d} n_i\left(e^{Z_{t+\Delta t,i}} - e^{Z_{t,i}}\right) = \sum_{i=1}^{d} n_i S_{t,i}\left(e^{X_{t+\Delta t,i}} - 1\right),$$

where $X_{t+\Delta t,i} = \ln(S_{t+\Delta t,i}/S_{t,i})$ is the logarithmic return of the ith stock during the period $[t, t + \Delta t]$. Let $w_{t,i} = n_i S_{t,i}/V_t$ be the weight of the ith stock in the portfolio or the proportion of the portfolio value invested in the ith stock at time t. The loss of holding the portfolio during the period $[t, t+\Delta t]$ is

$$L_t^{\Delta t} = -\Delta V_t = -V_t \sum_{i=1}^{d} w_{i,t}\left(e^{X_{t+\Delta t,i}} - 1\right).$$

Since $e^x - 1 \approx x$ when $x \approx 0$, the first-order approximation of the loss function is a linear function of the risk-factor changes, $L_{t,\text{linear}}^{\Delta t} = -V_t \sum_{i=1}^{d} w_{i,t} X_{t+\Delta t,i}$. □

Example 2.2 (European call option) *Consider a standard European call option on a non-dividend paying stock with maturity time T and exercise price K. Using the Black-Scholes option pricing formula (see Section 3.1), the value of a call option on a stock with price S_t at time t is expressed as*

$$C^{BS}(t, S_t; r, \sigma, K, T) = S_t \Phi(d_{1,t}) - Ke^{-r(T-t)}\Phi(d_{2,t}),$$

where Φ denotes the standard Gaussian distribution function, r is the continuously compounded risk-free interest rate, σ denotes the volatility of the underlying stock, and

$$d_{1,t} = \frac{\log(S_t/K) + (r + \frac{1}{2}\sigma^2)(T-t)}{\sigma\sqrt{T-t}}, \qquad d_{2,t} = d_{1,t} - \sigma(T-t).$$

Denote by $\mathbf{Z}_t = (\ln S_t, r_t, \sigma_t)'$ the vector of risk factors. The loss of holding the European call option during the period $[t, t + \Delta t]$ is

$$L_t^{\Delta t} = -C^{BS}(t + \Delta t, S_{t+\Delta t}; r_{t+\Delta}, \sigma_{t+\Delta}, K, T) + C^{BS}(t, S_t; r_t, \sigma_t, K, T),$$

and its first-order approximation is expressed as

$$L_{t,linear}^{\Delta t} = -\frac{\partial C^{BS}}{\partial t}\Delta t - \frac{\partial C^{BS}}{\partial S}S_t X_{t+\Delta t,1} - \frac{\partial C^{BS}}{\partial r}X_{t+\Delta t,2} - \frac{\partial C^{BS}}{\partial \sigma}X_{t+\Delta t,3},$$

where $X_{t+\Delta t,1} = \ln(S_{t+\Delta t}/S_t)$, $X_{t+\Delta t,2} = r_{t+\Delta t} - r_t$, and $X_{t+\Delta t,3} = \sigma_{t+\Delta t} - \sigma_t$ are three components of the risk-factor changes. The derivatives $\partial C^{BS}/\partial S$, $\partial C^{BS}/\partial r$, $\partial C^{BS}/\partial \sigma$, and $\partial C^{BS}/\partial t$ are called the delta, rho, vega, and theta of the option, respectively. These derivatives are often referred to as the Greeks; see Section 3.1. □

Delta-gamma approximation

For portfolios containing derivatives, portfolio value is often a nonlinear function of the risk factors and the linearized loss can be a very poor approximation to the true loss. Hence a second-order, or the delta-gamma approximation, is often used. In particular, given the value of the portfolio (2.2), the second-order approximation to the loss function $L_t^{\Delta t}$ is given by

$$L_{t,quad}^{\Delta t} := -\frac{\partial f}{\partial t}(t, \mathbf{Z}_t)\Delta t - \left(\frac{\partial f}{\partial \mathbf{Z}}(t, \mathbf{Z}_t)\right)' \mathbf{X}_{t+\Delta t} - \frac{1}{2}\mathbf{X}_{t+\Delta t}'\left(\frac{\partial^2 f}{\partial \mathbf{Z}\partial \mathbf{Z}'}(t, \mathbf{Z}_t)\right)\mathbf{X}_{t+\Delta t},$$
$$(2.4)$$

in which terms of order $(\Delta t)^2$ and $\mathbf{X}_t\Delta t$ are dropped from the second-order Taylor expansion since these terms tend to zero faster than Δt.

Example 2.3 (European call option) *Consider the European call option in Example 2.2 and the vector of risk factors $\mathbf{Z}_t = (\ln S_t, r_t, \sigma_t)'$ associated with the option. Given the loss function of holding the option during the period $[t, t + \Delta t]$,*

$$L_t^{\Delta t} = -C^{BS}(t + \Delta t, S_{t+\Delta t}; r_{t+\Delta}, \sigma_{t+\Delta}, K, T) + C^{BS}(t, S_t; r_t, \sigma_t, K, T),$$

its delta-gamma approximation is expressed as

$$L_{t,quad}^{\Delta t} = -\frac{\partial C^{BS}}{\partial t}\Delta t - \left(\frac{\partial C^{BS}}{\partial \mathbf{Z}}\right)' \mathbf{X}_{t+\Delta t} - \frac{1}{2}\mathbf{X}_{t+\Delta t}'\left(\frac{\partial^2 C^{BS}}{\partial \mathbf{Z}\partial \mathbf{Z}'}\right)\mathbf{X}_{t+\Delta t},$$

where $\mathbf{X}_{t+\Delta t} = (\ln(S_{t+\Delta t}/S_t), r_{t+\Delta t} - r_t, \sigma_{t+\Delta t} - \sigma_t)'$ *is the vector of risk-factor changes. When the change of interest rates are omitted (or* $r_{t+\Delta t} - r_t \approx 0$*), the delta-gamma approximation reduces to*

$$L_{t,quad}^{\Delta t} \approx -\frac{\partial C^{BS}}{\partial t}\Delta t - \frac{\partial C^{BS}}{\partial S}S_t X_{t+\Delta t,1} - \frac{\partial C^{BS}}{\partial \sigma}X_{t+\Delta t,3} - \frac{1}{2}\frac{\partial^2 C^{BS}}{\partial S^2}S_t^2 X_{t+\Delta t,1}^2$$
$$ - \frac{\partial^2 C^{BS}}{\partial S \partial \sigma}S_t X_{t+\Delta t,1}X_{t+\Delta t,3} - \frac{1}{2}\frac{\partial^2 C^{BS}}{\partial \sigma^2}X_{t+\Delta t,3}^2.$$

The derivative $\partial C^{BS}/\partial \sigma^2$ *is called the gamma of the option and is also one of the Greeks. If the change of the volatility is also omitted, the delta-gamma approximation can be further simplified to*

$$L_{t,quad}^{\Delta t} \approx -\frac{\partial C^{BS}}{\partial t}\Delta t - \frac{\partial C^{BS}}{\partial S}S_t X_{t+\Delta t,1} - \frac{1}{2}\frac{\partial^2 C^{BS}}{\partial S^2}S_t^2 X_{t+\Delta t,1}^2. \qquad \square$$

The delta-gamma approximations can also be very poor when the change of risk factors has significant third or higher order moments. In that case, more complicated methods need to be used to compute the loss function.

2.1.2   Value-at-Risk

Value-at-Risk (VaR) is one of the most important and widely used risk measures. It quantifies the maximum potential loss of a financial institution's position over a giving holding period with a given level of confidence. Consider a portfolio of assets and liabilities and a fixed time interval $[t, t + \Delta t]$. Denote by $L := L_t^{\Delta t}$ the loss of holding the portfolio during the period $[t, t + \Delta t]$ and let $F_L(l) = P(L \leq l)$ be its distribution function.

Definition 2.1 (Value-at-Risk) *The VaR of a portfolio with loss* L *at the confidence level* $\alpha \in (0,1)$ *is the minimum value that the probability of the loss exceeding* l *is no larger than* $1 - \alpha$*, i.e.,*

$$\mathrm{VaR}_\alpha(L) = \inf\{l \in \mathbb{R} : P(L > l) \leq 1 - \alpha\} = \inf\{l : F_L(l) \geq \alpha\}. \qquad (2.5)$$

In financial risk management, typical values for the confidence level α are 0.95 or 0.99. The time horizon Δt can be very different in calculating different types of risks. For example, in the calculation of market risk, Δt is usually one to ten days, while in the calculation of credit risk and operational risk, Δt is usually one year.

VaR is actually the quantile of the distribution of the loss function over the holding period. Specifically, consider a random variable Y and its distribution function $F_Y(y) = P(Y \leq y)$, its α-quantile is given by

$$q_\alpha(F_Y) := \inf\{y \in \mathbb{R} : F_Y(y) \geq \alpha\}.$$

When $F_Y(y)$ is continuous and strictly increasing, $q_\alpha(F_Y) = F_Y^{-1}(y)$. Hence, if the distribution function $F_L(l)$ is continuous and strictly increasing, its VaR can be expressed as $\mathrm{VaR}_\alpha(L) = q_\alpha(F_L)$.

Example 2.4 (VaR for Gaussian and student-t loss functions) *Consider the loss function L over the time period $[t, t+\Delta t]$. When the loss L follows a Gaussian distribution with mean μ and variance σ^2, i.e., $L \sim N(\mu, \sigma^2)$. Then given $\alpha \in (0,1)$, the VaR of the loss L at the confidence level α is $VaR_\alpha(L) = \mu + \sigma \Phi^{-1}(\alpha)$, where Φ is the cumulative distribution of a standard normal and $\Phi^{-1}(\alpha)$ is the α-quantile of Φ. If the loss L follows a Student-t distribution, $t(\nu, \mu, \sigma^2)$, such that $(L - \mu)/\sigma$ has a standard t distribution with ν degrees of freedom. Then $VaR_\alpha = \mu + \sigma t_\nu^{-1}(\alpha)$, where t_ν is the distribution function of a standard t distribution.* □

Example 2.5 (VaR of a stock portfolio) *Consider a portfolio of stocks. Denote by V_t the value of the portfolio at time t and by $w_{t,i}$ ($i = 1, \ldots, l$) the proportion of the portfolio value invested in the ith stock. Assume that the number of stocks remains same and there are no intermediate payments of income during the period $[t, t + \Delta t]$. Let $X_{t+\Delta t, i}$ be the logarithmic return of the ith stock during the period $[t, t + \Delta t]$, $\mathbf{w}_t = (w_{t,1}, \ldots, w_{t,d})'$ and $\mathbf{X}_{t+\Delta t} = (X_{t+\Delta t, 1}, \ldots, X_{t+\Delta t, d})'$. Example 2.1 shows that the linearized loss function of holding the portfolio during $[t, t + \Delta t]$ is $L_{t,linear}^{\Delta t} = -V_t \mathbf{w}_t' \mathbf{X}_{t+\Delta t}$. Suppose that $\mathbf{X}_{t+\Delta t}$ follows a multivariate normal distribution $N_d(\boldsymbol{\mu}, \boldsymbol{\Sigma})$ (see Section 2.2.1), then the linearized loss $L_{t,linear}^{\Delta t} \sim N(-V_t \mathbf{w}_t' \boldsymbol{\mu}, V_t^2 \mathbf{w}_t' \boldsymbol{\Sigma} \mathbf{w}_t)$ and the VaR of the portfolio during $[t, t + \Delta t]$ at the confidence level α is*

$$VaR_\alpha(L_{t,linear}^{\Delta t}) = -V_t \left(\mathbf{w}_t' \boldsymbol{\mu} + \sqrt{\mathbf{w}_t' \boldsymbol{\Sigma} \mathbf{w}_t} \, \Phi^{-1}(\alpha) \right).$$

We now consider the change in value of the portfolio over the k periods, $[t, t + \Delta t], \ldots, [t + (k-1)\Delta t, t + k\Delta t]$, and the corresponding VaR. Let $\mathbf{X}_{t+j\Delta t}$ be the logarithmic returns of the d stocks during the period $[t + (j-1)\Delta t, t + j\Delta t]$ ($j = 1, \ldots, k$) and suppose $\mathbf{X}_{t+\Delta t}, \ldots, \mathbf{X}_{t+k\Delta t}$ are independent and identically distributed as $N_d(\boldsymbol{\mu}, \boldsymbol{\Sigma})$. The linearized loss of holding the portfolio during the k-period is $L_{t,linear}^{k\Delta t} = -V_t \mathbf{w}_t' \sum_{j=1}^k \mathbf{X}_{t+j\Delta t} \sim N(-kV_t \mathbf{w}_t' \boldsymbol{\mu}, kV_t^2 \mathbf{w}_t' \boldsymbol{\Sigma} \mathbf{w}_t)$ and the k-period VaR of the portfolio during $[t, t + k\Delta t]$ at the confidence level α is

$$VaR_\alpha(L_{t,linear}^{k\Delta t}) = -V_t \left(k\mathbf{w}_t' \boldsymbol{\mu} + \sqrt{k\mathbf{w}_t' \boldsymbol{\Sigma} \mathbf{w}_t} \, \Phi^{-1}(\alpha) \right).$$

Note that if we further assume that $\boldsymbol{\mu} = \mathbf{0}$, we obtain $VaR_\alpha(L_{t,linear}^{k\Delta t}) = \sqrt{k} \, VaR_\alpha(L_{t,linear}^{\Delta t})$. So we have the multi-period VaR is the product of the one-period VaR and the square root of the number of periods, which is called the square-root-of-time rule. □

While VaR is widely used in banking regulations and internal risk management in banks, it has been undesirable mathematical characteristics such as a lack of subadditivity and convexity (Artzner et al., 1997, 1999); see Exercise 2.2. Kou et al. (2013) argued that subadditivity of VaR actually holds for wide classes of distributions of (L_1, L_2), and external risk measures should be robust to model misspecification and data quality. Because the mean is not a robust statistic whereas the median is, they propose replacing the conditional VaR or expected shortfall by the more robust *tail conditional median* (TCM)

$$\mathrm{TCM}_\alpha(L) = \mathrm{median}\big[L\,|\,L \geq \mathrm{VaR}_\alpha(L)\big], \tag{2.6}$$

which turns out to be equal to $\mathrm{VaR}\big(\tfrac{1}{2}(1+\alpha)\big)$ if F is continuous. Thus, $\mathrm{TCM}(\alpha)$ is VaR at a higher confidence level and incorporates tail information in a more robust way than the expected shortfall defined by (2.7).

2.1.3   Expected shortfall and coherent risk measures

The expected shortfall (ES), also known as the *conditional VaR*, is another risk assessment measure that quantifies the amount of tail risk. As an alternative measure of risk, ES is known to have better properties than VaR. Besides being used as a measure of the average magnitude of losses beyond the VaR threshold, ES is also used in optimizing or hedging a portfolio of financial instruments to reduce risk (Rockafellar and Uryasev, 2000, 2002).

Definition 2.2 (Expected shortfall) *For a loss L with $E|L| < \infty$ and distribution function F_L, the expected shortfall at confidence level $\alpha \in (0,1)$ is defined as*

$$ES_\alpha(L) = E(L|L \geq q_\alpha(F_L)) = \frac{1}{1-\alpha}\int_{q_\alpha(F_L)}^{\infty} x f_L(x)dx = \frac{1}{1-\alpha}\int_\alpha^1 q_u(F_L)du, \tag{2.7}$$

where $q_u(F_L)$ is the uth upper quantile of F_L.

The definitions of VaR and ES imply that they are related by

$$ES_\alpha(L) = \frac{1}{1-\alpha}\int_\alpha^1 \mathrm{VaR}_u(L)du \geq \mathrm{VaR}_\alpha.$$

This shows that ES is more conservative than VaR at the confidence level α and hence reflects tail risk better.

Example 2.6 (ES for Gaussian and Student-t loss distributions) *Suppose that the loss function $L \sim N(\mu, \sigma^2)$. Then*

$$ES_\alpha(L) = \frac{1}{1-\alpha}\int_{q_\alpha(F_L)}^{\infty} x f_L(x)dx = \frac{1}{1-\alpha}\int_{\mu+\sigma\Phi^{-1}(\alpha)}^{\infty} \frac{x}{\sqrt{2\pi}\sigma}\exp\Big\{-\frac{(x-\mu)^2}{2\sigma^2}\Big\}$$

$$= \frac{1}{1-\alpha}\int_{\Phi^{-1}(\alpha)}^{\infty} (\mu+\sigma y)\frac{1}{\sqrt{2\pi}}e^{-y^2/2}dy$$

$$= \frac{\mu}{1-\alpha}\Phi(y)\Big|_{\Phi^{-1}(\alpha)}^{\infty} + \frac{\sigma}{(1-\alpha)\sqrt{2\pi}}\int_{\Phi^{-1}(\alpha)}^{\infty} y e^{-y^2/2}dy$$

$$= \mu + \frac{\sigma}{(1-\alpha)\sqrt{2\pi}}\Big[-e^{-y^2/2}\Big]\Big|_{\Phi^{-1}(\alpha)}^{\infty} = \mu + \frac{\sigma}{1-\alpha}\phi(\Phi^{-1}(\alpha)),$$

where $\phi(\cdot)$ is the density of the standard Gaussian distribution. If the loss function L follows a Student-t distribution, $t(\nu, \mu, \sigma^2)$, then a similar argument shows that

$$ES_\alpha(L) = \mu + \frac{\left[\nu + (t_\nu^{-1}(\alpha))^2\right]g_\nu(t_\nu^{-1}(\alpha))}{(1-\alpha)(\nu-1)}\sigma,$$

where t_ν and g_ν are the distribution and the density functions of standard Student-t, respectively. □

Example 2.7 (VaR and ES of the SPY ETF) *An exchange-traded fund (ETF) is a type of investment fund that trades on stock exchanges, much like stocks. ETFs are designed to track the performance of a particular index, commodity, bond, or a basket of assets. Consider the daily log returns of the S&P 500 ETF Trust from January 4, 2000 to August 31, 2022, with $n = 5,702$ observations. Figure 2.1 shows the time series of daily adjusted closing price and their log returns of the SPY ETF and the histogram of daily log returns during the period. The sample mean and sample standard deviation of the daily returns of the SPY ETF during the sample period are $\widehat{\mu} = 2.479 \times 10^{-4}$ and $\widehat{\sigma} = 1.247 \times 10^{-2}$, respectively. Suppose an investor takes a long position of \$1 million in the SPY ETF and is interested in the tail risk of such a position. Assume that the daily log returns are independent and identically distributed (i.i.d.) and follow a normal distribution $N(\widehat{\mu}, \widehat{\sigma})$. The 95% VaR for a 1-day horizon is*

$$\$1,000,000 \times [-\widehat{\mu} + \widehat{\sigma}\Phi^{-1}(0.95)] = \$20,258.$$

The corresponding VaR for a 1-month horizon (21 trading days) is

$$\$1,000,000 \times [-21\widehat{\mu} + \sqrt{21}\widehat{\sigma}\Phi^{-1}(0.95)] = \$88,765.$$

Similarly, the corresponding ES for 1-day and 21-day horizons are

$$\$1,000,000 \times [-\widehat{\mu} + \frac{\widehat{\sigma}}{1-0.95}\phi(\Phi^{-1}(0.95))] = \$25,468,$$

and

$$\$1,000,000 \times [-21\widehat{\mu} + \frac{\sqrt{21}\widehat{\sigma}}{1-0.95}\phi(\Phi^{-1}(0.95))] = \$112,637.$$

respectively. □

In contrast to VaR, ES is always subadditive (see Exercise 2.3) and belongs to the class of coherent risk measures. Axioms of coherent risk measures were developed by Artzner et al. (1997, 1999), and they are expressed as follows.

Definition 2.3 *Consider a probability space $(\Omega, \mathcal{F}, P)$. Let $\mathcal{L}^0(\Omega, \mathcal{F}, P)$ be the set of all random variables on $(\Omega, \mathcal{F}, P)$ that are almost surely finite and $\mathcal{Y}$ a linear space of $\mathcal{L}^0(\Omega, \mathcal{F}, P)$. A risk measure $\varrho : \mathcal{Y} \to \mathbb{R}$ is called coherent if it satisfies the following axioms:*

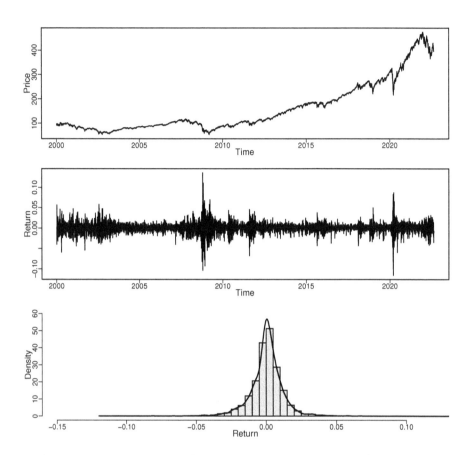

FIGURE 2.1: Time series of daily adjusted closing prices (top panel), log returns (middle panel), and the histogram of daily log returns (bottom panel) of the SPY ETF from January 4, 2000 to August 31, 2022.

(i) **Monotonicity**: $\varrho(L_1) \leq \varrho(L_2)$, if $L_1, L_2 \in \mathcal{Y}$ and $L_1 \leq L_2$ almost surely.

(ii) **Translation invariance**: $\varrho(L + b) = \varrho(L) + b$ for all $L \in \mathcal{Y}$ and $b \in \mathbb{R}$.

(iii) **Subadditivity**: $\varrho(L_1 + L_2) \leq \rho(L_1) + \rho(L_2)$ for all $L_1, L_2 \in \mathcal{Y}$.

(iv) **Positive homogeneity**: $\varrho(aL) = a\varrho(L)$ for all $L \in \mathcal{Y}$ and $a > 0$.

Artzner et al. (1999) argued that subadditivity is desirable for a risk measure because "a merger does not create extra risk." In modern portfolio theory, subadditivity ensures that the diversification principle holds because the subadditivity guarantees a lower risk measure for a diversified portfolio than a non-diversified portfolio. In banks' internal risk management, the subadditivity indicates that the overall risk of a financial firm is no more than the sum of risks of individual departments of the firm.

2.2 Multivariate models

The loss function $L_t^{\Delta t}$ defined in (2.2) involves the joint distribution of d risk factors. In order to characterize the joint behavior of risk factors, this section introduces several multivariate approaches. In particular, we first introduce multivariate normal and normal mixture distributions, and then present two dimension reduction methods, principal component analysis and factor analysis. We next introduce copula methods that model the joint distribution of a set of random variables via their marginal distributions. Moreover, since the normality assumption for risk factors is usually violated, several dependence measures are introduced to describe the concordance of random variables and/or their tail behavior.

2.2.1 Joint distributions of random variables

The distribution of a d-dimensional random vector $\mathbf{X} = (X_1, \ldots, X_d)'$ is characterized by its *joint density function* $f(x_1, \ldots, x_d)$ such that

$$F_{\mathbf{X}}(\mathbf{a}) = P(X_1 \leq a_1, \ldots, X_d \leq a_d) = \int_{-\infty}^{a_d} \cdots \int_{-\infty}^{a_1} f(x_1, \ldots, x_d) dx_1 \cdots dx_d$$

(2.8)

with $\mathbf{a} = (a_1, \ldots, a_d)'$, where $\mathbf{X}$ has a differentiable distribution function (which is the left-hand side of (2.8)). For the case of discrete $\mathbf{X}$,

$$P(\mathbf{X} = \mathbf{a}) = P(X_1 = a_1, \ldots, X_k = a_d).$$

Given the joint density function $f(x_1, \ldots, x_d)$ of $\mathbf{X}$, the density function f_i of the ith component X_i, called the *marginal density function*, can be obtained from $f(x_1, \ldots, x_d)$ by

$$f_i(x_i) = \int \cdots \int f(x_1, \ldots, x_d) dx_1 \cdots dx_{i-1} dx_{i+1} \cdots dx_d$$

in the continuous case and by summing over the other components of $\mathbf{X}$ in the discrete case. In general, for any $1 \leq i_1 < \cdots < i_j \leq d$, we can obtain from f the joint density function $f_{i_1, \ldots, i_j}$ of the subvector $(X_{i_1}, \ldots, X_{i_j})'$ of $\mathbf{X}$ by summing or integrating over the other components of $\mathbf{X}$. The random variables $X_1, \ldots, X_d$ are *mutually independent* if

$$f(x_{i_1}, \ldots, x_{i_k}) = f_{i_1}(x_{i_1}) \ldots f_{i_k}(x_{i_k}),$$

for any $1 \leq i_1 < \cdots < i_k \leq d$ and $k = 2, \ldots, d$. Moreover,

$$f(x_1, \ldots, x_d) = f_1(x_1) \prod_{i=2}^{d} f_i(x_i | x_1, \ldots, x_{i-1}),$$

where $f_i(x_i | x_1, \ldots, x_{i-1})$ is the *conditional density function* of X_i given $(X_1, \ldots, X_{i-1}) = (x_1, \ldots, x_{i-1})$.

The mean vector of $\mathbf{X} = (X_1, \ldots, X_d)'$ is defined as $E(\mathbf{X}) = (EX_1, \ldots, EX_d)'$. The covariance matrix of $\mathbf{X}$ is given by

$$\mathrm{Cov}(\mathbf{X}) = (\mathrm{Cov}(X_i, X_j))_{1 \leq i,j \leq d} = E\{(\mathbf{X} - E(\mathbf{X}))(\mathbf{X} - E(\mathbf{X}))'\}.$$

The mean $E(\mathbf{X})$ and covariance matrix $\mathrm{Cov}(\mathbf{X})$ have the following linearity (and bilinearity) properties. For nonrandom $d \times k$ matrix $\mathbf{A}$ and $d \times 1$ vector $\mathbf{B}$,

$$E(\mathbf{AX} + \mathbf{B}) = \mathbf{A}E(\mathbf{X}) + \mathbf{B}, \tag{2.9}$$

$$\mathrm{Cov}(\mathbf{AX} + \mathbf{B}) = \mathbf{A}\mathrm{Cov}(\mathbf{X})\mathbf{A}'. \tag{2.10}$$

The *correlation matrix* of $\mathbf{X}$ is the covariance matrix of the standardized vector $(X_1/\mathrm{Var}(X_1), \ldots, X_p/\mathrm{Var}(X_d))'$. Let $\mathbf{D} = \mathrm{diag}(\sqrt{\mathrm{Var}(X_1)}, \ldots, \sqrt{\mathrm{Var}(X_d)})$, then

$$\mathrm{Corr}(\mathbf{X}) = (\mathrm{Corr}(X_i, X_j))_{1 \leq i,j \leq d} = \mathbf{D}^{-1}\mathrm{Cov}(\mathbf{X})\mathbf{D}^{-1}.$$

Note that the (i, j)-th element of $\mathrm{Corr}(\mathbf{X})$ is

$$\mathrm{Corr}(X_i, X_j) = \mathrm{Cov}(X_i, X_j)/\sqrt{\mathrm{Var}(X_1)\mathrm{Var}(X_2)},$$

the pairwise linear correlation of X_i and X_j.

Multivariate normal distributions

An $d \times 1$ random vector $\mathbf{Z} = (Z_1, \ldots, Z_d)'$ is said to have the *d*-variate standard normal distribution if $Z_1, \ldots, Z_d$ are independent standard normal random variables. The joint density of $\mathbf{Z}$ at $\mathbf{Z} = \mathbf{z}$ is therefore

$$f(\mathbf{z}) = \prod_{i=1}^{d} \left\{ \frac{1}{\sqrt{2\pi}} \exp\left(-\frac{1}{2}z_i^2 \right) \right\} = (2\pi)^{-d/2} \exp(-\mathbf{z}'\mathbf{z}/2).$$

An $d \times 1$ random vector $\mathbf{X}$ is said to have a multivariate normal distribution if it is of the form $\mathbf{X} = \boldsymbol{\mu} + \mathbf{AZ}$, where $\mathbf{Z}$ is standard d-variate normal and $\boldsymbol{\mu}$ and $\mathbf{A}$ are $d \times 1$ and $d \times d$ nonrandom matrices.

Since $E(\mathbf{Z}) = \mathbf{0}$ and $\mathrm{Cov}(\mathbf{Z}) = \mathbf{I}$, the random vector $\mathbf{X} = \boldsymbol{\mu} + \mathbf{AZ}$ has mean $\boldsymbol{\mu}$ and covariance matrix $\boldsymbol{\Sigma} = \mathbf{AA}'$. Analogous to a univariate normal distribution that is determined by its mean and variance, we write $\mathbf{X} \sim N_d(\boldsymbol{\mu}, \boldsymbol{\Sigma})$. If $\boldsymbol{\Sigma}$ is nonsingular, then a change of variables applied to the density function of $\mathbf{Z}$ shows that the density function of $\mathbf{X}$ at $\mathbf{X} = \mathbf{x}$ is

$$f(\mathbf{x}) = \frac{1}{(2\pi)^{d/2}\sqrt{\det(\boldsymbol{\Sigma})}} \exp\left\{ -\frac{1}{2}(\mathbf{x} - \boldsymbol{\mu})'\boldsymbol{\Sigma}^{-1}(\mathbf{x} - \boldsymbol{\mu}) \right\}, \quad \mathbf{x} \in \mathbb{R}^d. \tag{2.11}$$

For the case $d = 1$, (2.11) reduces to the familiar normal density function

$$f(x) = \frac{1}{\sqrt{2\pi}\sigma} \exp\{-(x-\mu)^2/2\sigma^2\}, \qquad (2.12)$$

so Σ^{-1} and $\sqrt{\det(\Sigma)}$ in (2.11) are used to replace $1/\sigma^2$ and σ in (2.12). In the case $d = 2$, (2.11) can be written in the usual form of a bivariate normal density function:

$$f(x_1, x_2) = \frac{1}{2\pi\sigma_1\sigma_2\sqrt{1-\rho^2}} \exp\left\{-\left[\frac{(x_1-\mu_1)^2}{\sigma_1^2} - 2\rho\frac{(x_1-\mu_1)(x_2-\mu_2)}{\sigma_1\sigma_2}\right.\right.$$
$$\left.\left.+\frac{(x_2-\mu_2)^2}{\sigma_2^2}\right]\middle/[2(1-\rho^2)]\right\}, \qquad (2.13)$$

where $\sigma_i^2 = \text{Var}(X_i)$ and ρ is the correlation coefficient between X_1 and X_2.

Suppose that $\mathbf{X} \sim N_d(\boldsymbol{\mu}, \boldsymbol{\Sigma})$ is partitioned as

$$\mathbf{X} = \begin{pmatrix} \mathbf{X}_1 \\ \mathbf{X}_2 \end{pmatrix}, \quad \boldsymbol{\mu} = \begin{pmatrix} \boldsymbol{\mu}_1 \\ \boldsymbol{\mu}_2 \end{pmatrix}, \quad \boldsymbol{\Sigma} = \begin{pmatrix} \boldsymbol{\Sigma}_{11} & \boldsymbol{\Sigma}_{12} \\ \boldsymbol{\Sigma}_{21} & \boldsymbol{\Sigma}_{22} \end{pmatrix},$$

where $\mathbf{X}_1$ and $\boldsymbol{\mu}_1$ have dimension $r < d$ and $\boldsymbol{\Sigma}_{11}$ is a $r \times r$ matrix. Then the marginal distributions of $\mathbf{X}_1$ and $\mathbf{X}_2$ are also multivariate normal and are given by $\mathbf{X}_1 \sim N_r(\boldsymbol{\mu}_1, \boldsymbol{\Sigma}_{11})$, $\mathbf{X}_2 \sim N_{p-r}(\boldsymbol{\mu}_2, \boldsymbol{\Sigma}_{22})$. The conditional distribution of $\mathbf{X}_1$ given $\mathbf{X}_2 = \mathbf{x}_2$ is

$$N_r\left(\boldsymbol{\mu}_1 + \boldsymbol{\Sigma}_{12}\boldsymbol{\Sigma}_{22}^{-1}(\mathbf{x}_2 - \boldsymbol{\mu}_2), \; \boldsymbol{\Sigma}_{11} - \boldsymbol{\Sigma}_{12}\boldsymbol{\Sigma}_{22}^{-1}\boldsymbol{\Sigma}_{21}\right),$$

which can be shown by computing the conditional density function $f_{\mathbf{X}_1|\mathbf{X}_2}(\mathbf{x}_1|\mathbf{x}_2) = f(\mathbf{x})/f_{\mathbf{X}_2}(\mathbf{x}_2)$ via (2.11).

Estimates of mean vector and covariance matrix

Suppose there are n observations of a d-dimensional risk-factor return vector $\mathbf{X}_1, \ldots, \mathbf{X}_n$. Assume that these observations are independent and identically distributed and follow a distribution with mean vector $\boldsymbol{\mu}$ and covariance matrix $\boldsymbol{\Sigma}$. Standard estimators of $\boldsymbol{\mu}$ and $\boldsymbol{\Sigma}$ are given by the *sample mean vector* $\bar{\mathbf{X}}$ and the *sample covariance matrix* $\mathbf{S}$,

$$\bar{\mathbf{X}} = \frac{1}{n}\sum_{i=1}^{n}\mathbf{X}_i, \quad \mathbf{S} = \frac{1}{n-1}\sum_{i=1}^{n}(\mathbf{X}_i - \bar{\mathbf{X}})(\mathbf{X}_i - \bar{\mathbf{X}})'.$$

These two estimators are unbiased, that is, $E(\bar{\mathbf{X}}) = \boldsymbol{\mu}$ and $E(\mathbf{S}) = \boldsymbol{\Sigma}$.

Properties of $\bar{\mathbf{X}}$ and $\mathbf{S}$ can be obtained for some special multivariate distributions of $\mathbf{X}_i$. For example, if $\mathbf{X}_1, \ldots, \mathbf{X}_n$ follow independently a d-variate normal distribution $N_d(\boldsymbol{\mu}, \boldsymbol{\Sigma})$ with $d < n$ and positive definite $\boldsymbol{\Sigma}$. Then the sample mean vector $\bar{\mathbf{X}}$ and the sample covariance matrix $\mathbf{S}$ are independent. Moreover, since $\bar{\mathbf{X}}$ is a linear transformation of the multivariate normal vector

$(\mathbf{X}'_1, \ldots, \mathbf{X}'_n)'$, it is also normal. Making use of (2.9) and (2.10) to derive $E(\bar{\mathbf{X}})$ and $\mathrm{Cov}(\bar{\mathbf{X}})$, we then obtain $\bar{\mathbf{X}} \sim N(\boldsymbol{\mu}, \boldsymbol{\Sigma}/n)$. To obtain the distribution of the sample covariance matrix $\mathbf{S}$, we need the concept of Wishart distribution defined below.

Definition 2.4 *Suppose that vectors* $\mathbf{Z}_1, \ldots, \mathbf{Z}_n$ *follows independently d-variate normal distribution* $N(\mathbf{0}, \boldsymbol{\Sigma})$, *then the random matrix* $\mathbf{W} = \sum_{i=1}^{n} \mathbf{Z}_i \mathbf{Z}'_i$ *is said to have a Wishart distribution, denoted by* $W_d(\boldsymbol{\Sigma}, n)$.

This definition implies that the sample covariance matrix $\mathbf{S}$ follows a Wishart distribution with $n - 1$ degrees of freedom, i.e., $(n-1)\mathbf{S} \sim W_d(\boldsymbol{\Sigma}, n-1)$.

Besides the sample estimators, other estimators can also be obtained. For example, when $\mathbf{X}_1, \ldots, \mathbf{X}_n$ follow independently a d-variate normal distribution $N(\boldsymbol{\mu}, \boldsymbol{\Sigma})$, *maximum likelihood estimates* of $\boldsymbol{\mu}$ and $\boldsymbol{\Sigma}$ can be obtained, which are expressed as $\bar{\mathbf{X}}$ and $(n-1)\mathbf{S}/n$, respectively.

Properties of the estimators of $\boldsymbol{\mu}$ and $\boldsymbol{\Sigma}$ depend much on the true multivariate distribution of the observations, which is usually unknown in financial risk management. Hence, sample or maximum likelihood estimates are not necessarily the best estimators for problems in financial applications. In such cases, other types of distribution assumptions and estimators are available and should be considered.

Univariate and multivariate normality tests

Univariate or multivariate normal distributions are usually not a good description of financial risk-factor returns. The non-normality of the distributions of univariate and multivariate risk factors can be tested by univariate and multivariate tests.

A simple way to check the assumption of univariate normal distribution is to construct a *quantile-quantile* (Q-Q) plot of the quantiles of studentized residuals versus standard normal quantiles. The Q-Q plot of a distribution $F(X)$ versus another distribution $G(Y)$ plots the αth quantile x_α of F against the αth quantile y_α of G for $0 < \alpha < 1$. Suppose that the univariate samples $x_1, \ldots, x_n$ are normally distributed, then the Q-Q plot of studentized samples of $x_1, \ldots, x_n$ versus standard normal quantiles should lie approximately on a straight line.

Besides the Q-Q plot, other statistical tests are also commonly used to check univariate normality assumption. A common test is the Jarque and Bera test (Jarque and Bera, 1987) which is based on a distribution's skewness and kurtosis. The *skewness* of a random variable X with mean μ and variance σ^2 is defined as $\mathrm{sk}_1 = E(X - \mu)^3/\sigma^3$, which is 0 when X is symmetric. The *kurtosis* of X is defined as $\kappa_1 = E(X - \mu)^4/\sigma^4$, which is equal to 3 when X is normal. For a sample of n independent observations $x_1, \ldots, x_n$ from a distribution $F(X)$, the skewness and kurtosis of $F(X)$ can be estimated by the sample skewness $\widehat{\text{sk}}_1$ and sample kurtosis $\widehat{\kappa}_1$ given by

$$\widehat{\text{sk}}_1 = \frac{1}{n}\sum_{i=1}^{n}\frac{(x_i - \bar{x})^3}{\widehat{\sigma}^3}, \qquad \widehat{\kappa}_1 = \frac{1}{n}\sum_{i=1}^{n}\frac{(x_i - \bar{x})^4}{\widehat{\sigma}^4}, \qquad (2.14)$$

in which $\bar{x} = n^{-1}\sum_{i=1}^{n}x_i$ and $\widehat{\sigma} = n^{-1}\sum_{i=1}^{n}(x_i - \bar{x})^2$ are the sample mean and the sample variance of $x_1, \ldots, x_n$. Then to test for normality of $F(X)$ based on the sample skewness $\widehat{\text{sk}}_1$ and sample kurtosis $\widehat{\kappa}_1$, the *Jarque-Bera test* uses the test statistic

$$\text{JB} = n\left(\frac{\widehat{\text{sk}}_1^2}{6} + \frac{(\widehat{\kappa}_1 - 3)^2}{24}\right), \qquad (2.15)$$

which has an approximately χ_2^2-distribution under the null hypothesis that $F(X)$ is normal.

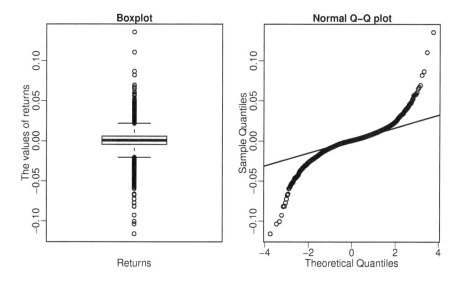

FIGURE 2.2: Normality assessment of SPY return: Boxplot (left) and normal Q-Q plot (right).

Example 2.8 (Normality test of the SPY return) *Consider the daily log return of the SPY ETF in Example 2.7. The sample skewness and sample kurtosis of the SPY log return is* $\widehat{\text{sk}}_1 = -0.2668$ *and* $\widehat{\kappa}_1 = 14.24$. *The Jarque-Bera test statistic of the sample is 30102 and the p-value of the Jarque-Bera test is less than* 2.2×10^{-16}. *This indicates that the daily log return of the SPY ETF during the sample period is significantly different from a normal distribution. Figure 2.2 shows the boxplot and the normal Q-Q plot of the SPY daily*

returns. In the boxplot, the median of the sample (the horizontal black line) is
6.730×10^{-4}, *which is slightly higher than the sample mean* $\widehat{\mu} = 2.479 \times 10^{-4}$.
The distance between the 75th and 25th quantiles is 1.068×10^{-2}, *which is*
smaller than $\widehat{\sigma} = 1.247 \times 10^{-2}$. *This indicates that the SPY returns are more*
concentrated than a normal distribution. In the normal Q-Q plot, the inverted
S-shaped curve of the points suggests that the more extreme empirical quan-
tiles of the data tend to be larger than the corresponding quantiles of a normal
distribution, indicating that the normal distribution is a poor model for these
returns.   □

To test for multivariate normality, one can use Mardia's test (Mardia,
1970, 1980) to check if the multivariate skewness and kurtosis are consistent
with a multivariate normal distribution. Mardia (1970) introduced the affine-
invariant skewness measure for a d-dimensional random vector $\mathbf{X}$ with mean
$\boldsymbol{\mu}$ and covariance matrix $\boldsymbol{\Sigma}$. Let $\mathbf{Y}$ be an independent copy of $\mathbf{X}$, the Mardia's
multivariate skewness and kurtosis are defined as

$$\mathrm{sk}_d = E\{[(\mathbf{X} - \boldsymbol{\mu})'\boldsymbol{\Sigma}^{-1}(\mathbf{Y} - \boldsymbol{\mu})]^3\}, \qquad \kappa_d = E\{[(\mathbf{X} - \boldsymbol{\mu})'\boldsymbol{\Sigma}^{-1}(\mathbf{Y} - \boldsymbol{\mu})]^2\},$$

Given a sample $\mathbf{x}_1, \dots, \mathbf{x}_n \in \mathbb{R}^d$, the sample estimate of sk_d and κ_d are given
by

$$\widehat{\mathrm{sk}}_d = \frac{n}{(n-1)^3} \sum_{i,j=1}^n m_{i,j}^3, \qquad \widehat{\kappa}_d = \frac{n}{(n-1)^2} \sum_{i=1}^n m_{i,i}^2, \qquad (2.16)$$

in which $m_{i,j} = (\mathbf{x}_i - \bar{\mathbf{x}})'\mathbf{S}^{-1}(\mathbf{x}_i - \bar{\mathbf{x}})$, and $\bar{\mathbf{x}}$ and $\mathbf{S}$ are the sample mean and
sample variance-covariance matrix of $\mathbf{x}_1, \dots, \mathbf{x}_n$. Then if $\mathbf{x}_1, \dots, \mathbf{x}_n$ comes
from a multivariate normal distribution, then

$$\frac{n}{6}\widehat{\mathrm{sk}}_d \sim \chi_k^2, \qquad [\widehat{\kappa}_d - d(d+2)]\sqrt{\frac{n}{8d(d+2)}} \sim N(0,1)$$

in which $k = d(d+1)(d+2)/6$ is the degree of freedom.

Example 2.9 (Normality test of 10 stock returns) *Consider the daily*
log returns of 10 stocks (AAPL, MSFT, AMZN, IBM, INTC, JPM, GE, CVX,
PFE, JNJ) from January 4, 2000 to August 31, 2022, with $n = 5,702$ *obser-*
vations. Table 2.1 lists the name, symbol, sample mean $\widehat{\mu}$, *sample standard*
deviation $\widehat{\sigma}$, *sample skewness, and sample kurtosis for these 10 stocks. The*
last column of the table shows the p-value of the Jarque-Bera tests of normality
for each of 10 stocks. Performing Mardia's multivariate tests on skewness and
kurtosis, one obtains that the p-value of both tests are less than 10^{-4}, *indicat-*
ing the multivariate normality assumption for the 10 stocks' daily returns is
not valid.   □

Normal mixture distributions

The random vector $\mathbf{X}$ is said to have a (multivariate) normal variance
mixture distribution if

$$\mathbf{X} = \boldsymbol{\mu} + \sqrt{W}\mathbf{A}\mathbf{Z}, \qquad (2.17)$$

TABLE 2.1: Statistics of 10 stocks' daily returns.

Name	Symbol	$\widehat{\mu} \times 10^4$	$\widehat{\sigma} \times 10^2$	$\widehat{\text{sk}}_1$	$\widehat{\kappa}_1$	p-value
Apple	AAPL	9.148	2.604	−4.045	115.726	0
Microsoft	MSFT	3.451	1.931	−0.164	12.510	0
Amazon	AMZN	5.867	3.157	0.427	15.700	0
IBM	IBM	1.199	1.657	−0.288	11.568	0
Intel	INTC	0.362	2.368	−0.558	12.853	0
JPMorgan	JPM	2.657	2.414	0.226	16.969	0
General Electric	GE	−1.857	2.120	−0.038	10.493	0
Chevron	CVX	3.796	1.764	−0.450	23.596	0
Pfizer	PFE	2.059	1.596	−0.153	8.367	0
Johnson Johnson	JNJ	3.214	1.226	−0.499	17.518	0

where the random vector $\mathbf{Z} \in \mathbb{R}^k$ follows a standard normal distribution $N_k(\mathbf{0}, \mathbf{I}_k)$, $W \geq 0$ is a non-negative scalar-valued random variable and independent of $\mathbf{Z}$, and $\mathbf{A} \in \mathbb{R}^{d \times k}$ and $\boldsymbol{\mu} \in \mathbb{R}^d$ are a matrix and a vector of constants, respectively.

In the above specification, we note that the conditional distribution $\mathbf{X}|W = w \sim N_d(\boldsymbol{\mu}, w\boldsymbol{\Sigma})$, where $\boldsymbol{\Sigma} = \mathbf{A}\mathbf{A}'$. Provided that $E|W| < \infty$, we have

$$E(\mathbf{X}) = \boldsymbol{\mu}, \quad \text{Cov}(\mathbf{X}) = E(W)\mathbf{A}E(\mathbf{Z}\mathbf{Z}')\mathbf{A}' = E(W)\boldsymbol{\Sigma}.$$

If we assume that $\boldsymbol{\Sigma}$ is positive definite and that the distribution of W has no point mass at zero, we may derive density of a normal variance mixture distribution. In particular, let $f_W(w)$ be the density function of W, the density of $\mathbf{X}$ at $\mathbf{X} = \mathbf{x}$ is given by

$$f(\mathbf{x}) = \frac{1}{(2\pi w)^{d/2}|\boldsymbol{\Sigma}|^{1/2}} \int \exp\left(-\frac{(\mathbf{x} - \boldsymbol{\mu})'\boldsymbol{\Sigma}^{-1}(\mathbf{x} - \boldsymbol{\mu})}{2w}\right) f_W(w)dw. \quad (2.18)$$

Note that if we assume W follows an inverse gamma distribution, $W \sim IG(\frac{\nu}{2}, \frac{\nu}{2})$, or equivalently, $\nu/W \sim \chi_\nu^2$, then $\mathbf{X}$ has a multivariate t distribution with ν degrees of freedom, whose density at $\mathbf{X} = \mathbf{x}$ is expressed as

$$f(\mathbf{x}) = \frac{\Gamma(\frac{\nu+d}{2})}{\Gamma(\frac{\nu}{2})(\pi\nu)^{d/2}|\boldsymbol{\Sigma}|^{1/2}} \exp\left(1 + \frac{(\mathbf{x} - \boldsymbol{\mu})'\boldsymbol{\Sigma}^{-1}(\mathbf{x} - \boldsymbol{\mu})}{\nu}\right)^{-(\nu+d)/2}.$$

Other types of normal mixture distributions can be similarly defined. For example, one may assume that the mean $\boldsymbol{\mu}$ in (2.17) is also a function of random variable W, then we obtain a normal mean-variance mixture distribution.

2.2.2 Principal component analysis

The aim of *principal component analysis* (PCA) is to reduce the dimensionality of highly correlated data by using a small number of principal components (or factors) to account for the variance of the original data. PCA is based on the *spectral decomposition theorem* in linear algebra, which says that any symmetric matrix $\mathbf{X} \in \mathbb{R}^{d \times d}$ can be written as

$$\mathbf{\Sigma} = \mathbf{\Gamma}\mathbf{\Lambda}\mathbf{\Gamma}', \tag{2.19}$$

where $\mathbf{\Lambda} = \mathrm{diag}(\lambda_1, \ldots, \lambda_d)$ is the diagonal matrix of eigenvalues of $\mathbf{\Sigma}$ with $\lambda_1 \geq \lambda_2 \geq \cdots \geq \lambda_d$, and $\mathbf{\Gamma} = (\mathbf{a}_1, \ldots, \mathbf{a}_d)$ is an orthogonal matrix of standardized eigenvectors $\mathbf{a}_1, \ldots, \mathbf{a}_d$ corresponding to $\lambda_1, \ldots, \lambda_d$, respectively, and satisfying that $\mathbf{a}_i \mathbf{a}_j' = \mathbf{a}_j' \mathbf{a}_i = 1_{\{i=j\}}$ for $i, j = 1, \ldots, d$. Here, $1_{\{i=j\}}$ is an indicator function whose value is 1 if $i = j$ and 0 otherwise.

Maximization of total variance of the projection

Given a random vector $\mathbf{X} = (X_1, \ldots, X_d)'$, we apply the above decomposition to its covariance matrix $\mathbf{\Sigma} = \mathrm{Cov}(\mathbf{X})$. Consider the linear combination $\mathbf{a}'\mathbf{x}$ with $\|\mathbf{a}\| = 1$ that has the largest variance over all such linear combinations. To maximize $\mathbf{a}'\mathbf{\Sigma}\mathbf{a} (= \mathrm{Var}(\mathbf{a}'\mathbf{x}))$ over $\mathbf{a}$ with $\|\mathbf{a}\| = 1$, introduce the Lagrange multiplier λ to obtain

$$\frac{\partial}{\partial a_i}\{\mathbf{a}'\mathbf{\Sigma}\mathbf{a} + \lambda(1 - \mathbf{a}'\mathbf{a})\} = 0 \quad \text{for } i = 1, \ldots, d. \tag{2.20}$$

The d equations in (2.20) can be written as the linear system $\mathbf{\Sigma}\mathbf{a} = \lambda\mathbf{a}$. Since $\mathbf{a} \neq \mathbf{0}$, this implies that λ is the largest eigenvalue of $\mathbf{\Sigma}$ and $\mathbf{a}$ is the corresponding eigenvector, and that $\lambda = \mathbf{a}'\mathbf{\Sigma}\mathbf{a}$. Let $\lambda_1 = \max_{\mathbf{a}:\|\mathbf{a}\|=1} \mathbf{a}'\mathbf{\Sigma}\mathbf{a}$ and $\mathbf{a}_1$ be the corresponding eigenvector with $\|\mathbf{a}_1\| = 1$. Next consider the linear combination $\mathbf{a}'\mathbf{x}$ that maximizes $\mathrm{Var}(\mathbf{a}'\mathbf{x}) = \mathbf{a}'\mathbf{\Sigma}\mathbf{a}$ subject to $\mathbf{a}_1'\mathbf{a} = 0$ and $\|\mathbf{a}\| = 1$. Introducing the Lagrange multipliers λ and η, we obtain

$$\frac{\partial}{\partial a_i}\{\mathbf{a}'\mathbf{\Sigma}\mathbf{a} + \lambda(1 - \mathbf{a}'\mathbf{a}) + \eta\mathbf{a}_1'\mathbf{a}\} = 0 \text{ for } i = 1, ..., d.$$

As in (2.20), this implies that the Lagrange multiplier λ is an eigenvalue of $\mathbf{\Sigma}$ with corresponding unit eigenvector $\mathbf{a}_2$ that is orthogonal to $\mathbf{a}_1$. Proceeding inductively in this way, we obtain the eigenvalue $\lambda_1 \geq \lambda_2 \geq \cdots \geq \lambda_d$ of $\mathbf{\Sigma}$ with the optimization characterization

$$\lambda_{k+1} = \max_{\mathbf{a}:\|\mathbf{a}\|=1, \mathbf{a}'\mathbf{a}_j=0 \text{ for } 1 \leq j \leq k} \mathbf{a}'\mathbf{\Sigma}\mathbf{a}. \tag{2.21}$$

The maximizer $\mathbf{a}_{k+1}$ of the right-hand side of (2.21) is an eigenvector corresponding to the eigenvalue λ_{k+1}.

The principal component

The principal component transformation of $\mathbf{X}$ is defined as

$$\mathbf{Y} = \mathbf{\Gamma}'(\mathbf{X} - \boldsymbol{\mu}),$$

and the ith element of $\mathbf{Y}$ is called the ith *principal component* of $\mathbf{X}$ and is given by

$$Y_i = \mathbf{a}_i'(\mathbf{X} - \boldsymbol{\mu}).$$

The eigenvalues λ_i, the eigenvectors $\mathbf{a}_i$, and the principal components have the following properties.

(i) $\lambda_i = \mathrm{Var}(\mathbf{a}_i^T \mathbf{x})$.

(ii) The elements of the eigenvectors $\mathbf{a}_i$ are called *factor loadings* in PCA, and the orthogonality of $\mathbf{a}_i$ implies that

$$\mathbf{I} = (\mathbf{a}_1, \ldots, \mathbf{a}_d)(\mathbf{a}_1, \ldots, \mathbf{a}_d)' = \mathbf{a}_1 \mathbf{a}_1' + \cdots + \mathbf{a}_d \mathbf{a}_d'.$$

More importantly, summing $\lambda_i \mathbf{a}_i \mathbf{a}_i' = \Sigma \mathbf{a}_i \mathbf{a}_i'$ over i, we obtain the following decomposition of $\boldsymbol{\Sigma}$ into d rank-one matrices:

$$\boldsymbol{\Sigma} = \lambda_1 \mathbf{a}_1 \mathbf{a}_1' + \cdots + \lambda_d \mathbf{a}_d \mathbf{a}_d'.$$

(iii) Since $\boldsymbol{\Sigma} = \mathrm{Cov}(\mathbf{X})$ and $\mathbf{X} = (X_1, \ldots, X_d)'$, $\mathrm{tr}(\boldsymbol{\Sigma}) = \sum_{i=1}^{d} \mathrm{Var}(X_i)$, we have

$$\lambda_1 + \cdots + \lambda_d = \sum_{i=1}^{d} \mathrm{Var}(X_i).$$

An important goal of PCA is to determine if the first few principal components can account for most of the overall variance $\sum_{i=1}^{d} \mathrm{Var}(X_i)$. This amounts to determining whether

$$\left(\sum_{i=1}^{k} \lambda_i \right) \Big/ \mathrm{tr}(\boldsymbol{\Sigma}) \text{ is near 1 for some small } k.$$

To apply PCA to data analysis, we suppose $\mathbf{X}_1, \ldots, \mathbf{X}_n$ are n independent observations from a multivariate population with mean $\boldsymbol{\mu}$ and covariance matrix $\boldsymbol{\Sigma}$. The mean vector $\boldsymbol{\mu}$ and covariance matrix $\boldsymbol{\Sigma}$ can be estimated by sample mean $\bar{\mathbf{X}} = \sum_{i=1}^{n} \mathbf{X}_i/n$ and the sample covariance matrix $\mathbf{S} = (n-1)^{-1} \sum_{i=1}^{n} (\mathbf{X}_i - \bar{\mathbf{X}})(\mathbf{X}_i - \bar{\mathbf{X}})'$, respectively. Let $\widehat{\mathbf{a}}_j = (\widehat{a}_{1j}, \ldots, \widehat{a}_{dj})^T$ be the eigenvector corresponding to the jth largest eigenvalue $\widehat{\lambda}_j$ of the sample covariance matrix $\mathbf{S}$.

Let $\mathbf{X}_i = (x_{i1}, \ldots, x_{id})'$, $i = 1, \ldots, n$, be the multivariate sample. Let $\mathbf{Z}_k$ be the k-th component of the n observations, i.e., $\mathbf{Z}_k = (x_{1k}, \ldots, x_{nk})'$, $1 \le k \le d$, and define

$$\mathbf{Y}_j = \widehat{a}_{1j} \mathbf{Z}_1 + \cdots + \widehat{a}_{dj} \mathbf{Z}_d, \qquad 1 \le j \le d,$$

where $\widehat{\mathbf{a}}_j = (\widehat{a}_{1j}, \ldots, \widehat{a}_{dj})'$ is the eigenvector corresponding to the jth largest eigenvalue $\widehat{\lambda}_j$ of the sample covariance matrix $\mathbf{S}$ with $\|\widehat{\mathbf{a}}_j\| = 1$. Since the matrix $\widehat{\mathbf{A}} := (\widehat{a}_{ij})_{1 \le i,j \le d}$ is orthogonal (i.e., $\widehat{\mathbf{A}}\widehat{\mathbf{A}}' = \mathbf{I}$), it follows that the observed data $\mathbf{X}_k$ can be expressed in terms of the "principal components" $\mathbf{Y}_j$ as

$$\mathbf{Z}_k = \widehat{a}_{k1}\mathbf{Y}_1 + \cdots + \widehat{a}_{kd}\mathbf{Y}_d. \tag{2.22}$$

Moreover, the sample correlation matrix of the transformed data y_{tj} is the identity matrix.

Example 2.10 (PCA of 10 stocks' daily returns) *Consider the daily log returns of 10 stocks from January 4, 2000 to August 31, 2022 in Example 2.9. One can perform the principal component analysis for the correlation matrix of 10 stocks' daily log returns. Table 2.2 shows the variance $\widehat{\lambda}_i$, the percentage of the overall variance ϱ_i explained by principal components $\mathbf{Y}_1, \dots, \mathbf{Y}_i$, and factor loadings of 10 principal components $\mathbf{Y}_1, \dots, \mathbf{Y}_{10}$. Note that the first five principal components explain almost the 90% overall variance of the sample.* □

TABLE 2.2: Variance and factor loadings of principal components of 10 stocks.

	$\mathbf{Y}_1$	$\mathbf{Y}_2$	$\mathbf{Y}_3$	$\mathbf{Y}_4$	$\mathbf{Y}_5$	$\mathbf{Y}_6$	$\mathbf{Y}_7$	$\mathbf{Y}_8$	$\mathbf{Y}_9$	$\mathbf{Y}_{10}$
$\widehat{\lambda}_i$	4.447	1.205	0.806	0.675	0.571	0.561	0.494	0.450	0.403	0.388
$\widehat{\varrho}_i$	0.444	0.556	0.667	0.778	0.889	1.000	1.000	1.000	1.000	1.000
	$\mathbf{a}_1$	$\mathbf{a}_2$	$\mathbf{a}_3$	$\mathbf{a}_4$	$\mathbf{a}_5$	$\mathbf{a}_6$	$\mathbf{a}_7$	$\mathbf{a}_8$	$\mathbf{a}_9$	$\mathbf{a}_{10}$
AAPL	0.289	0.374	0.150	0.385	0.751	0.017	0.163	0.079	0.082	0.035
MSFT	0.354	0.232	0.178	0.098	−0.287	−0.237	−0.395	-0.168	0.269	0.620
AMZN	0.253	0.407	0.315	−0.765	−0.008	−0.051	0.257	−0.026	−0.092	−0.093
IBM	0.344	0.064	−0.166	0.260	−0.458	0.195	0.663	0.271	0.113	0.096
INTC	0.344	0.302	0.055	0.293	−0.294	0.006	−0.368	0.014	−0.413	−0.557
JPM	0.351	−0.063	−0.408	−0.217	0.089	0.259	−0.238	−0.074	0.640	−0.335
GE	0.341	−0.134	−0.452	−0.152	0.177	0.324	−0.046	−0.252	−0.555	0.359
CVX	0.309	−0.289	−0.284	−0.069	0.110	−0.817	0.093	0.181	−0.081	−0.093
PFE	0.281	−0.462	0.411	−0.106	0.076	0.254	−0.238	0.624	−0.057	0.086
JNJ	0.276	−0.476	0.441	0.133	−0.006	0.009	0.229	−0.632	0.047	−0.173

2.2.3 Factor models and analysis

An alternative statistical method to determine factors is factor analysis. Letting $\mathbf{r} = (r_1, \dots, r_d)'$, $\boldsymbol{\alpha} = (\alpha_1, \dots, \alpha_d)'$, $\boldsymbol{\epsilon} = (\epsilon_1, \dots, \epsilon_d)'$, and $\mathbf{B}$ to be the $d \times k$ matrix whose ith row vector is $\mathbf{b}'_i$, we say that the random vector $\mathbf{X} = (X_1, \dots, X_d)'$ follows a k-factor model if

$$\mathbf{X} = \boldsymbol{\alpha} + \mathbf{B}\mathbf{f} + \boldsymbol{\epsilon}, \tag{2.23}$$

with $E\boldsymbol{\epsilon} = E\mathbf{f} = \mathbf{0}$ and $E(\mathbf{f}\boldsymbol{\epsilon}') = \mathbf{0}$. Although this looks like a regression model, the fact that the regressor $\mathbf{f}$ is unobservable means that least squares regression cannot be applied to estimate $\mathbf{B}$. Let $\mathbf{X}_i$, $i = 1, \dots, n$, be independent observations from (2.23) so that $\mathbf{X}_i = \boldsymbol{\alpha} + \mathbf{B}\mathbf{f}_i + \boldsymbol{\epsilon}_i$ and $E\boldsymbol{\epsilon}_i = E\mathbf{f}_i = \mathbf{0}$, $E(\mathbf{f}_i\boldsymbol{\epsilon}'_i) = \mathbf{0}$, $\text{Cov}(\mathbf{f}_i) = \boldsymbol{\Omega}$, and $\text{Cov}(\boldsymbol{\epsilon}_i) = \mathbf{V}$. Then

$$E(\mathbf{X}_i) = \boldsymbol{\alpha}, \qquad \text{Cov}(\mathbf{X}_i) = \mathbf{B}\boldsymbol{\Omega}\mathbf{B}' + \mathbf{V}. \tag{2.24}$$

The decomposition of the covariance matrix $\boldsymbol{\Sigma}$ of $\mathbf{X}_i$ in (2.24) is the essence of factor analysis, which separates variability into a *systematic* part due to the variability of certain unobserved factors, represented by $\mathbf{B\Omega B'}$, and an error (*idiosyncratic*) part, represented by $\mathbf{V}$. In terms of minimizing the risk of holding a portfolio of assets or liabilities, the covariance decomposition in (2.24) suggests that one might reduce the idiosyncratic risk of the portfolio by using diversification or hedging strategies, but not for the systematic risk of the portfolio.

Orthogonal factors and estimation

Factor analysis usually assumes a strict factor structure in which the factors account for all the pairwise covariances of the asset returns, so $\mathbf{V}$ is assumed to be diagonal; i.e., $\mathbf{V} = \mathrm{diag}(v_1, \ldots, v_d)$. Since $\mathbf{B}$ and $\boldsymbol{\Omega}$ are not uniquely determined by $\boldsymbol{\Sigma} = \mathbf{B\Omega B'} + \mathbf{V}$, the *orthogonal factor model* assumes that $\boldsymbol{\Omega} = \mathbf{I}$ so that $\mathbf{B}$ is unique up to an orthogonal transformation and $\mathbf{X} = \boldsymbol{\alpha} + \mathbf{Bf} + \boldsymbol{\epsilon}$ with $\mathrm{Cov}(\mathbf{f}) = \boldsymbol{\Omega}$. In particular, it yields

$$\mathrm{Cov}(\mathbf{X}, \mathbf{f}) = E\{(\mathbf{X} - \boldsymbol{\alpha})\mathbf{f}'\} = \mathbf{B\Omega} = \mathbf{B},$$

and the covariance decomposition in (2.24) can be reduced to

$$\mathrm{Var}(X_l) = \sum_{j=1}^{k} b_{lj}^2 + \mathrm{Var}(\epsilon_l), \qquad 1 \le l \le d,$$

$$\mathrm{Cov}(X_l, X_j) = \sum_{m=1}^{k} b_{lm} b_{jm}, \qquad 1 \le l, j \le d.$$

Assuming the observed $\mathbf{X}_i$ are independent $N(\boldsymbol{\alpha}, \boldsymbol{\Sigma})$, the likelihood function is

$$L(\boldsymbol{\alpha}, \boldsymbol{\Sigma}) = (2\pi)^{-dn/2} (\det \boldsymbol{\Sigma})^{-n/2} \exp\left\{ -\frac{1}{2} \sum_{i=1}^{n} (\mathbf{X}_i - \boldsymbol{\alpha})' \boldsymbol{\Sigma}^{-1} (\mathbf{X}_i - \boldsymbol{\alpha}) \right\},$$

with $\boldsymbol{\Sigma}$ constrained to be of the form $\boldsymbol{\Sigma} = \mathbf{BB'} + \mathrm{diag}(v_1, \ldots, v_d)$, in which $\mathbf{B}$ is $d \times k$. The MLE $\widehat{\boldsymbol{\alpha}}$ of $\boldsymbol{\alpha}$ is $\bar{\mathbf{X}} := n^{-1} \sum_{i=1}^{n} \mathbf{X}_i$, and we can maximize

$$-\frac{1}{2} n \log \det(\boldsymbol{\Sigma}) - \frac{1}{2} \mathrm{tr}(\mathbf{W}\boldsymbol{\Sigma}^{-1})$$

over $\boldsymbol{\Sigma}$ of the form above, where $\mathbf{W} = \sum_{i=1}^{n} (\mathbf{X}_i - \bar{\mathbf{X}})(\mathbf{X}_t - \bar{\mathbf{X}})'$. Iterative algorithms can be used to find the maximizer $\widehat{\boldsymbol{\Sigma}}$ and therefore also $\widehat{\mathbf{B}}$ and $\widehat{v}_1, \ldots, \widehat{v}_d$.

In factor analysis, the entries of the matrix $\widehat{\mathbf{B}}$ are called *factor loadings*. Since $\widehat{\mathbf{B}}$ is unique only up to orthogonal transformations, the usual practice is to multiply $\widehat{\mathbf{B}}$ by a suitably chosen orthogonal matrix $\mathbf{Q}$, called a *factor* *rotation*, so that the factor loadings have a simple interpretable structure. Letting $\widehat{\mathbf{B}}^* = \widehat{\mathbf{B}}\mathbf{Q}$, one choice of $\mathbf{Q}$ is a matrix that maximizes the *varimax criterion*

$$C = d^{-1} \sum_{j=1}^{k} \left[\sum_{i=1}^{d} \widehat{b}_{ij}^{*4} - \left(\sum_{i=1}^{d} \widehat{b}_{ij}^{*2} \right)^2 \Big/ d \right]$$

$$\propto \sum_{j=1}^{k} \mathrm{Var}\big(\text{squared loadings of the } j\text{th factor}\big).$$

Intuitively, maximizing C corresponds to spreading out the squares of the loadings on each factor as much as possible.

Since the values of the factors $\mathbf{f}_i$, $1 \le i \le n$, are unobserved, it is often of interest to impute these values, called *factor scores*, for model diagnostics. From the model $\mathbf{X} - \boldsymbol{\alpha} = \mathbf{B}\mathbf{f} + \boldsymbol{\epsilon}$ with $\mathrm{Cov}(\boldsymbol{\epsilon}) = \mathbf{V}$, the generalized least squares estimate of $\mathbf{f}$ when $\mathbf{B}$, $\mathbf{V}$, and $\boldsymbol{\alpha}$ are known is

$$\widehat{\mathbf{f}} = (\mathbf{B}'\mathbf{V}^{-1}\mathbf{B})^{-1}\mathbf{B}'\mathbf{V}^{-1}(\mathbf{X}_i - \boldsymbol{\alpha});$$

Bartlett (1937) therefore suggested estimating $\mathbf{f}_i$ by

$$\widehat{\mathbf{f}}_i = (\widehat{\mathbf{B}}'\widehat{\mathbf{V}}^{-1}\widehat{\mathbf{B}})^{-1}\widehat{\mathbf{B}}'\widehat{\mathbf{V}}^{-1}(\mathbf{X}_i - \bar{\mathbf{X}}).$$

Number of factors

The theory underlying factor analysis assumes that the number k of factors has been specified and does not indicate how to specify it. For the theory to be useful, however, k has to be reasonably small. The freedom in the choice of k has led to two ways in empirical work. One way is to repeat the estimation for several choices of k to see if the results are sensitive to increasing the number of factors. Another way involves more formal hypothesis testing that the k-factor model indeed holds when the $\mathbf{r}_t$ are independent $N(\boldsymbol{\alpha}, \boldsymbol{\Sigma})$. In such a case, the null hypothesis H_0 is that $\boldsymbol{\Sigma}$ is of the form $\mathbf{BB}' + \mathbf{V}$ with $\mathbf{V}$ diagonal and $\mathbf{B}$ being $d \times k$. The *generalized likelihood ratio* (GLR) statistic that tests this null hypothesis against unconstrained $\boldsymbol{\Sigma}$ is of the form

$$\Lambda = n\Big\{ \log \det \big(\widehat{\mathbf{B}}\widehat{\mathbf{B}}' + \widehat{\mathbf{V}}\big) - \log \det \big(\widehat{\boldsymbol{\Sigma}}\big) \Big\}, \qquad (2.25)$$

where $\widehat{\boldsymbol{\Sigma}} = n^{-1} \sum_{i=1}^{n} (\mathbf{X}_i - \bar{\mathbf{X}})(\mathbf{X}_i - \bar{\mathbf{X}})'$ is the unconstrained MLE of $\boldsymbol{\Sigma}$. Under H_0, Λ is approximately χ^2 with

$$\frac{1}{2}d(d+1) - \left\{ d(k+1) - \frac{1}{2}k(k-1) \right\} = \frac{1}{2}\big\{ (d-k)^2 - d - k \big\}$$

degrees of freedom. Bartlett (1954) has shown that the χ^2 approximation to the distribution of (2.25) can be improved by replacing n in (2.25) by $n - 1 - (2d + 4k + 5)/6$, which is often used in empirical studies. For example, Roll and Ross (1980) use this approach and conclude that there is an adequate number of factors in their empirical investigation of arbitrage pricing theory.

Example 2.11 (Factor analysis of 10 stock returns) *Consider the daily log returns of 10 stocks from January 4, 2000 to August 31, 2022 in Example 2.9. To fit the 10 stocks' daily log returns to the factor model (2.23) with $k = 4$ factors, one can first transform returns of each stock by $\tilde{\mathbf{x}}_i = (\mathbf{x}_i - \hat{\mu}_i)/\hat{\sigma}_i$, where $\hat{\mu}_i$ and $\hat{\sigma}$ are the sample mean and the sample standard deviation of stock i, respectively, during the sample period and given in Table 2.1. One can perform the factor analysis for transformed return series $\tilde{\mathbf{x}}_1, \ldots, \tilde{\mathbf{x}}_{10}$ to avoid the issue that one or more large variances of returns dominate the analysis. Table 2.3 shows the result of factor analysis with $k = 4$ by using the procedure explained above, including the factor loading matrix $\mathbf{B} = (\mathbf{b}_1, \ldots, \mathbf{b}_k)$, the diagonal matrix $\mathbf{V} = diag(v_1, \ldots, v_d)$, and the proportion of variance $\hat{\lambda}_i$ ($i = 1, \ldots, k$) explained by each factor. To check if the number of factors $k = 4$ is adequate, one can carry out the GLR test using the test statistic (2.25), which is given by $\Lambda = 35.53$ in this analysis. Since under H_0, Λ is approximately χ^2_{18}, the p-value of the GLR test is 0.0002, indicating that $k = 4$ is adequate.*

TABLE 2.3: Factor analysis of 10 stocks daily returns.

	$\mathbf{b}_1$	$\mathbf{b}_2$	$\mathbf{b}_3$	$\mathbf{b}_4$	v_i
AAPL	0.531	0.205	0.110	0.198	0.624
MSFT	0.629	0.248	0.244	0.235	0.428
AMZN	0.316	0.138	0.076	0.777	0.271
IBM	0.490	0.405	0.243	0.105	0.526
INTC	0.734	0.248	0.144	0.141	0.359
JPM	0.315	0.657	0.213	0.152	0.401
GE	0.268	0.674	0.242	0.103	0.405
CVX	0.235	0.471	0.362	0.056	0.588
PFE	0.152	0.272	0.600	0.092	0.534
JNJ	0.167	0.175	0.766	0.033	0.354
$\hat{\lambda}_i$	0.183	0.156	0.134	0.078	N/A

2.2.4 Copula

Although it is preferable to characterize the joint distribution of risk factors directly, it might be difficult to do so in practice. In such case, one may use copula methods to describe the joint distribution of risk factors via their marginal distributions.

A d-dimensional *copula* is a distribution function on $[0,1]^d$ with standard uniform marginal distributions. Let $C(\mathbf{u}) = C(u_1, \ldots, u_d)$ be the multivariate distribution function that is a copula. Then C is a mapping of the form $C : [0,1]^d \to [0,1]$, i.e., a mapping of the unit hypercube into the unit interval. Since $C(u_1, \ldots, u_d)$ is a distribution function, the following properties hold.

(i) $C(u_1, \ldots, u_d)$ is increasing in each component u_i.

(ii) $C(1, \ldots, 1, u_i, 1, \ldots, 1) = u_i$ for all $i \in \{1, \ldots, d\}$, $u_i \in [0, 1]$.

(iii) For all $(a_1, \ldots, a_d)$, $(b_1, \ldots, b_d) \in [0, 1]^d$ with $a_i \leq b_i$ we have

$$\sum_{i_1=1}^{2} \cdots \sum_{i_d=1}^{2} (-1)^{i_1 + \cdots + i_d} C(u_{1,i_1}, \ldots, u_{p,i_d}) \geq 0,$$

where $u_{j,1} = a_j$ and $u_{j,2} = b_j$ for all $j \in \{1, \ldots, d\}$.

The connection of marginal and joint distributions of a random vector is characterized by *Sklar's theorem* (Sklar, 1959) summarized below.

Theorem 2.1 (Sklar) *Let F be a joint distribution function with margins $F_1, \ldots, F_d$. Then there exists a copula $C : [0, 1]^d \to [0, 1]$ such that, for all $x_1, \ldots, x_d$ in $[-\infty, +\infty]$,*

$$F(x_1, \ldots, x_d) = C(F_1(x_1), \ldots, F_d(x_d)). \tag{2.26}$$

If the margins are continuous, then C is unique; otherwise C is uniquely determined on $\mathrm{Range}(F_1) \times \cdots \times \mathrm{Range}(F_d)$, where $\mathrm{Range}(F_i) = F_i([-\infty, +\infty])$ denotes the range of F_i. Conversely, if C is a copula and $F_1, \ldots, F_d$ are univariate distribution functions, then the function F defined in (2.26) is a joint distribution function with margins $F_1, \ldots, F_d$.

Note that (2.26) can be rewritten as

$$C(u_1, \ldots, u_d) = F(F_1^{-1}(u_1), \ldots, F_d^{-1}(u_d)), \tag{2.27}$$

in which $F_i^{-1}(y) = \inf\{x : F_i(x) \geq y\}$. Formulas (2.26) and (2.27) are fundamental in dealing with copulas.

Specifically, the copula function $C(u_1, \ldots, u_d)$ defined on $0 \leq u_i \leq 1$ ($1 \leq i \leq d$) can be used to provide a joint distribution function of $\mathbf{X} = (X_1, \ldots, X_d)'$ via

$$
\begin{aligned}
P\{X_1 \leq x_1, \ldots, X_d \leq x_d\} &= P\{F_1(X_1) \leq F_1(x_1), \ldots, F_d(X_d) \leq F_d(x_d)\} \\
&= C\big(F_1(x_1), \ldots, F_d(x_d)\big). \tag{2.28}
\end{aligned}
$$

A useful property of the copula of a distribution is its *invariance* under strictly increasing transformation of the marginals. In particular, let $(X_1, \ldots, X_d)'$ be a random vector with continuous marginals and copula C and let $F_1, \ldots, F_d$ be strictly increasing functions. It is easy to show that $(F_1(X_1), \ldots, F_d(X_d))$ also has copula C.

By definition, the value of a copula ranges from 0 to 1. The lower and upper bounds can be further improved by the *Fréchet bounds* for copulas. that is, for every copula $C(u_1, \ldots, u_d)$, we have the bounds

$$\max \left(\sum_{i=1}^{d} u_i + 1 - d, 0 \right) \leq C(u_1, \ldots, u_d) \leq \min(u_1, \ldots, u_d). \qquad (2.29)$$

Note that the above Fréchet bounds also implies bounds for joint and marginal distributions of a d-variate vector,

$$\max \left(\sum_{i=1}^{d} F_i(x_i) + 1 - d, 0 \right) \leq F(x_1, \ldots, x_d) \leq \min(F_1(x_1), \ldots, F_d(x_d)).$$

Commonly used copulas

There are three types of copulas: fundamental, implicit, and explicit copulas. Fundamental copulas include a number of important special dependence structures; implicit copulas are constructed from well-known multivariate distributions using Sklar's theorem; explicit copulas usually have simple closed-form expressions.

We first consider some fundamental copulas. The *independence copula* is defined as

$$C^{\mathrm{ind}}(u_1, \ldots, u_d) = \prod_{i=1}^{d} u_i. \qquad (2.30)$$

The *comonotonicity copula* is the Fréchet upper bound copula, that is,

$$C^{\mathrm{co}}(u_1, \ldots, u_d) = \min(u_1, \ldots, u_d). \qquad (2.31)$$

Note that this copula is the joint distribution function of the random vector $(U, \ldots, U)$, where $U \sim U(0, 1)$. The *countermonotonicity copula* is the two-dimensional Fréchet lower bound copula given by

$$C^{\mathrm{counter}}(u_1, u_2) = \max(u_1 + u_2 - 1, 0). \qquad (2.32)$$

This copula is the joint distribution function of the random vector $(U, 1 - U)$, where $U \sim U(0, 1)$.

Two commonly used implicit copulas include Gauss and t copulas. Specifically, if $\mathbf{X} \sim N_d(\boldsymbol{\mu}, \boldsymbol{\Sigma})$ is a Gaussian random vector ($\mathbf{R}$ is the correlation matrix of $\mathbf{X}$), then its copula is a so-called *Gauss copula* and given by

$$C_{\mathbf{R}}^{\mathrm{Ga}}(\mathbf{u}) = P(\Phi(X_1) \leq u_1, \ldots, \Phi(X_d) \leq u_d)) = \boldsymbol{\Phi}_{\mathbf{R}}(\Phi^{-1}(u_1), \ldots, \Phi^{-1}(u_d)),$$
$$(2.33)$$

where Φ denotes the standard univariate normal distribution function and $\boldsymbol{\Phi}_{\mathbf{R}}$ denotes the joint distribution function of $\mathbf{X}$. The Gauss copula can be expressed as an integral over the density of $\mathbf{X}$. In the case $d = 2$ and $|\rho| < 1$, we have

$$
\begin{aligned}
C_{\mathbf{R}}^{\mathrm{Ga}}(u_1, u_2) &= \Phi_2(\Phi^{-1}(u_1), \Phi^{-1}(u_2); \rho) \\
&= \int_{-\infty}^{\Phi^{-1}(u_1)} \int_{-\infty}^{\Phi^{-1}(u_2)} \frac{1}{2\pi\sqrt{1-\rho^2}} \exp\left\{ -\frac{s_1^2 - 2\rho s_1 s_2 + s_2^2}{2(1-\rho^2)} \right\} ds_1 ds_2,
\end{aligned}
\tag{2.34}
$$

where $\Phi_2(\cdot, \cdot; \rho)$ is the cumulative distribution function of two standard normally distributed random variables with correlation $\rho \in (-1, 1)$, and Φ is the cumulative distribution function of $N(0, 1)$. Figure 2.3 shows the perspective and contour plots of a bivariate Gauss copula with correlation $\rho = 0.7$.

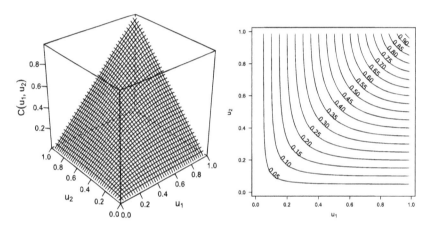

FIGURE 2.3: Perspective (left) and contour (right) plots of a bivariate Gauss copula with correlation $\rho = 0.7$.

Note that both the independence and comonotonicity copulas are special cases of the Gauss copula. If $\mathbf{R} = I_d$, we obtain the independence copula (2.30); if $\mathbf{R} = J_d$, the $d \times d$ matrix consisting entirely of ones, then we obtain comonotonicity (2.31). Also, for $d = 2, \rho = -1$ the Gauss copula is equal to the countermonotonicity copula (2.32). Thus in two dimensions the Gauss copula can be thought of as a dependence structure that interpolates between perfect positive and negative dependence, where the parameter ρ represents the strength of dependence.

Another important implicit copula is t-copula, which is a variant of the Gaussian copula. Suppose that $\mathbf{x} = (x_1, \ldots, x_d)'$ is a d-dimensional variable with $\mathrm{Var}(x_i) = 1$ for $1 \leq i \leq d$. Let $t_\nu(x)$ be the distribution function of a standard univariate t distribution with ν degrees of freedom,

$$
t_\nu(x) = \int_{-\infty}^{x} \frac{\Gamma(\frac{\nu+1}{2})}{\sqrt{\pi\nu}\Gamma(\frac{\nu}{2})} \left(1 + \frac{s^2}{\nu}\right)^{-\frac{\nu+1}{2}} ds,
$$

where $\Gamma(\cdot)$ is the Gamma function, and $\mathbf{t}_{\nu,\boldsymbol{\rho}}(\mathbf{x})$ be the joint distribution function of the vector with ν degrees of freedom, mean $\mathbf{0}$, and correlation matrix $\boldsymbol{\rho}$,

$$t_{\nu,\boldsymbol{\rho}}(x_1,\ldots,x_d) = \int_{-\infty}^{x_1} \cdots \int_{-\infty}^{x_d} \frac{\Gamma(\frac{\nu+d}{2})}{\Gamma(\frac{\nu}{2})(\nu\pi)^{d/2}|\boldsymbol{\rho}|^{1/2}} \left[1+\frac{1}{\nu}\mathbf{s}'\boldsymbol{\rho}^{-1}\mathbf{s}\right]^{-\frac{\nu+d}{2}} ds_1 \ldots ds_d.$$

Then the copula function

$$C_{\nu,\boldsymbol{\rho}}^t(u_1,\ldots,u_d) = \mathbf{t}_{\nu,\boldsymbol{\rho}}(t_\nu^{-1}(u_1),\ldots,t_\nu^{-1}(u_d)), \qquad (2.35)$$

is called a multivariate *Student t-copula* function. In the case $p = 2$, the distribution of a bivariate t distribution with degrees of freedom ν and correlation ρ is

$$t_{\nu,\rho}(x_1,x_2) = \int_{-\infty}^{x_1} \int_{-\infty}^{x_2} \frac{1}{2\pi\sqrt{1-\rho^2}}\left(1 + \frac{s^2+t^2-2\rho st}{\nu(1-\rho^2)}\right)^{-\frac{\nu+1}{2}} dsdt, \quad (2.36)$$

then a bivariate t copula is given by $C_{\nu,\rho}^t(u_1,u_2) = t_{\nu,\rho}(t_\nu^{-1}(u_1),t_\nu^{-1}(u_2))$. Figure 2.4 shows the perspective and contour plots of a bivariate t-copula with correlation $\rho = -0.4$ and degrees of freedom $\nu = 10$.

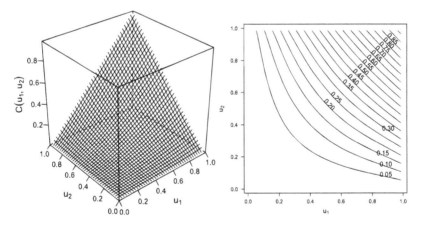

FIGURE 2.4: Perspective (left) and contour (right) plots of a bivariate t-copula with correlation $\rho = -0.4$ and degrees of freedom $\nu = 10$.

In contrast to implicit copulas that do not have simple analytical forms, explicit copulas can be expressed in closed forms. For example, a *Gumbel copula* is given by

$$C_\theta^{\text{Gu}}(u_1,\ldots,u_d) = \exp\left(-\left[(-\log u_1)^\theta + \cdots + (-\log u_d)^\theta\right]^{1/\theta}\right), \qquad \theta \geq 1. \tag{2.37}$$

A *Clayton copula* is expressed as

$$C_\theta^{\text{Cl}}(u_1, \ldots, u_d) = (u_1^{-\theta} + \cdots + u_d^{-\theta} - 1)^{-1/\theta}, \qquad \theta \geq 0. \tag{2.38}$$

In general, we can construct an *Archimedean copula* in the following way. Suppose ψ is a decreasing, continuous, convex function $\psi : [0, \infty) \to [0, 1]$ satisfying $\psi(0) = 1$ and $\lim_{t \to \infty} \psi(t) = 0$, then a p-variate Archimedean copula is defined as

$$C(u_1, \ldots, u_d) = \psi\big(\psi^{-1}(u_1) + \cdots + \psi^{-1}(u_d)\big). \tag{2.39}$$

Archimedean copulas are well-suited for modeling symmetric dependence structures of variables. They often have simpler computational algorithms than other copula families.

Copula densities

Copulas do not always have joint densities; the comonotonicity and countermonotonicity copulas are examples of copulas that are not absolutely continuous. However, the parametric copulas that we have met so far do have densities given by

$$c(u_1, \ldots, u_d) = \frac{\partial C(u_1, \ldots, u_d)}{\partial u_1 \cdots \partial u_d}. \tag{2.40}$$

For the implicit copulas of an absolutely continuous joint distribution function F with strictly increasing, continuous marginal distribution functions $F_1, \ldots, F_d$, the copula density is given by

$$c(u_1, \ldots, u_d) = \frac{f(F_1^{-1}(u_1), \cdots, F_d^{-1}(u_d))}{f_1(F_1^{-1}(u_1)) \cdots f_d(F_d^{-1}(u_d))}, \tag{2.41}$$

where f is the joint density of F, $f_1, \ldots, f_d$ are the marginal densities, and $F_1^{-1}, \cdots, F_d^{-1}$ are the ordinary inverses of the marginal distribution functions. For example, one can show that the density of the bivariate Gauss copula (2.34) is expressed as

$$c(u_1, u_2; \rho) = \frac{1}{\sqrt{1 - \rho^2}} \exp\left\{ -\frac{\rho^2(x_1^2 + x_2^2) - 2\rho x_1 x_2}{2(1 - \rho^2)} \right\}, \tag{2.42}$$

where $x_i = \Phi^{-1}(u_i)$ for $i = 1, 2$. Figure 2.5 shows the perspective and contour plots of densities of a bivariate Gauss copula with correlation $\rho = 0.7$ and a bivariate t-copula with correlation $\rho = 0.7$ (top) and degrees of freedom $\nu = 10$.

Conditional distributions of copulas

We now consider conditional distributions of copulas. We focus on two dimensions and suppose that (U_1, U_2) has distribution function C. Since a

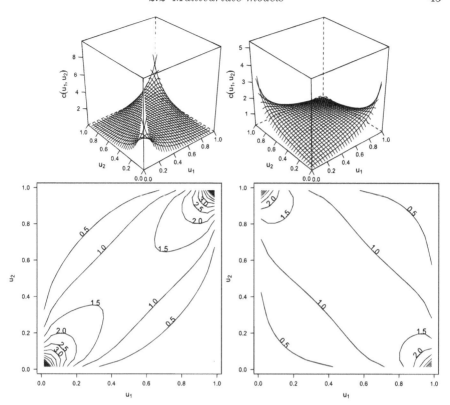

FIGURE 2.5: Perspective (left) and contour (right) plots of densities of a bivariate Gauss copula with correlation $\rho = 0.7$ and a bivariate t-copula with correlation $\rho = -0.4$ (top) and degrees of freedom $\nu = 10$ (bottom).

copula is an increasing continuous function in each argument,

$$
\begin{aligned}
C_{U_2|U_1}(u_2|u_1) &= P(U_2 \le u_2 | U_1 = u_1) \\
&= \lim_{\delta \to 0} \frac{C(u_1 + \delta, u_2) - C(u_1, u_2)}{\delta} = \frac{\partial}{\partial u_1} C(u_1, u_2),
\end{aligned}
\tag{2.43}
$$

where this partial derivative exists almost everywhere. A risk-management interpretation of the conditional distribution is the following. Suppose continuous risks (X_1, X_2) have the (unique) copula C. Then $1 - C_{U_2|U_1}(u_2|u_1)$ is the probability that X_2 exceeds its qth quantile given that X_1 attains its pth quantile. For a bivariate Gauss copula that is given by (2.34), its conditional distribution of U_2 given $U_1 = u_1$ is

$$
C_{U_2|U_1}(u_2|u_1) = \frac{\partial}{\partial u_1} C(u_1, u_2; \rho) = \Phi\left(\frac{\Phi^{-1}(u_2) - \rho \Phi^{-1}(u_1)}{\sqrt{1 - \rho^2}} \right).
\tag{2.44}
$$

Simulation of dependent variables

Let F be a d-variate joint distribution function with continuous marginals $F_1, \ldots, F_d$, and assume that F has a unique d-copula C. In order to simulate a vector $\mathbf{X} = (X_1, \ldots, X_d)'$ from F, it is sufficient to simulate a vector $\mathbf{U} = (U_1, \ldots, U_d)'$ from C. Then by Sklar's Theorem, $X_i = F_i^{-1}(U_i)$ $(i = 1, \ldots, d)$ have marginal distribution F_i and $\mathbf{X}$ has joint distribution F. We now consider a general procedure of simulating a sample from the copula C. For convenience, we assume that C is absolutely continuous.

Algorithm 2.1 (Simulation of copulas) *Assume the d-variate joint distribution F has continuous marginals $F_1, \ldots, F_d$ and a unique copula C.*

1. *Simulate $U_1 \sim U(0, 1)$ and denote by u_1 the simulated sample.*

2. *For $k = 2, \ldots, d$, simulate U_k conditional on $(u_1, \ldots, u_{k-1})$. Denote by $G_k(\cdot | u_1, \ldots, u_{k-1})$ the conditional distribution of U_k given $\{U_1 = u_1, \ldots, U_{k-1} = u_{k-1}\}$, which is expressed as*

$$
\begin{aligned}
G_k(u_k | u_1, \ldots, u_{k-1}) &= P(U_k \le u_k | U_1 = u_1, \ldots, U_{k-1} = u_{k-1}) \\
&= \frac{\partial_{u_1, \ldots, u_{k-1}} C(u_1, \ldots, u_k, 1, \ldots, 1)}{\partial_{u_1, \ldots, u_{k-1}} C(u_1, \ldots, u_{k-1}, 1, \ldots, 1)},
\end{aligned}
$$

where $\partial_{u_1, \ldots, u_{k-1}} C(u_1, \ldots, u_d)$ is the partial derivative of C with respect to $u_1, \ldots, u_{k-1}$. Simulate $U_k \sim U(0, 1)$ that is independent of $U_1, \ldots, U_{k-1}$ and take $u_k = G_k^{-1}(u_k | u_1, \ldots, u_{k-1})$.

3. *Let $\mathbf{x} = (x_1, \ldots, x_d)' := (F_1^{-1}(u_1), \ldots, F_d^{-1}(u_d))'$. Then $\mathbf{x}$ is a sample that has joint distribution F and marginals $F_1, \ldots, F_d$.*

Using the above algorithm, one can simulate explicit copulas that can be expressed in analytical forms. For example, Figure 2.6 shows the pairwise scatterplots of 1000 simulated points from the Clayton copula (2.38) with $d = 3$ and $\theta = 4$.

Algorithm 2.1 can be a little complicated, if the copula doesn't have an explicit form. For implicit copulas such as Gauss and t copulas, one can simulate them in a much simpler way. In particular, one can first simulate a random vector $\mathbf{X} = (X_1, \ldots, X_d)'$ that follows the distribution function $F(\mathbf{X})$, and then the random vector $\mathbf{U} = (F_1(X_1), \ldots, F_d(X_d))'$ has distribution function (or copula) C. This procedure can be used to simulate Gauss and t copulas. Figure 2.7 shows 500 simulated points from a bivariate Gauss copula with correlation 0.7 and a bivariate t-copula with correlation -0.4 and degrees of freedom $\nu = 10$.

Fitting copula to data

Suppose that the random vector $\mathbf{X} = (X_1, \ldots, X_d)'$ has joint distribution function F and continuous marginal distribution $F_1, \ldots, F_d$. Then by Sklar's

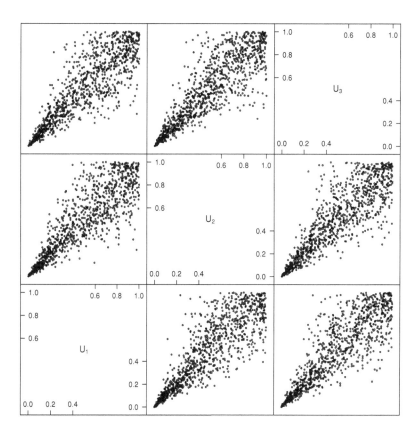

FIGURE 2.6: Pairwise scatterplots of 1000 simulated points from a trivariate Clayton copula with $\theta = 4$.

theorem, there exists a unique representation $F(x) = C(F_1(x_1), \ldots, F_d(x_d))$. Let $\mathbf{x}_1, \ldots, \mathbf{x}_n \in \mathbb{R}^d$ be observed samples that draw independently and identically from the distribution function F. We consider the estimation of a parametric copula C by the following two-step procedure.

The first step is to estimate margins and construct a *pseudo-sample* observations from the copula and the original sample $\mathbf{x}_1, \ldots, \mathbf{x}_n$. To do this, one needs to estimate the marginal distribution F_i $(i = 1, \ldots, d)$ using the univariate samples $x_{1,i}, \ldots, x_{n,i}$. There are two methods to obtain the marginal estimate $\widehat{F}_i$.

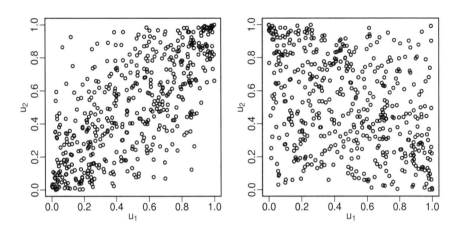

FIGURE 2.7: 500 simulated points from a bivariate Gauss copula with correlation 0.7 (left) and a bivariate t-copula with correlation -0.4 and degrees of freedom $\nu = 10$ (right).

(i) **Nonparametric estimation**. The marginal distribution F_i is estimated by

$$\widehat{F}_{i,n}(x) = \frac{1}{n+1} \sum_{t=1}^{n} I_{\{x_{t,i} \leq x\}}. \qquad (2.45)$$

Note that this is different from the usual empirical distribution function, as the denominator of the sum is $n+1$ instead of n. This guarantees that the pseudo-copula data $\widehat{F}_{i,n}(x_{t,i}) \in (0,1)$ so that the copula density can be evaluated at all pseudo-samples.

(ii) **Parametric estimation**. If the marginal distribution F_i is assumed to be a parametric distribution, one can use the maximum likelihood method to fit the F_i to the univariate samples $x_{1,i}, \ldots, x_{n,i}$. The parametric univariate distributions are usually chosen from distributions with heavier tails than normal, such as Student t, lognormal, or extreme value distributions.

After the estimates of the marginal distribution function $\widehat{F}_1, \ldots, \widehat{F}_d$ are obtained, one can construct the pseudo-sample from the copula, $\{\widehat{\mathbf{U}}_1, \ldots, \widehat{\mathbf{U}}_n\}$, in which

$$\widehat{\mathbf{U}}_t = (\widehat{U}_{t,1}, \ldots, \widehat{U}_{t,d})' = (\widehat{F}_1(x_{t,1}), \ldots, \widehat{F}_d(x_{t,d}))'.$$

The second step of the estimation procedure is to estimate the copula parameters by maximum likelihood from the pseudo-sample. Assume that the copula is parameterized by a finite dimensional vector $\boldsymbol{\theta}$, so that this parametric copula is expressed as $C_{\boldsymbol{\theta}}$. Denote by $c_{\boldsymbol{\theta}}$ the copula density of $C_{\boldsymbol{\theta}}$ as in (2.40). The *maximum likelihood estimate* (MLE) of $\boldsymbol{\theta}$ can be obtained by maximizing

$$\ln L(\boldsymbol{\theta}; \mathbf{X}) = \sum_{i=1}^{n} \ln c_{\boldsymbol{\theta}}(\mathbf{x}_i). \tag{2.46}$$

Since it is usually difficult to get an analytical solution that maximizes the log-likelihood (2.46), numerical algorithm is often used to maximize (2.46).

Following the two-step procedure, one can fit a number of copula models to data and select the best fitted model based on some information criterion (such as the Akaike information criterion).

Example 2.12 (Trivariate copulas for stock returns) *Consider the daily log returns of JPMorgan, Amazon, and Microsoft from January 4, 2000 to August 31, 2022. We use the above two-step procedure to fit a trivariate copula to these three stock returns. The first step is estimate the marginal distributions by the nonparametric method (2.45) and compute the pseudo-sample from the copula $\mathbf{U}_t = (U_{t,1}, U_{t,2}, U_{t,3})' = (U_{t,\text{JPM}}, U_{t,\text{AMZN}}, U_{t,\text{MSFT}})'$ that is shown in Figure 2.8.*

The second step is to fit the pseudo-sample to a parametric copula. We consider three copulas: the Gauss copula $C_{\mathbf{R}}^{\text{Ga}}$ (2.33), the t-copula $C_{\nu,\boldsymbol{\rho}}^t$ (2.35), and the Clayton copula C_{θ}^{Cl} (2.38) with $d = 3$. Table 2.4 shows the estimated parameters, their standard errors (in parenthesis), and maximized log-likelihood for three copula models. The t-copula in Table 2.4 shows the highest maximized log-likelihood among three models, suggesting a better goodness-of-fit than the other two copulas. □

TABLE 2.4: Estimated trivariate copulas for daily returns of three stocks.

Copula	Coefficients	Log-likelihood
$C_{\mathbf{R}}^{\text{Ga}}$	$R_{1,2} = 0.359$ (.011), $R_{1,3} = 0.448$ (.010) $R_{2,3} = 0.483$ (.009)	1485.0
$C_{\nu,\boldsymbol{\rho}}^t$	$\rho_{1,2} = 0.379$ (.013), $\rho_{1,3} = 0.468$ (.011) $\rho_{2,3} = 0.513$ (.011), $\widehat{\nu} = 3.584$ (.140)	2016.5
C_{θ}^{Cl}	$\widehat{\theta} = 0.615$ (0.015)	1321.9

2.2.5 Dependence measures

Dependence of random variables can be described from different perspectives, for example, linear correlation, concordance, and tail dependence. These perspectives are associated with different measures. The conventional linear correlation is commonly used in statistics and many applications, but it is only natural in the context of multivariate normal or more generally, elliptical models. Embrechts et al. (2002) discussed two fallacies caused by the linear

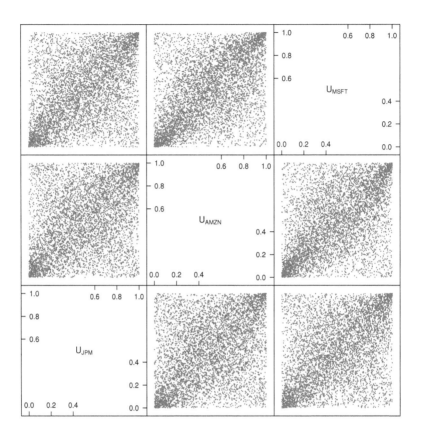

FIGURE 2.8: Pairwise scatterplots of a pseudo-sample from a copula for trivariate JPMorgan, Amazon and Microsoft daily log-returns in Example 2.12.

correlation. The first is that the marginal distributions and pairwise correlations of a random vector determine its joint distribution, and the second is that, for given univariate distributions F_1 and F_2 and any correlation value $\rho \in [-1, 1]$, one can always construct a joint distribution F with marginals F_1 and F_2 and correlation ρ.

In the follows, we introduce dependence measures that describe the concordance of random variables or their tails. For two points (x_1, x_2) and $(\tilde{x}_1, \tilde{x}_2)$ in $\mathbb{R}^2$, they are *concordant* if $(x_1 - \tilde{x}_1)(x_2 - \tilde{x}_2) > 0$ and *discordant* if $(x_1 - \tilde{x}_1)(x_2 - \tilde{x}_2) < 0$. Three measures, Kendall's tau, Spearman's rho, and tail dependence can be used to characterize such concordance.

Kendall's tau

Kendall's coefficient was first introduced by Fechner around 1900 and redis-covered by Kendall (1938), according to Nelsen (1991). It is a rank correlation of measuring concordance of bivariate random vectors.

Definition 2.5 *Given two random vectors (X_1, X_2) and (Y_1, Y_2) that are in-dependent and identically distributed with copula C, Kendall's tau, denoted by ρ_τ, is given by*

$$\rho_\tau = P[(X_1 - Y_1)(X_2 - Y_2) > 0] - P[(X_1 - Y_1)(X_2 - Y_2) < 0]. \quad (2.47)$$

For random variables with continuous distribution functions, Kendall's tau depends only on the unique copula C of (X_1, X_2) and an explicit formula can be obtained.

Theorem 2.2 *Suppose that random variables X_1 and X_2 have continuous marginal distributions and unique copula C. Kendall's tau for X_1 and X_2 is*

$$\rho_\tau(X_1, X_2) = 4 \int_0^1 \int_0^1 C(u_1, u_2) dC(u_1, u_2) - 1, \quad (2.48)$$

and

$$\rho_\tau(X_1, X_2) = 1 - 4 \int_0^1 \int_0^1 \frac{\partial C(u_1, u_2)}{\partial u_1} \frac{\partial C(u_1, u_2)}{\partial u_2} du_1 du_2. \quad (2.49)$$

Theorem 2.2 can be shown as follows. For (2.48), consider marginal dis-tribution functions F_1 and F_2 and $(\tilde{X}_1, \tilde{X}_2)$, which is an independent copy of (X_1, X_2). Using the interchangeability of the pairs (X_1, X_2) and $(\tilde{X}_1, \tilde{X}_2)$, we have

$$\begin{aligned}
\rho_\tau(X_1, X_2) &= 4P(X_1 < \tilde{X}_1, X_2 < \tilde{X}_2) - 1 \\
&= 4P(F_1(X_1) < F_1(\tilde{X}_1), F_2(X_2) < F_2(\tilde{X}_2)) - 1.
\end{aligned} \quad (2.50)$$

Let $U_i = F_i(X_i)$ and $\tilde{U}_i = F_i(\tilde{X}_i)$ for $i = 1, 2$. Since the distribution function of (U_1, U_2) and $(\tilde{U}_1, \tilde{U}_2)$ is $C(u_1, u_2)$, we obtain

$$\rho_\tau(X_1, X_2) = 4E[P(U_1 < \tilde{U}_1, U_2 < \tilde{U}_2)] - 1$$

$$= 4 \int_0^1 \int_0^1 P(U_1 < u_1, U_2 < u_2) dC(u_1, u_2) - 1 = 4 \int_0^1 \int_0^1 C(u_1, u_2) dC(u_1, u_2) - 1.$$

The equivalence of (2.48) and (2.49) can be obtained by using the following identify

$$\int_0^1 \int_0^1 C(u_1, u_2) dC(u_1, u_2) + \int_0^1 \int_0^1 \frac{\partial C(u_1, u_2)}{\partial u_1} \frac{\partial C(u_1, u_2)}{\partial u_2} du_1 du_2 = \frac{1}{2}.$$

Spearman's rho

Another type of rank correlation is Spearman's rho, which is defined as follows.

Definition 2.6 *Given three random vectors (X_1, X_2), (Y_1, Y_2), and (Z_1, Z_3) that are independent and identically distributed with copula C, Spearman's rho, denoted by ρ_S, is*

$$\rho_S = 3[P(X_1 - Y_1)(X_2 - Z_3) > 0) - P((X_1 - Y_1)(X_2 - Z_3) < 0)].$$

Spearman's rho and Kendall's tau are closely related. For three independent and identically distributed random vectors (X_1, X_2), (Y_1, Y_2), and (Z_1, Z_3), we may write

$$\rho_S(X_1, X_2) = 6P((X_1 - Y_1)(X_2 - Z_3) > 0) - 3$$
$$= 6P[(U_1 - \tilde{U}_1)(U_2 - \tilde{U}_3) > 0)] - 3, \qquad (2.51)$$

where $U_i = F_i(X_i)$, $i = 1, 2$, $\tilde{U}_1 = F_1(\tilde{Y}_1)$, and $\tilde{U}_3 = F_2(Z_3)$. Conditional on U_1 and U_2, we have

$$\rho_S(X_1, X_2) = 6E[P[(U_1 - \tilde{U}_1)(U_2 - \tilde{U}_3) > 0)|U_1, U_2]] - 3$$
$$= 6E(U_1U_2 + (1 - U_1)(1 - U_2)) - 3.$$

Since $E(U_i) = 1/2$, $i = 1, 2$, we have $\rho_S(X_1, X_2) = 12\mathrm{Cov}(U_1, U_2)$. Since $\mathrm{Var}(U_i) = 1/12$, $i = 1, 2$, the above argument implies the following theorem.

Theorem 2.3 *If X_1 and X_2 have continuous marginal distributions F_1 and F_2, then $\rho_S(X_1, X_2) = \mathrm{Corr}(F_1(X_1), F_2(X_2))$.*

For random variables with continuous distribution functions, an explicit formula can be obtained for Spearman's rho. In particular, suppose that random variables X_1 and X_2 have continuous marginal distributions and unique copula C. then Spearman's rho for X_1 and X_2 is given by

$$\rho_S(X_1, X_2) = 12 \int_0^1 \int_0^1 [C(u_1, u_2) - u_1 u_2] du_1 du_2. \qquad (2.52)$$

Example 2.13 (Tau and rho for the Gauss copula) *Let $\mathbf{X} = (X_1, X_2)$ have a bivariate distribution with the Gauss copula C_ρ^{Ga} and continuous margins. To compute Kendall's tau and Spearman's rho for $\mathbf{X}$, we note that the rank correlation ρ_τ and ρ_S is a copula property, hence we can simply assume that $\mathbf{X} \sim N(\mathbf{0}, \mathbf{R})$, where $\mathbf{R}$ is a correlation matrix with off-diagonal element ρ. Denote $\mathbf{Y} = \tilde{\mathbf{X}} - \mathbf{X}$, where $\tilde{\mathbf{X}}$ is an independent copy of $\mathbf{X}$. It is easy to see that $\mathbf{Y} \sim N(\mathbf{0}, 2\mathbf{R})$. For Kendall's tau, formula (2.50) implies*

$$\rho_\tau(X_1, X_2) = 4P(Y_1 > 0, Y_2 > 0) - 1.$$

To show this, consider the spherical transformation $Y_1 = R\cos\theta$ and $Y_2 = R(\rho\cos\theta + \sqrt{1-\rho^2}\sin\theta)$ where R is a positive radial random variable and θ is an independent uniformly distributed angle on $[\pi,\pi)$. Let $\phi = \arcsin\rho$. Then

$$P(Y_1 > 0, Y_2 > 0) = P(\cos\theta > 0, \sin(\theta + \phi) > 0) = \frac{1}{4} + \frac{\arcsin\rho}{2\pi}.$$

This shows that

$$\rho_\tau(X_1, X_2) = \frac{2}{\pi}\arcsin\rho. \tag{2.53}$$

To calculate Spearman's rho for $\mathbf{X}$, let $\mathbf{Z} = (Z_1, Z_2)'$ where Z_1, Z_2 are two independent standard normal random variables. Let $Y_i = X_i - Z_i$ $(i = 1, 2)$. Formula (2.51) implies that

$$\rho_S(X_1, X_2) = 3\big(2P(Y_1 Y_2 > 0) - 1\big) = 3(4P(Y_1 > 0, Y_2 > 0) - 1)$$

Note that now $\mathbf{Y} = (Y_1, Y_2)' \sim N(\mathbf{0}, \mathbf{R} + I_2)$, hence $\mathrm{corr}(Y_1, Y_2) = \rho/2$ and $P(Y_1 > 0, Y_2 > 0) = \frac{1}{4} + (2\pi)^{-1}\arcsin(\rho/2)$. This shows that

$$\rho_S(X_1, X_2) = \frac{6}{\pi}\arcsin\left(\frac{\rho}{2}\right). \tag{2.54}$$

Formulas (2.53) and (2.54) link the rank correlations directly to the correlation of the Gauss copula, hence can be used to calibrate the correlation of the Gauss copula in empirical studies. □

Tail dependence

The coefficients of tail dependence measure the concordance in the tail or extreme values of a bivariate distribution and are defined as follows.

Definition 2.7 *Let X_1 and X_2 be random variables with distribution functions F_1 and F_2. The coefficient of upper tail dependence of X_1 and X_2 is*

$$\lambda_u(X_1, X_2) = \lim_{q \to 1^-} P(X_2 > F_2^{-1}(q)|X_1 > F_1^{-1}(q)), \tag{2.55}$$

provided the limit exists in $[0,1]$. Analogously, the coefficient of lower tail dependence of X_1 and X_2 is

$$\lambda_l(X_1, X_2) = \lim_{q \to 0^+} P(X_2 \le F_2^{-1}(q)|X_1 \le F_1^{-1}(q)), \tag{2.56}$$

If $\lambda_u = 0$ (or $\lambda_l = 0$), X_1 and X_2 are *asymptotically independent* in the upper (or lower) tail. Otherwise, X_1 and X_2 have extremal dependence in the upper (or lower) tail. If the marginal distributions F_1 and F_2 are continuous so that the copula C of the bivariate function is unique, simple expressions can be obtained for λ_u and λ_l. Specifically, we have

$$\lambda_l = \lim_{q \to 0^+} \frac{C(q,q)}{q}, \qquad \lambda_u = \lim_{q \to 1^-} \frac{1 - 2q + C(q,q)}{1 - q}. \tag{2.57}$$

When the copula $C(u_1, u_2)$ is exchangeable (i.e., $C(u_1, u_2) = C(u_2, u_1)$ for all $u_1, u_2 \in [0, 1]$), applying L'Hôpital's rule, we obtain that

$$\lambda_l = \lim_{q \to 0^+} P(U_2 \le q | U_1 = q) + \lim_{q \to 0^+} P(U_1 \le q | U_2 = q) = 2 \lim_{q \to 0^+} P(U_2 \le q | U_1 = q).$$

Example 2.14 (Asymptotic independence of the Gauss copula) *Consider the bivariate Gauss copula C_ρ^{Ga} as in (2.34). Let $X_i = \Phi^{-1}(U_i)$ $(i = 1, 2)$, then (X_1, X_2) has a bivariate Gaussian distribution with standard margins and correlation ρ. Hence,*

$$\lambda_l = 2 \lim_{q \to 0^+} P(\Phi^{-1}(U_2) \le \Phi^{-1}(q) | \Phi^{-1}(U_1) = \Phi^{-1}(q)) = 2 \lim_{x \to \infty} P(X_2 \le x | X_1 = x).$$

Since the conditional distribution $X_2 | (X_1 = x)$ is a normal distribution with mean ρx and variance $1 - \rho^2$, it can be calculated that

$$\lambda_l = 2 \lim_{x \to \infty} \Phi\left(x \sqrt{\frac{1 - \rho}{1 + \rho}}\right) = 0.$$

The radial symmetry of the copula implies that $\lambda_u = \lambda_l = 0$. This shows that the Gauss copula is asymptotically independent in the tail.   □

Example 2.15 (Tail dependence of the t copula) *Consider the bivariate t copula (2.36). Let $X_i = t_\nu^{-1}(U_i)$ $(i = 1, 2)$, where t_ν is the distribution function of a univariate t distribution with ν degrees of freedom. To calculate the coefficient of lower tail dependence λ_l, we note that the conditional distribution given $X_1 = x$ is expressed as*

$$\left(\frac{\nu + 1}{\nu + x^2}\right)^{1/2} \frac{X_2 - \rho x}{\sqrt{1 - \rho^2}} \sim t_{\nu + 1}.$$

Making use of this result and using an argument analogous to Example 2.14, one can show that

$$\lambda_l^t = 2 t_{\nu + 1}\left(-\sqrt{\frac{(\nu + 1)(1 - \rho)}{(1 - \rho)}}\right).$$

By the radial symmetry of the copula, we have $l_u^t = l_l^t$. This shows that the bivariate t-copula is asymptotically dependent in both the upper and lower tails.   □

2.3   Time series models

To characterize the temporal aspects of risk factors, this section introduces univariate and multivariate time-series models. We start with basic concepts of time series analysis and then go over univariate ARMA models and volatility models. After that, multivariate linear time series models and volatility models are presented.

2.3.1 Weakly stationary series

A time series $\{x_t\}$ is said to be *stationary* (or *strictly stationary*) if the joint distribution of $(x_{t+h_1},\dots,x_{t+h_m})$ does not depend on t for any integer $m \geq 1$ and any intergers $h_1,\dots,h_m$. This indicates that shifting the time origin forward or backward t lags has no effect on the joint distribution of $(x_{h_1},\dots,x_{h_m})$ for any $m \geq 1$. This concept of stationarity is a little restrictive for time series analysis, hence in practice it is often useful to define stationary process in a weak sense by making use of the first two moments. Specifically, a time series $\{x_1, x_2, x_3, \dots\}$ is said to be *weakly stationary* (or *covariance stationary*) if Ex_t and $\mathrm{Cov}(x_t, x_{t+k})$ do not depend on $t \geq 1$. For a weakly stationary sequence $\{x_t\}$, $\mu := Ex_t$ is called its *mean* and $\gamma(h) := \mathrm{Cov}(x_t, x_{t+h})$, $h \geq 0$, is called the *autocovariance function*. If $\{x_t\}$ is stationary and $E|x_1| < \infty$, then $\lim_{n\to\infty} n^{-1} \sum_{t=1}^{n} x_t = \mu$.

By definition of the weakly stationary time series, the correlation coefficient $\mathrm{Corr}(x_t, x_{t+h})$ at lag h also does not depend on t and is given by $\rho(h) := \gamma(h)/\gamma(0)$, $h \geq 0$, which is called the *autocorrelation function* (ACF) of $\{x_t\}$. Suppose that $\sum_{h=1}^{\infty} |\gamma(h)| < \infty$. Given the sample $\{x_1,\dots,x_n\}$ from a weakly stationary time series, a consistent estimate of μ is the sample mean $\bar{x} = n^{-1} \sum_{t=1}^{n} x_n$ by the weak law of large numbers for weakly stationary sequences. The method of moments estimates the autocovariance function $\gamma(h)$ by

$$\widehat{\gamma}_h = \frac{1}{n-h} \sum_{t=1}^{n-h} (x_{t+h} - \bar{x})(x_t - \bar{x}) \tag{2.58}$$

and the autocorrelation function $\rho(h)$ by $\widehat{\rho}_h = \widehat{\gamma}_h/\widehat{\gamma}_0$ for $h = 1, 2, \dots$. Here, $\widehat{\gamma}_h$ and $\widehat{\rho}_h$ are referred to as the sample autocovariance function and the *sample ACF*, respectively.

Besides the ACF, another kind of measure that also describes the serial dependence of $\{x_t\}_{t\geq 1}$ is the *partial ACF* (PACF), which characterizes the dependence of two samples by removing the linear effect of all samples between these two. In particular, let x_h^{h-1} be the regression of x_h on $\{x_{h-1},\dots,x_1\}$, which is $x_h^{h-1} = \beta_0 + \beta_1 x_{h-1} + \cdots + \beta_{h-1} x_1$ ($\beta_0 = 0$ is the mean of x_t). Then the PACF of the process $\{x_t\}$ is defined as

$$\gamma_1^p = \mathrm{Corr}(x_1, x_0), \qquad \gamma_h^p = \mathrm{Corr}(x_h - x_h^{h-1}, x_0 - x_0^{h-1})$$

for $h \geq 2$. In practice, the PACF γ_h^p can be estimated as follows for the time series sample $\{x_1,\dots,x_n\}$. For each $h \geq 1$, regress x_t on $\{1, x_{t-1},\dots,x_{t-h}\}$ for $t = h+1,\dots,n$. Then γ_h^p can be estimated by the regression coefficient of x_{t-h}, we denote it by $\widehat{\gamma}_h^p$ and call it the *sample PACF*.

Tests of independence

Suppose x_t are independent and identically distributed, and their second moment exists, i.e., $Ex_t^2 < \infty$. Then $\rho_j = 0$ for all $j \geq 1$. Moreover, $\widehat{\rho}_h$ is asymptotically $N(0, 1/n)$ as $n \to \infty$, for any fixed $h \geq 1$, and $\widehat{\rho}_1,\dots,\widehat{\rho}_h$

are asymptotically independent. This property has been used to test the null hypothesis $\rho_h = 0$ with approximate significance level α. The test rejects H_0 if $|\hat{\rho}_h| \geq z_{1-\alpha/2}/\sqrt{n}$, where z_q is the qth quantile of the standard normal distribution. In addition, to test the null hypothesis $\rho_1 = \cdots = \rho_h = 0$, a widely used test statistic is the *Box-Pierce statistic* $Q^*(m)$ or the *Ljung-Box statistic* $Q(m)$, where

$$Q^*(m) = n \sum_{h=1}^{m} \hat{\rho}_h^2, \qquad Q(m) = n(n+2) \sum_{h=1}^{m} \frac{\hat{\rho}_h^2}{n-h}.$$

Both $Q^*(m)$ and $Q(m)$ are asymptotically χ_m^2 as $n \to \infty$ when the x_t are i.i.d. Therefore the null hypothesis is rejected if $Q(m)$ or $Q^*(m)$ exceeds the $1-\alpha$ quantile $\chi_{m;1-\alpha}^2$ of the χ_m^2-distribution. For a moderate sample size m, $Q(m)$ is better approximated by χ_m^2 than $Q^*(m)$.

2.3.2   Univariate ARMA processes

Consider a time series x_t with mean $\mu = E(x_t)$. Let $y_t = x_t - \mu$ be the series after removing the non-zero mean. Then the process $\{x_t\}$ is a zero-mean autoregressive and moving average (ARMA) process with orders p and q, denoted as ARMA(p, q), if it is weakly stationary and satisfies

$$y_t = x_t - \mu, \qquad y_t = \sum_{i=1}^{p} \phi_i y_{t-i} + u_t + \sum_{j=1}^{q} \psi_j u_{t-j}, \qquad (2.59)$$

where the innovations u_t are independent and identically distributed random variables with mean 0 and variance σ^2.

Stationary and invertible conditions

Under the stationary condition that all roots of equation $1 - \sum_{i=1}^{p} \phi_i z^i = 0$ lie outside the unit circle, the ARMA(p, q) process (2.59) is weakly stationary and has a representation of the form

$$y_t = \sum_{i=0}^{\infty} \alpha_i u_{t-i}, \qquad (2.60)$$

where the α_i are coefficients that satisfy $a_0 = 1$ and $\sum_{i=1}^{\infty} |\alpha_i| < \infty$. Then given the representation (2.60), the autocovariance and autocorrelation of the ARMA process (2.59) can be expressed as, for $h = \pm 1, \pm 2, \ldots$,

$$\gamma(h) = \sum_{i=0}^{\infty} \alpha_i \alpha_{i+|h|}, \qquad \rho(h) = \frac{\sum_{i=0}^{\infty} \alpha_i \alpha_{i+|h|}}{\sum_{i=0}^{\infty} \alpha_i^2}. \qquad (2.61)$$

Under the invertibility condition that all roots of equation $1 + \sum_{j=1}^{q} \psi_j z^j = 0$ lie outside the unit circle, the ARMA(p, q) process (2.59) is invertible, i.e., the innovation u_t can be recovered from an (infinite) linear sum of observations,

$$u_t = y_t + \sum_{j=1}^{\infty} \theta_j y_{t-j},$$

where the θ_j are coefficients that satisfy $\sum_{j=1}^{\infty} |\theta_j| < \infty$. In practice, the ARMA models we fit to real data are both invertible and stationary.

Yule-Walker equation for ACF

When $p = 0$ or $\phi_1 = \cdots = \phi_p = 0$ in (2.59), the ARMA(p, q) process reduces to a MA(q) process, $x_t = u_t + \sum_{j=1}^{q} \psi_j u_{t-j}$. Corresponding to this, the autocovariance and autocorrelation (2.61) become

$$\gamma(h) = \sum_{i=0}^{q-h} \phi_i \phi_{i+h}, \qquad \rho(h) = \frac{\sum_{i=0}^{q-h} \phi_i \phi_{i+h}}{\sum_{i=0}^{q} \phi_i^2}, \qquad h = 1, 2, \ldots.$$

When $q = 0$ or $\psi_1 = \cdots = \psi_q = 0$ in (2.59), the ARMA(p, q) process becomes an AR(p) process, $x_t = \sum_{i=1}^{p} \phi_i x_{t-i} + u_t$. For this AR$(p)$ process or the general ARMA(p, q) process, the ACF $\rho(h)$ can be calculated via the *Yule-Walker equation*. Specifically, multiplying through equation (2.59) by x_{t-h}, take expectation and divide by σ^2, assuming that the variance of x_t is finite. Then, using the fact $\rho(h) = \rho(-h)$ for all h, we obtain

$$\rho(h) = \phi_1 \rho(h-1) + \cdots + \phi_p \rho(h-p), \qquad h \geq \max(p, q+1).$$

This set of difference equation is called the Yule-Walker equations and has the general solution

$$\rho(h) = A_1 \pi_1^{-|h|} + \cdots + A_p \pi_p^{-|h|},$$

where $\pi_1, \ldots, \pi_p$ are the roots of the equation $1 - \sum_{i=1}^{p} \phi_i y^i = 0$, and the constants $A_1, \ldots, A_p$ are chosen to satisfy the initial conditions for $\rho(0), \ldots, \rho(p-1)$.

Example 2.16 *Consider the AR(2) process $x_t = x_{t-1} - \frac{1}{2} x_{t-2} + u_t$, where u_t are independent and identically distributed standard normal random variables. Since the roots of the equation $1 - z + \frac{1}{2} z^2 = 0$ are complex numbers, $1 \pm i$, which are outside the unit circle, the process is stationary. To find the ACF process, we note that the first Yule-Walker equation gives $\rho(1) = \rho(0) = \frac{1}{2} \rho(-1)$, then $\rho(-1) = \rho(1)$ gives $\rho(1) = 2/3$. For $k \geq 2$, the Yule-Walker equations are $\rho(h) = \rho(h-1) - \frac{1}{2} \rho(h-2)$. Then for $h = 0, 1, 2, \ldots$, the ACF of the process can be expressed as*

$$\rho(h) = A_1 (1+i)^{-h} + A_2 (1-i)^{-h} = 2^{-h/2} \left(A_1 e^{-i \frac{\pi h}{4}} + A_2 e^{i \frac{\pi h}{4}} \right).$$

Using $\rho(0) = 1$ and $\rho(1) = 2/3$, some algebra gives

$$\rho(h) = 2^{-h/2}(\cos\frac{\pi h}{4} + \frac{1}{3}\sin\frac{\pi h}{4}),$$

for $h = 0, 1, 2, \ldots$. □

Parameter estimation and order determination

Consider the ARMA(p, q) model (2.59) with $u_t \sim N(0, \sigma^2)$. Let $\boldsymbol{\theta} = (\mu, \phi_1, \ldots, \phi_p, \psi_1, \ldots, \psi_q)'$. The log-likelihood function is given by

$$l(\boldsymbol{\theta}, \sigma) = -\frac{n}{2}\log(2\pi\sigma^2) - \frac{1}{2\sigma^2}\sum_{t=1}^{n}\big(x_t - \mu - \phi_1(x_{t-1} - \mu) - \cdots - \phi_p(x_{t-p} - \mu)$$

$$-\psi_1 u_{t-1} - \cdots - \psi_q u_{t-q}\big)^2, \tag{2.62}$$

in which $u_1, \ldots, u_{n-1}$ can be retrieved recursively from (2.59) for a given $\boldsymbol{\theta}$ under the initial condition $x_0 = \cdots = x_{-p+1} = 0 = u_0 = \cdots = u_{-q+1}$.

An important practical issue is the choice of the order (p, q) of an ARMA model. The information criteria *Akaike information criterion* (AIC) and *Bayesian information criterion* (BIC) can be used to address this issue,

$$\text{AIC}(d) = -\frac{2}{n}l(\widehat{\boldsymbol{\theta}}, \widehat{\sigma}) + \frac{2d}{n}, \tag{2.63}$$

$$\text{BIC}(d) = -\frac{2}{n}l(\widehat{\boldsymbol{\theta}}, \widehat{\sigma}) + \frac{d\log n}{n}, \tag{2.64}$$

where n is the sample size, $d = p + q + 1$, $l(\boldsymbol{\theta}, \sigma)$ is the log-likelihood function of the ARMA(p, q) model, and $(\widehat{\boldsymbol{\theta}}, \widehat{\sigma})$ is the MLE. The model selection procedure is to choose the model with the smallest AIC or BIC. Note that we can replace $d = p + q + 1$ by $d = p + q$ in (2.63) or (2.64), as is done by some software packages, since we are basically comparing the information criteria of candidate ARMA(p, q) models. Whereas BIC is concerned with choosing the correct p and q when the data are indeed generated by a finite-order ARMA model, the AIC is designed to select the model that predicts best when the underlying model has infinite order.

Prediction in ARMA models

In the case that $\psi_1 = \cdots = \psi_q = 0$, the ARMA$(p, q)$ model becomes an AR(p) model, the conditional expectation $E(x_{t+1}|x_t, x_{t-1}, \ldots)$ is $\phi_1 x_t + \cdots + \phi_p x_{t-p+1}$. It is the minimum-variance one-step-ahead forecast $\widehat{x}_{t+1|t}$ of x_{t+1} based on the current and past observations $x_t, x_{t-1}, \ldots$. In general, we can assume the initial conditions, $x_0 = \cdots = x_{-p+1} = 0 = u_0 = \cdots = u_{-q+1}$ for an ARMA(p, q), so that the one-step-ahead forecasts can be written as

$$x_t(1) = E(x_{t+1}|x_t, x_{t-1}, \ldots) = \mu + \sum_{i=1}^{p}\phi_i(x_{t-i+1} - \mu) + \sum_{j=1}^{q}\psi_j u_{t-j+1}.$$

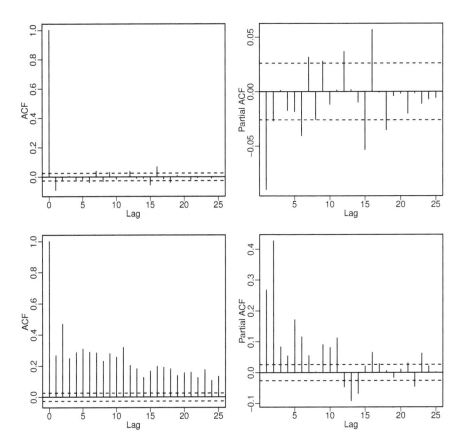

FIGURE 2.9: The ACF and PACF of the SPY daily log returns (top two) and squared log returns (bottom two).

The h-steps-ahead forecasts can be similarly obtained, that is,

$$x_t(h) = E(x_{t+h}|x_t, x_{t-1}, \dots) = \mu + \sum_{i=1}^{p} \phi_i(x_t(h-i) - \mu) + \sum_{j=1}^{q} \psi_j u_t(h-j),$$

where

$$x_t(h-i) = \begin{cases} E(x_{t+h-i}|x_t, x_{t-1}, \dots) & h > i \\ x_{t+h-i} & h \le i \end{cases}, \qquad u_t(h-j) = \begin{cases} 0 & h > j \\ u_{t+h-j} & h \le j \end{cases}.$$

Example 2.17 (ARMA models for SPY returns) *Consider the daily log return of the SPY ETF in Example 2.7. The top two panels of Figure 2.9 show that the ACF and PACF of the return series are significantly different from 0, and hence an ARMA model seems appropriate. Denote by* x_t

the SPY daily log return at time t. We fit an ARMA(1,1) model to the return series, and obtain the following estimated model

$$x_t = 0.000255_{(.00015)} + 0.2043_{(0.1608)}x_{t-1} + u_t - 0.2970_{(0.1575)}u_{t-1}, \quad u_t \sim N(0, 0.0124^2),$$

in which the standard errors of the estimated coefficients are shown in paren-thesis of the subscript. One can further use the fitted ARMA(1,1) model to calculate the 1-day and 2-day ahead predictions and their standard errors, which gives 0.0014 (0.0124) and 0.0005 (0.0125), respectively. Since the ob-served daily log returns at September 1 and 2, 2022 are 0.0031 and −0.011, which are interior points of the 95% prediction interval, it seems the fitted model performs well. However, the fitted model misses a salient feature of the return series, time varying volatility. The bottom two panel of Figure 2.9 show the temporal dependence of the squared returns r_t^2 over time, which involves time-varying volatilities discussed in the next section. □

2.3.3   GARCH and stochastic volatility models

Many financial time series display time-varying volatilities, which is a dif-ferent feature from the linear time series. There are generally two types of volatility models, the GARCH type and the stochastic volatility type. Most time series volatility models belong to the GARCH type, meaning that there is only one source of randomness in the model. In such case, the likelihood of the model has an analytical form with recursive representation of volatilities, and then the maximum likelihood method can be used to make inference on the model. In contrast to the GARCH type volatility models, the stochas-tic volatility type models have more than one source of randomness. In such a case, the likelihood of the model involves n-fold integrals and it is pro-hibitive to use the maximum likelihood method due to the computational cost. However, since stochastic volatility time series models are usually the discrete-time analogs of continuous-time stochastic volatility models, they are still widely used among financial practitioners. In this section, we introduce several GARCH type models and one stochastic volatility model.

ARCH(k) models

To better describe the idea, we may represent u_t having mean zero in the form

$$u_t = \sigma_t \epsilon_t, \tag{2.65}$$

where $\{\epsilon_t\}$ denotes a sequence of i.i.d. random variables with zero mean and unit variance, and σ_t may be thought of as the local conditional standard deviation of the process. The ϵ_t may have a normal distribution but this as-sumption is not necessary for much of what follows. For example, one may assume that ϵ_t follow a standardized Student t-distribution. Let x_ν be a Stu-dent t-distribution with $\nu > 2$ degrees of freedom. Then $\epsilon_t = x_\nu / \sqrt{\nu/(\nu-2)}$

has a *standardized Student t-distribution* with variance 1 and probability density function

$$f(\epsilon) = \frac{\Gamma\big((\nu+1)/2\big)}{\Gamma(\nu/2)\sqrt{(\nu-2)\pi}}\left(1+\frac{\epsilon^2}{\nu-2}\right)^{-(\nu+1)/2}. \qquad (2.66)$$

In any case the unconditional distribution of u_t generated by a non-linear model will not generally be normal but rather fat-tailed (or leptokurtic). Suppose we additionally assume that the square of σ_t depends on the most recent value of the derived series by

$$\sigma_t^2 = \omega + \sum_{i=1}^{k}\alpha_i u_{t-i}^2, \qquad (2.67)$$

where the parameters ω and $\alpha_1,\dots,\alpha_k$ are nonnegative and satisfy $\alpha_1+\cdots+\alpha_k < 1$ to ensure that σ_t^2 is non-negative. A model for u_t satisfying equations (2.65) and (2.67) is called an *autoregressive conditionally heteroscedastic model of order k* or ARCH(k) (Engle, 1982). The adjective "autoregressive" arises because the value of σ_t^2 depends on past values of the derived series, albeit in squared form. Since ARCH(k) models and their inference can be viewed as special cases of GARCH models, we present them together with GARCH models that introduces next.

GARCH(k, h) models

The ARCH model has been generalized to allow the variance to depend on past values of σ_t^2 as well as on past values of u_t^2. A derived variable satisfying equation (2.65) is said to follow a *generalized ARCH* or GARCH (Bollerslev, 1986) model of order (k, h) when the local conditional variance is given by

$$\sigma_t^2 = \omega + \sum_{i=1}^{k}\alpha_i u_{t-i}^2 + \sum_{j=1}^{h}\beta_j \sigma_{t-j}^2, \qquad (2.68)$$

where $\omega \geq 0$ and $\alpha_i, \beta_j \geq 0$ for all i, j.

The GARCH(k, h) model (2.68) can be considered as an ARMA model of volatility with martingale difference innovations. Let $\beta_i = 0$ if $i > h$, $\alpha_j = 0$ if $j > k$, and $\eta_t = u_t^2 - \sigma_t^2$. Note that

$$E(\eta_t) = E(\sigma_t^2\epsilon_t^2) - E(\sigma_t^2) = E(\epsilon_t^2 E(\sigma_t^2|\mathcal{F}_{t-1})) - E(\sigma_t^2) = 0.$$

Then η_t is a martingale difference. Moreover,

$$u_t^2 = \omega + \sum_{j=1}^{\max(h,k)}(\alpha_j+\beta_j)u_{t-j}^2 + \eta_t - \sum_{i=1}^{h}\beta_i\eta_{t-i}.$$

Hence, the same invertibility and stationarity assumptions of ARMA models apply to GARCH models. For example, to ensure that u_t is covariance stationary, it is required that all roots of $1 - \sum_{j=1}^{\max(h,k)}(\alpha_j + \beta_j)z^j = 0$ lie outside the unit circle. Since the α's and β's are usually assumed to be nonnegative, then we have $\sum_{j=1}^{k}\alpha_j + \sum_{i=1}^{h}\beta_i < 1$. Under these conditions, the unconditional variance of u_t is given by

$$E(u_t^2) = \frac{\omega}{1 - \sum_{i=1}^{h}\beta_i - \sum_{j=1}^{k}\alpha_j}.$$

GARCH models can capture the volatility clustering and high leptokurtic features of return series. To illustrate this, we examine the GARCH(1, 1) case,

$$u_t = \sigma_t \epsilon_t, \qquad \sigma_t^2 = \omega + \alpha u_{t-1}^2 + \beta \sigma_{t-1}^2, \tag{2.69}$$

where $\omega > 0$, $\alpha, \beta > 0$, and $\alpha + \beta < 1$. The model implies that a large u_{t-1}^2 or σ_{t-1}^2 will lead to a large σ_t^2, which in turn will give rise to a large $u_t^2 = \sigma_t^2 \epsilon_t^2$. This is consistent with the volatility clustering observed in financial time series. Moreover, even when ϵ_t is standard normal, GARCH(1, 1) can still be highly leptokurtic since

$$\frac{E(u_t^4)}{[\mathrm{Var}(u_t)]^2} = \frac{3[1 - (\alpha + \beta)^2]}{1 - (\alpha + \beta)^2 - 2\alpha^2} > 3$$

when $(\alpha + \beta)^2 + 2\alpha^2 < 1$.

We next consider the inference of GARCH(k, h) models. Let $\boldsymbol{\theta} = (\omega, \alpha_1, \ldots, \alpha_k, \beta_1, \ldots, \beta_h)'$. Assuming that the ϵ_t are i.i.d. $N(0, 1)$, the log-likelihood function in the GARCH(1, 1) model (2.68) is given by

$$l(\boldsymbol{\theta}) = -\frac{1}{2}\sum_{t=1}^{n}\log\sigma_t^2 - \sum_{t=1}^{n}\frac{u_t^2}{2\sigma_t^2} - \frac{n}{2}\log(2\pi), \tag{2.70}$$

in which the σ_t^2 can be computed recursively by (2.68) when the initial values $\sigma_0^2, \ldots, \sigma_{1-h}^2$ and the value of $\boldsymbol{\theta}$ are given. Then MLE $\hat{\boldsymbol{\theta}}$ is the solution of

$$0 = \nabla l(\boldsymbol{\theta}) = -\frac{1}{2}\sum_{t=1}^{n}\left[\frac{1}{\sigma_t^2} - \frac{u_t^2}{\sigma_t^4}\right]\nabla\sigma_t^2, \tag{2.71}$$

in which $\nabla\sigma_t^2$ can be evaluated recursively by

$$\nabla\sigma_t^2 = (1, u_{t-1}^2, \ldots, u_{t-k}^2, \sigma_{t-1}^2, \ldots, \sigma_{t-h}^2)' + \sum_{i=1}^{h}\beta_i\nabla\sigma_{t-i}^2.$$

A variant of the GARCH model above assumes that the ϵ_t follow a standardized Student t-distribution with density function given by (2.66), in which

$\nu \geq 2$ is treated as an unknown parameter. The underlying motivation is to allow a heavier tail in the distribution of u_t^2 and to use the data to determine how heavy the tail should be. In this case, the likelihood function is

$$l(\boldsymbol{\theta}, \nu) = n \log \left(\frac{\Gamma((\nu+1)/2)}{\Gamma(\nu/2)\sqrt{(\nu-2)\pi}} \right) - \frac{1}{2} \sum_{t=1}^{n} \log \sigma_t^2 - \frac{\nu+1}{2} \sum_{t=1}^{n} \log \left(1 + \frac{u_t^2}{(\nu-2)\sigma_t^2} \right).$$

ARCH and GARCH models can be easily used to forecast volatilities of a time series. Take the GARCH(1,1) model (2.69) as an example. Let $\sigma_t^2(m) = E(\sigma_{t+m}^2 | u_t, u_{t-1}, \ldots)$ be the m-step-ahead forecast of σ_{t+m}^2 at the forecast origin t. Replacing time index t by $t+1$ in (2.69) and taking expectations on both sides yields that

$$\sigma_t^2(1) = E(\sigma_{t+1}^2 | u_t, u_{t-1}, \ldots) = \omega + \alpha u_t^2 + \beta \sigma_t^2, \qquad \lambda = \alpha + \beta,$$

where σ_t can be estimated by fitting the model (2.69) to observations $u_1, \ldots, u_t$. Then the m-step-ahead forecast of σ_{t+m}^2 can be calculated as follows

$$\sigma_t^2(m) = \omega(1 - \lambda^m)/(1 - \lambda) + (\alpha u_t^2 + \beta \sigma_t^2)\lambda^{m-1}.$$

Similarly, for the GARCH(h, k) model (2.68), we can obtain the following one-step and k-step ahead forecasts

$$\sigma_t^2(m) = \omega + \sum_{j=1}^{k} \alpha_j u_t^2(m-j) + \sum_{j=1}^{h} \beta_i \sigma_t^2(m-i),$$

where the $(m-i)$-steps-ahead forecast $u_t^2(m-i)$ are defined as $u_t^2(m-j) = \sigma_t^2(m-j)$ for $k > j$, and $u_t^2(m-j) = u_{t+k-j}^2$ for $k \leq j$.

Exponential GARCH models

ARCH and GARCH models characterize the conditional variance of a time series as a linear function of the squared past innovations and observations, hence positive and negative shocks on the series have the same effects on volatility. In practice, it is well known that the asset prices respond differently to positive and negative shocks, hence various extensions of ARCH and GARCH models have been introduced to account for the asymmetric effect of positive and negative shocks on volatilities.

One such kind of volatility models are the *exponential GARCH* (Nelson, 1991). An EGARCH(h, k) model for the series u_t is given by $u_t = \sigma_t \epsilon_t$ and

$$u_t = \sigma_t \epsilon_t, \quad \log(\sigma_t^2) = \omega + \sum_{i=1}^{h} \beta_i \log(\sigma_{t-i}^2) + \sum_{j=1}^{k} f_j(\epsilon_{t-j}), \qquad (2.72)$$

where the ϵ_t are i.i.d. with mean 0 and $f_j(\epsilon) = \alpha_j \epsilon + \gamma_j(|\epsilon| - E|\epsilon|)$. Note that the random variable $f_j(\epsilon_t)$ is the sum of two zero-mean random variables $\alpha_j \epsilon_t$ and $\gamma_j(|\epsilon_t| - E|\epsilon_t|)$. We can rewrite $f_j(\epsilon_t)$ as

$$
f_j(\epsilon_t) = \begin{cases} (\alpha_j + \gamma_j)\epsilon_t - \gamma_j E|\epsilon_t|, & \text{if } \epsilon_t \geq 0, \\ (\alpha_j - \gamma_j)\epsilon_t - \gamma_j E|\epsilon_t|, & \text{if } \epsilon_t < 0, \end{cases}
$$

which shows the asymmetry of the volatility response to positive and negative returns. Since (2.72) represents $\log(\sigma_t^2)$ in ARMA form with innovations $f_j(\epsilon_{t-j})$, σ_t^2 and therefore u_t also are stationary if the zeros of $1 - \beta_1 z - \cdots - \beta_h z^h$ lie outside the unit circle.

The EGARCH model can be estimated as follows. Let $\boldsymbol{\theta} = (\omega, \alpha_1, \ldots, \alpha_k, \gamma_1, \ldots, \gamma_k, \beta_1, \ldots, \beta_h)'$. Given initial values $\sigma_0^2, \ldots, \sigma_{1-h}^2$ and assuming that ϵ_t are standard normal, the log-likelihood function $l(\boldsymbol{\theta})$ has the same form as (2.70), where σ_t^2 is defined recursively by (2.72) and $E|\epsilon_t| = \sqrt{2/\pi}$. The MLE $\hat{\boldsymbol{\theta}}$ is the solution of (2.71) in which $\nabla \sigma_t^2$ can be calculated recursively by the following

$$
\begin{aligned}
\nabla \sigma_t^2 &= \nabla \left\{ \exp \left(\omega + \sum_{j=1}^{k} \left[\alpha_j \epsilon_{t-j} + \gamma_j \left(|\epsilon_{t-j}| - \sqrt{\frac{2}{\pi}} \right) \right] + \sum_{i=1}^{h} \beta_i \log \sigma_{t-i}^2 \right) \right\} \\
&= \sigma_t^2 \left\{ \left(1, \epsilon_{t-1}, \ldots, \epsilon_{t-k}, |\epsilon_{t-1}| - \sqrt{2/\pi}, \ldots, |\epsilon_{t-k}| - \sqrt{2/\pi}, \right. \right. \\
&\qquad \left. \left. \log \sigma_{t-1}^2, \ldots, \log \sigma_{t-h}^2 \right)' + \sum_{i=1}^{h} \frac{\beta_i}{\sigma_{t-i}^2} \nabla \sigma_{t-i}^2 \right\}.
\end{aligned}
$$

To see how to use EGARCH model to forecast volatilities, we take the EGARCH(1, 1) model with standard normal ϵ_t as an example, for which $\log(\sigma_t^2) = \omega + \beta \log(\sigma_{t-1}^2) + f(\epsilon_{t-1})$, or equivalently

$$
\sigma_t^2 = \left(\sigma_{t-1} \right)^{2\beta} \exp \left\{ \omega + f(\epsilon_{t-1}) \right\}, \tag{2.73}
$$

where $f(\epsilon) = \alpha \epsilon + \gamma(|\epsilon| - \sqrt{2/\pi})$. Define $g(c) = E\{ \exp[c(\omega + f(\epsilon))] \}$ for $\epsilon \sim N(0,1)$ and $c > 0$. Let Φ be the standard normal distribution function, then $g(c)$ is expressed as

$$
\begin{aligned}
g(c) &= e^{c\omega} \int_{-\infty}^{\infty} e^{-\epsilon^2/2} \exp \left\{ c\alpha\epsilon + c\gamma(|\epsilon| - \sqrt{2/\pi}) \right\} d\epsilon / \sqrt{2\pi} \\
&= e^{c(\omega - \gamma\sqrt{2/\pi})} \left\{ e^{(\alpha+\gamma)^2 c^2/2} \Phi[c(\gamma + \alpha)] + e^{(\alpha-\gamma)^2 c^2/2} \Phi[c(\gamma - \alpha)] \right\}.
\end{aligned}
$$

Let $\mathcal{F}_t = \{\epsilon_s : s \leq t\}$. In view of (2.73), using some algebra, we obtain the k-step prediction of the volatility

$$
\sigma_t^2(k) = g(1)g(\beta) \ldots g(\beta^{k-1}) \left(\sigma_t^2 \right)^{\beta^k}.
$$

Asymmetric power ARCH models

Besides the EGARCH model, another kind of extensions that account for the asymmetric effect of positive and negative shocks on volatilities are *asymmetric power ARCH* (or APARCH/APGARCH) models (Ding et al., 1993). Let ϵ_t be a sequence of i.i.d. variables with mean 0 and variance 1. An APGARCH(h, k) model for the series u_t is given by $u_t = \sigma_t \epsilon_t$ and

$$\sigma_t^\delta = \omega + \sum_{i=1}^h \alpha_i \big(|\epsilon_{t-j}| - \gamma_i \epsilon_{t-i} \big)^\delta + \sum_{j=1}^k \beta_j \sigma_{t-j}^\delta, \tag{2.74}$$

in which $\omega > 0$, $\alpha_j \geq 0$, $\beta_i \geq 0$, and $|\gamma_i| \leq 1$. The APARCH models are a very general class of models, which include as special cases the TS-GARCH model (Taylor, 1986; Schwert, 1989), the GJR-GARCH model (Glosten et al., 1993), the T-ARCH model (Zakoian, 1994), the N-ARCH model (Higgins and Bera, 1992), and the Log-ARCH model (Geweke, 1986; Pantula, 1986).

The APARCH model can be estimated via the maximum likelihood method. Let $\boldsymbol{\theta} = (\omega, \alpha_1, \ldots, \alpha_k, \gamma_1, \ldots, \gamma_k, \beta_1, \ldots, \beta_h)'$. Given initial values $\sigma_0^2, \ldots, \sigma_{1-h}^2$ and assuming that ϵ_t are standard normal, the log-likelihood function $l(\boldsymbol{\theta})$ has the same form as (2.70), where σ_t^2 is defined recursively by (2.74). The MLE $\widehat{\boldsymbol{\theta}}$ is the solution of (2.71) in which $\nabla \sigma_t^2$ can be calculated recursively by the following

$$
\begin{aligned}
\nabla \sigma_t^\delta \;=\; & \Big(1, \big(|\epsilon_{t-1}| - \gamma_i \epsilon_{t-1} \big)^\delta, \ldots, \big(|\epsilon_{t-k}| - \gamma_i \epsilon_{t-k} \big)^\delta, \\
& -\alpha_1 \delta \big(|\epsilon_{t-1}| - \gamma_1 \epsilon_{t-1} \big)^{\delta-1} \epsilon_{t-1}, \ldots, -\alpha_k \delta \big(|\epsilon_{t-k}| - \gamma_k \epsilon_{t-k} \big)^{\delta-1} \epsilon_{t-1}, \\
& \log \sigma_{t-1}^2, \ldots, \log \sigma_{t-h}^2 \Big)' + \sum_{i=1}^h \beta_i \nabla \sigma_{t-i}^\delta.
\end{aligned}
$$

The forecasting of volatilities by the APARCH model can be done analogously as the GARCH and EGARCH models. Denote by $\mathcal{F}_t$ the set of observations up to time t, in view of (2.74), the m-step ahead of forecast of σ_t^2 or the conditional expectation of σ_{m+t}^δ given $\mathcal{F}_t$ can be recursively calculated by

$$E(\sigma_{t+m}^\delta | \mathcal{F}_t) = \omega + \sum_{i=1}^h \alpha_i E\big((|\epsilon_{t+m-j}| - \gamma_i \epsilon_{t+m-i})^\delta | \mathcal{F}_t \big) + \sum_{j=1}^k \beta_j E(\sigma_{t+m-j}^\delta | \mathcal{F}_t).$$

Example 2.18 (GARCH, EGARCH, and APARCH models for SPY returns) *Consider the daily log return of the SPY ETF in Examples 2.7 and 2.17. The bottom two panels of Figure 2.9 show the ACF and PACF of the squared SPY daily returns. Comparing to the ACF and PACF of the SPY returns, which are shown in the top two panels of Figure 2.9, we notice that the ACF and PACF of squared returns are even stronger than those of the original return series. This indicates the volatility has a significant temporal dependence. Denote by x_t the SPY return series. Suppose that $x_t = \mu + \sigma_t \epsilon_t$, where ϵ_t are i.i.d. standard normal random variables. We assume*

TABLE 2.5: Estimated GARCH, EGARCH, and APARCH models for the SPY returns.

	μ	ω	α_1	β_1	γ_1	δ
GARCH(1,1)	6.75e-4	2.42e-6	0.1283	0.8552		
	(1.05e-4)	(7.30e-7)	(.0099)	(.0103)		
EGARCH(1,1)	3.06e-4	−0.2786	−0.1495	0.9696	0.1643	
	8.59e-5	(.0011)	(.0071)	(.0003)	(.0043)	
APARCH(1,1)	2.20e-4	3.45e-4	0.0972	0.8948	0.9542	0.9883
	(9.23e-5)	(1.53e-4)	(.0071)	(.0071)	(.0081)	(.0830)

that σ_t follows three volatility models: GARCH(1,1) in (2.69), EGARCH(1,1) in (2.72), and APARCH(1,1) in (2.74). Table 2.5 shows the estimated parameters and their standard errors (in parenthesis) using the maximum likelihood method.   □

Stochastic volatility models

Besides the above GARCH type volatility models, another type of volatility models are the stochastic volatility models. For illustration purpose, we consider the following *stochastic volatility* (SV) model

$$u_t = \sigma_t \epsilon_t, \quad \sigma_t^2 = e^{h_t}, \quad h_t = \phi_0 + \phi_1 h_{t-1} + \cdots + \phi_p h_{t-p} + \eta_t, \qquad (2.75)$$

which has AR(p) dynamics for $\log \sigma_t^2$. The ϵ_t and η_t in (2.75) are assumed to be independent normal random variables with $\epsilon_t \sim N(0,1)$ and $\eta_t \sim N(0,\sigma^2)$. We usually assume that roots of $1 - \phi_1 y - \cdots - \phi_p y^p = 0$ lie outside the unit circle so that the log volatility process h_t is stationary. A complication of the SV model is that unlike in usual AR(p) models, the h_t in (2.75) is an unobserved state undergoing AR(p) dynamics, while the observations are u_t such that $u_t | h_t \sim N(0, e^{h_t})$.

To make inference on the SV model (2.75), we note that, since the likelihood function of $\boldsymbol{\theta} = (\sigma, \phi_0, \ldots, \phi_p)^T$ involves n-fold integrals, it is prohibitively difficult to compute the MLE by numerical integration for usual sample sizes. Usually the SV models can be estimated by the *quasi-maximum likelihood* (QML) method or the *Markov chain Monte Carlo* (MCMC) method. Here, we introduce the QML method for the SV model, and the MCMC method for the SV model is introduced in Section 7.4.1.

To fix the ideas, we consider the following the case with $p = 1$, $\phi_0 = \omega$ and $\phi_1 = \phi$. Letting $y_t = \log u_t^2$, note that $y_t = h_t + \log \epsilon_t^2$, where $\log \epsilon_t^2$ is distributed like $\log \chi_1^2$, which has mean -1.27. Let $\xi_t = \log \epsilon_t^2 - E(\log \chi_1^2)$. Note that this is a linear state-space model with unobserved states h_t and observations y_t satisfying

$$h_t = \omega + \phi h_{t-1} + \eta_t, \qquad y_t = h_t + E(\log \chi_1^2) + \xi_t.$$

Let $h_{t|t-1}$ denote the minimum-variance linear predictor of h_t and $h_{t-1|t-1}$ denote the best linear estimate of h_{t-1} based on observations up to time $t-1$. Then by Theorem 6.11, the Kalman filter gives the recursion

$$h_{t|t-1} = \omega + \phi h_{t-1|t-1}, \quad h_{t|t} = h_{t|t-1} + V_{t|t-1}\left[y_t - E(\log \chi_1^2) - h_{t|t-1}\right]/v_t,$$
$$V_{t|t-1} = \phi^2 V_{t-1|t-1} + \sigma^2, \quad V_{t|t} = V_{t|t-1}(1 - V_{t|t-1}/v_t),$$

where $v_t = V_{t|t-1} + \text{Var}(\log \chi_1^2)$ is the variance of $e_t := y_t - E(\log \chi_1^2) - h_{t|t-1}$ and $V_{t|t} = \text{Var}(h_{t|t})$. If ξ_t were normal (with mean 0 and variance $\text{Var}(\log \chi_1^2)$), then e_t would be $N(0, v_t)$ and therefore the log-likelihood function would be given by

$$l(\omega, \phi, \sigma^2) = -\frac{1}{2}\sum_{t=1}^{n} \log v_t - \frac{1}{2}\sum_{t=1}^{n} e_t^2/v_t \qquad (2.76)$$

up to additive constants that do not depend on (ω, ϕ, σ^2). By a result of Dunsmuir (1979), the estimator that maximizes (2.76) is still consistent and asymptotically normal. However, since the log-likelihood (2.76) is a quasi-likelihood, the QML estimator that maximizes (2.76) has been shown to have a larger mean squared error than the estimates from the MCMC method.

2.3.4 Multivariate ARMA processes

This section considers multivariate ARMA models for multiple series of financial risk-factor data. The idea of the multivariate ARMA models is similar to that of univariate ARMA models in Section 2.3.2 except that a new concept, cointegration, arises for the joint movement of multiple time series.

A d-dimensional time series $\mathbf{x}_t = (x_{t1}, \cdots, x_{td})'$ is called *weakly stationary* (or *stationary in the weak sense*) if its mean vectors and covariance matrices are time-invariant and therefore can be denoted by

$$\boldsymbol{\mu} = E(\mathbf{x}_t), \quad \boldsymbol{\Gamma}_h = E[(\mathbf{x}_t - \boldsymbol{\mu}_t)(\mathbf{x}_{t+h} - \boldsymbol{\mu}_{t+h})'] = (\Gamma_{h,ij})_{1 \leq i,j \leq d}. \qquad (2.77)$$

Letting $\mathbf{D} = \text{diag}\{\sqrt{\Gamma_{0,11}}, \cdots, \sqrt{\Gamma_{0,dd}}\}$, the *cross-correlation matrix* of $\mathbf{x}_t$ is given by

$$\boldsymbol{\rho}_0 = (\rho_{0,ij})_{1 \leq i,j \leq d} = \mathbf{D}^{-1}\boldsymbol{\Gamma}_0\mathbf{D}^{-1},$$

where $\rho_0(i,j)$ is the correlation coefficient between x_{ti} and x_{tj}. Similarly, the lag-h cross-correlation matrix of $\mathbf{x}_t$ is given by

$$\boldsymbol{\rho}_h = (\text{Corr}(x_{t,i}, x_{t+h,j}))_{1 \leq i,j \leq d} = \mathbf{D}^{-1}\boldsymbol{\Gamma}_h\mathbf{D}^{-1}.$$

Given observations $\{\mathbf{x}_1, \ldots, \mathbf{x}_n\}$ from a weakly stationary time series, we can estimate its mean $\boldsymbol{\mu}$ by $\bar{\mathbf{x}} = n^{-1}\sum_{t=1}^{n} \mathbf{x}_t$ and its lag-h cross-covariance matrix $\boldsymbol{\Gamma}_h$ by

$$\widehat{\boldsymbol{\Gamma}}_h = \frac{1}{n}\sum_{t=h+1}^{n} (\mathbf{x}_t - \bar{\mathbf{x}})(\mathbf{x}_{t-h} - \bar{\mathbf{x}})'.$$

Let $\mathbf{u}_t$ be i.i.d. d-variate Gaussian random variables with mean $\mathbf{0} \in \mathbb{R}^d$ and variance-covariance matrix $\boldsymbol{\Sigma}$. The process $\{\mathbf{x}_t\}$ is a zero-mean *vector ARMA* (VARMA) if it is a covariance-stationary process and satisfies

$$\mathbf{x}_t - \sum_{i=1}^{p} \boldsymbol{\Phi}_i \mathbf{x}_{t-i} = \mathbf{u}_t + \sum_{j=1}^{q} \boldsymbol{\Psi}_j \mathbf{u}_{t-j},$$

where $\boldsymbol{\Phi}_i$ $(i = 1, \ldots, p)$ and $\boldsymbol{\Psi}_j$ $(j = 1, \ldots, q)$ are parameter matrices in $\mathbb{R}^{d \times d}$. $\{\mathbf{x}_t\}$ is a VARMA(p, q) process with mean $\boldsymbol{\mu}$ if the centered series $\{\mathbf{x}_t - \boldsymbol{\mu}\}$ is a zero-mean VARMA(p, q) process.

When the VARMA process $\{\mathbf{x}_t\}$ is covariance-stationary, it has a *moving average* (MA) representation of the form

$$\mathbf{x}_t = \sum_{i=0}^{\infty} \boldsymbol{\alpha}_i \mathbf{u}_{t-i}, \tag{2.78}$$

where $\{\boldsymbol{\alpha}_i = (\alpha_{i,jk}) \in \mathbb{R}^{d \times d}\}_{i \geq 0}$ is a sequence of matrices with absolutely summable components, i.e., $\sum_{i=0}^{\infty} |\alpha_{i,jk}| < \infty$ for any $j, k \in \{1, \ldots, d\}$. With the MA representation (2.78), the covariance matrices $\boldsymbol{\Gamma}_h$ in (2.77) is given by

$$\boldsymbol{\Gamma}_h = \mathrm{Cov}\Big(\sum_{i=0}^{\infty} \boldsymbol{\alpha}_i \mathbf{u}_{t+h-i}, \sum_{j=0}^{\infty} \boldsymbol{\alpha}_j \mathbf{u}_{t-j} \Big) = \sum_{i,j=0}^{\infty} \boldsymbol{\alpha}_{i+h} E(\mathbf{u}_{t-i} \mathbf{u}'_{t-j}) \boldsymbol{\alpha}'_j = \sum_{i=0}^{\infty} \boldsymbol{\alpha}_{i+h} \boldsymbol{\Sigma}_{\mathbf{u}} \boldsymbol{\alpha}'_i. \tag{2.79}$$

Vector autoregressive (VAR) processes

An important class of multivariate regression models with stochastic regressors is the autoregressive model

$$\mathbf{x}_t = \boldsymbol{\Phi}_1 \mathbf{x}_{t-1} + \cdots + \boldsymbol{\Phi}_p \mathbf{x}_{t-p} + \mathbf{u}_t, \tag{2.80}$$

where $\mathbf{u}_t$ are i.i.d. $d \times 1$ random vectors with mean $\mathbf{0}$ and covariance matrix $\boldsymbol{\Sigma}$. Let $\boldsymbol{\Phi}(B) = \mathbf{I} - \boldsymbol{\Phi}_1 B - \cdots - \boldsymbol{\Phi}_p B^p$, where B is the backshift operator defined by $B\mathbf{x}_t = \mathbf{x}_{t-1}$. Then (2.80) can be expressed as $\boldsymbol{\Phi}(B)\mathbf{x}_t = \boldsymbol{\mu} + \boldsymbol{\epsilon}_t$. The model (2.80) is covariance stationary if all the roots of $\det(\boldsymbol{\Phi}(\lambda)) = 0$ lie outside the unit circle. It can be shown that the least squares estimate of $(\boldsymbol{\Phi}_1, \ldots, \boldsymbol{\Phi}_p)$ is consistent when (2.80) is stationary.

For a VAR process with order 1, $\mathbf{x}_t = \boldsymbol{\Phi}_1 \mathbf{x}_{t-1} + \mathbf{u}_t$ the MA representation (2.78) can be written as $\mathbf{x}_t = \sum_{i=0}^{\infty} \boldsymbol{\Phi}_1^i \mathbf{u}_{t-i}$, and the covariance matrix function of the process follows from (2.77) and (2.79) and is expressed as

$$\boldsymbol{\Gamma}_h = \sum_{i=0}^{\infty} \boldsymbol{\alpha}_{i+h} \boldsymbol{\Sigma}_{\mathbf{u}} \boldsymbol{\alpha}'_i = \boldsymbol{\Phi}_1^h \boldsymbol{\Gamma}_0, \qquad h = 0, 1, 2, \ldots.$$

Example 2.19 (VAR(1) model for three stocks) *Consider the daily log returns of JPMorgan, Amazon, and Microsoft in Example 2.12. Let* $\mathbf{x}_t = (x_{t,1}, x_{t,2}, x_{t,3})' = (x_{t,\text{JPM}}, x_{t,\text{AMZN}}, x_{t,\text{MSFT}})'$ *and suppose that the return series* $\mathbf{x}_t$ *follows the VAR process (2.80) with* $p = 1$. *Using the ML method, we can obtain the following estimated coefficients with standard errors in parenthesis*

$$\widehat{\mathbf{\Phi}}_1 = \begin{pmatrix} -0.0899_{(.0150)} & -0.0613_{(.0201)} & -0.0707_{(.0120)} \\ 0.0075_{(.0126)} & -0.0002_{(.0252)} & 0.0218_{(.0107)} \\ -0.0027_{(.0202)} & 0.0587_{(.0285)} & -0.0569_{(.0161)} \end{pmatrix},$$

and the estimated covariance matrix of innovations

$$\widehat{\mathbf{\Sigma}} = \begin{pmatrix} 5.783 & 2.445 & 2.071 \\ 2.445 & 9.934 & 2.631 \\ 2.071 & 2.631 & 3.681 \end{pmatrix} \times 10^{-4}.$$

Unit root and cointegration

Another way to write the VAR(p) model (2.80) is to consider the lag-1 difference of $\mathbf{x}$'s,

$$\Delta\mathbf{x}_t = (\mathbf{B}_1 - \mathbf{I})\mathbf{x}_{t-1} - \sum_{j=1}^{p-1} \mathbf{B}_{j+1}\Delta\mathbf{x}_{t-j} + \mathbf{u}_t, \tag{2.81}$$

where $\mathbf{B}_j = \sum_{i=j}^{p} \mathbf{\Phi}_i$. Letting $\mathbf{\Phi}(B) = \mathbf{I} - \mathbf{\Phi}_1 B - \cdots - \mathbf{\Phi}_p B^p$, assume that

$$\det(\mathbf{\Phi}(z)) \text{ has all zeros at 1 or outside the unit circle.} \tag{2.82}$$

Let $\mathbf{\Pi} = \mathbf{B}_1 - \mathbf{I} = -\mathbf{\Phi}(1)$. If rank($\mathbf{\Pi}$) = k and $\boldsymbol{\mu} = \mathbf{0}$, then $\mathbf{\Phi}(1)$ is nonsingular and the VAR(p) process $\mathbf{\Phi}(B)\mathbf{x}_t = \boldsymbol{\mu}_0 + \boldsymbol{\epsilon}_t$ is stationary in view of (2.82). If rank($\mathbf{\Pi}$) = 0, then $\sum_{i=1}^{p} \mathbf{\Phi}_i = \mathbf{I}$, implying that

$$\Delta\mathbf{x}_t = \boldsymbol{\mu} - \sum_{j=1}^{p-1} \mathbf{B}_{j+1}\Delta\mathbf{x}_{t-j} + \mathbf{u}_t, \tag{2.83}$$

and therefore $\Delta\mathbf{x}_t$ is a VAR($p-1$) process. The remaining case $0 < $ rank($\mathbf{\Pi}$) = $r < d$ implies that there exist $d \times r$ matrices $\boldsymbol{\alpha}$ and $\boldsymbol{\beta}$ such that

$$\mathbf{\Pi} = \boldsymbol{\alpha}\boldsymbol{\beta}', \quad \text{rank}(\boldsymbol{\alpha}) = \text{rank}(\boldsymbol{\beta}) = r, \tag{2.84}$$

and therefore the VAR(p) model (2.81) can be written as

$$\Delta\mathbf{x}_t = \boldsymbol{\mu} + \boldsymbol{\alpha}\boldsymbol{\beta}'\mathbf{x}_{t-1} - \sum_{j=1}^{p-1} \mathbf{B}_{j+1}\Delta\mathbf{x}_{t-j} + \boldsymbol{\epsilon}_t, \tag{2.85}$$

noting that $\mathbf{B}_1 = \mathbf{\Pi} - \mathbf{I}$. Comparison of (2.85) with (2.83) shows that (2.85) has an extra term $\boldsymbol{\alpha\beta}'\mathbf{x}_{t-1}$, which is called the *error correction term*. Accordingly, (2.85) is called an *error correction model* (ECM). Since $\boldsymbol{\alpha}$ has rank r, there exists a $k \times (k-r)$ matrix $\mathbf{A}$ such that $\mathbf{A}'\boldsymbol{\alpha} = \mathbf{0}$. Pre-multiplying (2.81) by $\mathbf{A}'$ shows that the $(d-r)$-dimensional series $\mathbf{A}'\mathbf{x}_t$ has no error correction term (or equivalently rank($\mathbf{A}'\boldsymbol{\Phi}$) = 0). In fact, det($\boldsymbol{\Phi}(z)$) has $d-r$ zeros equal to 1 and r zeros outside the unit circle by (2.82).

Definition 2.8 *(i) A multivariate time series $\mathbf{x}_t$ is said to be unit-root non-stationary if it is nonstationary but $\Delta\mathbf{x}_t$ is stationary in the weak sense.*

(ii) A nonzero $d \times 1$ vector $\mathbf{b}$ is called a cointegration vector of a unit-root nonstationary time series $\mathbf{x}_t$ if $\mathbf{b}'\mathbf{x}_t$ is weakly stationary.

(iii) A multivariate time series is said to be cointegrated if all its components are unit-root nonstationary and there exists a cointegration vector. If the linear space of cointegrating vectors (with $\mathbf{0}$ adjoined) has dimension $r > 0$, then the time series is said to be cointegrated with order r.

The column vectors of $\boldsymbol{\beta} = (\boldsymbol{\beta}_1, \ldots, \boldsymbol{\beta}_r)$ have the property that $\boldsymbol{\beta}_i'\mathbf{x}_t$ is weakly stationary and are therefore cointegrating vectors. An economic interpretation of a cointegrated multivariate time series $\mathbf{x}_t$ is that its components have some common trends that result in $\boldsymbol{\beta}_i'\mathbf{x}_t$ having long-run equilibrium for $1 \le i \le r$ even though the individual components $x_{t,i}$ are nonstationary and have variances diverging to ∞. In particular, if $\beta_{ji} \ne 0$, then linear regression of $x_{t,j}$ on the other components of $\mathbf{x}_t$ would not be spurious even though $\mathbf{x}_t$ is unit-root nonstationary.

2.3.5   Multivariate volatility models

This section introduces several commonly used multivariate volatility models. Since they have the form of GARCH models, they are also called *multivariate GARCH* or MGARCH models. Let $\boldsymbol{\epsilon}_t$ be i.i.d. d-variate normal random vectors with mean $\mathbf{0}$ and covariance $\mathbf{I}_d$. The process $\{\mathbf{u}_t\}$ is said to be a MGARCH process if it is strictly stationary and satisfies

$$\mathbf{u}_t = \boldsymbol{\Sigma}_t^{1/2}\boldsymbol{\epsilon}_t, \qquad t = 1, 2, \ldots, \tag{2.86}$$

where $\boldsymbol{\Sigma}_t^{1/2} \in \mathbb{R}^{d \times d}$ is the Cholesky factor of a positive-definite matrix $\boldsymbol{\Sigma}_t$ that is a functional of observations $\{\mathbf{u}_s\}_{s \le t-1}$. Let $\mathcal{F}_t$ be the set of functionals of observations $\{\mathbf{u}_s\}_{s \le t-1}$ (i.e., the set of information up to time $t-1$). Equation (2.86) implies that $\mathrm{E}(\mathbf{u}_t|\mathcal{F}_{t-1}) = 0$ and $\mathrm{Cov}(\mathbf{u}_t|\mathcal{F}_{t-1}) = \boldsymbol{\Sigma}_t$. Denote by $\sigma_{t,lj}$ $(1 \le l, j \le d)$ the (l,j)-th element of the matrix $\boldsymbol{\Sigma}_t$ and $\sigma_{t,l} = \sqrt{\sigma_{t,ll}}$ the conditional standard deviation (or volatility) of the lth component series $\{u_{t,l}\}$. Let $\mathbf{D}_t$ be the diagonal matrix diag($\sigma_{t,1}, \ldots, \sigma_{t,d}$) and $\mathbf{R}_t$ be the correlation matrix of $\boldsymbol{\Sigma}_t$. Then $\boldsymbol{\Sigma}_t = \mathbf{D}_t\mathbf{R}_t\mathbf{D}_t$. The diagonal matrix $\mathbf{D}_t$ is called the *volatility matrix*, and the $\mathbf{R}_t$ is referred to as the *conditional correlation matrix*.

Constant conditional correlation GARCH model

The process $\mathbf{u}_t$ is a *constant conditional correlation GARCH* (CCC-GARCH) process if its conditional covariance matrix is of the form $\boldsymbol{\Sigma}_t = \mathbf{D}_t \mathbf{R}_c \mathbf{D}_t$, in which the correlation matrix $\mathbf{R}_c$ is a constant, positive-definite correlation matrix, $\mathbf{D}_t = \operatorname{diag}(\sigma_{t,1}, \ldots, \sigma_{t,d})$, and $\sigma_{t,l}$ ($l = 1, \ldots, d$) satisfy the univariate GARCH(k_l, h_l) model

$$\sigma_{t,l}^2 = \omega_l + \sum_{i=1}^{k_l} \alpha_{l,i} u_{t-i,l}^2 + \sum_{j=1}^{h_l} \beta_{l,j} \sigma_{t-j,l}^2, \tag{2.87}$$

where $\omega_l > 0, \alpha_{l,1}, \ldots, \alpha_{l,k_l}, \beta_{l,1}, \ldots, \beta_{l,h_l} \geq 0$, and $\sum_{i=1}^{k_l} \alpha_{l,i} + \sum_{j=1}^{h_l} \beta_{l,j} < 1$. These conditions imply that the lth component series $\{u_{t,l}\}$ is covariance-stationary, hence the d-variate process $\{\mathbf{u}_t\}$ is covariance-stationary.

The CCC-GARCH model suggests an ad-hoc but simple two-step method of estimation procedure. First, one can fit an univariate GARCH model to the component series of $\mathbf{u}_t$ and denote the estimated volatility of the each component series as $\hat{\sigma}_{t,l}^2$ ($l = 1, \ldots, d$) and $\widehat{\mathbf{D}}_t = \operatorname{diag}(\hat{\sigma}_{t,1}, \ldots, \hat{\sigma}_{t,d})$. Second, estimate the conditional correlation matrix $\mathbf{R}_c$ from the standardized residual series $\hat{\mathbf{v}}_t = \widehat{\mathbf{D}}_t^{-1} \mathbf{u}_t$.

Dynamic conditional correlation GARCH model

A *Dynamic conditional correlation GARCH* (DCC-GARCH) model extends the CCC-GARCH model and allows conditional correlations evolving dynamically. Specifically, the process $\mathbf{u}_t$ is a DCC-GARCH process if its conditional covariance matrix is of the form $\boldsymbol{\Sigma}_t = \mathbf{D}_t \mathbf{R}_t \mathbf{D}_t$, in which (i) $\mathbf{D}_t = \operatorname{diag}(\sigma_{t,1}, \ldots, \sigma_{t,d})$, $\sigma_{t,l}$ ($l = 1, \ldots, d$) satisfy the univariate GARCH model (2.87), (ii) the positive-definite conditional correlation process $\mathbf{R}_t$ is the correlation matrix of the process

$$\mathbf{R}_t = \operatorname{Corr}\left\{ \left(1 - \sum_{i=1}^{k} a_i - \sum_{j=1}^{h} b_j\right) \mathbf{R}_c + \sum_{i=1}^{k} a_i \mathbf{v}_{t-i} \mathbf{v}_{t-i}' + \sum_{j=1}^{h} b_j \mathbf{R}_{t-j} \right\}, \tag{2.88}$$

where $\mathbf{R}_c$ is a positive-definite correlation matrix, $\mathbf{v}_t = \mathbf{D}_t^{-1} \mathbf{u}_t$, $\alpha_1, \ldots, \alpha_k, \beta_1, \ldots, \beta_l \geq 0$ and $\sum_{i=1}^{h} \alpha_i + \beta_{j=1}^{k} \beta_j < 1$.

The DCC-GARCH model can be estimated by the following three-step method. First, fit a univariate GARCH model to the component series to get the volatility matrix $\widehat{\mathbf{D}}_t$ and get an estimated standardized residual process $\hat{\mathbf{v}}_t = \widehat{\mathbf{D}}_t^{-1} \mathbf{u}_t$. Second, estimate $\mathbf{R}_c$ by the sample correlation matrix of the process $\hat{\mathbf{v}}_t$. Third, use the estimate $\widehat{\mathbf{R}}_c$ and the residual process $\hat{\mathbf{v}}_t$ to fit the model (2.88).

Example 2.20 (DCC-GARCH models for stock returns) *Consider the daily log returns of JP Morgan, Amazon, and Microsoft in Examples 2.12 and 2.19. Let $\mathbf{x}_t = (x_{t,1}, x_{t,2}, x_{t,3})' = (x_{t,\text{JPM}}, x_{t,\text{AMZN}}, x_{t,\text{MSFT}})'$. Suppose that* *the return series* $\mathbf{x}_t$ *follows the MGARCH process* (2.86). *We fit a DCC-GARCH(1, 1) model to* $\mathbf{x}_t$. *Table 2.6 shows the estimated model parameters and their standard errors (in parenthesis). Figure 2.10 shows the estimated volatility and conditional correlation of daily returns of the three stocks.* □

TABLE 2.6: Estimated DCC-GARCH(1, 1) model parameters for three stock returns.

	μ	ω	α	β	a_1	b_1
JPMorgan	7.58e-4	3.64e-6	0.0935	0.9008	0.0783	0.9910
	(1.93e-4)	(2.57e-6)	(.0220)	(.0229)	(.0026)	(.0036)
Amazon	1.09e-3	2.60e-6	0.0187	0.9781		
	(3.03e-4)	(1.24e-6)	(.0012)	(.0001)		
Microsoft	7.45e-4	0.00001	0.0859	0.8949		
	(1.93e-4)	(1.93e-4)	(.0150)	(.0353)		

BEKK-GARCH models

Comparing to the CCC-GARCH and DCC-GARCH models, a more straightforward way of modeling the conditional covariance matrix $\mathbf{\Sigma}_t$ is to vectorize $\mathbf{\Sigma}_t$ and specify explicitly a dynamic structure for the vectorized series. Denote by "vech" the *vector half* operator, i.e., $\mathrm{vech}(\mathbf{\Sigma}_t) = (\sigma_{t,11}, \ldots, \sigma_{t,1d}, \sigma_{t,22}, \ldots, \sigma_{t,2d}, \ldots, \sigma_{t,dd})' \in \mathbb{R}^{d(d+1)/2}$. A vectorized MGARCH model is given by

$$\mathrm{vech}(\mathbf{\Sigma}_t) = \mathrm{vech}(\widetilde{\mathbf{C}}) + \sum_{i=1}^{k} \widetilde{\mathbf{A}}_i \mathrm{vech}(\mathbf{u}_{t-i}\mathbf{u}'_{t-i}) + \sum_{j=1}^{q} \widetilde{\mathbf{B}}_j \mathrm{vech}(\mathbf{\Sigma}_{t-j}), \quad (2.89)$$

in which matrices $\widetilde{\mathbf{A}}_i, \widetilde{\mathbf{B}}_j, \widetilde{\mathbf{C}} \in \mathbb{R}^{(d(d+1)/2) \times (d(d+1)/2)}$. The general form of the MGARCH model (2.89) has $(1 + (k + h)d(d + 1)/2)d(d + 1)/2 \sim O(d^4)$ parameters, since there are too many parameters to estimate, the MGARCH model (2.89) has very limited usage in practice.

To overcome the weakness, Engle and Kroner (1995) introduced the BEKK (named after Baba, Engle, Kroner, and Kraft) model. In particular, for the process $\mathbf{u}_t$ as in (2.86), the covariance matrix $\mathbf{\Sigma}_t$ is expressed as

$$\mathbf{\Sigma}_t = \mathbf{C} + \sum_{j=1}^{k} \mathbf{A}'_j \mathbf{u}_{t-j} \mathbf{u}'_{t-j} \mathbf{A}_j + \sum_{i=1}^{h} \mathbf{B}_i \mathbf{\Sigma}_{t-i} \mathbf{B}'_i, \quad (2.90)$$

in which $\mathbf{C}$ (required to be symmetric and positive definite) and $\mathbf{A}_j, \mathbf{B}_i$ are $d \times d$ coefficient matrices. The BEKK model only involves $d(d+1)/2 + (k+h)d^2$ number of parameters, which is much smaller than that of the MGARCH model (2.89). However, when dimension d increases, the number of parameters in the BEKK model can still be very large. In such a case, a sparse representation of the BEKK model should be considered (Yao et al., 2024).

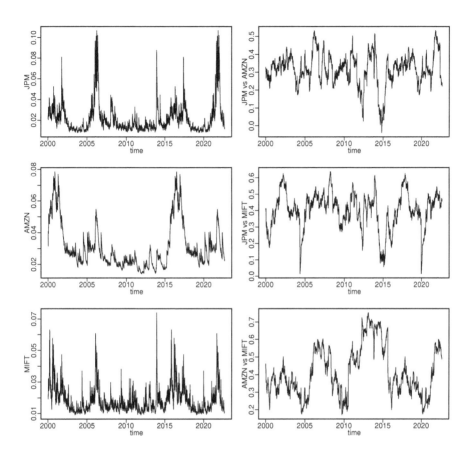

FIGURE 2.10: The estimated volatility (left) and correlation (right) in the DCC-GARCH(1,1) model for JPMorgan, Amazon, and Microsoft daily returns.

Generalized orthogonal GARCH models

When the dimension d of the vector increases, the number of parameters in the aforementioned MGARCH models increases rapidly and some dimension reduction method such as principal component analysis can be used to reduce the number of parameters in the model. One type of such models are the *generalized orthogonal GARCH* (GO-GARCH) models (van der Weide, 2002), which are described as follows. Let $\mathbf{u}_t$ be the observed series as in (2.86). Assume that $\mathbf{u}_t$ is a linear combination of d uncorrelated components $\mathbf{f}_t$,

$$\mathbf{u}_t = \mathbf{\Gamma}\mathbf{f}_t,$$

where $\{\mathbf{f}_t = (f_{t,1}, \ldots, f_{t,d})'\}$, $f_{t,i}$ $(i = 1, \ldots, d)$ is unobserved and have unit variance, and the mixing matrix $\boldsymbol{\Gamma} \in \mathbb{R}^{d \times d}$ is constant and invertible. Then the conditional covariance of $\mathbf{f}_t$ at time t can be represented by a diagonal matrix $\mathbf{H}_t$, i.e., $\mathrm{Cov}(\mathbf{f}_t) = \mathbf{H}_t = \mathrm{diag}(\nu_{t,1}^2, \ldots, \nu_{t,d}^2)$. For $l = 1, \ldots, d$, the ith component series $f_{t,i}$ is described by a GARCH(1,1) process

$$\nu_{t,l}^2 = (1 - \alpha_l - \beta_l) + \alpha_l f_{t-1,l}^2 + \beta_i \nu_{t-1,l}^2,$$

where $\alpha_l, \beta_l > 0$, $\alpha_l + \beta_l < 1$, and $\nu_{0,l}^2 = 1$ equals the (l, l)-th element of the unconditional covariance matrix. Hence, the conditional covariance of $\mathbf{u}_t$ is expressed as $\boldsymbol{\Sigma}_t = \boldsymbol{\Gamma} \mathbf{H}_t \boldsymbol{\Gamma}'$.

Example 2.21 (GO-GARCH models for stock returns) *Consider the daily log returns of JP Morgan, Amazon, and Microsoft in Examples 2.12, 2.19, and 2.20. We still denote by $\mathbf{x}_t = (x_{t,\mathrm{JPM}}, x_{t,\mathrm{AMZN}}, x_{t,\mathrm{MSFT}})'$. Suppose that the return series $\mathbf{x}_t$ follows the MGARCH process (2.86). We fit the above GO-GARCH model to $\mathbf{x}_t$ and obtain the following estimate*

$$\widehat{\boldsymbol{\Gamma}} \approx \begin{pmatrix} -0.0241 & -0.0006 & 0.0012 \\ -0.0085 & -0.0089 & 0.0291 \\ -0.0083 & -0.0174 & 0.0014 \end{pmatrix}, \quad \begin{array}{lll} \widehat{\mu}_1 = 2.95e-4, & \widehat{\alpha}_1 = 0.0921, & \widehat{\beta}_1 = 0.9024, \\ \widehat{\mu}_2 = 5.87e-4, & \widehat{\alpha}_2 = 0.0436, & \widehat{\beta}_2 = 0.9449, \\ \widehat{\mu}_3 = 3.38e-4, & \widehat{\alpha}_3 = 0.0103, & \widehat{\beta}_3 = 0.9884. \end{array}$$

Figure 2.11 shows estimated volatilities $\sigma_{t,i}$ of JPMorgan, Amazon, and Microsoft returns and estimated volatilities $\nu_{t,i}$ of uncorrelated components $\mathbf{f}_t$. □

2.4   Extreme value theory and quantile regression

This section introduces extreme value theory which studies the limiting laws of extreme values in large samples and can be used to describe the extremal behavior of financial risk factors, and quantile regression which models the dependence of the quantile of a response variable on a set of explanatory variables.

2.4.1   Maxima and minima

The *generalized extreme value* (GEV) family of distributions arises from the asymptotic distribution of the extreme order statistics. Assume that $x_1, \ldots, x_n$ are i.i.d. samples drawn from a distribution without replacement. Denote by $x_{(1)} = \min\{x_1, \ldots, x_n\}$ and $x_{(n)} = \max\{x_1, \ldots, x_n\}$ the minima and maxima of the sample. Since the x_t are i.i.d.,

$$P(x_{(1)} \geq x) = P(x_t \geq x \text{ for all } 1 \leq t \leq n) = P^n(x_1 \geq x).$$

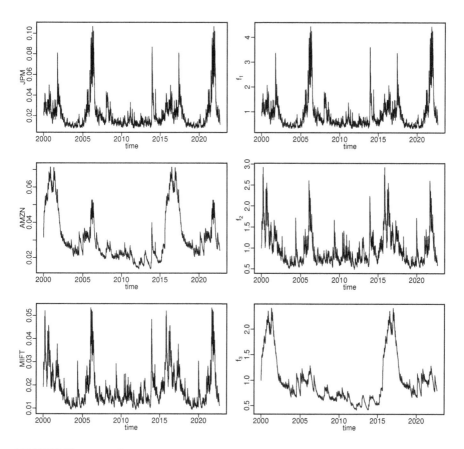

FIGURE 2.11: The estimated volatility of JPMorgan, Amazon, and Microsoft daily returns (left) and uncorrelated components $\mathbf{f}_t$ (right).

Using this result, one can show that, if there exist constants $\alpha_n > 0$ and β_n such that $(x_{(1)} - \beta_n)/\alpha_n$ converges in distribution, the limiting distribution function F must be of the form

$$F(x) = \begin{cases} 1 - \exp[-(1 + cx)^{1/c}] & \text{if } c \neq 0, \\ 1 - \exp[-\exp(x)] & \text{if } c = 0, \end{cases} \tag{2.91}$$

for $x < -1/c$ if $c < 0$ and for $x > -1/c$ for $c > 0$. The case $c = 0$ in (2.91) is referred to as the *Gumbel* family, while $c < 0$ corresponds to the *Fréchet* family and $c > 0$ the *Weibull* family. The parameter c is called the *shape parameter* of the GEV distribution, $-1/c$ is called the *tail index*, and β_n and α_n are called the *location* and *scale* parameters, respectively, for normalizing the sample minimum.

For a given sample, there is only a single minimum or maximum, and (α_n, β_n, c) cannot be estimated with only one observation ($x_{(1)}$ or $x_{(n)}$). One way to circumvent this difficulty is to partition the sample into subsamples and to apply extreme value theory to the subsamples. Specifically, divide a sample of size n into k nonoverlapping subsamples, each with m observations:

$$\{x_1, \ldots, x_m\}, \{x_{m+1}, \ldots, x_{2m}\}, \ldots, \{x_{(k-1)m+1}, \ldots, x_{km}\},$$

assuming for simplicity that $n = mk$. Let $M_{m,i} = \min\{x_{(i-1)m+j} : 1 \leq j \leq m\}$ for $i = 1, \ldots, k$. For sufficiently large m, we can apply extreme value theory to each subsample to conclude that the subsample minima $M_{m,i}$, $i = 1, \ldots, k$, can be regarded as a subsample of k observations from a GEV distribution after renormalization; i.e., $(M_{m,i} - \beta)/\alpha$ has distribution function (2.91). In this way, the parameters α, β, c can be estimated by maximizing the likelihood function

$$L(\alpha, \beta, c) = \prod_{i=1}^{k} \left\{ \frac{1}{\alpha} f\left(\frac{M_{m,i} - \beta}{\alpha}\right) \right\},$$

where $f(x)$ is the derivative of the distribution function $F(x)$ in (2.91).

2.4.2   Threshold exceedance

Instead of subsamples of prespecified sizes, an alternative approach is to use exceedance of the samples over some prespecified threshold. Consider n i.i.d. samples $x_1, \ldots, x_n$ drawn from a distribution without replacement. If there exist constants $\alpha_n > 0$ and β_n such that $(x_{(n)} - \beta_n)/\alpha_n$ converges in distribution, then the limiting distribution function F has the following form

$$F(x) = \begin{cases} \exp[-(1 - cx)^{1/c}] & \text{if } c \neq 0, x < 1/c \\ \exp[-\exp(x)] & \text{if } c = 0. \end{cases}$$

In this case, for η_n so chosen that $\alpha_n - c(\eta_n - \beta_n) \to \psi > 0$, $P\{r_i \leq x + \eta_n | r_i > \eta_n\}$ can be shown to converge to

$$G_{c,\psi}(x) = \begin{cases} 1 - (1 - cx/\psi)^{1/c} & \text{if } c \neq 0, \\ 1 - \exp(-x/\psi) & \text{if } c = 0. \end{cases} \tag{2.92}$$

in which $\psi := \alpha - c(\eta - \beta) > 0$, $x > 0$ when $c < 0$ and $0 < x \leq \psi/c$ when $c > 0$. The function $G_{c,\psi}$ is the distribution function of the *generalized Pareto* (GP) distribution. The parameters ψ and c can be estimated from the subsample of returns x_i that exceed η_n by the method of maximum likelihood.

Example 2.22 (VaR for GEV and GP distributions) *Let L be the loss of holding a portfolio during a period. If $(L - \beta)/\alpha$ has a GEV distribution (2.91). Then the $(1 - p) \times 100\%$ VaR at the confidence level $1 - p$ is given by*

$$\mathrm{VaR}_{1-p}(L) = \begin{cases} \beta - \dfrac{\alpha}{c}\left\{1 - \left(-\log p\right)^c\right\} & \text{if } c \neq 0, \\ \beta + \alpha \log\left(-\log p\right) & \text{if } c = 0. \end{cases}$$

If $r - \eta$ follows a GP distribution (2.92) conditional on $r > \eta$, then the $(1 - p) \times 100\%$ VaR at the confidence level $1 - p$ is

$$\mathrm{VaR}_{1-p}(L) = \begin{cases} \eta + (\psi/c)(1 - p^c) & \text{if } c \neq 0, \\ \eta - \psi \log p & \text{if } c = 0. \end{cases} \qquad \square$$

2.4.3 Quantile regression

Quantile regression is a type of regression analysis that estimates the conditional quantiles of response variable across values of the predictor variables. Different from the linear regression analysis that estimates the conditional mean of the response variables, the quantile regression estimates are more robust against outliers of the response variables and provides a way to model the quantiles directly using the predictor variables.

Let $q_\alpha(Y)$ be the α-th quantile of the random variable Y. To understand quantile regression, we first define a loss function $\rho_\alpha(u) = (\alpha - 1_{\{u<0\}})u$, where $1_{\{\cdot\}}$ is an indicator function, and argue that $q_\alpha(Y)$ minimizes the expected loss of $Y - u$ with respect to u,

$$q_\alpha(Y) = \arg\min_u E\big[\rho_\alpha(Y - u)\big].$$

To see this, we note that

$$E\rho_\alpha(Y - u) = (\alpha - 1)\int_{-\infty}^{u}(y - u)dF_Y(y) + \alpha\int_{u}^{\infty}(y - u)dF_Y(y).$$

Differentiating with respect to u yields

$$0 = (1 - \alpha)\int_{-\infty}^{u}dF_Y(y) - \alpha\int_{u}^{\infty}dF(x) = F_Y(u) - \alpha.$$

This indicates that any element of $\{y : F_Y(y) = \alpha\}$ minimizes expected loss, hence the solution is given by $u^* = F_Y^{-1}(y) = q_\alpha(Y)$. If instead, n i.i.d. samples $y_1, \ldots, y_n$ are observed, we replace the distribution F_Y by the empirical distribution function $F_n(y) = n^{-1}\sum_{i=1}^{n} 1_{\{y_i \leq y\}}$, then the αth sample quantile minimizes the expected loss

$$\int \rho_\alpha(y - u)dF_n(y) = n^{-1}\sum_{i=1}^{n} \rho_\alpha(y_i - u).$$

The above argument provides an important clue to extend linear regression $E(Y|\mathbf{X}) = \boldsymbol{\beta}'\mathbf{X}$ to the quantile regression model

$$q_\alpha(Y|\mathbf{X}) = \boldsymbol{\beta}'\mathbf{X}, \qquad (2.93)$$

with covariate vector $\mathbf{x} \in \mathbb{R}^p$, by solving the optimization problem

$$\min_{\boldsymbol{\beta} \in \mathbb{R}^p} \sum_{i=1}^{n} \rho_\alpha(y_i - \boldsymbol{\beta}'\mathbf{x}_i). \qquad (2.94)$$

The optimization problem (2.94) can be reformulated into a linear programming problem by introducing extra $2n$ slack parameters $\mathbf{u}, \mathbf{v} \in \mathbb{R}_+^n$:

$$\min_{\boldsymbol{\beta} \in \mathbb{R}^p, \mathbf{u}, \mathbf{v} \in \mathbb{R}_+^n} \left\{ \tau \mathbf{1}_n' \mathbf{u} + (1 - \tau) \mathbf{1}_n' \mathbf{v} \,\middle|\, \boldsymbol{\beta}' \mathbf{x}_i + u_i - v_i = y_i, i = 1 \ldots, n \right\},$$

in which $\mathbb{R}_+^n = [0, \infty)^n$. One can use simplex methods or interior point methods to solve this linear program and get optimal solution. Statistical inferences can be done for quantile regression. Compare to regular linear regressions, quantile regression is more robust against outliers. It also has advantages in modeling weaker relationship between predictors and response where variance dispersion or heteroscedastic errors exist.

Example 2.23 (Quantile regression of SPY returns) *Consider the daily log returns of the SPY ETF and JP Morgan, Amazon, Microsoft stocks from January 4, 2000 to August 31, 2022. Denote by y_t the SPY return and $\mathbf{x}_t = (x_{t,1}, x_{t,2}, x_{t,3})' = (x_{t,\mathrm{JPM}}, x_{t,\mathrm{AMZN}}, x_{t,\mathrm{MSFT}})'$ the JP Morgan, Amazon, Microsoft stock returns at time t, respectively. By choosing the quantile $\alpha = 0.02, 0.04, \ldots, 0.98$, we carry out the quantile regression analysis of the SPY returns versus returns of JP Morgan, Amazon, Microsoft stocks. Figure 2.12 shows the estimated quantile regression coefficients and 95% confidence intervals for each quantile regression. In particular, for $\alpha = 0.99$, the estimated regression model is expressed as*

$$q_{0.99}(y_t) = 0.0170_{(.0010)} + 0.2483_{(.0451)} x_{t,1} + 0.0688_{(.0308)} x_{t,2} + 0.1813_{(.0448)} x_{t,3} + \epsilon_t,$$

in which the standard errors of the estimated coefficient are shown in parenthesis and the estimated standard deviation of ϵ_t is 0.0068.   □

Supplements and problems

This chapter introduces basic concepts in financial risk management and commonly used multivariate models, time series models, and extreme value theory. McNeil et al. (2015) provide a comprehensive and detailed introduction on these topics. Roncalli (2020) also introduces these statistical methods after he explains several types of financial risk and their regulatory criteria. Jorion (2006) describe extensively quantitative methods and implementation of using VaR to assess different types of financial risks.

Our introduction on statistical methods only covers basic results on each specific topic. Interested readers can find more discussion on these topics in other reference. For multivariate methods, Bilodeau and Brenner (1999), Timm (2002), and Rencher and Christensen (2012) provide standard textbook treatment on multivariate distribution and dimension reduction methods, and

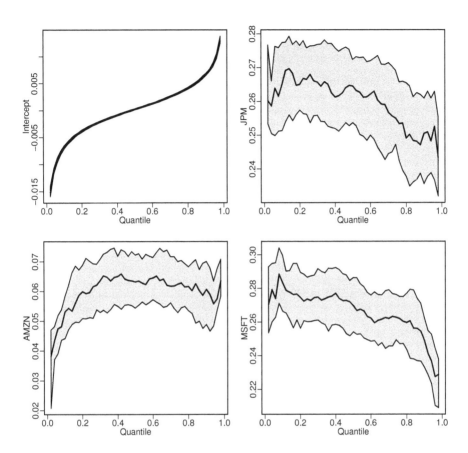

FIGURE 2.12: The estimated quantile regression coefficients and 95% confidence intervals for returns of SPY versus JPMorgan, Amazon, and Microsoft.

Nelsen (2006) describes comprehensively the theory of copula and its application to statistics, probability, and actuarial science. Besides, Cherubini et al. (2004) and Cherubini et al. (2011) give detailed discussion on the applications of copula and dynamic copula methods in finance.

There are many textbook treatment on time series models and analysis. For example, Chatfield and Xing (2019) introduce the topics with R examples at the undergraduate level. Shumway and Stoffer (2017) present a balanced and comprehensive treatment of both time and frequency domain methods on statistical time series analysis. Graduate level treatment on time series models and analysis with financial applications can be found in Tsay (2010) and Tsay (2013). Moreover, Francq and Zakoian (2019) provide a detailed survey of GARCH models and their applications in finance. Our introduction on time series models doesn't involve nonlinear time series, we refer interested readers to Fan and Yao (2003), Douc et al. (2014), and Tsay and Chen (2019).

The extreme value theory presented in Section 2.4 is a very elementary introduction to the topic. General treatments on basic theoretical framework of extreme value models and related statistical inferential techniques can be found in Embrechts et al. (1997), Coles (2001), and Haan and Ferreira (2006). Moreover, introductions to multivariate extreme value theory and extreme values in time series can be introduced in Beirlant et al. (2004) and McNeil et al. (2015).

1. **(Alternative expression for ES)**. Show that, for $0 < \alpha < 1$, we have

$$\text{ES}_\alpha = \frac{1}{1-\alpha} E[(L - q_\alpha(L))^+] + q_\alpha(L)$$

$$= \frac{1}{1-\alpha} \{ E(L; L > q_\alpha(L)) + q_\alpha(L)[1 - \alpha - P(L > q_\alpha(L))] \}.$$

(Hint: Use $q_\alpha(L) = F_L^{-1}(\alpha)$ and $E[(L - q_\alpha(L))^+] = E(L; L > q_\alpha(L)) - q_\alpha(L)P(L > q_\alpha(L))$.)

2. **(Non-subadditivity of VaR)**. Consider two independent and identically distributed random variables L_1 and L_2. For $i = 1, 2$, L_i has a mixture distribution $L_i = \epsilon_i + \eta_i$, where $\epsilon_i \sim N(0, 1)$ and η_i takes values 0 and -10 with probability 0.991 and 0.009, respectively. Show that $\text{VaR}(L_1 + L_2) > \text{VaR}(L_1) + \text{VaR}(L_2)$.

3. **(Subadditivity of ES)**. For any $L_1, L_2 \in \mathcal{Y}$ and $\alpha \in (0, 1)$, show that

$$\text{ES}_\alpha(L_1) + \text{ES}_\alpha(L_2) \geq \text{ES}_\alpha(L_1 + L_2).$$

(Hint: Let $I_i = I_{\{L_i \geq q_\alpha(L_i)\}}$ for $i = 1, 2$ and $I_{12} = I_{\{L_1+L_2 \geq q_\alpha(L_1+L_2)\}}$ and show $L_i(I_i - I_{12}) \geq q_\alpha(L_i)(I_i - I_{12})$ for $i = 1, 2$.)

4. Consider a portfolio of two zero-coupon bonds, whose default times are independent. The probability density function of (L_1, L_2) is given in Table 2.7. Compute (1) the probability distribution of $L = L_1 + L_2$, (2) $\text{VaR}_{99\%}(L)$, and (3) $\text{ES}_{99\%}(L)$.

5. **(Fréchet bounds)** For copula $C(u_1, \ldots, u_p)$, show the following bounds

$$\max \left(\sum_{i=1}^p u_i + 1 - p, 0 \right) \leq C(u_1, \ldots, u_p) \leq \min(u_1, \ldots, u_p).$$

6. Consider a bivariate Gauss copula that is given by (2.34). Show that its density is given by (2.42) and its conditional distribution of U_2 given $U_1 = u_1$ is given by (2.44).

TABLE 2.7: Density function of (L_1, L_2)

	$L_1 = 0$	$L_1 = 50$	$L_2 = 100$	
$L_2 = 0$	95.2%	1.4%	0.8%	97.4%
$L_2 = 50$	1.5%	0.1%	0.1%	1.7%
$L_2 = 100$	0.7%	0.1%	0.1%	0.9%
	97.4%	1.6%	1%	

7. Consider the bivariate t copula

$$C_{\nu,\rho}^t(u_1, u_2) = \int_{-\infty}^{x_1} \int_{-\infty}^{x_2} \frac{1}{2\pi\sqrt{1-\rho^2}} \left(1 + \frac{s^2 + t^2 - 2\rho st}{\nu(1-\rho^2)}\right)^{-\frac{\nu+1}{2}} ds dt,$$

where $x_i = t_\nu^{-1}(u_i)$ for $i = 1, 2$. Show that its copula density is

$$c_{\nu,\rho}(u_1, u_2) = \rho^{-\frac{1}{2}} \frac{\Gamma(\frac{\nu+2}{2})\Gamma(\frac{\nu}{2})}{\Gamma(\frac{\nu+1}{2})^2} \left[\frac{1 + \frac{x_1^2 + x_2^2 - 2\rho x_1 x_2}{\nu(1-\rho^2)}}{\left(1 + \frac{x_1^2}{\nu}\right)\left(1 + \frac{x_1^2}{\nu}\right)}\right]^{-\frac{\nu+2}{2}},$$

and its conditional distribution of U_2 given $U_1 = u_1$ is

$$C_{U_2|U_1}(u_2|u_1) = t_{\nu+1}\left(\sqrt{\frac{\nu+1}{\nu+x_1^2}} \frac{x_2 - \rho x_1}{\sqrt{1-\rho^2}}\right).$$

8. (1) Consider the bivariate Gumbel copula

$$C_\theta^{Gu}(u_1, u_2) = \exp\left(-\left[(-\log u_1)^\theta + (-\log u_2)^\theta\right]^{1/\theta}\right), \qquad \theta \geq 1.$$

Show that if $\theta = 1$, it reduces to the independence copula, and if $\theta \to \infty$, it converges to the two-dimensional comonotonicity copula.

(2) Consider the bivariate Clayton copula

$$C_\theta^{Cl}(u_1, u_2) = (u_1^{-\theta} + u_2^{-\theta} - 1)^{-1/\theta}, \qquad \theta \geq 0.$$

Show that if $\theta \to 0$, it converges to the independent copula, and if $\theta \to \infty$, it converges to the two-dimensional comonotonicity copula.

9. Consider the d-dimensional Clayton copula $C_\theta^{Cl}(u_1, \ldots, u_d)$ as in (2.38). Show its density is given by

$$c_\theta^{Cl}(u_1, \ldots, u_d) = (u_1 \ldots u_d)^{-\theta-1}(u_1^{-\theta} + \cdots + u_d^{-\theta} - 1)^{-\frac{1}{\theta}-d} \prod_{i=0}^{d-1}(1 + i\theta).$$

10. (**Scatterplot of simulated copula**) Consider the following copulas

    (1) A bivariate Gauss copula (2.34) with $\rho = 0.5$.
    (2) A bivariate t copula (2.35) with $\nu = 10$ degrees of freedom and correlation $\rho = 0.4$.
    (3) A three dimensional Gumbel copula (2.37) with $\theta = 2$.
    (4) A four dimensional Clayton copula (2.38) with $\theta = 1$.

    For each case, simulate 1000 random vector and give their pairwise scatterplots.

11. (**Comparison of simulated copulas**) Consider the following sets of copulas.

    (1) Four dimensional Gumbel copula (2.37) with $\theta = 3, 10, 25$.
    (2) Four dimensional Clayton copula (2.38) with $\theta = 2, 8, 16$.

    For each set of copulas, simulate 1000 random vector for each θ and compare their joint and marginal distributions.

12. (**Attainable correlations of lognormal random variables**) Let $Z \sim N(0, 1)$ and consider two random variables $X_1 = e^Z$ and $X_2 = e^{\sigma Z}$. Compute the correlation of X_1 and X_2 and show that the minimum and maximum of the correlations are

$$\rho_{\min} = \frac{e^{-\sigma} - 1}{\sqrt{(e - 1)(e^{-\sigma} - 1)}}, \qquad \rho_{\max} = \frac{e^{\sigma} - 1}{\sqrt{(e - 1)(e^{-\sigma} - 1)}}.$$

13. Consider the Marshall-Olkin copula (Marshall and Olkin, 1967a,b) that is characterized by parameters m and n and defined $C^{MO}(u_1, u_2) = \min(u_1^{1-m} u_2, u_1 u_2^{1-n})$. Compute its copula density and show that its Kendall's tau, Spearman's rho, and the coefficient of upper tail dependence are given by $mn/(m + n - mn)$, $3mn/(2m + 2n - mn)$, and $\min(m, n)$, respectively.

14. (**Tail dependence of Gumbel and Clayton copulas**) Consider the Gumbel copula (2.37) and the Clayton copula (2.38) with $d = 2$. Show that their upper and lower tail dependence are given by $\lambda_u^{\mathrm{Gu}} = 2 - 2^{1/\theta}$, $\lambda_l^{\mathrm{Gu}} = 0$, $\lambda_u^{\mathrm{Cl}} = 0$, and $\lambda_l^{\mathrm{Cl}} = 2^{-1/\theta}$.

15. Let Z_t be independent and identically distributed standard Gaussian random variables. Show that if the AR(2) process $X_t = \alpha_1 X_{t-1} + \alpha_2 X_{t-2} + Z_t$ is weakly stationary, α_1 and α_2 satisfy conditions $\alpha_1 + \alpha_2 < 1$, $\alpha_1 - \alpha_2 > -1$, and $\alpha_2 > -1$.

16. Consider the AR(2) process

$$X_t = \frac{2}{a(a+2)}X_{t-1} + \frac{1}{a(a+2)}X_{t-2} + u_t,$$

where u_t are i.i.d. $N(0,1)$ random variables and the constant a takes real values. (a) Find the condition under which the process X_t is stationary. (b) Find the ACF of the process X_t.

17. Consider the ARMA(1,1) process

$$X_t = \alpha X_{t-1} + Z_t + \beta Z_{t-1},$$

where $|\alpha| < 1$, $|\beta| < 1$, $\alpha, \beta \neq 0$, and Z_t are i.i.d. standard Gaussian random variables. Show that the ACF of this process is

$$\rho(k) = \frac{(1+\alpha\beta)(\alpha+\beta)}{1+2\alpha\beta+\beta^2}\alpha^{k-1}, \quad k \geq 1.$$

18. Consider the GARCH(1,1) process (2.69), in which $w, \alpha, \beta > 0$, $\alpha + \beta < 1$, and ϵ_t are i.i.d. random variables with mean 0 and variance 1. Suppose that the fourth moment of the process exist and denote by κ_u and κ_ϵ the kurtosis of u_t and ϵ_t, respectively. Show that

$$\kappa_u = \frac{(1-(\alpha+\beta)^2)\kappa_\epsilon}{1-(\alpha+\beta)^2-(\kappa_\epsilon-1)\alpha^2}.$$

19. (**ACF of GARCH(1,1) process**) Consider the GARCH(1,1) process (2.69), in which $w, \alpha, \beta > 0$, $\alpha + \beta < 1$, $1 - 3\alpha^2 - \beta^2 - 2\alpha\beta > 0$, and ϵ_t are i.i.d. normal random variables with mean 0 and variance 1. Show that

(1) The ACF of u_t^2 satisfies

$$\rho_{u^2}(h) = (\alpha+\beta)\rho_{u^2}(h-1), \quad h = 2, 3, \ldots,$$

(2) The ACF of u_t^2 at lag 1 is given by

$$\rho_{u^2}(1) = \frac{\alpha(1-\beta^2-\alpha\beta)}{1-\beta^2-2\alpha\beta}.$$

20. (**Maxima of exponential and Pareto distributions**) Suppose that $x_1, \ldots, x_n$ are i.i.d. samples drawn from the distribution $F(x)$. Let $x_{(n)} = \{x_1, \ldots, x_n\}$ be the maxima of the sample. Choosing normalizing sequence c_n and d_n, we then have $P((x_{(n)} - d_n)/c_n \leq x) = F^n(c_n x + d_n)$.

(1) If the distribution $F(x)$ is an exponential distribution function, i.e., $F(x) = 1 - \exp(-\beta x)$ for $\beta > 0$ and $x \geq 0$. Choosing normalizing sequence $c_n = 1/\beta$ and $d_n = (\log n)/\beta$, show that

$$\lim_{n \to \infty} P((x_{(n)} - d_n)/c_n \leq x) = \exp(-e^{-x}), \quad x \in \mathbb{R}.$$

(2) If the distribution $F(x)$ is a Pareto distribution function, i.e., $F(x) = 1 - (\kappa/(\kappa + x))^\alpha$ for $\alpha > 0$, $\kappa > 0$ and $x \geq 0$. Choosing normalizing sequence $c_n = \kappa n^{1/\alpha}/\alpha$ and $d_n = \kappa n^{1/\alpha} - \kappa$, show that

$$\lim_{n \to \infty} P((x_{(n)} - d_n)/c_n \leq x) = \exp\left(-(1 + (x/\alpha))^{-\alpha}\right), \quad 1 + (x/\alpha) > 0.$$

Chapter 3

Financial derivatives and their pricing theory

A *derivative* is a financial instrument having a value derived from or contingent on the values of more basic underlying variables. Since the price of a derivative is usually a nonlinear function of risk factors, the value change or the loss of holding a portfolio containing derivatives is also a nonlinear function of risk factors. In such a case, evaluating or hedging financial risk due to the value change of a derivative involves the step of calculating the derivative price. This chapter introduces the pricing theory of options, interest rates models, and valuation of credit derivatives. To avoid technical complexity, all stochastic processes considered in this chapter are assumed to be diffusion processes and have no jumps.

Options are financial derivatives that give the holder the right, but not the obligation, to buy (call option) or sell (put option) a specific asset at a predetermined price (the strike price) on or before a specified expiration date. The underlying assets for options include but are not limited to stocks, bonds, commodities, or even a currency. Options are traded both on exchanges and in the over-the-counter market. Section 3.1 introduces the Black-Scholes-Merton option pricing theory, which is fundamental in derivative pricing. Two option pricing problems are introduced, one is for European options and the other is for American options. To relax the assumption of constant volatility in the Black-Scholes-Merton model, Section 3.2 introduces stochastic volatility models that allow the volatility of underlying assets to change randomly.

To manage and hedge interest rate risk, interest rate derivatives are widely used by financial institutions, corporations, and investors to hedge against adverse movements in interest rates. To price interest rate derivatives that are contingent on future interest rates, stochastic interest rate models are developed to understand the relationship between interest rates and the prices of derivative instruments. Section 3.3 introduces several interest rate models such as short rate models, LIBOR and swap market models, and the Heath-Jarrow-Morton models. The section also explains how to price some interest rate derivatives based on introduced interest rate models.

Interest rate models in Section 3.3 focus solely on modeling the behavior of interest rates over time and do not incorporate the issuer's credit risk. To evaluate such credit risk, one can estimate the probability of default or the obligor's creditworthiness by analyzing related factors and historical data, which usually involves satistical models and methods of survival analysis. Section 3.4 presents statistical models of survival analysis and their inference methods. In particular, it introduces basic concepts on survivor functions, proportional hazards models, additive hazard models, and accelerated failure time models. Besides statistical approaches, two kinds of mathematical frameworks have been developed to characterize the obligor's credit default risk. One is to use structural approach to describe the relationship between obligors' asset value process and default barriers, and the other is to model directly the obligor's hazard as a stochastic process and evaluate the probability of default. Section 3.5 explains the main ideas and models for these two approaches.

Besides default risk, obligor's credit ratings and their rating migrations also play an important role in credit risk management. Section 3.6 introduces statistical models on dynamics of the obligors' credit rating migrations and their inference methods, which includes time-homogeneous Markov chain models and piecewise time-homogeneous Markov chain models. Moreover, Section 3.6 also introduces the modulated Markov and semi-Markov rating transitions models which incorporate the obligors' (or firms') heterogeneity into the analysis of credit rating migrations.

An important development in derivative markets since the 1990s has been the growth of credit derivatives. Credit derivatives allow companies to trade credit risks in much the same way that they trade market risks. By buying and selling credit derivative contracts, banks and other financial institutions can actively manage their portfolios of credit risks and get protected in case the borrower fails to repay the debt. Retail banks are typically net buyers of protection against credit events, other investors such as hedge funds and investment banks often act as both sellers and buyers of credit protection. Section 3.7 introduces two kinds of credit derivatives, credit default swaps (CDS) and collateralized debt obligations. The evaluation of the CDS spread and the structure of CDO tranches are explained in the section.

One key assumption in the models and methods presented in Sections 3.4–3.6 is that all factors related to credit risk can be observed, which is certainly not true in practice. Section 3.8 removes this assumption and introduces frailty variables as common factors. As a result, the unconditional distribution of credit events among the set of firms are dependent due to the common factors, but these credit events are conditional independent given the common factors. Section 3.8 presents two mixture models (Bernoulli mixture model and Poisson mixture model) and factor copulas to explain how static frailties are used in credit risk analysis. Models of dynamic frailty and correlated default are also presented.

3.1 Black-Scholes option pricing theory

The Black-Scholes-Merton theory for pricing and hedging options is of fundamental importance in the development of financial derivatives. Two equivalent approaches can be used to derive the Black-Scholes formulas for the option price. One is to construct a self-financing replication portfolio and derive the partial differential equation that the option price satisfies, and the other is to construct a risk-neutral measure under which the discounted option price becomes a martingale.

3.1.1 European options

A *call (put) option* gives the holder the right to buy (sell) the underlying asset (e.g., stock) at a certain date, known as the *expiration date* or *maturity*, at a certain price, which is called the *strike price*. Let K denote the strike price of an option with expiration date T and let S be the current price of the underlying asset. Let S_t denote the asset price at time t and denote $x_+ = \max(x, 0)$. Suppose that S_t is a *geometric Brownian motion* (GBM) with drift μ and volatility σ,

$$dS_t = \mu S_t dt + \sigma S_t dW_t^{\mathbb{P}}, \qquad (3.1)$$

where $W_t^{\mathbb{P}}$ is a standard Brownian motion on a physical measure probability space. Denote by $g(S_T)$ the payoff of an option, where $g(S_T) = (K - S_T)_+$ for a put option and $g(S_T) = (S_T - K)_+$ for a call option.

To use the self-financing replication strategy to construct an differential equation for the option, one usually assumes the following. First, the market has a risk-free asset with constant interest rate r. Second, continuous hedging can occur and there are no transaction costs. Third, short selling is allowed, and the asset is perfectly divisible.

There are usually two ways to find the prices of options, one is to construct a self-financing portfolio via dynamic hedging and derive a partial differential equation that the value of the option satisfies, and the other is to construct a risk-neutral measure via the Girsanov Theorem so that the discounted value of the option becomes a martingale process. In the following, we first show these two pricing procedures for underlying assets paying no dividends and extend the result to the ones paying dividend.

Self-financing replication portfolio and partial differential equation approach

Consider the case where the asset pays no dividends. Let $f(t, S)$ be the option price at time t when $S_t = S$. Consider a portfolio Π that at time t holds 1 unit of the option and $-\Delta$ units of the asset. Then $\Pi_t = f(t, S_t) - S_t \Delta$.

Since $dS_t = \mu S_t dt + \sigma S_t dW_t$, where W_t is a standard Brownian motion on some probability space, applying Ito's formula to $df(t, S_t)$ yields

$$d\Pi_t = \sigma S_t \left(\frac{\partial f}{\partial S} - \Delta \right) dw_t + \left(\mu S_t \frac{\partial f}{\partial S} + \frac{1}{2} \sigma^2 S^2 \frac{\partial^2 f}{\partial S^2} + \frac{\partial f}{\partial t} - \theta S_t \Delta \right) dt. \quad (3.2)$$

The portfolio Π_t becomes risk-free if $\Delta = \partial f / \partial S$, for which the coefficient of dW_t in (3.2) is equal to 0. In this case, Π_t should have the same return as the risk-free asset (i.e., $d\Pi_t = r\Pi_t dt$) because otherwise there are arbitrage opportunities, and (3.2) reduces to the partial differential equation (PDE)

$$\frac{\partial f}{\partial t} + rS \frac{\partial f}{\partial S} + \frac{1}{2} \sigma^2 S^2 \frac{\partial^2 f}{\partial S^2} = rf \quad \text{for} \quad 0 \le t < T. \quad (3.3)$$

The boundary condition of the PDE is $f(T, S) = g(S)$, where $g(S)$ is the payoff function of an option. The PDE (3.3) with the boundary condition $f(T, S) = g(S)$ can be solved explicitly for f, yielding the Black-Scholes formulas for the option prices $c_t = c(t, S_t)$ and $p_t = p(t, S_t)$:

$$c(t, S) = S\Phi(d_1) - Ke^{-r(T-t)}\Phi(d_2), \quad (3.4a)$$

$$p(t, S) = Ke^{-r(T-t)}\Phi(-d_2) - S\Phi(-d_1), \quad (3.4b)$$

where $\Phi(\cdot)$ is the standard normal cumulative distribution function and

$$d_1 = \frac{\log(S/K) + (r + \sigma^2/2)(T - t)}{\sigma\sqrt{T - t}}, \quad d_2 = d_1 - \sigma\sqrt{T - t}.$$

Let $p_t = p(t, S_t)$ and $c_t = c(t, S_t)$. It is easy to see from (3.4a) and (3.4b) that the following *put-call parity* holds,

$$S_t e^{-q(T-t)} + p_t - c_t = Ke^{-r(T-t)}.$$

Risk-neutral measure and martingale approach

Another way to obtain the solution is to construct a risk-neutral measure $\mathbb{Q}$ by using the Girsanov theorem. Then under the risk neutral measure $\mathbb{Q}$, S_t follows a GBM given below

$$dS_t = rS_t dt + \sigma S_t dW_t^{\mathbb{Q}},$$

where $W_t^{\mathbb{Q}}$ is Brownian motion under the measure $\mathbb{Q}$ and is defined by

$$dW_t^{\mathbb{Q}} = \frac{\mu - r}{\sigma} dt + dW_t^{\mathbb{P}}.$$

Then applying the Feynman-Kac formula, we obtain the martingale solution

$$f(t, S_t) = E^{\mathbb{Q}}\left[e^{-r(T-t)} g(S_T) \right].$$

Note that the dynamics of S_t under measure $\mathbb{Q}$ implies that $S_T = S_t e^{(r-\sigma^2/2)(T-t)+\sigma(W_T^{\mathbb{Q}}-W_t^{\mathbb{Q}})}$, so that S_T is log-normally distributed under $\mathbb{Q}$. Then given the payoff function $g(S_T) = (K - S_T)_+$ for a put option and $g(S_T) = (S_T - K)_+$ for a call option, some algebra yields the option prices (3.4a) and (3.4b).

Underlying asset with dividends

When the asset pays dividends at rate q, the self-financing replication strategy can be modified to yield the PDE

$$\frac{\partial f}{\partial t} + (r - q)S\frac{\partial f}{\partial S} + \frac{1}{2}\sigma^2 S^2 \frac{\partial^2 f}{\partial S^2} = rf \quad \text{for } 0 \le t < T. \tag{3.5}$$

Given the boundary condition $f(T, S) = g(S)$, where g is the payoff function of an option, the PDE (3.5) can be solved explicitly for f, yielding the following option prices $c_t = c(t, S_t)$ and $p_t = p(t, S_t)$:

$$c(t, S) = Se^{-q(T-t)}\Phi(d_1) - Ke^{-r(T-t)}\Phi(d_2), \tag{3.6a}$$
$$p(t, S) = Ke^{-r(T-t)}\Phi(-d_2) - Se^{-q(T-t)}\Phi(-d_1), \tag{3.6b}$$

where $\Phi(\cdot)$ is the standard normal cumulative distribution function and

$$d_1 = \frac{\log(S/K) + (r - q + \sigma^2/2)(T - t)}{\sigma\sqrt{T - t}}, \quad d_2 = d_1 - \sigma\sqrt{T - t}.$$

Same results can be obtained by using the risk-neutral measure and the martingale approach.

For the martingale approach, the dynamics of the stock price can be shown to satisfy

$$dS_t = (r - q)S_t dt + \sigma S_t dW_t^{\mathbb{Q}}.$$

The total gain process of holding the asset plus accumulated dividends is a $\mathbb{Q}$-martingale. The dynamics of S_t under measure $\mathbb{Q}$ is

$$S_T = S_t e^{(r - q - \sigma^2/2)(T-t) + \sigma(W_T^{\mathbb{Q}} - W_t^{\mathbb{Q}})}.$$

The martingale solution of the option price has the form

$$f(t, S_t) = E^{\mathbb{Q}}\left[e^{-(r-q)(T-t)}g(S_T)\right],$$

then applying the Feynman-Kac formula yields the call and put option prices (3.6a) and (3.6b).

Option Greeks

Option Greeks are measures of the sensitivity of an option's price to its determining parameters and are widely used in risk management of option portfolios; see Example 2.3 that uses the delta-gamma approximation to approximate the loss function of holding a European call option. Given an option (or option portfolio), the first and second derivatives of the value of the portfolio V with respect to the components $\mathbf{S}$ are called the *delta* and the *gamma* of the portfolio, respectively. For the European call option price (3.6a), the delta and the gamma are given by

$$\frac{\partial c}{\partial S}(t, S) = e^{-q(T-t)}\Phi(d_1), \quad \frac{\partial^2 c}{\partial S^2}(t, S) = \frac{\phi(d_1)}{S_t\sigma\sqrt{T - t}}, \tag{3.7a}$$

in which $\phi(x) = (2\pi)^{-1/2}\exp(x^2/2)$ is the density function of the standard normal distribution. For a portfolio involving derivatives, it is called *delta-neutral* if its delta is zero.

The sensitivity of option prices to other factors are defined as follows. The first derivative of the option value with respect to the time, the volatility of the underlying, and the interest rate are called the portfolio's *theta*, *vega* and *rho*, respectively. For the European call option (3.6a), they are expressed as

$$\frac{\partial c}{\partial t}(t, S) = -\frac{S_t\phi(d_1)\sigma e^{-q(T-t)}}{2\sqrt{T-t}} + qS_t\Phi(d_1)e^{-q(T-t)} - rKe^{-r(T-t)}\Phi(d_2),$$
(3.7b)

$$\frac{\partial c}{\partial\sigma}(t, S) = Se^{-q(T-t)}\phi(d_1)\sqrt{T-t},$$
(3.7c)

$$\frac{\partial c}{\partial r}(t, S) = (T-t)Ke^{-r(T-t)}\Phi(d_2);$$
(3.7d)

see Exercise 3.3.

3.1.2   American options

An American option, similar to a European option, gives the holder (buyer) the right, but not the obligation, to buy or sell an underlying asset at a strike price on or before a predetermined expiration date. The key difference between American and European options is that American options can be exercised at any time prior to or on the expiration date, whereas European options can only be exercised at expiration.

Merton (1974a) extended the Black-Scholes theory for pricing European options to American options. The Black-Scholes PDE (3.5) still holds in the continuation region $\mathcal{C}$ of (t, S_t) before exercise, and the exercise boundary $\partial\mathcal{C}$ is determined by the *free boundary condition* $\partial f/\partial S = 1$ (or -1) for a call (or put) option. Optimal exercise of the option occurs when the asset price exceeds or falls below an exercise boundary $\partial\mathcal{C}$ for a call or put option, respectively. Unlike the explicit formula (3.6a) or (3.6b) for European options, there is no closed-form solution of the free-boundary PDE, and numerical methods such as finite differences are needed to compute American option prices under this theory. The free-boundary PDE can also be represented probabilistically as the value function of the optimal stopping problem

$$f(t, S) = \sup_{\tau\in\mathcal{T}_{t,T}} E^{\mathbb{Q}}[e^{-r(\tau-t)}g(S_\tau)|S_t = S],$$
(3.8)

where $\mathcal{T}_{t,T}$ denotes the set of stopping times τ whose values are between t and T, and $E^{\mathbb{Q}}$ is expectation with respect to the *risk-neutral measure* $\mathbb{Q}$ under which S_t is GBM with drift $r - q$ and volatility σ. Cox et al. (1975) proposed to approximate GBM by a binomial tree with root node S_0 at time 0, so that (3.8) can be approximated by a discrete-time and discrete-state optimal stopping problem that can be computed by backward induction; see Hull

(2021, Chapter 13) for an introduction to the binomial tree method. Besides finite-difference and tree methods, Monte carlo methods are also widely used to price American options; see, for example, Longstaff and Schwartz (2001).

Early exercise boundary

Denote $f(t, S)$ by $C(t, S)$ for an American call option, and by $P(t, S)$ for an American put option. Jacka (1991) and Carr et al. (1992) derived the decomposition formula

$$
\begin{aligned}
P(t, S) \;=\; & p(t, S) + K\rho e^{\rho u} \int_u^0 \left\{ e^{-\rho s} \Phi\left(\frac{\bar{z}(s) - z}{\sqrt{s - u}} \right) \right. \\
& \left. - \mu e^{-(\mu \rho s + u/2) + z} \Phi\left(\frac{\bar{z}(s) - z}{\sqrt{s - u}} - \sqrt{s - u} \right) \right\} ds \quad (3.9)
\end{aligned}
$$

and a similar formula relating $C(t, S)$ to $c(t, S)$, where $\bar{z}(u)$ is the early exercise boundary $\partial \mathcal{C}$ under the transformation

$$\rho = r/\sigma^2, \; \mu = q/r, \; u = \sigma^2(t - T), \; z = \log(S/K) - (\rho - \mu\rho - 1/2)u. \quad (3.10)$$

Ju (1998) found that the *early exercise premium*, $P(t, S) - p(t, S)$ in (3.9), can be computed in closed form if $\partial \mathcal{C}$ is a piecewise exponential function that corresponds to a piecewise linear $\bar{z}(u)$. By using such an assumption, Ju (1998) reported numerical studies showing that his method with three equally spaced pieces substantially improves previous approximations to option prices in both accuracy and speed. AitSahlia and Lai (2001) introduced the transformation (3.10) to reduce GBM to Brownian motion, which can be approximated by a symmetric Bernoulli random walk, and to transform the early exercise boundary $\partial \mathcal{C}$ to $\bar{z}(u)$ in the new coordinate system. They developed a corrected random walk approximation to compute by backward induction the optimal stopping boundary $\bar{z}(\cdot)$, which their numerical results show can indeed be well approximated by a piecewise linear function with a few pieces. The integral obtained by differentiating that in (3.9) with respect to S also has a closed-form expression when $\bar{z}(\cdot)$ is piecewise linear, and approximating $\bar{z}(\cdot)$ by a linear spline that uses a few unevenly spaced knots gives a fast and reasonably accurate method for computing the delta $\Delta = \partial P/\partial S$ of an American put. Similar results also hold for American call options on dividend-paying stocks; American calls on stocks that do not pay dividends are optimally exercised at maturity.

3.1.3 Implied volatility

The interest rate r in the Black-Scholes formula (3.4a) or (3.4b) for the price of a European option is usually taken to be the yield of a short-maturity Treasury bill at the time when the contract is initiated. The parameter in

(3.4a) or (3.4b) that cannot be directly observed is σ. Equating (3.4a) or (3.4b) to the actual price of the option yields a nonlinear equation in σ whose solution is called the *implied volatility* of the underlying asset. Traders calculate implied volatilities from actively traded options on a stock and use them to price over-the-counter options on the same stock and to calculate the option's delta for hedging applications. The implied volatilities computed from call and put options with the same strike price K and time to maturity $T - t$ should be equal because the put–call parity relationship (3.1.1) holds for both the Black-Scholes price pair $\left(p_t^{\mathrm{BS}}, c_t^{\mathrm{BS}}\right)$ and the market price pair $\left(p_t^{\mathrm{M}}, c_t^{\mathrm{M}}\right)$, from which it follows that $c_t^{\mathrm{BS}} - c_t^{\mathrm{M}} = p_t^{\mathrm{BS}} - p_t^{\mathrm{M}}$ and therefore the equation $c_t^{\mathrm{BS}} = c_t^{\mathrm{M}}$ gives the same solution for σ as $p_t^{\mathrm{BS}} = p_t^{\mathrm{M}}$.

Smiles, skews, and surfaces

A call option, whose payoff function is $(S - K)_+$, is said to be *in the money*, *at the money*, or *out of the money* according to whether $S_t > K$, $S_t = K$, or $S_t < K$, respectively. Puts have the reverse terminology since the payoff function is $(K - S)_+$. According to the Black-Scholes theory, the σ in (3.4a) and (3.4b) is the volatility of the underlying asset and therefore does not vary with K and T. However, for some equity options, a *volatility skew* is observed (i.e., the implied volatility is a decreasing function of the strike price K). The *volatility smile* is common in foreign currency options, for which the implied volatility is relatively low for at-the-money options and becomes higher as the option moves into the money or out of the money, giving the "smile" shape of the implied volatility curve as a function of K (with minimum around $K = S$). Moreover, implied volatilities also tend to vary with time to maturity.

The implied volatilities of options on an underlying asset, therefore, are often quoted as a function of K and T. *Volatility surfaces*, usually presented in the form of a table, provide the volatilities for pricing an option on the asset with any strike price and any maturity. Volatility surfaces play an important role in options trading and risk management. Traders can use volatility surfaces to identify mispriced options, construct volatility-based trading strategies, and hedge against volatility risk. Risk managers may also use volatility surfaces to assess and manage the overall volatility exposure of their options portfolios.

3.2   Stochastic volatility models

This section introduces briefly three types of stochastic volatility models. Each of these models offers a different perspective on how to model and understand asset price dynamics and volatility behavior, providing valuable tools for option pricing, risk management, and hedging in financial markets.

3.2.1 The local volatility model

Unlike the Black-Scholes model, which assumes constant volatility, the *local volatility* model accounts for the volatility smile or skew observed in options markets. The local volatility model provides an exact fit to all European option prices on any given day. It assumes that the *risk-neutral process* of the asset price has the more general form

$$dS_t = (r_t - q_t)S_t dt + \sigma(t, S_t)S_t dW_t, \qquad (3.11)$$

which replace the constant σ in the Black-Scholes theory by a function $\sigma(t, S_t)$. Dupire (1979) has shown that the function $\sigma(t, S)$ is given analytically by

$$\frac{\sigma^2(T, K)}{2} = \left\{ \frac{\partial c_T}{\partial T} + q_T c_T + K(r_T - q_T)\frac{\partial c_T}{\partial K} \right\} \Big/ K^2 \frac{\partial^2 c_T}{\partial K^2}, \qquad (3.12)$$

where c_T is the market price of a European call option with strike price K and maturity T.

There are mainly two methods to price a European option under the risk-neutral model (3.11), finite-difference approximations and tree-based approaches. Andersen and Brotherton-Ratcliffe (1997) used finite difference approximations of (3.12) to recalibrate the model (3.11) daily to the market prices of standard European options. By recalibrating daily, one can use the model to closely match observed option prices, thereby ensure that the model remains consistent with market conditions over time. An alternative approach, proposed by Derman and Kani (1994) and Rubinstein (1994), approximates (3.11) by an *implied tree*, which is a discrete-time Markov chain approximation to (3.11) in the form of a binomial tree that is recalibrated daily to the market prices of standard options. Since this approach focuses exclusively on "in-sample" fitting, which uses the sample consisting of all European option prices on a given day, its disadvantage is that the in-sample correctness does not extend to out-of-sample forecasting of future European option prices or pricing of American options that can be exercised at any time prior to the expiration date T.

3.2.2 The constant elasticity of variance model

Whereas the local volatility model changes the constant σ in the Black-Scholes theory by a function $\sigma(t, S_t)$, the *constant elasticity of variance* (CEV) model replaces σ by σS^α, imposing an additional parameter for the Black-Scholes model. Specifically, Cox and Ross (1976) assumed that the risk-neutral process of the asset price follows the CEV model

$$dS_t = (r - q)S_t dt + \sigma S_t^\alpha dW_t.$$

Here, a value of $\alpha = 1$ results in constant volatility, while values greater than or less than 1 imply increasing or decreasing volatility, respectively, as the asset price changes.

Cox and Ross (1976) showed that the formulas (3.4a) and (3.4b) for $c(t, S)$ and $p(t, S)$ can be modified by replacing $\Phi(d_1)$ and $\Phi(d_2)$ by the distribution functions of certain noncentral χ^2-distributions:

$$
\begin{aligned}
c(t, S) &= Se^{-q(T-t)}\left[1 - \chi^2(a; b+2, c)\right] - Ke^{-r(T-t)}\chi^2(c; b, a), \\
p(t, S) &= Ke^{-r(T-t)}\left[1 - \chi^2(c; b, a)\right] - Se^{-q(T-t)}\chi^2(a; b+2, c),
\end{aligned}
$$

in the case $\alpha > 1$ and

$$
\begin{aligned}
c(t, S) &= Se^{-q(T-t)}\left[1 - \chi^2(c; -b, a)\right] - Ke^{-r(T-t)}\chi^2(a; 2-b, c), \\
p(t, S) &= Ke^{-r(T-t)}\left[1 - \chi^2(a; 2-b, c)\right] - Se^{-q(T-t)}\chi^2(c; -b, c),
\end{aligned}
$$

in the case $0 < \alpha < 1$, where K is the strike price and T is the expiration date of the option, $\chi^2(\cdot; \nu, \lambda)$ is the distribution function of the noncentral chi-square distribution (defined below) with ν degrees of freedom and noncentrality parameter λ, and

$$
v = \frac{\sigma^2}{2(r-q)(\alpha-1)}\left\{e^{2(r-q)(\alpha-1)(T-t)} - 1\right\},
$$

$$
a = \frac{\left[Ke^{-(r-q)(T-t)}\right]^{2(1-\alpha)}}{(1-\alpha)^2 v}, \quad b = \frac{1}{1-\alpha}, \quad c = \frac{S^{2(1-\alpha)}}{(1-\alpha)^2 v};
$$

see Hull (2021, Section 27.3). The parameters α and σ of the CEV model can be estimated by nonlinear least squares, minimizing over (α, σ) the sum of squared differences between the model prices and the market prices.

3.2.3 The stochastic volatility model

The continuous-time *stochastic volatility* (SV) model under the risk-neutral measure is $dS_t/S_t = (r-q)dt + \sigma_t dW_t$, in which $v_t := \sigma_t^2$ is modeled by

$$
dv_t = \alpha(v^* - v_t)dt + \beta v_t^\xi d\widetilde{W}_t, \tag{3.13}
$$

where $\widetilde{W}_t$ is Brownian motion that is independent of W_t. For this SV model, Hull and White (1987) have shown that the price of a European option is given by $\int_0^\infty b(w)g(w)dw$, where $b(w)$ is the Black-Scholes price in which σ is replaced by w, and w is the average variance rate during the life of the option, which is a random variable with density function g determined by the stochastic dynamics (3.13) for v_t. Although there is no analytic formula for g, Hull and White (1987) have used this representation of the option price to develop closed-form approximations to the model price of the form $\int_0^\infty b(w)g(w)dw$. The parameters α, β, v^*, and ξ in (3.13) can be estimated by minimizing the sum of squared differences between the model prices and the market prices. The SV model has been used to account for the volatility smile associated with the Black-Scholes prices. Heston (1993) introduced a specific case of the stochastic volatility model where $\xi = 1/2$. This special case simplifies the model and allows for a closed-form solution for the prices of European call and put options.

3.3 Stochastic interest rates

To price interest rate derivatives that are contingent on future interest rates, stochastic models of interest rate dynamics are often used. This section introduces different types of interest rates and commonly used models for short rates and forward rates.

3.3.1 Types of interest rates

An interest rate can be considered as the rate of return for a "risk-free" asset (e.g., Treasury bills, bonds). If the asset has cash value P_0 at time 0 and the interest rate at time t is r_t, then the cash value of the asset at time t is given by

$$P_t = P_0 \exp\left(\int_0^t r_s ds \right).$$

In the above formula, P_t can be regarded as the value of a bank account at time $t \geq 0$, and r_s is commonly called the *short rate* (or *instantaneous spot rate*); the short rate can be changed on a daily basis by the bank.

Risk-free bonds

A *zero-coupon bond* pays a specific amount, called the *face* (or *par*) *value*, at maturity without intermediate coupon payments. U.S. Treasury bills (T-bills) are zero-coupon bonds with fixed terms to maturity of 13, 26, and 52 weeks. U.S. Treasury notes (T-notes) are semiannual coupon bonds with maturities of 1 to 10 years; U.S. Treasury bonds (T-bonds) are semiannual coupon bonds with maturities of more than 10 years.

The *n-year zero rate* (also called *zero-coupon yield*) is the *yield* (i.e., interest rate) of a zero-coupon bond that matures in n years. For a coupon-bearing bond, the yield is the interest rate implied by its payment structure. Let $A = $ face value of bond, $B = $ bond price, $C_j = $ coupon payment at time t_j ($j = 1, \ldots, J$), and n denote the number of years to maturity. Under continuous compounding, the bond's *yield to maturity* (or simply *yield*) is given by the equation

$$B = Ae^{-yn} + \sum_{j=1}^{J} C_j e^{-yt_j}.$$

Bond prices are often quoted in two different forms in the market. The *dirty* price is the actual amount paid in return for the full amount of all future coupon payments and the principal. It is the sum of the *clean* price and the *accrued interest*.

The price at time $t \leq T$ of a zero-coupon bond with face value 1 and maturity date T is denoted by $P(t,T)$. Clearly $P(T,T) = 1$. The *spot rate*

$R(t,T)$ at time t of the bond is the yield (under continuous compounding) given by

$$R(t,T) = -\frac{\log P(t,T)}{T-t} \tag{3.14}$$

or, equivalently

$$P(t,T) = \exp\big\{-(T-t)R(t,T)\big\}.$$

For a coupon-bearing bond, the *par yield* at time t specifies the coupon rate $\rho(t,T)$ that causes the price of the bond, issued at t and maturing at T, to equal its par value. For example, in the case of a bond that pays coupons annually and matures in T years, the par yield $\rho(t,T)$ in year $t<T$ is given by

$$\rho(t,T)\sum_{s=t+1}^{T} P(t,s) + P(t,T) = 1.$$

Solving the above equation yields that $\rho(t,T) = \{1-P(t,T)\}/\sum_{s=t+1}^{T} P(t,s)$.

Forward rates

A *forward rate agreement* is a contract at the current time t for a loan between the expiration date T_1 of the contract and the maturity date T_2 of the loan. The contract gives its holder a loan at time T_1 with a fixed rate of simple interest for the period $T_2 - T_1$, to be paid at time T_2 besides the principal; the holder also receives at time T_2 an interest payment based on the forward rate $F(t,T_1,T_2)$ that is defined as

$$F(t,T_1,T_2) = \frac{1}{T_2 - T_1}\log\frac{P(t,T_1)}{P(t,T_2)}. \tag{3.15}$$

In the case $T_1 = t$, the forward rate becomes the spot rate $R(t,T_2)$ defined by (3.14). The *instantaneous forward rate* at time t is defined as

$$f(t,T) = \lim_{\delta\to 0} F(t,T,T+\delta) = -\frac{1}{P(t,T)}\frac{\partial P(t,T)}{\partial T}. \tag{3.16}$$

Hence, the price $P(t,T)$ of a zero-coupon bond can be expressed as

$$P(t,T) = \exp\left[-\int_t^T f(t,u)du\right]. \tag{3.17}$$

The short rate r_t is related to the instantaneous forward rate by

$$r(t) = \lim_{T\to t} f(t,T) = f(t,t).$$

LIBOR (London InterBank Offered Rate)

The LIBOR is an interest rate at which banks lend to one another in the London interbank market. It serves as a benchmark for short-term interest rates globally and is widely used in financial markets for pricing various financial instruments, including interest rate swaps, floating-rate loans, and derivatives. The LIBOR rates apply to loans with various maturities, such as 1 day, 1 month, 3 months, 6 months, 1 year, and up to 5 years. The LIBOR rate is an annualized, simple rate of interest that will be delivered at the end of a specified period. The LIBOR *zero curve* up to 1 year is determined by the 1-month, 3-month, 6-month, and 12-month LIBOR rates. Eurodollar futures contracts can be used to extend the zero curve to longer maturities.

There are therefore different types of forward rates, and a major distinction can be made between interbank rates (LIBOR) and government (U.S. Treasury, Japanese Treasury, etc.) rates. The same notations $P(t, T)$, $R(t, T)$, $f(t, T)$, however, are used to refer to the rates in different sectors in what follows. In the bond options market, the forward rate is often associated with LIBOR, for which simple, instead of continuously compounded, interest rates are used. In particular, the *forward LIBOR rate* is defined as

$$F(t, T_1, T_2) = \frac{1}{T_2 - T_1} \left[\frac{P(t, T_1)}{P(t, T_2)} - 1 \right], \tag{3.18}$$

in which $P(t, T)$ is given by the LIBOR term structure. Note that the last equality in (3.16) holds under either (3.15) or (3.18).

During the past decade, LIBOR faced significant integrity issues following the manipulation scandal that surfaced in the late 2000s. Concerns about the reliability and sustainability of LIBOR as a reference rate grew over time. In response to the shortcomings of LIBOR, financial regulators and industry stakeholders embarked on a concerted effort to identify and promote alternative reference rates that are more robust, transparent, and based on actual transactions rather than expert judgment. In the US, the *Alternative Reference Rates Committee* (ARRC) recommended the *Secured Overnight Financing Rate* (SOFR) as the preferred alternative to USD LIBOR. SOFR is based on overnight Treasury repurchase agreement transactions and is considered a more representative measure of short-term borrowing costs. Similarly, in the UK, the Financial Conduct Authority (FCA) endorsed the *Sterling Overnight Index Average* (SONIA) as the replacement for GBP LIBOR. SONIA is based on overnight unsecured transactions in the sterling money markets.

Swap rates and interest rate swaps

The *interest rate swap* is an extension of the forward rate agreement involving two financial institutions. In an interest rate swap, there are two parties involved, often referred to as the "fixed-rate payer" (say, financial institution A) and the "floating-rate payer" (say, financial institution B). Institution A agrees to pay institution B cash flows equal to the interest at a predetermined fixed rate while institution B agrees to pay institution A cash flows equal to the interest at a floating rate on a *notional principal* (which is not exchanged between institutions A and B) for a prespecified period of time. The floating rate is usually LIBOR.

Under a swap contract that is initialized at time $T = T_0$, there are swap payments at times $T_1, \ldots, T_M$ with $T_i - T_{i-1} = \tau_i$. The contract specifies that one party pays a fixed rate $\tau_i K$ at time T_i and receives from the other party the floating rate $\tau_i L(T_{i-1}, T_{i-1}, T_i)$, where $L(t, T_{i-1}, T_i) = \{P(t, T_{i-1})/P(t, T_i) - 1\}/\tau_i$ is used to denote the forward LIBOR rate (3.18). The party that pays the fixed rate is called the *payer* and the other party, paying the floating rate, is called the *receiver* of the swap. To ensure no arbitrage, K has to be chosen at time $t \leq T$ so that the swap has value 0, i.e.,

$$K = \frac{P(t, T_0) - P(t, T_M)}{\sum_{i=1}^{M} P(t, T_i)\tau_i}, \tag{3.19}$$

noting that $\sum_{i=0}^{M-1} \{P(t, T_i)/P(t, T_{i+1}) - 1\} P(t, T_{i+1}) = P(t, T_0) - P(t, T_M)$. The value of K given by (3.19) is called the *forward swap rate* and is denoted by $s(t, T_0, T_M)$. For the particular case $t = T_0$ and $\tau_i = \tau$, it is called the *swap rate* with maturity $M\tau$ and is denoted by s_M. Combining (3.19) with (3.18) yields an alternative expression for the forward swap rate that links it to the underlying forward rates:

$$s(t, T_0, T_M) = \sum_{i=0}^{M-1} w_i F(t, T_i, T_{i+1}) \tag{3.20}$$

with

$$w_i = P(t, T_{i+1})\tau_{i+1} \left/ \left\{ \sum_{j=0}^{M-1} P(t, T_{j+1})\tau_{j+1} \right\} \right. .$$

3.3.2   Short rate models

A typical way of modeling the short rate (or the instantaneous spot rate) r_t is to assume r_t follows the diffusion process

$$dr_t = m(t, r_t)dt + \nu(t, r_t)dW_t.$$

Let $\mu(t, r) = m(t, r)/r$, $\sigma(t, r) = \nu(t, r)/r$, so $dr_t/r_t = \mu(t, r_t)dt + \sigma(t, r_t)dW_t$. Consider a portfolio consisting of two bonds with different maturities. Using Ito's formula and hedging arguments similar to those of Section 3.1, one can show that in the absence of arbitrage,

$$\lambda(t, r) = \frac{\mu(t, r) - r}{\sigma(t, r)}$$

is the same for all derivatives dependent on r_t (e.g., bonds with the same maturities). $\lambda(t, r)$ is called the *market price of risk*. Moreover, the price $g(t, r)$ of an interest rate derivative satisfies the Black-Scholes-type PDE

$$\frac{\partial g}{\partial t} + (m - \lambda \nu)\frac{\partial g}{\partial r} + \frac{\nu^2}{2}\frac{\partial^2 g}{\partial r^2} = rg, \quad 0 \le t < T. \tag{3.21}$$

In the case of a bond with par value 1 ($g = P$), we obtain the terminal condition $g(T, r) = 1$. The solution of (3.21) with this terminal condition can be expressed as the expectation

$$E^{\mathbb{Q}}\left\{\exp\left(-\int_t^T r_s ds\right)\Big| r_t = r\right\}. \tag{3.22}$$

Here, $E^{\mathbb{Q}}$ denotes the expectation under which r_t is the diffusion process

$$dr_t = (m - \lambda\nu)dt + \nu dW_t,$$

or $\mathbb{Q}$ is the "equivalent martingale measure."

Unlike the Black-Scholes model and pricing theory, which are commonly used in the equity options market, there is no single dominant model with a specific choice of $m(t, r)$ and $s(t, r)$ in (3.21) that is widely accepted for pricing various derivative securities in the interest rate market. Instead, one encounters a plethora of models and risk-neutral measures that are different from the physical (real-world) measure. Rebonato (2004) gives an overview of the historical development of these models. Another modeling issue for interest rate derivatives is that unlike the relatively short maturity of equity options, the interest rate market considers bonds with maturities as long as 10–30 years. Because of anticipated changes in the economy over such a long period, there are no convincing models that can actually capture the real-world interest rate movements 10–30 years in the future. What one can model at best are the perceived movements that are reflected in the prices of interest rate derivatives.

Vasicek, Cox-Ingersoll-Ross, and Hull-White models

When r_t is a Gaussian process under the risk-neutral measure, the expectation (3.22) has a closed-form expression in view of the formula for moment generating functions of normal random variables. Empirical studies have shown that interest rates fluctuate around some level. Motivated by this *mean-reverting* property, Vasicek (1977) and Hull and White (1990) propose the following models for r_t

$$\text{Vasicek (1977):} \qquad dr_t = \kappa(\theta - r_t)dt + \sigma dW_t, \tag{3.23}$$
$$\text{Hull and White (1990):} \qquad dr_t = (\theta_t - \kappa r_t)dt + \sigma dW_t, \tag{3.24}$$

in which θ_t is a nonrandom function of t such that $P(0,T)$ agrees with the term structure at time 0. Note that (3.23) is a special case of (3.24) with $\theta_t = \kappa\theta$, but this choice often fails to match the initial term structure exactly.

A conceptual difficulty with Gaussian process models for r_t is that $\{r_t < 0\}$ has positive probability. The following (non-Gaussian) modification of the Vasicek model proposed by Cox et al. (1985) (abbreviated by CIR) has positive r_t when $2\kappa\theta > \sigma^2$:

$$dr_t = \kappa(\theta - r_t)dt + \sigma\sqrt{r_t}dW_t. \tag{3.25}$$

The price $p_0(t,T)$ of a zero-coupon bond with par value 1 given by the expectation (3.22) has a closed-form expression for the models (3.23)–(3.25):

$$p_0(t,T) = \alpha(t,T)e^{-\beta(t,T)r_t}. \tag{3.26}$$

The functions α and β in (3.26) are given below.

(i) For the Vasicek model, $\beta(t,T) = (1 - e^{-\kappa(T-t)})/\kappa$,

$$\alpha(t,T) = \exp\left\{\left(\theta - \frac{\sigma^2}{2\kappa^2}\right)[\beta(t,T) - (T-t)] - \frac{\sigma^2}{4\kappa}\beta^2(t,T)\right\}.$$

(ii) For the CIR model, letting $h = \sqrt{\kappa^2 + 2\sigma^2}$,

$$\alpha(t,T) = \left[\frac{2he^{(\kappa+h)(T-t)/2}}{2h + (\kappa+h)[e^{(T-t)h} - 1]}\right]^{2\kappa\theta/\sigma^2},$$

$$\beta(t,T) = \frac{2(e^{(T-t)h} - 1)}{2h + (\kappa+h)[e^{(T-t)h} - 1]}.$$

(iii) For the Hull-White model, $\beta(t,T) = (1 - e^{-\kappa(T-t)})/\kappa$,

$$\log\alpha(t,T) = \log\frac{p_0(0,T)}{p_0(0,t)} - \beta(t,T)\frac{\partial}{\partial t}\log P(0,t) - \frac{\sigma^2(1 - e^{-2\kappa t})}{4\kappa}\beta^2(t,T).$$

Provided with the zero coupon bond price (3.26), the spot rate $R(t,T)$ defined by (3.14) is an affine function of the short rate r_t:

$$R(t,T) = -\frac{\log\alpha(t,T)}{T-t} + r_t\frac{\beta(t,T)}{T-t}.$$

Moreover, the instantaneous forward rate $f(t,T)$ defined by (3.16) is also an affine function of r_t,

$$f(t,T) = -\frac{\partial}{\partial T}\log p_0(t,T) = -\frac{\partial}{\partial T}\log\alpha(t,T) + r_t\frac{\partial}{\partial T}\beta(t,T).$$

One-factor-model-based bond option prices

Provided with the preceding short-rate models, Jamshidian (1989) has derived explicit formulas for the prices of European options on bonds. In the Vasicek model, the price $Z(t, T, \widetilde{T}, K)$ at time t of a European option with strike K, maturity T, and written on a zero-coupon bond that matures at time $\widetilde{T}$ and has par value 1 is

$$
\begin{aligned}
Z_c(t, T, \widetilde{T}, K) &= p_0(t, \widetilde{T})\Phi(h) - Kp_0(t, T)\Phi(h - \sigma_P), &\text{(3.27a)} \\
Z_p(t, T, \widetilde{T}, K) &= Kp_0(t, T)\Phi(\sigma_P - h) - p_0(t, \widetilde{T})\Phi(-h), &\text{(3.27b)}
\end{aligned}
$$

in which

$$
\sigma_P = \sigma\sqrt{\frac{1 - e^{-2\kappa(T-t)}}{2\kappa}}\beta(T, \widetilde{T}), \quad h = \frac{1}{\sigma_P}\log\frac{p_0(t, \widetilde{T})}{p_0(t, T)K} + \frac{\sigma_P}{2}.
$$

The formulas (3.27a) and (3.27b) still hold for the Hull-White model, with the same definition of σ_P but with h redefined as

$$
h = \frac{1}{\sigma_P}\log\frac{p_0(0, \widetilde{T})}{p_0(0, T)K} + \frac{\sigma_P}{2}.
$$

In the CIR model, the price at time t of a European call option with strike K, maturity T, and written on a zero-coupon bond that matures at time $\widetilde{T}$ and has par value 1 is

$$
\begin{aligned}
Z_c(t, T, \widetilde{T}, K) = p_0(t, \widetilde{T})\chi^2\left(2\mu[\rho + \psi + \beta(T, \widetilde{T})]; \frac{4\kappa\theta}{\sigma^2}, \frac{2\rho^2 r_t e^{h(T-t)}}{\rho + \psi + \beta(T, \widetilde{T})}\right) \\
- Kp_0(t, T)\chi^2\left(2\mu[\rho + \psi]; \frac{4\kappa\theta}{\sigma^2}, \frac{2\rho^2 r_t e^{h(T-t)}}{\rho + \psi}\right)
\end{aligned}
$$

when the short rate at time t is r_t, where

$$
\rho = \frac{2h}{\sigma^2 e^{h(T-t)} - 1}, \quad \psi = \frac{\kappa + h}{\sigma^2}, \quad \mu = \frac{\ln(\alpha(T, \widetilde{T})/K)}{\beta(T, \widetilde{T})},
$$

and $\chi^2(\cdot; \nu, \lambda)$ is the distribution function of the noncentral chi-square distribution with ν degrees of freedom and noncentrality parameter λ. The price of a European put option can be found by the *put–call parity* relation

$$
Z_c + Kp_0(t, T) = Z_p + p_0(t, \widetilde{T}).
$$

Multifactor affine yield models

The one-factor models (3.23), (3.24), and (3.25) only rely on one source of randomness dW_t in the short rate r_t. In these models, the yields for different maturities are perfectly correlated. A simple way to introduce additional sources of randomness is to replace dW_t by $d\mathbf{W}_t$, where $\mathbf{W}_t$ is d-dimensional standard Brownian motion. Then given a d-dimensional diffusion process of the form $d\mathbf{x}_t = \mathbf{B}\mathbf{x}_t dt + \Sigma^{1/2} d\mathbf{W}_t$, the short rate process is represented as $r_t = \mu + \boldsymbol{\theta}' d\mathbf{x}_t$. For example, consider the two-factor short rate model $r_t = x_t + y_t$ with factors x_t and y_t given by

$$dx_t = -ax_t dt + \sigma dW_{t,1}, \qquad dy_t = -by_t dt + \nu dW_{t,2},$$

where $a, b, \sigma, \widetilde{\sigma}$ are positive constants and $(W_{t,1}, W_{t,2})$ is a two-dimensional Brownian motion. Then the price of a zero-coupon bond maturing at time T can be expressed as

$$p_0(t,T) = \exp\left\{ -\frac{1 - e^{-a(T-t)}}{a} x_t - \frac{1 - e^{-b(T-t)}}{b} y_t + \frac{1}{2}\alpha(t,T) \right\},$$

where

$$
\begin{aligned}
\alpha(t,T) &= \frac{\sigma^2}{a^2}\left[T - t + \frac{2}{a}e^{-a(T-t)} - \frac{1}{2a}e^{-2a(T-t)} - \frac{3}{2a} \right] \\
&\quad + \frac{\nu^2}{b^2}\left[T - t + \frac{2}{b}e^{-b(T-t)} - \frac{1}{2b}e^{-2b(T-t)} - \frac{3}{2b} \right].
\end{aligned}
$$

Empirical studies using principal component analysis for yield curves suggest that usually $d = 2$ or 3 suffices for the number of factors.

3.3.3   Forward rate dynamics

Instead of modeling the stochastic dynamics of the short rate process r_t, an alternative approach is to model the instantaneous forward rate $f(t,T)$, noting that $P(t,T)$ can be retrieved from $f(t,\cdot)$ via (3.17). An even more fundamental entity is the forward rate (3.15) that involves a three-dimensional time index (t, T_1, T_2), and therefore a stochastic model for (3.15) is a random field rather than a stochastic process such as r_t that involves one-dimensional time t. However, because the relevant dates in interest rate derivatives are typically fixed at $T_1, \ldots, T_M$, one only has to work with a vector-valued process $(F_1(t), \ldots, F_M(t))$, where $F_i(t) = F(t, T_i, T_{i+1})$. In the next, we first describe Black's model for the dynamics of $F_i(t)$ and the associated Black-Scholes type formulas that are commonly used in the interest rate market for pricing caps/floors and swaptions. Then we present the LIBOR and swap market models that resolve certain inconsistencies in Black's model and provide theoretical justifications and extensions of the standard market formulas for pricing caps/floors and swaptions.

Standard market formulas based on Black's model of forward prices

The market convention of pricing commonly traded interest rate derivatives is based on Black's formula (Black, 1976). First consider a European option with strike price K and expiration date T on a zero-coupon bond maturing at time $\widetilde{T} > T$. Black's formula yields the option price for a call and a put option

$$
\begin{aligned}
Z_c(t, T, \widetilde{T}, K) &= p_0(t, \widetilde{T}) \big[F_t \Phi(d_1) - K\Phi(d_2) \big] & \text{(3.28a)} \\
Z_p(t, T, \widetilde{T}, K) &= p_0(t, \widetilde{T}) \big[K\Phi(-d_2) - F_t \Phi(-d_1) \big] & \text{(3.28b)}
\end{aligned}
$$

at time t, where

$$
d_1 = \frac{\log(F_t/K) + \sigma^2(T-t)/2}{\sigma\sqrt{T-t}}, \quad d_2 = d_1 - \sigma\sqrt{T-t},
$$

under the assumption that F_t follows the geometric Brownian motion $dF_t/F_t = \mu dt + \sigma dW_t$. Besides this GBM assumption on the forward price F_t of the bond (with maturity date $\widetilde{T}$) at the expiration date T of the option, (3.28a) and (3.28b) also assume that the interest rate is a deterministic function so that the price of a forward contract is the same as that of a futures contract with the same delivery date.

Interest rate caps and floors are two popular interest rate derivatives used by interest rate borrowers and lenders to manage interest rate risk associated with floating-rate loans. An interest rate *cap* is designed to provide insurance against the rate of interest on a floating-rate loan rising above a specified level R^*, called the *cap rate*. The cap rate represents the maximum interest rate that the interest rate borrower is willing to pay on the loan. If the reference rate (usually LIBOR) exceeds the cap rate at any reset date during the life of the cap, the cap buyer receives a payment from the cap seller. In an interest rate cap, the floating rate of the loan is periodically reset to equal LIBOR at dates $T_1, \ldots, T_M$. At each reset date T_i during the life of the cap with expiration date $\widetilde{T}(> T_M)$, if LIBOR exceeds R^*, the cap's payoff at date T_{i+1} is LIBOR minus R^*, and there is no payoff if LIBOR falls below R^*. Note that the payment dates (when payoffs from the cap are assessed) are $T_1, \ldots, T_M$. In a similar way (but with R^* minus LIBOR as the payoff), an interest rate *floor* with *floor rate* R^* provides insurance against LIBOR falling below R^*.

Consider an interest rate cap described above. Denote by $F_i(t)$ the forward rate $F(t, T_i, T_{i+1})$, where the T_i ($i = 1, \ldots, M$) are the reset dates. Let $\tau_i = T_{i+1} - T_i$, which is called a *tenor*. The cap can be viewed as a portfolio of M *caplets* that are call options with payoff $N\tau_i(F_i - R^*)_+$ at time T_{i+1}, in which $F_i = F(T_i, T_i, T_{i+1})$, R^* is the cap rate, and N the notional principal of the cap; see Hull (2021, Section 29.2). Note that F_i is the interest rate observed at time T_i for the period between T_i and T_{i+1}; i.e., $F_i = F(T_i, T_i, T_{i+1})$. Assuming GBM dynamics $dF_i/F_i = \mu_i dt + \sigma_i dW_t^{(i)}$, the value at time $t \le T_i$ of the ith caplet is given by (3.28a) as

$$
Z_i(t) = N\tau_i p_0(t, T_{i+1})\{F_i(t)\Phi(d_{1i}) - R^*\Phi(d_{2i})\}, \qquad \text{(3.29)}
$$

where $d_{1i} = \{\log(F_i(t)/R^*) + \sigma_i^2(T_i - t)/2\}/\{\sigma_i\sqrt{T_i - t}\}$, $d_{2i} = d_{1i} - \sigma_i\sqrt{T_i - t}$. The cap price at time t (before the first reset date) is the sum $\sum_{i=1}^{M} Z_i(t)$ of the prices of these caplets. The standard market price of floors is similar, using put options in place of the call options.

LIBOR market model and swap market model

LIBOR market and swap market models are introduced by Brace et al. (1997) and Jamshidian (1997), respectively. They assume arbitrage-free stochastic processes for forward rates or swap rates, respectively and are consistent with Black's model, which is a variant of the Black-Scholes model in Section 3.1.

Consider the probability measure $\mathbb{P}_{T_i}$ that is forward risk-neutral with respect to the numeraire $p_0(t, T_i)$. Here, a numeraire is a measure or unit of value used to standardize the pricing and valuation of financial instruments or assets. Note that the forward rate $F_i(t)$ satisfies $F_i(t)p_0(t, T_{i+1}) = \{p_0(t, T_i) - p_0(t, T_{i+1})\}/\tau_i$. Suppose that, under $\mathbb{P}_{T_i}$, $F_i(t)$ is a driftless GBM defined by

$$dF_i(t) = \nu_i(t)F_i(t)dW_i(t), \qquad t \leq T_i, \tag{3.30}$$

in which $\nu_i(t)$ is a deterministic function. Then $\{F_i(t), t \leq T_i\}$ is a martingale under $\mathbb{P}_{T_i}$, and it follows from (3.30) that $F_i(T_i)$ is lognormal under $\mathbb{P}_{T_i}$, with

$$\mathrm{Var}_{\mathbb{P}_{T_i}}\left[\log F_i(T_i)\big|\mathcal{F}_t\right] = \int_t^{T_i} \nu_i^2(u)du,$$

$$\mathbb{E}_{\mathbb{P}_{T_i}}\left[\log F_i(T_i)\big|\mathcal{F}_t\right] = \log F_i(t) - \frac{1}{2}\int_t^{T_i} \nu_i^2(u)du.$$

The price of the ith caplet is given by

$$N\tau_i p_0(t, T_{i+1})E^{\mathbb{P}_{T_i}}\left[(F_i(T_i) - R^*)_+\right] = N\tau_i p_0(t, T_{i+1})\{F_i(t)\Phi(d_{1i}) - R^*\Phi(d_{2i})\}, \tag{3.31}$$

where N is the notional principal and d_{1i} and d_{2i} are the same as those in Black's formula (3.29) when the ith caplet volatility σ_i satisfies

$$\sigma_i^2 = \frac{1}{T_i - t}\int_t^{T_i} \nu_i^2(u)du, \qquad 1 \leq i \leq M;$$

i.e., σ_i^2 is the average instantaneous variance $\nu_i^2(\cdot)$ over the time interval (t, T_i).

Swaptions (or swap options) are options on interest rate swaps giving the holder the right to enter into an interest rate swap at a certain time T in the future. A European payer swaption gives the holder the right to pay fixed rate R^* and receive floating rate (LIBOR) at times $T_1 \leq \cdots \leq T_M$ with $T_1 > T$ (which is the swaption's expiration date). Let $T_0 = T$, and denote the forward swap rate $s(t, T_0, T_M)$ in (3.20) simply by $s(t)$ for $t \leq T$. The swaption is exercised at its expiration date T only when $s(T) > R^*$. Therefore the payoff of this swaption is

$$N \sum_{i=1}^{M} \tau_i \big(s(T) - R^*\big)_+ P(T, T_i),$$

where N is the notional principal. Consider the probability measure $\mathbb{Q}$ that is forward risk-neutral with respect to the numeraire $N \sum_{i=1}^{M} \tau_i p_0(t, T_i)$, which is the present value of the interest rate swap on the notional principal N. Suppose that, under $\mathbb{Q}$, $s(t)$ follows a driftless geometric Brownian motion $ds(t) = \varsigma(t)s(t)dW_t$, where $\varsigma(t)$ is a deterministic function. Then an argument similar to that in (3.31) yields

$$N \sum_{i=1}^{M} \tau_i p_0(t, T_i) E_{\mathbb{Q}} \big[(s(T) - R^*)_+ \big] = \Big\{ N \sum_{i=1}^{M} \tau_i p_0(t, T_i) \Big\} \Big\{ s(t) \Phi(d_1) - R^* \Phi(d_2) \Big\},$$

$$(3.32)$$

where

$$d_1 = \frac{\log(s(t)/R^*) + \sigma^2(T - t)/2}{\sigma \sqrt{T - t}},$$

$$d_2 = d_1 - \sigma \sqrt{T - t}, \qquad \sigma^2 = \frac{1}{T - t} \int_t^T \varsigma^2(u) du.$$

The swaption price (3.32) is the same as the widely used Black's formula for swaptions. Note that the swap market model expresses the swaption volatility σ as the square root of the average instantaneous variance $\varsigma^2(\cdot)$ over the interval (t, T).

Incompatibility between the LIBOR and swap market models

The above LIBOR and swap market models are not compatible, as it is challenging to reconcile these interest rate models used in financial markets. Intuitively, LIBOR is a benchmark interest rate used for short-term inter-bank lending, while swap rates represent the fixed interest rates exchanged in interest rate swap contracts. LIBOR and swap rates are closely related, but have differences in market conventions, liquidity, and underlying assumptions. These differences can lead to discrepancies between the LIBOR and swap rates, and pose challenges for modeling and pricing interest rate derivatives that rely on both LIBOR and swap rates.

Specifically, the LIBOR market model retrieves Black's formula (3.29) for pricing caplets by assuming that $F_i(\cdot)$ follows GBM and hence lognormal, whereas the swap market model retrieves Black's formula (3.32) for pricing swaptions by assuming that $s(T_0)(= s(T_0, T_0, T_M))$ is lognormal. However, these two assumptions preclude each other since, by (3.20), $s(T_0) = \sum_{i=0}^{M-1} w_i(T_0) F_i(T_0)$, in which $w_i(t) = p_0(t, T_{i+1}) \tau_i / \big\{ \sum_{j=0}^{M-1} p_0(t, T_{j+1}) \tau_j \big\}$ is a nonlinear function of $F_i(T_0)$. One can use Ito's formula and the LIBOR market model to derive the stochastic dynamics of $s(t)$, from which it follows that the instantaneous swap rate volatility $\varsigma(t)$ is related to the instantaneous volatilities $\nu_i(t)$ in (3.30) by

$$\varsigma^2(u) = \sum_{i=1}^{M}\sum_{j=1}^{M} \nu_i(t)\nu_j(t)\rho_{ij}(t)\psi_i(t)\psi_j(t),$$

where $\rho_{ij}(t)$ is the instantaneous correlation between the forward rates $F_i(t)$ and $F_j(t)$ and

$$\psi_j(t) = \frac{w_j(t) + \sum_{i=1}^{M} F_i(t)dw_j(t)/dF_j(t)}{\sum_{i=1}^{M} w_i(t)F_i(t)};$$

see Rebonato (2002, p.175) for detailed discussion.

In recent years, concerns about the reliability and integrity of LIBOR as a benchmark rate have led to efforts to transition to alternative reference rates, such as the *Secured Overnight Financing Rate* (SOFR) in the United States and the *Sterling Overnight Index Average* (SONIA) in the UK. These transitions further underscore the importance of addressing the incompatibility between different interest rate models in financial markets. Market participants need to ensure that their models and frameworks are aligned with the new reference rates and market conventions to maintain consistency and accuracy in pricing and risk management.

The HJM models of the instantaneous forward rate

The HJM framework is a class of interest rate models that describe the evolution of the entire yield curve, including the instantaneous forward rates, over time. These models were introduced by Heath et al. (1992) and have become a fundamental tool for understanding and modeling interest rate dynamics in financial markets. The HJM model assumes that the instantaneous forward rate process $f(t,T)$ for every given maturity T follows a k-factor model of the form

$$df(t,T) = \sum_{i=1}^{k} \sigma_i(t,T)s_i(t,T)dt + \sum_{i=1}^{k} \sigma_i(t,T)dW_i(t) \qquad (3.33)$$

under an equivalent martingale measure, where $\mathbf{W}(t) = (W_1(t),\ldots,W_k(t))'$ is k-dimensional Brownian motion. They showed that, in the absence of arbitrage,

$$s_i(t,T) = \int_t^T \sigma_i(t,u)du.$$

The initial condition $f(0,T)$ in the stochastic differential equation (3.33) is the market forward rate curve. For the one-factor case $k = 1$, we can rewrite (3.33) as

$$f(t,T) = f(0,T) + \int_0^t \sigma(u,T)s(u,T)du + \int_0^t \sigma(u,T)dW(u),$$

which yields the following representation of the short rate $r(t) = f(t,t)$:

$$r(t) = f(0,t) + \int_0^t \sigma(u,t)s(u,t)du + \int_0^t \sigma(u,t)dW(u).$$

An important advantage of (3.33) over short-rate models is that it can include multiple factors naturally. Although a number of multifactor short-rate models have been proposed, most of them involve decomposing $r(t)$ into a sum of unobservable state variables $x_i(t)$ that are used to introduce additional Brownian motions $W_i(t)$ as sources of randomness but do not have a physical interpretation. For $k \geq 2$ or for general volatility functions $\sigma(t,T)$ in the case $k = 1$, the short rate $r(t)$ in the HJM models is non-Markovian, making it difficult to use tree methods to compute prices of interest rate derivatives because they lead to non-recombining trees. Monte Carlo simulations are needed to compute these prices in HJM models.

3.4 Survival analysis and intensity process

Interest rate models introduced in the preceding section do not involve the issuer's credit risk. When the issuer's credit risk should be taken into consideration, one needs to model the occurrence of default. In such a case, statistical methods and models in survival analysis are often used. This section introduces concepts in survival analysis and commonly used models that analyze the failture time and its dependence upon explanatory variables.

3.4.1 Survivor function and its estimate

Let τ be the failure time of an individual from a homogeneous population. The *survival function* of τ is

$$S(t) = P(\tau > t) = 1 - F(t),$$

where $F(t) = P(\tau \leq s)$. Clearly, $S(t)$ is a non-increasing left-continuous function of t with $S(0) = 1$ and $\lim_{t\to\infty} S(t) = 0$. If τ is absolutely continuous, the *probability density function* of τ is $f(t) = -dS(t)/dt$, and the *hazard* (or *intensity*) function is

$$\lambda(t) = \lim_{h \to 0+} \frac{P(t \leq \tau < t+h \mid \tau \geq t)}{h} = f(t)/S(t). \qquad (3.34)$$

Integrating (3.34) with respect to t yields

$$S(t) = \exp\left[-\int_0^t \lambda(u)du\right] = \exp[-\Lambda(t)],$$

where $\Lambda(t) = \int_0^t \lambda(s)ds$ is called the *cumulative hazard function*. Without assuming F to be continuous, we can define the cumulative hazard function by the Stieltjes integral

$$\Lambda(t) = \int_0^t \frac{dF(s)}{1 - F(s-)},$$

in which $F(s-) = \lim_{x \uparrow s} F(x)$ is the left limit of $F(x)$ at s. Regarding the indicator functions $1_{\{\tau \le t\}}, t \ge 0$, as a stochastic process, an important property of Λ is that

$$\{1_{\{\tau \le t\}} - \Lambda(\tau \wedge t), t \ge 0\} \text{ is a martingale,} \qquad (3.35)$$

where $\tau \wedge t = \min(\tau, t)$.

Failure time distributions

Some parametric models have been used for failure time distributions, which are listed below. These distributions admit analytical form expressions for tail probabilities and hence simple formulas for survivor and hazard functions. For convenience, we denote by $Y = \log \tau$ the log failure time.

The *one-parameter exponential distribution* assumes the hazard function is constant, $\lambda(t) = \lambda > 0$, over the range of τ. The exponential distribution has the *memoryless property*, means that the conditional probability of failure in a time interval of specified length is the same regardless of how long the individual has been on study. The survivor and density functions of τ are $S(t) = e^{-\lambda t}$ and $f(t) = \lambda e^{-\lambda t}$, respectively.

Generalizing the exponential distribution to allow for a power dependence of the hazard over time, we obtain the two-parameter *Weibull distribution*. The survivor and hazard functions of a Weibull distribution with shape parameter γ and scale parameter $1/\lambda$ are given by

$$S(t) = \exp[-(\lambda t)^\gamma], \qquad \lambda(t) = \lambda \gamma (\lambda t)^{\gamma - 1}, \qquad \lambda, \gamma > 0,$$

respectively. The hazard function is monotone decreasing for $\gamma < 1$, increasing for $\gamma > 1$, and reduces to the constant exponential hazard if $\gamma = 1$. The left panel of Figure 3.1 shows the hazard functions of the Weibull distribution with $\lambda = 0.8$ and $\gamma = 1, 2.5, 0.5$, respectively. Note that the function $\log[-\log S(t)] = \gamma(\log t + \log \lambda)$ is straight line of $\log t$ with slope γ and intercept $\gamma \log \lambda$, this can be used to estimate parameters γ and λ from a sample estimate of the survivor function when the Weibull distribution is assumed. The transformed variable $W := (\log \tau + \log \lambda)\gamma$ has the probability density function $\exp(w - e^w)$ for $w \in (-\infty, \infty)$, which is an extreme value (minimum) distribution; see Section 2.4.

A *log-normal distribution* of τ with mean $\mu = -\log \lambda$ and variance σ^2 can be obtained by assuming $\log \tau = \mu + \sigma W$ and $W \sim N(0, 1)$. The survivor function of τ is $S(t) = 1 - \Phi((\log \lambda t)/\sigma)$, and its hazard function is $f(t)/S(t)$

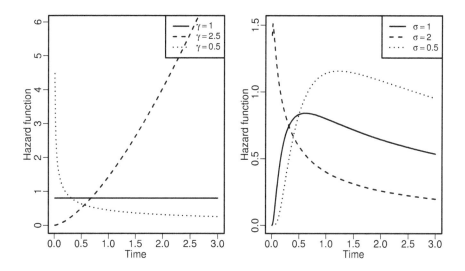

FIGURE 3.1: Hazard functions of the Weibull distribution (left) with $\lambda = 0.8$ and the log-normal distribution (right) with $\mu = 0$.

with the density function

$$f(t) = \frac{1}{\sqrt{2\pi}\sigma t} \exp\left[-\frac{(\log \lambda t)^2}{2\sigma^2}\right].$$

The hazard function has value 0 at $t = 0$, increases to a maximum and then decreases, approaching zero as t becomes large. The right panel of Figure 3.1 shows the hazard functions of the log-normal distribution with mean $\mu = 0$ and $\sigma = 1, 2, 0.5$, respectively. As shown in the figure, the hazard function has value 0 at $t = 0$ and increases to a maximum and then decreases. The log-normal model is simple to apply if there is no censoring.

Censoring, life tables and product-limit estimators

In applications of survival analysis to biomedical and econometric data, the τ_i may not be fully observable because of censoring. Some subjects (or firms) may not fail during the observation period or may have been lost in follow-up during the period; the data on these individuals are said to be *right-censored*. In the rest of the discussion, we assume that the right-censoring mechanism is *independent*, that is, the failure time distribution under censoring at each time $t > 0$ is the same as that with no censoring.

One way to estimate the survivor function is to use a life table. A life table is a summary of the survival data grouped into convenient intervals. Suppose, for example, that the data are grouped into intervals $I_1, \ldots, I_k$, and for $j = 1, \ldots, k$, the interval I_j can be expressed as

$$I_j = (b_0 + \cdots + b_{j-1}, b_0 + \cdots + b_j).$$

By convention, we denote $b_0 = 0$. The life table then presents the number of failures and censored survival times falling in each interval. In particular, Denote n_j the number of subjects alive at the beginning of I_j, d_j the number of deaths during I_j, and l_j the number lost to follow-up during I_j. Let

$$p_j = P(\text{failure in } I_j \mid \text{survival to enter } I_j),$$

then the standard life-table estimator of p_j is

$$\widehat{p}_j = d_j/(n_j - l_j) \text{ if } n_j > 0, \text{ and } 0 \text{ if } n_j = 0. \tag{3.36}$$

Then for the survivor function at the end of I_j, i.e., $P(\text{survival at time } t_k) = (1 - p_1) \cdots (1 - p_k)$, the corresponding life-table estimator is

$$\widehat{S}(b_1 + \cdots + b_j) = (1 - \widehat{p}_1) \cdots (1 - \widehat{p}_j). \tag{3.37}$$

With the same spirit, Altman (1989) and others have developed mortality tables for loans and bonds by using methods in actuarial science. These mortality tables (i.e., life tables) are used by actuaries to set premiums for life insurance policies.

Nonparametric maximum likelihood estimation

We now consider the nonparametric maximum likelihood estimate of the survival function. Let $t_1, \ldots, t_k$ be the observed failure times in a sample of size $n = n_0$ from a homogeneous population with (unknown) survival function $S(t)$. Suppose that d_j items fail at t_j and m_j items are censored in the interval $[t_j, t_{j+1})$ at times $t_{j1}, \ldots, t_{jm_j}$, $j = 0, \ldots, k$, where $t_0 = 0$ and $t_{k+1} = \infty$. Let $n_j = \sum_{i=j}^{k}(m_i + d_i)$ be the number of items at risk at a time just prior to t_j. Then the contribution to the likelihood of a censored survival time at t_{jl} is $P(\tau > t_{jl}) = S(t_{jl})$ and the probability of failure at t_j is $P(\tau = t_j) = S(t_j^-) - S(t_j)$. The likelihood of the data is expressed as

$$L = \prod_{j=0}^{k} \left\{ [S(t_j^-) - S(t_j)]^{d_j} \prod_{l=1}^{m_j} S(t_{jl}) \right\}. \tag{3.38}$$

To maximize the above likelihood function, we note that the maximum likelihood estimate $\widehat{S}(t)$ is a discrete survival function with hazard components $\widehat{\lambda}_1, \ldots, \widehat{\lambda}_k$ at $t_1, \ldots, t_k$, and

$$\widehat{S}(t_j) = \prod_{i=1}^{j}(1 - \widehat{\lambda}_i), \qquad \widehat{S}(t_j^-) = \prod_{i=1}^{j-1}(1 - \widehat{\lambda}_i),$$

Plug the above into the likelihood function (3.38), we obtain that

$$L = \prod_{j=1}^{k} \left\{ \lambda_j^{d_j} \prod_{i=1}^{j-1} (1 - \lambda_i)^{d_j} \prod_{i=1}^{j} (1 - \lambda_i)^{m_j} \right] \right\} = \prod_{j=1}^{k} \lambda_j^{d_j} (1 - \lambda_j)^{n_j - d_j}. \quad (3.39)$$

The above likelihood is maximized by $\widehat{\lambda}_j = d_j/n_j$ $(j = 1, \ldots, k)$. Hence, the Kaplan-Meier (Kaplan and Meier, 1958) estimate or the *product limit estimate* of the survival function is

$$\widehat{S}(t) = \prod_{t_j \leq t} \frac{n_j - d_j}{n_j}. \quad (3.40)$$

Corresponding to this, the cumulative hazard function $\Lambda(t)$ can be estimated by the *Nelson-Aalen estimator*,

$$\widehat{\Lambda}(t) = \sum_{t_j \leq t} \widehat{\lambda}_j = \sum_{t_j \leq t} d_j/n_j. \quad (3.41)$$

To obtain an asymptotic distribution of $\widehat{S}(t)$ at a given time t, we may treat the likelihood function (3.39) as a likelihood with parameters $\lambda_1, \ldots, \lambda_k$ and estimate the asymptotic variance of $\widehat{\lambda}_j$ by $d_j(n_j - d_j)/n_j^3$ using standard likelihood method. Then an asymptotic variance estimate of $\log \widehat{S}(t)$ is

$$\widehat{\mathrm{Var}}[\log \widehat{S}(t)] = \sum_{t_j \leq t} \frac{1}{(1 - \widehat{\lambda}_j)^2} \widehat{\mathrm{Var}}(1 - \widehat{\lambda}_j) = \sum_{t_j \leq t} \frac{d_j}{n_j(n_j - d_j)},$$

and hence the *Greenwood's formula* (Greenwood, 1926) is obtained for the asymptotic variance of $\widehat{S}(t)$

$$\widehat{\mathrm{Var}}[\widehat{S}(t)] = [\widehat{S}(t)]^2 \sum_{t_j \leq t} \frac{d_j}{n_j(n_j - d_j)}. \quad (3.42)$$

Then an approximate 95% confidence interval for $S(t)$ is $\widehat{S}(t) \pm 1.96[\widehat{\mathrm{Var}}(\widehat{S}(t))]^{1/2}$.

Example 3.1 (Kaplan-Meier estimate of the survivor function on bank default) *Consider a sample of the Federal Deposit Insurance Corporation (FDIC) Call Reports (i.e., Reports of Condition and Income) data from 2002 to 2011. A call report is a regulatory report that must be filed by banks in the U.S. on a quarterly basis with the FDIC. It contains information about the bank's financial health. By examining banks' call reports, one can get some insight on the healthiness of the U.S. banking system. The sample contains 335 California banks from the fourth quarter of 2002 to the fourth quarter of 2011, and 39 among these banks failed during that period. Figure 3.2 shows the number of events and the Kaplan-Meier estimate of the survivor function with approximate 95% confidence intervals over time.* □

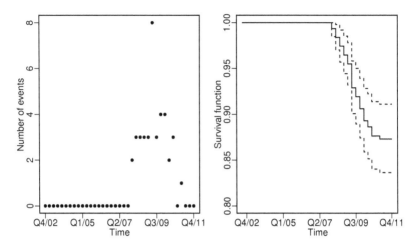

FIGURE 3.2: The number of events (left) and the Kaplan-Meier estimate of the survivor function with approximate 95% confidence intervals (right) for the California bank failure data.

Product-limit estimate of survival function

Recall that the actuarial (life-table) estimate (3.37) and the Kaplan-Meier estimate (3.40) has the form of

$$\widehat{S}(t) = \prod_{s<t}\left(1 - \frac{\Delta N(s)}{Y(s)}\right), \qquad (3.43)$$

where we again use the notation $0/0$. Since $N(s)$ has at most n jumps, the product in (3.43) is finite. Note that $\widehat{S}(t) = \prod_{s<t}(1 - \Delta\widehat{\Lambda}(s))$ by (3.46) and (3.43), which is an estimate of the survivor function

$$S(t) = \prod_{s<t}(1 - d\Lambda(s)) \text{ for all } t \text{such that } \Lambda(t) < \infty. \qquad (3.44)$$

The product in (3.44) is called the *product integral* of the nondecreasing right-continuous function $\Lambda(\cdot)$ on $[0, t]$. It is defined by the limit of $\prod\{1 - [\Lambda(t_i) - \Lambda(t_{i-1})]\}$, where $0 = t_0 < t_1 < \cdots < t_m = t$ is a partition of $[0, t]$ and the product is in the natural order from left to right, as the mesh size $\max_{1\le i\le m}|t_i - t_{i-1}|$ approaches 0. Moreover, it can be shown that the Kaplan-Meier estimate (3.43) has the following asymptotic distribution

$$(\widehat{S}(t) - S(t)) \bigg/ \left[\widehat{S}(t)\left\{\int_0^t \frac{I_{\{Y(s)>0\}}}{Y^2(s)}dN(s)\right\}^{1/2}\right] \xrightarrow{\mathcal{D}} N(0,1),$$

where $\xrightarrow{\mathcal{D}}$ denotes convergence in distribution; see Kalbfleisch and Prentice (2002, Section 5.6.1).

Nelson-Aalen estimator of the cumulative hazard function

Instead of discretizing the failure times τ_i as in life tables, we can regard $n_j - l_j$ as the number of subjects at risk at an observed failure time s, and generalize $\widehat{p}_j$ in (3.36) to

$$\widehat{p}_j = \frac{\text{the number of failures at } s}{\text{the number of subjects at risk at } s}.$$

"Loss to follow-up" corresponds to right censoring, which can be formulated by introducing censoring variables c_i that indicate the time of loss to follow-up. The censoring variable c_i can also indicate the length of the observation period. The observations, therefore, are $(T_i, \delta_i), i = 1, \cdots, n$, where $T_i = \min(\tau_i, c_i)$ and $\delta_i = I_{\{\tau_i \leq c_i\}}$ is the censoring indicator that indicates whether T_i is an actual failure time or is censored. Subject i is "at risk" at time s if $T_i \geq s$ (i.e., has not failed and not been lost to follow-up prior to s). Let

$$Y(s) = \sum_{i=1}^{n} I_{\{T_i \geq s\}}, \qquad N(s) = \sum_{i=1}^{n} I_{\{T_i \leq s, \, \delta_i = 1\}}. \qquad (3.45)$$

Note that $Y(s)$ is the risk set size and $\Delta N(s) = N(s) - N(s-)$ is the number of observed deaths at time s, and that $\Delta N(s)/Y(s)$ is the analog of $\widehat{p}_j$. Moreover, the Nelson-Aalen estimator (3.41) of the cumulative hazard function $\Lambda(t)$ for the case of fully observable τ_i can be readily extended to

$$\widehat{\Lambda}(t) = \sum_{s \leq t} \frac{\Delta N(s)}{Y(s)} = \int_0^t \frac{I_{\{Y(s) > 0\}}}{Y(s)} dN(s) \qquad (3.46)$$

for the censored data $(T_i, \delta_i), 1 \leq i \leq n$. It is can be shown that the following asymptotic normality holds for $\widehat{\Lambda}(t)$,

$$(\widehat{\Lambda}(t) - \Lambda(t)) \Big/ \left\{ \int_0^t \frac{I_{\{Y(s) > 0\}}}{Y^2(s)} dN(s) \right\}^{1/2} \xrightarrow{\mathcal{D}} N(0, 1),$$

as $n \to \infty$; see Kalbfleisch and Prentice (2002, Section 5.6.1).

3.4.2 Proportional hazards models

We have focused so far on the estimation of the survival distribution of a failure time τ. In applications, one often wants to use a model for τ to predict future failures from a vector $\mathbf{x}(t)$ of predictors based on current and past observations. The predictor $\mathbf{x}(t)$ is called a time-varying (or time-dependent) covariate. When $\mathbf{x}(t) = \mathbf{x}$ does not depend on t, it is called a time-independent

(or baseline) covariate. In practice, some predictors may be time-independent
while other components of $\mathbf{x}(t)$ may be time-varying.

Since prediction of future default from $\mathbf{x}(t)$ is relevant only if $\tau > t$ (i.e.,
if default has not occurred at or before t), one is interested in modeling the
conditional distribution of τ given $\tau > t$, e.g., by relating the hazard function
$\lambda(t)$ to $\mathbf{x}(t)$. Noting that $\lambda(t)dt$ is the probability that $\tau \in [t, t+dt]$, an analog
of the logistic regression model is the *proportional hazards* (or *Cox regression*)
model

$$\lambda(t) = \lambda_0(t) \exp(\boldsymbol{\beta}'\mathbf{x}(t)), \tag{3.47}$$

in which $\lambda_0(t)$ is the time-varying analog of the intercept parameter α in
logistic regression. Putting $\mathbf{x}(t) = 0$ in (3.47) shows that $\lambda_0(\cdot)$ is also a hazard
function; it is called the *baseline hazard*. Whereas logistic regression involves
binary outcomes and their covariates, Cox regression fits (3.47) to the observed
data $(T_i, \delta_i, \mathbf{x}_i(t)), t \leq T_i, 1 \leq i \leq n$.

Parametric approach and maximum likelihood

The parameters of (3.47) include the regression coefficients $\boldsymbol{\beta}$ and the baseline hazard function $\lambda_0(\cdot)$. Since the function $\lambda_0(t)$ is an infinite-dimensional
parameter and there are only n observations, one way to overcome difficulties
caused by the infinite-dimensional parameter $\lambda_0(t)$ is to specify a parametric
failure time model for $\lambda_0(t)$ up to a parameter vector $\boldsymbol{\theta}$ so that the parameters
of (3.47) reduce to $(\boldsymbol{\beta}, \boldsymbol{\theta})$, which can be estimated by the maximum likelihood
method.

The baseline hazard $\lambda_0(t)$ can be assumed to as the hazard of a parametric distribution, such as the exponential, the Weibull, or the log-normal
distribution in Section 3.4.1. Once a parametric model for the baseline hazard
$\lambda_0(t)$, say $\lambda_0(t; \boldsymbol{\theta})$, is specified, we can proceed to derive the log-likelihood
of the data. For notational simplicity, assume that the covariates are time-
invariant. Note that an observed failure contributes a factor $f_i(T_i; \boldsymbol{\theta})$ to the
likelihood, where $f_i = \lambda_i S_i$ is the density function of the τ_i, and a censored
failure contributes a factor $S_i(T_i; \boldsymbol{\theta})$ to the likelihood. Since

$$\log S_i(t; \boldsymbol{\theta}) = -\Lambda_i(t; \boldsymbol{\theta}), \quad \Lambda_0(t; \boldsymbol{\theta}) = \int_0^t \lambda_0(u; \boldsymbol{\theta})du, \quad \Lambda_i(t; \boldsymbol{\theta}) = \Lambda_0(t; \boldsymbol{\theta}) \exp(\boldsymbol{\beta}'\mathbf{x}_i),$$

where $\Lambda_i(t; \boldsymbol{\theta})$ is the cumulative hazard function of τ_i, the log-likelihood function can be written as

$$\begin{aligned}
l(\boldsymbol{\beta}, \boldsymbol{\theta}) &= \sum_{i=1}^{n} \{\delta_i \log f_i(T_i) + (1 - \delta_i) \log S_i(T_i)\} \\
&= \sum_{i=1}^{n} \left\{ \delta_i [\log \lambda_0(T_i; \boldsymbol{\theta}) + \boldsymbol{\beta}'\mathbf{x}_i(T_i)] - \Lambda_0(T_i; \boldsymbol{\theta})e^{\boldsymbol{\beta}'\mathbf{x}_i(T_i)} \right\}.
\end{aligned}$$

Then the maximum likelihood estimates $(\widehat{\boldsymbol{\beta}}, \widehat{\boldsymbol{\theta}})$ can be obtained by solving the
equation $\partial l(\boldsymbol{\beta}, \boldsymbol{\theta})/\partial \boldsymbol{\beta} = 0$ and $\partial l(\boldsymbol{\beta}, \boldsymbol{\theta})/\partial \boldsymbol{\theta} = 0$.

Semiparametric approach and partial likelihood

Instead of assuming a parametric model to estimate the baseline hazard function $\lambda_0(\cdot)$, Cox (1972, 1975) introduced a semiparametric method to estimate the finite-dimensional parameter $\boldsymbol{\beta}$ in the presence of an infinite-dimensional nuisance parameter $\lambda_0(\cdot)$. This method is *semiparametric* in the sense of being nonparametric in $\lambda_0(\cdot)$ but parametric in $\boldsymbol{\beta}$ (in terms of $\boldsymbol{\beta}'\mathbf{x}_t$ rather than the more general $\sum \beta_j g_j(x_j(t))$ in generalized additive models). Cox's *partial likelihood* method decomposes the likelihood function into two factors, with one involving only $\boldsymbol{\beta}$ and the other involving both $\boldsymbol{\beta}$ and Λ. It estimates $\boldsymbol{\beta}$ by maximizing the partial likelihood, which is the first factor that only involves $\boldsymbol{\beta}$. The partial likelihood method is described as follows.

Order the observed censored failure times as $\tau_{(1)} < \cdots < \tau_{(m)}$, with $m \le n$. Let C_j denote the set of censored T_i's in the interval $[\tau_{(j-1)}, \tau_{(j)})$ and let (j) denote the individual failing at $\tau_{(j)}$, noting that with probability 1 there is only one failure at $\tau_{(j)}$ because the failure time distributions have density functions. Let $R_{(j)} = \{i : T_i \ge \tau_{(j)}\}$ denote the risk set at $\tau_{(j)}$. Then the conditional probability that the individual fails at $\tau_{(j)}$ is

$$P\{(j)|C_j, (l), C_l, 1 \le l \le j - 1\} = P\{(j) \text{ fails at } \tau_{(j)}|R_{(j)}, \text{ one failure at } \tau_{(j)}\}$$

$$= \exp\left[\boldsymbol{\beta}'\mathbf{x}_{(j)}(\tau_{(j)})\right] \Big/ \sum_{i \in R_{(j)}} \exp\left[\boldsymbol{\beta}'\mathbf{x}_{(i)}(\tau_{(j)})\right].$$

The partial likelihood is defined as the product of the above probability at all observed censored failure times,

$$\prod_{j=1}^{m} P\{(j)|C_j, (l), C_l, 1 \le l \le j - 1\}.$$

Ignoring the other factors $P\{C_{j+1}|(l), C_l, 1 \le l \le j\}$ in the likelihood function, the so-called Cox's regression estimator $\widehat{\boldsymbol{\beta}}$ maximizes the partial log-likelihood

$$l_p(\boldsymbol{\beta}) = \sum_{j=1}^{m} \left\{ \boldsymbol{\beta}'\mathbf{x}_{(j)}(\tau_{(j)}) - \log\left(\sum_{i \in R_{(j)}} \exp\left[\boldsymbol{\beta}'\mathbf{x}_{(i)}(\tau_{(j)})\right] \right) \right\}. \tag{3.48}$$

Equivalently, $\widehat{\boldsymbol{\beta}}$ is the solution of $\partial l_p(\boldsymbol{\beta})/\partial \boldsymbol{\beta} = 0$.

Let $w_i(\boldsymbol{\beta}) = \exp\left(\boldsymbol{\beta}'\mathbf{x}_i(\tau_{(j)})\right)/\sum_{l \in R_{(j)}} \exp\left[\boldsymbol{\beta}'\mathbf{x}_l(\tau_{(j)})\right]$ for $i \in R_{(j)}$. The first and second derivative of the partial likelihood with respect to $\boldsymbol{\beta}$ are expressed as

$$\frac{\partial}{\partial \boldsymbol{\beta}} l_p(\boldsymbol{\beta}) = \sum_{j=1}^{m} \left\{ \mathbf{x}_{(j)}(\tau_{(j)}) - \sum_{i \in R_{(j)}} w_i(\boldsymbol{\beta})\mathbf{x}_{(i)}(\tau_{(j)}) \right\}, \tag{3.49}$$

$$-\left(\frac{\partial^2}{\partial \beta_k \partial \beta_h} l_p(\boldsymbol{\beta}) \right)_{k,h} = \sum_{j=1}^{m} \left\{ \sum_{i \in R_{(j)}} w_i(\boldsymbol{\beta})\mathbf{x}_i(\tau_{(j)})\mathbf{x}_i'(\tau_{(j)}) \right.$$

$$\left. - \left[\sum_{i \in R_{(j)}} w_i(\boldsymbol{\beta})\mathbf{x}_i(\tau_{(j)}) \right] \cdot \left[\sum_{i \in R_{(j)}} w_i(\boldsymbol{\beta})\mathbf{x}_i(\tau_{(j)}) \right] \right\}$$

$$\tag{3.50}$$

Since $\sum_{i \in R_{(j)}} w_i(\boldsymbol{\beta}) = 1$, we can interpret the term $\bar{\mathbf{x}}(\tau_{(j)}) := \sum_{i \in R_{(j)}} w_i(\boldsymbol{\beta})\mathbf{x}_i(\tau_{(j)})$ in (3.49) and (3.50) as a weighted average of covariates over the risk set. Each summand in (3.49) therefore compares the covariate at an observed failure to its weighted average over the risk set. Moreover, each summand in (3.50) can be expressed as a sample covariance matrix of the form

$$\sum_{i \in R_{(j)}} w_i(\boldsymbol{\beta})\{\mathbf{x}_i(\tau_{(j)}) - \bar{\mathbf{x}}(\tau_{(j)})\}\{\mathbf{x}_i(\tau_{(j)}) - \bar{\mathbf{x}}(\tau_{(j)})\}'.$$

Making use of martingale theory, Cox's regression estimator $\widehat{\boldsymbol{\beta}}$ can be shown to satisfy the usual asymptotic properties of maximum likelihood estimates even though partial likelihood is used; see Kalbfleisch and Prentice (2002, Section 5.7). In particular, it can be shown that as $n \to \infty$,

$$(-\ddot{l}_p(\boldsymbol{\beta}))^{1/2}(\widehat{\boldsymbol{\beta}} - \boldsymbol{\beta}_0) \text{ has a limiting standard normal distribution,}$$

where $\ddot{l}_p(\boldsymbol{\beta})$ is the Hessian matrix of second partial derivatives $(\partial^2/\partial\beta_k\partial\beta_h)$ $l_p(\boldsymbol{\beta})$, given by (3.50). One can perform usual likelihood inference (Lai and Xing, 2008a, Section 2.4.2), treating the partial likelihood as a likelihood function and apply likelihood-based selection of covariates similar to that for generalized linear models in Section 4.4.4. Moreover, even though $\widehat{\boldsymbol{\beta}}$ is based on partial likelihood, it has been shown to be asymptotically efficient. Computation of $\widehat{\boldsymbol{\beta}}$ and related likelihood inference can usually be carried out by standard statistical softwares.

Estimation of baseline hazard

When there are no covariates, $\Lambda = \Lambda_0$ can be estimated by the Nelson-Aalen estimator (3.46). Note that (3.46) has jumps only at uncensored observations and that $Y(s)$ is the sum of 1's over the risk set $\{i : T_i \geq s\}$ at s. When τ_i has hazard function $\exp[\boldsymbol{\beta}'\mathbf{x}_i(s)]\lambda_0(s)$, we modify $Y(s)$ to

$$Y(s) = \sum_{i \in R_{(j)}} \exp(\boldsymbol{\beta}'\mathbf{x}_i(s)) \text{ at } s = \tau_{(j)}, \tag{3.51}$$

using the same notation as in Section 3.4.1. The Breslow estimator of Λ_0 in the proportional hazards regression model (3.47) is again given by (3.46) but with $Y(s)$ defined by (3.51).

Regression diagnostics

Similar to generalized linear models, we can use the assumed proportional hazards regression model (3.47) to derive *generalized residuals* for regression diagnostics. Letting $\Lambda_i(t) = \int_0^t \exp(\boldsymbol{\beta}'\mathbf{x}_i(s))d\Lambda_0(s)$, note that $\Lambda_i(\tau_i)$ has an exponential distribution with mean 1 since

$$P\{\Lambda_i(\tau_i) > t\} = P\{\tau_i > \Lambda_i^{-1}(t)\} = S_i(\Lambda_i^{-1}(t)) = \exp\{-\Lambda_i(\Lambda_i^{-1}(t))\} = e^{-t}.$$

Therefore, $(\Lambda_i(T_i), \delta_i), 1 \leq i \leq n$, is a sample of size n from the exponential distribution with mean 1. If (3.47) is the correct model, plotting $\widehat{\Lambda}_i(T_i)$ against a particular predictor x_{ik} or against $\widehat{\boldsymbol{\beta}}' \mathbf{x}_i$ should not reveal any systematic pattern, where $\widehat{\Lambda}_i$ replaces $(\boldsymbol{\beta}, \Lambda_0)$ in Λ_i by $(\widehat{\boldsymbol{\beta}}, \widehat{\Lambda}_0)$. It is not obvious, however, from the residual plots how to correct for departures from the assumed regression model (3.47). *Martingale residuals* have been introduced as an alternative, using the fact that

$$M_i(t) = N_i(t) - \int_0^t 1_{\{T_i \geq s\}} \exp[\boldsymbol{\beta}' \mathbf{x}_i(s)] d\Lambda_0(s)$$

is a martingale under (3.47), where $N_i(t) = 1_{\{T_i \leq t, \delta_i = 1\}}$ is an indicator process. Replacing t by $\max_{1 \leq i \leq n} T_i$ and $(\boldsymbol{\beta}, \Lambda_0)$ by $(\widehat{\boldsymbol{\beta}}, \widehat{\Lambda}_0)$ yields the martingale residuals $\widehat{M}_i$. One can plot $\widehat{M}_i$ versus x_{ik} or $\widehat{\boldsymbol{\beta}}' \mathbf{x}_i$ to detect departure from the martingale (zero-mean) pattern that suggest ways to enhance the model.

Example 3.2 (Cox regression analysis on bank default) *Consider the California bank failure data in Example 3.1. We now study the dependence of banks' survivor probabilities on some covariates by fitting a proportional hazard models (3.47) with unknown baseline $\lambda_0(t)$ to the sample. Three covariates are considered here, i.e., $\mathbf{x}_t = $ (eqpct, p3assetpct, nclnlspct)$_t$, where eqpct is the percentage of the bank's total equity in the bank's total assets, p3assetpct is the percentage of the bank's assets 30 to 89 days pass due in the bank's total assets, and nclnlspct is the percentage of the bank's noncurrent loans and leases in the bank's total assets. Table 3.1 shows the estimated regression coefficients and their standard errors in the Cox model (3.47). All three covariates are significant with p-value less than 0.1%. Figure 3.3 shows the estimated survivor probabilities with 95% confidence intervals and the estimated baseline hazard function in the Cox regression model. Note that the estimated survivor probabilities are very closed to 1 during the sample period.* □

TABLE 3.1: Estimated coefficients in the Cox regression model on bank default.

	β	$\exp(\beta)$	$\text{se}(\beta)$	z-statistics	p-value
eqpct	-0.8467	0.4288	0.0950	-8.9096	<2e-16
p3assetpct	0.2378	1.2685	0.0902	2.6380	0.00834
nclnlspct	0.1626	1.1765	0.0333	4.8851	1.03e-6

3.4.3 Additive hazard models

The proportional hazards model (3.47) characterizes the impact of covariates on hazard functions in a multiplicative way and has been a popular choice

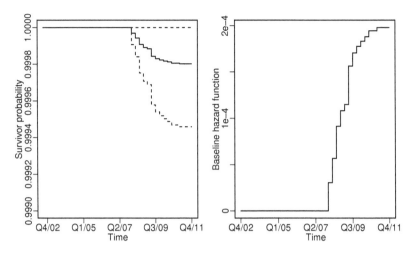

FIGURE 3.3: Estimated survivor probabilities with 95% confidence intervals (left) and the estimated baseline hazard function (right) in the Cox regression model.

for the regression analysis of survival data. Another way to model the effect of covariates on hazard is to consider the additive effect.

Time varying coefficients

The first additive hazard model is the *additive hazard regression* introduced by Aalen (1980),

$$\lambda(t) = \lambda_0(t) + \boldsymbol{\beta}(t)'\mathbf{x}(t), \tag{3.52}$$

in which $\lambda_0(t)$ is the unknown time-dependent intercept, $\mathbf{x}(t) \in \mathbb{R}^p$ is the vector of covariates (possibly time-dependent), and $\boldsymbol{\beta}(t) = (\beta_1(t), \dots, \beta_p(t))'$ is a time-dependent vector of coefficients. Denote $B_k(t) = \int_0^t \beta_k(u)du$ the cumulative effect of $\beta_k(t)$, $k = 1, \dots, p$, and $\mathbf{B}(t) = (B_1(t), \dots, B_p(t))'$. Then an estimate of $\boldsymbol{\beta}(t)$ can be obtained by differencing an estimate of $\mathbf{B}(t)$. Define $Y(t)$ and $N(t)$ as in (3.45), then

$$M_i(t) = N_i(t) - \int_0^t Y_i(u)\lambda(u)du = N_i(t) - \int_0^t Y_i(u)[\lambda_0(t) + \boldsymbol{\beta}(t)'\mathbf{x}(t)]du,$$

is a martingale difference under (3.52). Let $\mathbf{X}(t)$ be the matrix in which the ith row is given by $(Y_i(t), Y_i(t)\mathbf{x}'(t))$ and $\mathbf{N}(t) = (N_1(t), \dots, N_n(t))'$. Then the cumulative effect $\mathbf{B}(t)$ at time t can be estimated by

$$\widehat{\mathbf{B}}(t) = (\mathbf{X}(t)'\mathbf{X}(t))^{-1}\mathbf{X}(t)\mathbf{N}(t)$$

if $\mathbf{X}(t)$ has full rank.

Time invariant coefficients

Aalen's additive model (3.52) uses time-varying coefficients and the estimates are inherently non-parametric. Lin and Ying (1994) studied the case of time-invariant coefficients

$$\lambda(t) = \lambda_0(t) + \boldsymbol{\beta}'\mathbf{x}(t), \qquad (3.53)$$

and developed an inference procedure using estimating equations. In particular, let $Y_i(t) = I_{\{T_i \geq t\}}$, $N_i(t) = I_{\{T_i \leq t, \delta_i = 1\}}$, and

$$\bar{\mathbf{x}}(t) = \sum_{i=1}^{n} \mathbf{x}_i(t) Y_i(t) \Big/ \sum_{i=1}^{n} Y_i(t),$$

where $0/0$ is considered as 0. Lin and Ying (1994) proposed the following estimating function

$$U(\boldsymbol{\beta}) = \sum_{i=1}^{n} \int \{\mathbf{x}_i(t) - \bar{x}(t)\}\{dN_i(t) - \boldsymbol{\beta}'\mathbf{x}_i(t)Y_i(t)dt\},$$

which is based on the simple fact that, when $\boldsymbol{\beta}$ takes the true parameter value, $U(\boldsymbol{\beta})$ is a martingale integral and therefore has mean 0. Since U is linear in $\boldsymbol{\beta}$, solving the estimating equation $U(\boldsymbol{\beta}) = 0$ gives us an estimate of $\boldsymbol{\beta}$, which can be expressed as,

$$\widehat{\boldsymbol{\beta}} = \left[\sum_{i=1}^{n} \int \{\mathbf{x}_i(t) - \bar{\mathbf{x}}(t)\}^{\otimes 2} Y_i(t) dt \right]^{-1} \sum_{i=1}^{n} \int \{\mathbf{x}_i(t) - \bar{\mathbf{x}}(t)\} dN_i(t).$$

where $\mathbf{a}^{\otimes 2} = \mathbf{a}\mathbf{a}'$. Applying the standard counting-process martingale theory, it can be shown that $\widehat{\boldsymbol{\beta}}$ is asymptotically normal with mean $\boldsymbol{\beta}$ and with a variance-covariance matrix consistently estimated $\widehat{\mathbf{A}}^{-1}\widehat{\mathbf{V}}\widehat{\mathbf{A}}^{-1}$, where

$$\widehat{\mathbf{A}} = \sum_{i=1}^{n} \int \{\mathbf{x}_i(t) - \bar{\mathbf{x}}(t)\}^{\otimes 2} Y_i(t) dt, \qquad \widehat{\mathbf{V}} = \sum_{i=1}^{n} \int \{\mathbf{x}_i(t) - \bar{\mathbf{x}}(t)\}^{\otimes 2} dN_i(t).$$

Example 3.3 (Additive hazard models for bank default) *Consider the California bank failure data in Examples 3.1 and 3.2. We now fit additive hazard models (3.52) and (3.53) to the data with the three covariates $\mathbf{x}_t = ($eqpct, p3assetpct, nclnlspct$)_t$. Figure 3.4 shows the estimated cumulative coefficient $\widehat{\mathbf{B}}(t)$ for model (3.52) with covariates $(1, \mathbf{x}'_t)$. Note that the plots indicate that, with the evolvement of time, p3assetpct and nclnlspct cause higher risk whereas eqpct leads to lower risk. For the additive hazard model (3.53) with covariates $\mathbf{x}_t$, Table 3.2 shows the estimated coefficients and their standard errors in model (3.53), which indicates that the additive effect of eqpct in (3.53) is not significant. Note that both additive models show that p3assetpct and nclnlspct have positive additive effects on hazard functions of bank failure.* □

TABLE 3.2: Estimated coefficients in the additive hazard model 3.53 on bank default.

	β	$se(\beta)$	z-statistics	p-value
eqpct	-2.097e-05	2.053e-05	-1.021	0.3071
p3assetpct	4.769e-03	2.072e-03	2.301	0.0214
nclnlspct	7.053e-03	1.538e-03	4.586	4.51e-06

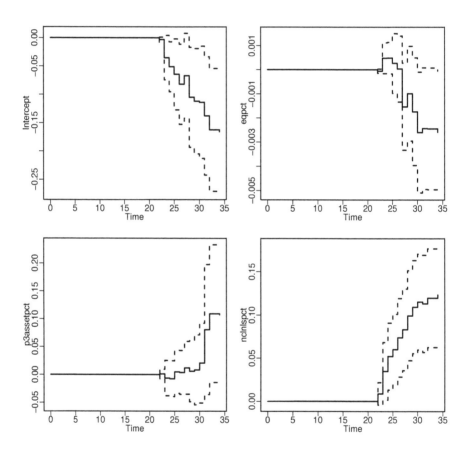

FIGURE 3.4: Estimated cumulative coefficients $\widehat{\mathbf{B}}(t)$ in the additive hazard model (3.52) for the bank failure data. The x-axis is the index of quarters starting from the fourth quarter in 2002.

Frailty and additive-multiplicative hazard

   In survival analysis and actuarial science, frailty variables are often used to account for unobserved heterogeneity or individual-specific characteristics that may influence the hazard or risk of experiencing an event. The additive hazard regression models (3.52) and (3.53) can be easily extended to incorporate the frailty effect. Let $\xi(t)$ be a possibly time-dependent frailty process which is independent of the covariate process $\mathbf{x}$. For example, an additive frailty model can be expressed as

$$\lambda(t) = \lambda_0(t) + \xi(t) + \boldsymbol{\beta}'\mathbf{x}(t).$$

An inference procedure can be obtained by extending the procedure in Lin and Ying (1994). An interesting feature of the above additive frailty model is that, when marginalized over ξ, it is still an additive hazards model. Moreover, the ideas of multiplicative and additive hazard models (3.47) and (3.52) can be combined to obtain an *additive-multiplicative hazard model* (Lin and Ying, 1995).

$$\lambda(t) = Y(t)\Big(\boldsymbol{\alpha}(t)'\mathbf{x}(t) + \lambda_0(t)\exp\{\boldsymbol{\beta}'\mathbf{z}(t)\}\Big),$$

in which $\boldsymbol{\alpha}(t)$ is a time-varying component and $\boldsymbol{\beta}$ is the regression coefficients. Inference on $\boldsymbol{\alpha}(t)$, $\boldsymbol{\beta}$ and $\lambda_0(t)$ can be made by using the procedure in Lin and Ying (1995) and Martinussen and Scheike (2002).

3.4.4 Accelerated failure time models

Let τ be a failure time and $\mathbf{z}$ a covariate vector that doesn't depend on time. The *accelerated failure time* (ADT) model assumes that

$$\log(\tau) = -\boldsymbol{\beta}'\mathbf{x}_t + \epsilon, \tag{3.54}$$

where $\boldsymbol{\beta}$ is the p-dimensional regression coefficient and ϵ is a residual term with unspecified distribution. This specification leads to the following hazard function

$$\lambda(t) = \lambda_0[t\exp(\boldsymbol{\beta}'\mathbf{x}_t)]\exp(\boldsymbol{\beta}'\mathbf{x}_t), \tag{3.55}$$

where $\lambda_0(t)$ is the hazard associated with the unspecified error distribution $\exp(\epsilon)$. Because the covariates acts multiplicatively on time, the failure time is accelerated or decelerated relative to $\lambda_0(t)$.

Let C be the censoring time for T and $\tilde{T} = \min\{T, C\}$, $\Delta = 1_{\{T \leq C\}}$ is an indicator variable for censoring. Consider (3.54) as the model for the intensity of the counting process $N(t) = 1_{\{\tilde{T} \leq t\}}\Delta$. Assume that n i.i.d. counting processes are being observed subject to this model. Define the time-transformed counting process

$$N^*(t) = (N_1(t\exp(-\boldsymbol{\beta}'\mathbf{x}_1)), \ldots, N_n(t\exp(-\boldsymbol{\beta}'\mathbf{x}_n)))$$

with associated at risk process $Y_i^*(t; \boldsymbol{\beta}) = Y_i(t \exp(-\boldsymbol{\beta}' \mathbf{x}_i))$, $i = 1, \ldots, n$. This suggest that $\Lambda_0(t) = \int_0^t \lambda_0(s) ds$ should be estimated by the Breslow-type estimator

$$\widehat{\Lambda}_0(t; \boldsymbol{\beta}) = \int_0^t \frac{d\widetilde{N}^*(s)}{S_0^*(s; \boldsymbol{\beta})}, \qquad d\widetilde{N}^*(s) = \sum_{i=1}^n dN_i^*(t), \ S_0^*(s; \boldsymbol{\beta}) = \sum_{i=1}^n Y_i^*(t; \boldsymbol{\beta}).$$

To estimate $\boldsymbol{\beta}$, one can replace $d\Lambda_0(s)$ in the efficient score function for $\boldsymbol{\beta}$ by the above estimator and obtain an estimating equation. Then an efficient estimator of $\boldsymbol{\beta}$ can be derived from the estimating equation; see detailed discussion in Martinussen and Scheike (2006, Section 8.1).

3.5   Credit default risk

Interest rate models in Section 3.3 do not involve the issuer's credit risk. When the borrower has the possibility of default, the credit risk (or simply, the default risk) of the borrower needs to be valuated. The valuation of corporate default risk dates back to Merton (1974b) and Black and Cox (1976) that viewed the bonds and stocks issued by a firm as contingent claims on the assets of the firm and applied option-pricing techniques to evaluate corporate liabilities. This method establishes the connection between an economic-pricing model and a statistical model describing default. In this section, we first introduce structural models that reply on option-pricing techniques and then intensity models that make use of the idea of survival analysis. More detailed discussion on these models can be found in Duffie and Singleton (2003) and Lando (2004).

3.5.1   Structural models of default risk

Structural models attempt to derive the relationship from some underlying economic theory. Structural models for the default of a firm assume that default is triggered when the firm's asset and debt values enter some region whose boundary is called a "default barrier."

Merton's model of debt and equity values

As in the Black-Scholes model, Merton (1974b) assumes a market with continuous trading, no transaction costs, unlimited short selling and perfectly divisible assets, and with a risk-free asset that has constant interest rate r. The asset value A_t of the firm is assumed to follow a geometric Brownian motion (GBM)

$$dA_t = \mu A_t dt + \sigma A_t dW_t. \tag{3.56}$$

Suppose at time 0 the firm has issued two kinds of contingent claims: equity and debt, in which debt is assumed to be a zero-coupon bond with face value D and maturity date T. We can regard the firm as being run by equity owners. At maturity date T of the bond, the payoffs to debt, $V_d(T)$, and equity, $V_e(T)$ are given as

$$V_e(T) = (A_T - D)^+, \qquad V_d(T) = \min(D, A_T).$$

That is, if $A_T < D$ at maturity T, bond holders receive a recovery of A_T instead of the promised payment D.

To value the debt and equity prior to the maturity date T, note that the equity value at time t is given by the Black-Scholes formula

$$V_e(t) = E^Q \left\{ e^{-r(T-t)} V_e(T) | A_t \right\} = A_t \Phi(d_1) - De^{-r(T-t)} \Phi(d_2), \qquad (3.57)$$

where Q is the risk-neutral measure under which the drift of the GBM is r instead of μ, and

$$d_1 = \frac{\log(A_t/D) + (r + \sigma^2/2)(T-t)}{\sigma \sqrt{T-t}}, \qquad d_2 = d_1 - \sigma \sqrt{T-t}.$$

Furthermore, since the firm's assets are equal to the value of debt plus equity, i.e., $V_t = V_e(t) + V_d(t)$, the debt value at time t is

$$V_d(t) = E^Q \left\{ e^{-r(T-t)} \min(A_T, D) | A_t \right\} = A_t - V_e(t), \qquad (3.58)$$

note that $\min(A_t, D) = A_t - (A_t - D)^+$.

Credit spread

An important characteristic of a defaultable bond is the difference between its yield and the yield of an equivalent default-free bond, i.e., the *credit spread*. Given the above argument, the yield at date t of the corporate bond (or the firm's debt value) is expressed as

$$y(t, T) = \frac{1}{T-t} \log \frac{D}{V_d(t)},$$

then the credit spread, or the difference between the yield of this bond and a corresponding treasury bond with interest r, is given by

$$s(t, T) = y(t, T) - r = \frac{1}{T-t} \log \frac{D}{V_d(t)} - r.$$

The Merton model allows the credit spread curve $s(t, T)$ to be hump shaped or monotonically decreasing (in cases where the firm's value is smaller than the face value of debt). Furthermore, it can be verified that

$$\lim_{t \to T} s(t, T) = \begin{cases} +\infty & \text{if } V_T < D, \\ 0 & \text{if } V_T > D. \end{cases}$$

This result indicates that, if a firm is not in financial distress at a date very close to the maturity date T, it is very unlikely to default. In other words, if the asset value remains above the default threshold close to maturity, the probability of default is low. However, empirical evidence suggests that short-term credit spreads are often non-zero even when firms are not in immediate financial distress or close to default. This contradicts with the prediction of Merton's model and is considered as a major shortcoming of Merton's approach. The Black and Cox's default barrier model (discussed later in this section) suffer from the same shortcoming.

One possible explanation for the discrepancy between model predictions and empirical evidence is the presence of jumps or discontinuous movements in the firm's asset value. These jumps are caused by unexpected events, news, or shocks that affect the firm's financial condition. Introducing jumps in the dynamics of the firm's value can lead to more realistic modeling of credit risk and better alignment with observed credit spread behavior. Mason and Bhattacharya (1981) introduced this approach as a potential remedy for the shortcomings of Merton's and Black and Cox's models.

KMV's credit monitor model and distance to default

Empirical studies show that firms generally do not default when their asset value falls below the book value of their total liabilities. This suggests that the default point (or the asset value at which the firm will actually default) should lie somewhere between short-term and total liabilities, and hence introduce the concept of *distance to default*.

In the 1990s, the KMV Corporation of San Francisco (now merged into Moody's KMV) introduced its Credit Monitor Model for default prediction using Merton's basic framework. To begin with, suppose a firm's asset value A_t follows a geometric Brownian motion (3.56), then it can be expressed by

$$A_t = A_0 \exp\{(\mu - \frac{1}{2}\sigma^2)t + \sigma W_t\}.$$

Therefore, the probability of default on the debt at date t is

$$P(A_T < D|\mathcal{F}_t) = P\left\{(\mu - \frac{\sigma^2}{2})T + \sigma W_T < \log(\frac{D}{A_0})\Big|\mathcal{F}_t\right\}$$

$$= \Phi\left(-\frac{\log(A_t/D) + (\mu - \frac{1}{2}\sigma^2)(T - t)}{\sigma\sqrt{T - t}}\right).$$

This leads to the definition of the KMV's distance to default at time t, which is defined by

$$DD = \frac{\log(A_t/D) + (\mu - \frac{1}{2}\sigma^2)(T - t)}{\sigma\sqrt{T - t}},$$

so that the probability of default at date t under Merton's model is $\Phi(-DD)$. Since $\sigma\sqrt{T - t}$ is the standard deviation of $\log A_T$ conditional on $\mathcal{F}_t$, DD can be regarded as the distance of the logged firm value from the firm's logged liability, in the unit of 'standard deviation.' Instead of using the normal distribution for $\log A_T$ to predict the default probability, the KMV Credit Monitor Model uses a large proprietary database of firms and firm defaults to construct an EDF (expected default frequency) score for a firm with κ standard deviations away from the default (i.e., $DD = \kappa$) at the beginning of a future period. The idea is to stratify the firms based on their DD's and estimate empirically the default probability of the firm from its associated stratum. The firm's DD is evaluated from its balance sheet and option models for its debt.

Black-Cox framework of default barriers

Black and Cox (1976) extend the Merton model by allowing default to occur prior to the maturity of the bond; specifically, when the asset value A_t hits a non-random, continuous boundary $g(t) \leq D$. They also allow the equity owners to be paid dividends, at rate q, so that under the risk-neutral measure $\mathbb{Q}$ the GBM for A_t has drift $r - q$, that is, A_t follows the process

$$dA_t = (r - q)A_t dt + \sigma A_t dW_t^{\mathbb{Q}}.$$

Let $\tau = \inf\{0 \leq t \leq T : A_t = g(t)\}$. At time $\min(\tau, T)$, the payoff to equity holders is

$$V_e(\min(\tau, T)) = (A_T - D)^+ 1_{\{\tau > T\}},$$

and the payoff to bond holders is

$$V_d(\min(\tau, T)) = \min(A_T, D)I_{\{\tau > T\}} + g(\tau)1_{\{\tau \leq T\}},$$

in which the first summand reflects that the bond holders receive $\min(A_T, D)$ at the bond's maturity if default has not occurred up to that time, while the second summand reflects that the bond holders receive the recovery of $A_\tau = g(\tau)$ upon default. Then the value of $V_e(\min(\tau, T))$ and $V_d(\min(\tau, T))$ at time t can be evaluated via their discounted expectation under the risk-neutral measure.

3.5.2 Intensity models of defaultable claims

The previous section assumes a deterministic hazard function for defaultable claims, which is not realistic. We now extend the discussion to the case of *stochastic hazard* processes. Consider a probability space $(\Omega, \mathcal{F}, \mathbb{P})$. Let $\{\mathbf{x}_t \in \mathbb{R}^d\}$ be a process of state variables defined on $(\Omega, \mathcal{F}, \mathbb{P})$ and $\{\mathcal{F}_t\}_{t \geq 0}$ the filtration (i.e., the information set) generated by the process X, i.e., $\mathcal{F}_t = \sigma\{\mathbf{x}_s; 0 \leq s \leq t\}$. Let $Y_t = 1_{\{\tau \leq t\}}$ $(t \geq 0)$ be the jump or default indicator process, $\{\mathcal{H}_t; t \geq 0\}$ the filtration generated by the process $\{Y_t\}$, i.e., $\mathcal{H}_t = \sigma\{Y_u; u \leq t\}$. We consider a new filtration defined by

$$\mathcal{G}_t = \mathcal{F}_t \vee \mathcal{H}_t, \qquad t \geq 0,$$

meaning that $\mathcal{F}_t$ is the smallest σ-algebra that contains $\mathcal{G}_t$ and $\mathcal{H}_t$. In the context of credit risk models, $\mathcal{G}_t$ contains information about the state variables and the occurrence or non-occurrence of default up to time t and is usually assumed observable for econometricians or investors. Note that the default time τ is $\mathcal{H}_t$ stopping time and hence also a $\mathcal{G}_t$ stopping time, but τ is not necessary a stopping time with respect to the filtration $\mathcal{F}_t$ generated by the state variable process.

Doubly stochastic random time

A random time τ is said to be *doubly stochastic* if there exists a positive $\mathcal{F}_t$-adapted process $\lambda(t)$ such that $\Lambda(t) = \int_0^t \lambda(s)ds$ is strictly increasing and finite for every $t > 0$ and such that, for all $t \geq 0$,

$$P(\tau > t|\mathcal{F}_\infty) = \exp\left(-\int_0^t \lambda(s)ds\right) = \exp\left(-\Lambda(t)\right).$$

The above equation means that, given information of state variable $\mathbf{x}_t$ up to time t, the future information of $\{\mathbf{x}_s\}_{s>t}$ does not contain any extra information on the probability that default occurs before time t. Doubly stochastic random times are also known as *conditional Poisson* or *Cox* random times. Actually, since (3.5.2) implies that $(\tau > t|\mathcal{F}_\infty)$ is $\mathcal{F}_t$-measurable, we have

$$P(\tau > t|\mathcal{F}_\infty) = P(\tau > t|\mathcal{F}_t).$$

Based on the definition, a doubly stochastic random time τ can be explicitly constructed. Specifically, let ζ be an exponential random variable with mean 1 that is independent of $\mathcal{F}_\infty$, and $\lambda(\mathbf{x}_t)$ be a positive $\mathcal{F}_t$ measurable process such that $\int_0^t \lambda(\mathbf{x}_u)du$ is strictly increasing and finite for all $t > 0$. Define the random time τ by

$$\tau = \inf\left\{t : \int_0^t \lambda(\mathbf{x}_u)du \geq \zeta\right\}, \tag{3.59}$$

then τ is doubly stochastic with $\mathcal{F}_t$ measurable intensity process $\lambda(\mathbf{x}_t)$. Let $Y_t := 1_{\{\tau \leq t\}}$, we then call the jump process $\{Y_t\}_{t\geq 0}$ a doubly stochastic process or *Cox process*.

Intensity of doubly stochastic random times

The intensities of doubly stochastic random times have nice properties, which extend the martingale property (3.35) in Section 3.4.1. In particular, let τ be a doubly stochastic random time defined by (3.59), and $\lambda(t)$ is the associated intensity process, then $M_t := Y_t - \Lambda(\tau \wedge t)$, $t \geq 0$, is a $\mathcal{G}_t$ martingale. According to the definition of τ and $\mathcal{G}_t$, we can show that, for any integrable random variable Z,

$$E(1_{\{\tau>t\}}Z|\mathcal{G}_t) = 1_{\{\tau>t\}}\frac{E(1_{\{\tau>t\}}Z|\mathcal{F}_t)}{P(\tau > t|\mathcal{F}_t)}. \tag{3.60}$$

Furthermore, if $T > t$ and Z is $\mathcal{F}_T$-measurable, we have

$$E(1_{\{\tau > T\}} Z | \mathcal{G}_t) = 1_{\{\tau > t\}} E\left(\exp\left(- \int_t^T \lambda(\mathbf{x}_s) ds \right) Z \Big| \mathcal{F}_t \right). \qquad (3.61)$$

Prices of three types of contingent claims

Let $(\Omega, \mathcal{F}, \mathcal{F}_t, \mathbb{Q})$ be a filtered probability space, where $\mathbb{Q}$ is the equivalent martingale measure. Denote τ the default time of a firm and $Y_t = 1_{\{\tau \leq t\}}$ be the associated default indicator process. As before, we define $\mathcal{H}_t = \sigma(\{Y_s : s \leq t\})$ and $\mathcal{G}_t = \mathcal{F}_t \vee \mathcal{H}_t$. Suppose that the market for credit products is complete, so that we can use the martingale measure $\mathbb{Q}$ as the pricing measure for defaultable securities. Then given a financial claim with pay-off H and maturity $T \geq t$, its price at time t is expressed as

$$H_t = E^{\mathbb{Q}}\left(\exp\left(- \int_t^T r(\mathbf{x}_s) ds \right) H \Big| \mathcal{G}_t \right). \qquad (3.62)$$

Finally, we assume that, under $\mathbb{Q}$, the default time τ is a doubly stochastic random time with state-variable process $\mathbf{x}_t$, filtration $\mathcal{F}_t = \sigma(\{\mathbf{x}_s; 0 \leq s \leq t\})$, and hazard process $\lambda(\mathbf{x}_t)$.

We now derive the price of a zero-coupon defaultable bond. Consider a zero-coupon bond, with maturity date T and par value 1, issued by a firm at time 0. Assume that there is a short-rate process $r(\mathbf{x}_s)$ under the risk-neutral measure $\mathbb{Q}$ such that the default-free bond price is given by

$$p(0, T) = E^{\mathbb{Q}}\left\{ \exp\left(- \int_0^T r(\mathbf{x}_s) ds \right) \right\}. \qquad (3.63)$$

Using equation (3.60) and assuming zero recovery at default, we can express the price of the defaultable bond at time 0 as

$$
\begin{aligned}
p_1(0, T) &= E^{\mathbb{Q}}\left[1_{\{\tau > T\}} \exp\left(- \int_0^T r(\mathbf{x}_s) ds \right) \Big| \mathcal{G}_0 \right] \\
&= E^{\mathbb{Q}}\left\{ 1_{\{\tau > 0\}} E^{\mathbb{Q}}\left[\exp\left(- \int_0^T [\lambda(\mathbf{x}_t) + r(\mathbf{x}_s)] ds \right) \Big| \mathcal{F}_0 \right] \right\} \qquad (3.64) \\
&= E^{\mathbb{Q}}\left\{ \exp\left(- \int_0^T (r + \lambda)(\mathbf{x}_s) ds \right) \right\}.
\end{aligned}
$$

Thus, (3.64) replaces the short rate $r(\mathbf{x}_s)$ in the default-free bond pricing formula by the default-adjusted short rate $(r + \lambda)(\mathbf{x}_s)$.

The pricing formula (3.64) for the zero-coupon defaultable bond with zero recovery can be extended to other contingent claims. Here, we consider three types of contingent claims and provide their pricing formulas.

(i) A contingent claim with a promised payment of $p(\mathbf{x}_T)$ at time T if there is no default. The actual payment at time T is $f(\mathbf{x}_T)1_{\{\tau>T\}}$.

$$E^{\mathbb{Q}}\left[\exp\left(-\int_t^T r(\mathbf{x}_s)ds\right)p(\mathbf{x}_T)1_{\{\tau>T\}}\Big|\mathcal{G}_t\right]$$
$$= 1_{\{\tau>t\}}E^{\mathbb{Q}}\left[\exp\left(-\int_t^T (r+\lambda)(\mathbf{x}_s)ds\right)p(\mathbf{x}_T)\Big|\mathcal{F}_t\right]. \tag{3.65a}$$

(ii) A contingent claim paying $g(\mathbf{x}_t)1_{\{\tau>t\}}$, $0 \leq t \leq T$, continuously until default or the maturity date T in the case of no default.

$$E^{\mathbb{Q}}\left[\int_t^T g(\mathbf{x}_s)1_{\{\tau>s\}}\exp\left(-\int_t^s r(\mathbf{x}_u)du\right)ds\Big|\mathcal{G}_t\right]$$
$$= 1_{\{\tau>t\}}E^{\mathbb{Q}}\left[\int_t^T g(\mathbf{x}_s)\exp\left(-\int_t^s (r+\lambda)(\mathbf{x}_u)du\right)ds\Big|\mathcal{F}_t\right]. \tag{3.65b}$$

(iii) A contingent claim with payment $h(\mathbf{x}_\tau)$ at the default date $\tau \leq T$ (T is the maturity date).

$$E^{\mathbb{Q}}\left[\exp\left(-\int_t^s r(\mathbf{x}_u)du\right)h(\mathbf{x}_\tau)1_{\{t<\tau\leq T\}}\Big|\mathcal{G}_t\right]$$
$$= 1_{\{\tau>t\}}E^{\mathbb{Q}}\left[\int_t^T h(\mathbf{x}_s)\lambda(\mathbf{x}_s)\exp\left(-\int_t^s (r+\lambda)(\mathbf{x}_u)du\right)ds\Big|\mathcal{F}_t\right]. \tag{3.65c}$$

Affine models

To evaluate the conditional expectations on the right-hand side of equations (3.65a)–(3.65c), most models assume that defaults are doubly stochastic random time and $\lambda(\mathbf{x}_t)$ and $r(\mathbf{x}_t)$ are functions of p-dimensional Markovian state variable process $\mathbf{x}_t$. Since $\mathbf{x}_t$ is a Markov process, conditional expectations on the right-hand side of equations (3.65a)-(3.65c) are given by some functions $f(t, \mathbf{x}_t)$. It is well known that under some additional regularity assumptions the function f can be computed as solution of a parabolic *partial differential equation* (PDE), which is called the *Feynman-Kac formula*. In particular, when $\mathbf{x}_t$ follows an *affine jump diffusion*, then the conditional expectation on the right-hand side of (3.65a) has the form

$$f(t, \mathbf{x}) = \exp\left[\alpha(t, T) + \boldsymbol{\beta}(t, T)'\mathbf{x}\right] \tag{3.66}$$

for some deterministic functions $\alpha : [0, T] \to \mathbb{R}$ and $\boldsymbol{\beta} : [0, T] \to \mathbb{R}^p$; moreover, α and $\boldsymbol{\beta}$ are determined by a $(p+1)$-dimensional *ordinary differential equation* (ODE) system that is easily solved numerically.

For illustration purpose, we consider the case here where the state variable process is given by a one-dimensional diffusion. Assume that the state variable process x_t is the unique solution of the stochastic differential equation

$$dx_t = \mu(x_t)dt + \sigma(x_t)dW_t, \qquad x_0 = x \in D \subset \mathbb{R},$$

where W_t is a standard, one-dimensional Brownian motion on some filtered probability space $(\Omega, \mathcal{F}, \mathcal{F}_t, \mathbb{P})$, μ and σ are continuous functions from D to $\mathbb{R}$. Suppose that the function $f : [0, T] \times D \to \mathbb{R}$ is continuous, once continuously differentiable in t and twice continuously differentiable in x on $[0, T) \times D$, and that f solves the terminal-value problem

$$
\begin{cases}
\dfrac{\partial f}{\partial t} + \dfrac{1}{2}\sigma^2(x)\dfrac{\partial^2 f}{\partial x^2} + \mu(x)\dfrac{\partial f}{\partial x} + h(x) = R(x)f, & (t, x) \in [0, T) \times D, \\
f(T, x) = g(x), & x \in D.
\end{cases}
\tag{3.67}
$$

where functions $R(x)$, $g(x)$, and $f(x)$ are functions defined on D. If f is bounded or, more generally, $\max_{0 \le t \le T} f(t, x) \le C(1 + x^2)$ for $x \in D$, then

$$E\left[\exp\left(-\int_t^T R(x_s)ds\right)g(x_T) + \int_t^T h(x_s)\exp\left(-\int_t^s R(x_u)du\right)ds\,\Big|\,\mathcal{F}_t\right] = f(t, x_t).$$

Consider the special case $h \equiv 0$ and $g(x) = e^{ux}$, $ux \le 0$ for $x \in D$. Suppose that R, μ, and σ^2 are affine function of x, i.e., there are constants $\rho_0, \rho_1, k_0, k_1, h_0$, and h_1 such that

$$R(x) = \rho_0 + \rho_1 x \ge 0, \quad \mu(x) = k_0 + k_1 x, \quad \sigma^2(x) = h_0 + h_1 x \ge 0.$$

then the solution of the PDE (3.67) has the form (3.66). Suppose that $\widetilde{f}$ is a solution of (3.67) and has the form

$$\widetilde{f}(t, x) = \exp\left[\alpha(t, T) + \beta(t, T)x\right].$$

Substitute the above form into equation (3.67). We can obtain the following ODE system for $\alpha(t, T)$ and $\beta(t, T)$,

$$\beta'(t, T) = \rho_1 - k_1\beta(t, T) - \frac{1}{2}h_1\beta^2(t, T), \quad \beta(T, T) = u, \tag{3.68a}$$

$$\alpha'(t, T) = \rho_0 - k_0\beta(t, T) - \frac{1}{2}h_0\beta^2(t, T), \quad \alpha(T, T) = 0, \tag{3.68b}$$

where $\alpha'(t, T)$ and $\beta'(t, T)$ are the derivatives of α and β with respect to t. The ODE (3.68a) for $\beta(t, T)$ is a so-called *Ricatti equation*. Actually, the ODE system (3.68a) and (3.68b) has a unique solution $(\alpha(t, T), \beta(t, T))$. If there is some C such that $\beta(t, T)x \le C$ for all $t \in [0, T]$ and $x \in D$, we have

$$E\left[\exp\left(-\int_t^T R(x_s)ds\right)e^{ux}\,\Big|\,\mathcal{F}_t\right] = \exp\left[\alpha(t, T) + \beta(t, T)x\right].$$

Note that the above discussion extends the price of default-free bond for different interest rate models in Section 3.3.2.

3.6   Credit rating migration

Credit rating migration, also known as credit rating transition, refers to the movement of a borrower's credit rating from one category to another over time. Credit ratings are assessments provided by credit rating agencies that evaluate the creditworthiness of borrowers, including corporations, governments, and other entities, by assigning them a rating based on their ability to repay debt obligations. Credit rating migration occurs when the creditworthiness of a borrower changes, leading to an upgrade or downgrade in their credit rating.

This section introduces probabilistic and statistical models that describe and predict the probability of changes in credit ratings over time. We first assume all firms are homogeneous, and introduce time-homogeneous and piecewise time-homogeneous Markov chain models for credit rating migration. Then we relax the homogeneity assumption by assuming firms' rating transitions may depend on firm's healthiness and economic environment, and introduce modulated Markov and semi-Markov models for credit rating migrations.

3.6.1   Time-homogeneous Markov chains

To present the idea, we fix an underlying probability space $(\Omega, \mathcal{G}, \mathbb{P})$ and a finite set $\mathcal{S} = \{1, \ldots, K\}$, which is the state space for all considered Markov chains in this section.

Discrete-time Markov chains

Let $\{\phi_t\}_{t=0,1,\ldots}$ be a sequence of random variables on $(\Omega, \mathcal{G}, \mathbb{P})$ with values in $\mathcal{S}$. The process ϕ is said to be a Markov chain if, for every time n and every combination of states $\{i_0, i_1, \ldots, i_n\}$, we have

$$P(\phi_{n+1} = j | \phi_0 = i_0, \ldots, \phi_{n-1} = i_{n-1}, \phi_n = i) = P(\phi_{n+1} = j | \phi_n = i). \quad (3.69)$$

The Markov chain is said to be *time-homogeneous* if $P(\phi_n = i_n | \phi_{n-1} = i_{n-1})$ does not depend on n, and then we define the one-period transition probability from i to j as

$$p_{ij} = P(\phi_{n+1} = j | \phi_n = i) \quad \text{for any } n.$$

We represent these transition probabilities in a matrix

$$P = \begin{pmatrix} p_{11} & p_{12} & \cdots & p_{1K} \\ p_{21} & p_{22} & \cdots & p_{2K} \\ \vdots & \vdots & \ddots & \vdots \\ p_{K,1} & p_{K,2} & \cdots & p_{K,K} \end{pmatrix}, \quad (3.70)$$

where $\sum_{j=1}^{K} p_{ij} = 1$ for all i. Given the above definition, the n-step transition probabilities can be obtained by taking the nth power of the one-period transition matrix, i.e.,

$$p_{ij}^{(n)} := P(\phi_n = j | \phi_0 = i) = (P^n)_{ij},$$

where $(P^n)_{ij}$ is the ijth element of the matrix obtained by raising P to the power n.

A *time inhomogeneous* Markov chain still satisfies (3.69), but the transition probability matrix P is not independent of calendar time. Hence, we can write

$$P(t, t+1) = \begin{pmatrix} p_{11}(t, t+1) & p_{12}(t, t+1) & \cdots & p_{1K}(t, t+1) \\ p_{21}(t, t+1) & p_{22}(t, t+1) & \cdots & p_{2K}(t, t+1) \\ \vdots & \vdots & \ddots & \vdots \\ p_{K,1}(t, t+1) & p_{K,2}(t, t+1) & \cdots & p_{K,K}(t, t+1) \end{pmatrix}$$

for the transition-probability matrix between times t and $t+1$ and we now have the following connection between multi-period and one-period transition probabilities

$$P(t, u) = P(t, t+1)P(t+1, t+2)\cdots P(u-1, u).$$

Estimation of discrete-time transition matrix

In credit risk management, it is convenient to assume that the credit quality (or rating) of obligors (or firms) follows a Markov chain which is characterized by a credit rating transition matrix. Assume that a rating transition process of an obligor follows a K-state discrete time Markov chain process with the finite state space $\mathcal{S} = \{1, \ldots, K\}$, which consists of K different rating classes, and state 1 denotes the best credit rating class and state K represents the default (or absorb) state. Then the $K \times K$ credit rating transition matrix (3.70) becomes

$$P = \begin{pmatrix} p_{11} & p_{12} & \cdots & p_{1K} \\ p_{21} & p_{22} & \cdots & p_{2K} \\ \cdots & & & \\ p_{K-1,1} & p_{K-1,2} & \cdots & p_{K-1,K} \\ 0 & 0 & \cdots & 1 \end{pmatrix}, \tag{3.71}$$

in which $p_{ii} = 1 - \sum_{j \neq i} p_{ij}$ for all i and p_{ij} represents the actual probability of moving to state j from initial rating state i in one time step. Usually, the estimates of credit rating transition matrices published by rating agencies usually use a discrete-time setting. Specifically, suppose there are N_i firms in a given rating category i at the beginning of the period and that N_{ij} firms migrate to the category j at the end of the period. The transition probability p_{ij} can be estimated by nonparametric maximum likelihood methods, and the estimate is given by

$$\widehat{p}_{ij} = N_{ij}/N_i, \qquad \text{for } j \neq i. \tag{3.72}$$

However, there are serious difficulties with this estimate in practice. First, if no transition from rating class i to j is observed during the period, the estimated transition probability $\hat{p}_{ij}$ becomes 0. Second, one year is the shortest period for which most rating agencies report transition probabilities or intensities, while in practical risk management, it is important to be able to analyze the risk of credit events and credit transition over a shorter period of time, such as a month. To address such concern, a continuous time homogeneous Markov framework is usually assumed for the rating transition process.

Continuous-time Markov chains

The definition of the discrete-time can be extended to continuous-time Markov chains by making the time parameter continuous. Specifically, for a stochastic process ϕ which is indexed by a continuous parameter $t \in [0, \infty)$ with values in $\mathcal{S}$, the Markov property holds if, for $i, j \in \mathcal{S}$,

$$P(\phi_t = j | \phi_{s_0} = i_0, \dots, \phi_{s_{n-1}} = i_{n-1}, \phi_s = i) = P(\phi_t = j | \phi_s = i) \qquad (3.73)$$

in which $s_0 < s_1 < \cdots < s_{n-1} < s$. For a continuous time Markov chain, the probability transition matrices satisfy that

$$P(s, t) = P(s, u)P(u, t), \qquad \text{for } s < u < t.$$

A time-homogeneous Markov chain on $\mathcal{S}$ can be characterized by its associated *generator matrix*, i.e., a $K \times K$ matrix Λ for which

$$P(0, t) = \exp(\Lambda t), \qquad t > 0, \qquad (3.74)$$

where the exponential of the matrix Λt is defined as

$$\exp(t\Lambda) := \sum_{k=0}^{\infty} \frac{\Lambda^k t^k}{k!}.$$

Provided that $P(0, t)$ is a probability transition matrix (i.e., rows sum to 1), the elements of its generator matrix $\Lambda = (\lambda_{ij})_{1 \leq i,j \leq K}$, defined by (3.74), satisfy that, for each $i \in \mathcal{S}$,

$$\lambda_{ii} = -\sum_{j \neq i} \lambda_{ij}, \qquad \lambda_{ij} \geq 0 \text{ for } 1 \leq i \neq j \leq K.$$

When a continuous homogeneous Markov chain is assumed for the rating transition process, as the default state (or the Kth state) is an absorbing state, the last row of the generator matrix Λ becomes all zeros, i.e., $\lambda_{(K,j)} = 0$, for $1 \leq j \leq K$.

To extend an homogeneous Markov chain to a time-inhomogeneous chain, one can assume that the transition probability matrix $P(s, t)$ during the period (s, t) has the product integral representation

$$P(s, t) = \prod_{s < u \leq t} (I_K + d\Lambda(u)), \qquad (3.75)$$

where I_K is the $K \times K$ identity matrix and $\Lambda(u) = (\Lambda_{ij}(u))_{1 \le i,j \le K}$ is the *intensity measure* of the Markov chain satisfying

$$\lambda_{ii}(u) = -\sum_{j \ne i} \lambda_{ij}(u), \qquad \lambda_{ij}(u) \ge 0 \text{ for } 1 \le i \ne j \le K.$$

Likelihood and Bayesian inference of generator matrices

Given observations of one-year transition probabilities, one expects to find a generator matrix Λ consistent with the one-year transition matrix P such that $P = \exp(\Lambda)$. Such problem is known as the *embedding problem* for Markov chains, and an discussion of this issue can be found in Israel et al. (2001). In general, the problem doesn't have a solution, and even it has, the solution is not necessarily unique. One way to avoid the embedding problem is to estimate rating transition based on continuous-time data. Denote, for the period $(0, t)$, $K_{ij}^{(0,t)}$ the number of transitions from category i to category j, $S_i^{(0,t)}$ the amount of time that firms spend in category i, then the likelihood of observed firms' rating transitions over the period $(0, t)$ is expressed as

$$\exp\left\{ \sum_{i=1}^{K-1} \left[\sum_{j \ne i} K_{ij}^{(0,t)} \log \lambda_{ij} - \left(\sum_{j \ne i} \lambda_{ij} + 1 - K \right) S_i^{(0,t)} \right] \right\}$$
$$\propto \prod_{i \ne j} \lambda_{ij}^{K_{ij}^{(0,t)}} \exp\left(-\lambda_{ij} S_i^{(0,t)} \right). \tag{3.76}$$

Then the elements in Λ can be estimated by their maximum likelihood estimators

$$\widehat{\lambda}_{ij,\mathrm{ml}} = K_{ij}^{(0,t)} \Big/ S_i^{(0,t)}, \qquad (i,j) \in \mathcal{K}, \tag{3.77}$$

in which $\mathcal{K} = \{(i,j) | i \ne j, 1 \le i \le K - 1, 1 \le j \le K\}$; see Küchler and Sørensen (1997, Section 3.4). The generator matrix Λ can also be estimated by the Bayes method. Assuming the off-diagonal elements λ_{ij} follow independently a Gamma(α_{ij}, β_i) prior distribution with the density function

$$g(\lambda_{ij}) = \frac{\beta_{ij}^{\alpha_{ij}}}{\Gamma(\alpha_{ij})} \lambda_{ij}^{\alpha_{ij}-1} \exp\left(-\lambda_{ij}\beta_{ij} \right), \qquad (i,j) \in \mathcal{K}, \tag{3.78}$$

one can show that the posterior distribution of λ_{ij} is Gamma$(\alpha_{ij} + K_{ij}^{(0,t)}, \beta_i + S_i^{(0,t)})$, then the element λ_{ij} can be estimated their posterior mean of the Gamma distribution, i.e.,

$$\widehat{\lambda}_{ij,\mathrm{Bayes}} = \frac{\alpha_{ij} + K_{ij}^{(0,t)}}{\beta_i + S_i^{(0,t)}}.$$

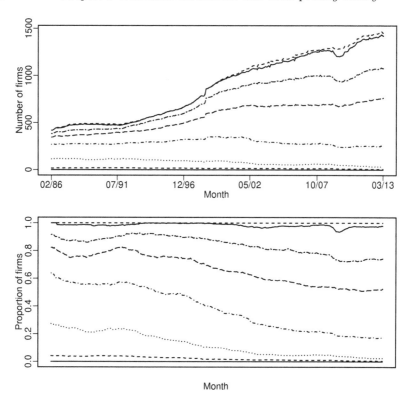

FIGURE 3.5: Number of firms (top) and their proportion (bottom) in each rating category from February 1986 to March 2013. The regions bounded by two adjacent curves represent rating categories $\mathcal{AAA}$, $\mathcal{AA}$, $\mathcal{A}$, $\mathcal{BBB}$, $\mathcal{BB}$, $\mathcal{B}$, and $\mathcal{CCC}$.

Example 3.4 (Credit rating transitions) *Consider a sample of Standard & Poor's monthly credit ratings of firms starting from February 1986 and ending March, 2013. The rating categories in the sample have been re-grouped into $K = 8$ rating categories, $\mathcal{AAA}$, $\mathcal{AA}$, $\mathcal{A}$, $\mathcal{BBB}$, $\mathcal{BB}$, $\mathcal{B}$, $\mathcal{CCC}$, and $\mathcal{D}$ (the default category). There are totally 1,817 firms in the sample. Figure 3.5 shows the number of firms and their proportion in each rating category during the sample period. We estimate the rating transition transition matrix in both discrete and continuous time-homogeneous Markov models. Table 3.3 shows the estimated rating transition and generator matrices of credit ratings for the sample. In particular, part (a) is the maximum likelihood estimate of one-year rating transition probability matrix in discrete time whose elements are given by (3.72). The blank entries in the estimated transition probability matrix indicate that no rating transitions were observed between two rating categories during the sample period. Part (b) of Table 3.3 is the maximum likelihood estimate of*

3.6 Credit rating migration

TABLE 3.3: Transition and generator matrices of credit rating during 1986–2013

		$\mathcal{AAA}$	$\mathcal{AA}$	$\mathcal{A}$	$\mathcal{BBB}$	$\mathcal{BB}$	$\mathcal{B}$	$\mathcal{CCC}$	$\mathcal{D}$
	$\mathcal{AAA}$	.99516	.00464	.00020					
	$\mathcal{AA}$	.00037	.99082	.00840	.00037		.00005		
	$\mathcal{A}$		.00104	.99285	.00587	.00015	.00007	.00001	
(a)	$\mathcal{BBB}$			.00232	.99374	.00371	.00020	.00002	.00001
	$\mathcal{BB}$			.00005	.00525	.98705	.00736	.00025	.00004
	$\mathcal{B}$			.00011	.00019	.00776	.98540	.00643	.00011
	$\mathcal{CCC}$				.00018	.00213	.03439	.95249	.01081
	$\mathcal{AAA}$	−.05810	.05568	.00242					
	$\mathcal{AA}$	.00443	−.11019	.10078	.00443		.00055		
	$\mathcal{A}$		.01246	−.08578	.07047	.00178	.00089	.00018	
(b)	$\mathcal{BBB}$			.02780	−.07517	.04454	.00243	.00027	.00013
	$\mathcal{BB}$			.00064	.06299	−.15534	.08831	.00298	.00043
	$\mathcal{B}$			.00129	.00232	.09312	−.17518	.07717	.00129
	$\mathcal{CCC}$				.00213	.02553	.41269	−.57011	.12976
	$\mathcal{AAA}$	.94367	.05122	.00483	.00025	7.4e-6	1.6e-5	6.9e-7	4.4e-8
	$\mathcal{AA}$	.00407	.89635	.09149	.00729	2.4e-4	5.4e-4	2.5e-6	2.0e-6
	$\mathcal{AA}$	.00003	.01130	.91928	.06517	.00304	.00101	1.7e-4	1.7e-5
(c)	$\mathcal{BBB}$	2.4e-7	1.6e-4	.02569	.92977	.03991	.00395	3.6e-4	1.6e-4
	$\mathcal{BB}$	9.2e-9	7.0e-6	.00141	.05638	.86094	.07589	.00465	7.2e-4
	$\mathcal{B}$	1.1e-8	7.4e-6	.00122	.00475	.08025	.85463	.05399	.00515
	$\mathcal{CCC}$	1.2e-9	1.1e-6	2.5e-4	.00288	.03240	.28903	.57580	.09964
	$\mathcal{AAA}$	.97140	.02670	.00184	5.8e-5	1.2e-6	4.0e-6	1.2e-7	4.5e-9
	$\mathcal{AA}$	.00212	.94657	.04800	.00296	6.0e-5	2.7e-4	7.0e-6	3.4e-7
	$\mathcal{A}$	6.6e-6	.00593	.95840	.03388	.00122	4.7e-4	8.7e-5	4.3e-6
(d)	$\mathcal{BBB}$	3.1e-8	4.1e-5	.01336	.96368	.02107	.00162	1.6e-4	7.4e-5
	$\mathcal{BB}$	9.6e-10	1.4e-6	5.2e-4	.02978	.92656	.04087	.00198	2.8e-4
	$\mathcal{B}$	1.4e-9	1.9e-6	6.2e-4	.00181	.04317	.92051	.03216	.00173
	$\mathcal{CCC}$	8.7e-11	1.4e-7	6.8e-5	.00123	.01484	.17206	.75521	.05658

the generator matrix of rating transition probability matrices whose elements are given by (3.77). The blank entries in the generator matrices also indicate no rating transition were observed between the two rating categories during the sample period. Parts (c) and (d) of Table 3.3 show the estimated one-year and half-year rating transition probability matrices $\widehat{P}(0,1)$ and $\widehat{P}(0,0.5)$ using the estimated generator matrix in part (b) and equation (3.74). In contrast to the estimated rating transition probability matrix in part (a), no entries in parts (c) and (d) are exactly zero due to the advantage of continuous-time Markov chain setting. For parts (c) and (d), all staying (or transition, respectively) probabilities in $\widehat{P}(0,1)$ are higher (or lower, respectively) than those in $\widehat{P}(0,0.5)$. $\qquad\square$

3.6.2   Piecewise time-homogeneous continuous-time Markov chains

The assumptions of time homogeneity and Markovian behavior of the rating process in the preceding section have been challenged by recent studies on the presence of various non-Markovian behavior such as industry heterogeneity, rating drifts, and time variations. Assume that a rating transition process of a firm follows a K-state hon-homogeneous continuous time Markov process, and the process is characterized by a transition probability matrix $P(s,t)$ over the period (s,t). Suppose that there are n rating transitions observed over the period (s,t). For a transition time t_i in (s,t), denote $\Delta N_{kj}(t_i)$ the number of transitions observed from state k to state j at time t_i, $\Delta N_k(t_i) = \sum_{1 \le i \le K, i \ne k} \Delta N_{kj}(t_i)$, and $Y_k(t_i)$ the number of firms in state k right before time t_i. Then the product-limit representation of a time-inhomogeneous Markov chain (3.75) can be estimated by

$$\widehat{P}(s,t) = \prod_{s < u \le t} \left(I_K + \Delta \widehat{\Lambda}(u) \right) = \prod_{i=1}^{n} \left(I_K + \Delta \widehat{\Lambda}(t_i) \right), \qquad (3.79)$$

in which

$$\Delta \widehat{\Lambda}(t_i) = \begin{pmatrix} -\frac{\Delta N_1(t_i)}{Y_1(t_i)} & \frac{\Delta N_{12}(t_i)}{Y_1(t_i)} & \frac{\Delta N_{13}(t_i)}{Y_1(t_i)} & \cdots & \frac{\Delta N_{1K}(t_i)}{Y_1(t_i)} \\ \frac{\Delta N_{21}(t_i)}{Y_2(t_i)} & -\frac{\Delta N_2(t_i)}{Y_2(t_i)} & \frac{\Delta N_{23}(t_i)}{Y_2(t_i)} & \cdots & \frac{\Delta N_{2K}(t_i)}{Y_2(t_i)} \\ \vdots & \vdots & & \cdots & \vdots \\ \frac{\Delta N_{K-1,1}(t_i)}{Y_{K-1}(t_i)} & \frac{\Delta N_{K-1,2}(t_i)}{Y_{K-1}(t_i)} & \cdots & -\frac{\Delta N_{K-1}(t_i)}{Y_{K-1}(t_i)} & \frac{\Delta N_{K-1,K}(t_i)}{Y_{K-1}(t_i)} \\ 0 & 0 & \cdots & \cdots & 0 \end{pmatrix}.$$

In the matrix above, the kth diagonal element counts the fraction of the exposed firms $Y_k(t_i)$ leaving the state at time t_i, and the kj'th off-diagonal element count the fraction of transitions from the kth category to the jth category in the number of exposed firms at time t_i. Note that the variable Y has incorporated the case of censoring for which there is no change in the estimator at the time of a censoring event. Furthermore, the last row in $\Delta \widehat{\Lambda}(t_i)$ is zero because the Kth state (i.e., default state) is absorbent. The product-limit estimator (3.79) converges asymptotically the true transition probability matrix $P(s,t)$ when the number of sample n gets larger; see more details in Andersen et al. (1995, Section 4.4).

Continuous-time setting

Though the product-limit estimator (3.79) has nice statistical properties, it is not very helpful in credit risk management since the observed rating records are not frequent enough during small time intervals. Xing et al. (2012b) proposed a piecewise time-homogeneous continuous-time Markov chain model in which the generator matrices $\Lambda(t)$ are piecewise constant and jumps in $\Lambda(t)$ indicate changes of patterns of credit rating dynamics. The discretization of the model allows efficient and recursive estimators of time-varying generator matrices and is practically applicable in analysis of time-inhomogeneous rating transitions.

Specifically, assume that the generator matrices $\Lambda(t)$ are piecewise constant and the changes in $\Lambda(t)$ follow a Poisson process $\{N_\Lambda(t); t \geq 0\}$ with a constant rate η. Then the duration between two adjacent structural breaks follows an exponential distribution with mean $1/\eta$. The generator matrix between two adjacent change times is constant and it is expressed as $\Lambda(t) = Q_{N_\Lambda(t)}$ at time t, where $Q_1, Q_2, \ldots$ are independent and identically distributed random generator matrices such that the off-diagonal elements $\lambda^{(i,j)}$ follow independently a Gamma(α_{ij}, β_i) prior distribution with the density function

$$g(\lambda^{(i,j)}) = \frac{\beta_i^{\alpha_{ij}}}{\Gamma(\alpha_{ij})} \left[\lambda^{(i,j)}\right]^{\alpha_{ij}-1} \exp(-\lambda^{(i,j)}\beta_i), \qquad (i,j) \in \mathcal{K}, \qquad (3.80)$$

in which $\mathcal{K} = \{(i,j)|i \neq j, 1 \leq i \leq K-1, 1 \leq j \leq K\}$. This assumption indicates that the time-dependent credit rating transition matrix $P(s,t)$ during the period (s,t) can be characterized in the following way.

If $\Lambda(t)$ undergoes M changes during the period (s,t), say the change times are $\tau_1 < \cdots < \tau_M$, the transition matrix during the period (s,t) can be characterized as

$$P(s,t) = \prod_{k=1}^{M+1} P(\tau_{k-1}, \tau_k) = \prod_{k=1}^{M+1} \exp\left(\int_{\tau_{k-1}}^{\tau_k} \Lambda(u)du\right) = \prod_{k=1}^{M+1} \exp\left[(\tau_k - \tau_{k-1})\Lambda(\tau_k-)\right],$$

in which $\tau_0 = s, \tau_{M+1} = t$. Note that the exponent in the above equation usually can not be simplified to $P(s,t) = \exp\left[\int_{\tau_0}^{\tau_{M+1}} \Lambda(u)du\right]$, because the components $P(\tau_0, \tau_1), \ldots, P(\tau_M, \tau_{M+1})$ may not commute. If $\Lambda(t)$ has no changes over the period (s,t), then $P(s,t)$ becomes homogeneous and can be characterized by

$$P(s,t) = \exp\left(\int_s^t \Lambda(u)du\right) = \exp\left[(t-s)\Lambda(t-)\right].$$

In such a case, since we assume that the elements of $\Lambda(t)$ follow the conjugate Gamma priors (3.80), the posterior distribution of entries $\lambda_{s,t}^{(i,j)}$ $(i \neq j)$ in $\Lambda(t)$ given transition history over the period (s,t) is Gamma$(K_{s,t}^{(i,j)} + \alpha_{ij}, S_{s,t}^{(i)} + \beta_i)$, where $K_{s,t}^{(i,j)}$ denotes the number of transitions from category i to category j and $S_{s,t}^{(i)}$ is the amount of time spent in category i; see Section 3.6.1.

Statistical inference

Suppose that there are n realizations of a Markov chain with generator matrices $\Lambda(t)$ during the period $(0, T)$ and firms' rating migrations from state i to state j at the period (s,t) are conditional independent given the generator matrix $\Lambda(t)$. The inference on piecewise-constant generator matrices $\Lambda(t)$ can be made as follows.

First, consider an evenly spaced partition for the period $(0, T)$, $0 = t_0 < t_1 < \cdots < t_L = T$, and assume that changes in $\Lambda(t)$ can only happen at the times $t_1, \ldots, t_L$. Let $I_1 = 1$ and $I_l = N_\Lambda(t_l-) - N_\Lambda(t_{l-1}-)$ for $l = 2, \ldots, L$. Then $I_l = 1$ if $\Lambda(t)$ are same at periods (t_{l-2}, t_{l-1}) and (t_{l-1}, t_l), and $\{I_l\}$ is a sequence of independent and identically distributed Bernoulli random variables with success probability $p = 1 - \exp(-\eta T / L)$. It is reasonable to assume that there is at most one structural break at time t_l and all changes in $\Lambda(t)$ can be identified if the partition of $(0, T)$ is fine enough.

Denote by $\mathcal{Y}_{s,t}$ the observed rating transitions over the period (s, t). Let $R_l = \max\{t_{m-1} | I_m = 1, m \le l\}$ be the most recent time of change in $\Lambda(t)$ up to time t_{l-1}. Then the conditional distribution of $\lambda^{(i,j)}_{t_{m-1}, t_l}$ given $R_l = t_{m-1}$ and $\mathcal{Y}_{t_{m-1}, t_l}$ is $\mathrm{Gamma}(K^{(i,j)}_{t_{m-1}, t_l} + \alpha_{ij}, S^{(i)}_{t_{m-1}, t_l} + \beta_i)$. Let $p_{m,l} = P(R_l = t_{m-1} | \mathcal{Y}_{t_{m-1}, t_l})$. It can be shown that the posterior distribution of $\lambda^{(i,j)}_{t_{l-1}, t_l}$ given $\mathcal{Y}_{(0, t_l)}$ can be expressed as a mixture of Gamma distributions,

$$\lambda^{(i,j)}_{t_{l-1}, t_l} | \mathcal{Y}_{(0, t_l)} \sim \sum_{m=1}^{l} p_{m,l} \mathrm{Gamma}(K^{(i,j)}_{t_{m-1}, t_l} + \alpha_{ij}, S^{(i)}_{t_{m-1}, t_l} + \beta_i), \qquad (3.81)$$

in which the mixture weight can be calculated recursively by $p_{m,l} = p^*_{m,l} / \sum_{m=1}^{l} p^*_{m,l}$ and

$$p^*_{m,l} = \begin{cases} p\pi_{00}/\pi_{ll} & m = l, \\ (1 - p)p_{m,l-1}\pi_{m,l-1}/\pi_{m,l} & m < l. \end{cases} \qquad (3.82)$$

The terms $\pi_{m,l}$ and $\pi_{0,0}$ in the above equation are expressed as

$$\pi_{00} = \prod_{i,j \in \mathcal{K}} \frac{\beta_i^{\alpha_{i,j}}}{\Gamma(\alpha_{ij})} \qquad \pi_{ml} = \prod_{i,j \in \mathcal{K}} \frac{(S^{(i)}_{t_{m-1}, t_l} + \beta_i)^{(K^{(i,j)}_{t_{m-1}, t_l} + \alpha_{ij})}}{\Gamma(K^{(i,j)}_{t_{m-1}, t_l} + \alpha_{ij})}.$$

The forward estimates of generator matrices (3.81) also provide the probability that the most recent change-point occurs in the period $[t_{h-1}, t_l]$ up to time t_l:

$$P(R_l \in [t_{h-1}, t_l] | \mathcal{Y}_{0, t_l}) = \sum_{m=h}^{l} p_{m,l}. \qquad (3.83)$$

Analogous to the Gamma mixture distribution (3.81), one can obtain the posterior distribution of $\lambda^{(i,j)}_{t_{l-1}, t_l}$ given $\mathcal{Y}_{(t_l, T)}$. Combining these two Gamma mixtures, we obtain the posterior distribution of $\lambda^{(i,j)}_{t_{l-1}, t_l}$ given $\mathcal{Y}_{(0, T)}$ for $1 \le l < L$,

$$\lambda^{(i,j)}_{t_{l-1}, t_l} | \mathcal{Y}_{(0, T)} \sim \sum_{1 \le m \le l \le k \le L} q_{mlk} \mathrm{Gamma}(K^{(i,j)}_{t_{m-1}, t_k} + \alpha_{ij}, S^{(i)}_{t_{m-1}, t_k} + \beta_i), \qquad (3.84)$$

in which q_{mlk} is the probability that the most recent times of change up to and down to t_l are t_m and t_k, respectively. Some algebra shows that $q_{mlk} = q_{mlk}^* / \sum_{1 \le m \le l \le k \le L} q_{mlk}^*$,

$$q_{mlk}^* = \begin{cases} pp_{m,l} & m \le l = k, \\ (1-p)p_{m,l}\widetilde{p}_{k,l+1}\pi_{m,l}\pi_{l+1,k} / (\pi_{m,k}\pi_{00}) & m \le l < k, \end{cases}$$

and $\widetilde{p}_{k,l+1} = \widetilde{p}_{k,l+1}^* / \sum_{k=l+1}^{L} \widetilde{p}_{k,l+1}^*$,

$$\widetilde{p}_{k,l+1}^* = \begin{cases} p\pi_{00} / \pi_{l+1,l+1} & k = l+1, \\ (1-p)\widetilde{p}_{k,l+2}\pi_{l+2,k} / \pi_{l+1,k} & k > l+1. \end{cases}$$

Then the generator matrix $\Lambda(t)$ at period (t_{l-1}, t_l) can be estimated by its posterior mean,

$$\widehat{\lambda}_{t_{l-1},t_l}^{(i,j)} = \sum_{1 \le m \le l \le k \le L} \frac{\pi_{mlk}(K_{t_{m-1},t_k}^{(i,j)} + \alpha_{ij})}{S_{t_{m-1},t_k}^{(i)} + \beta_i}, \qquad l = 1, \ldots, L-1,$$

and

$$\widehat{\lambda}_{t_{L-1},t_L}^{(i,j)} = \sum_{m=1}^{L} \frac{p_{m,l}(K_{t_{m-1},t_L}^{(i,j)} + \alpha_{ij})}{S_{t_{m-1},t_L}^{(i)} + \beta_i}, \qquad l = L.$$

Making use of the above estimates, the one-period probability transition matrix $P(t_{l-1}, t_l)$ can be estimated by $\widehat{P}(t_{l-1}, t_l) = \exp\left[(t_l - t_{l-1})\widehat{\Lambda}(t_l)\right]$ and the multi-period probability transition matrix $P(t_{m-1}, t_k)$ can be estimated by

$$\widehat{P}(t_{m-1}, t_k) = \prod_{l=m}^{k} P(t_{l-1}, t_l) = \prod_{l=m}^{k} \exp\left[(t_l - t_{l-1})\widehat{\Lambda}(t_l)\right].$$

3.6.3 Modulated Markov and semi-Markov rating transitions

In contrast to the default event in credit risk analysis which usually occurs once during the lifetime of a firm, firms' rating transitions are recurrent events and happens multiple times due to changes of firms' financial healthiness and economic environment. Additionally, due to the fact that the number of rating categories is usually more than one, rating transitions are also events involving *competing risks*. Here, competing risk refers to the situation where individuals or subjects are exposed to multiple mutually exclusive causes of failure, and the occurrence of one type of failure precludes the occurrence of other types of failures. The Markov chain models in the previous section assume homogeneous firms in the analysis, which is clearly not practical. To study rating transition risk for heterogeneous firms, modulated Markov and semi-Markov models can be used. This section introduces first the statistical concept of recurrent events and competing risk, and then modulated Markov and semi-Markov models for firms' rating transition intensities.

Recurrent events

For convenience, consider the case that there is only one type of event for a firm, or specifically, transition to a specific non-absorbing rating category $\mathcal{R}$. Suppose that, starting from a well-defined time origin, one observes a point process $(\tau_1, \tau_2, \dots)$ for a firm, where $\tau_i < C$ is the time to the firm's ith transition to $\mathcal{R}$, $i = 1, 2, \dots$, and C is the total follow-up time that right-censors the point process. There may also be a baseline covariate process $\mathbf{x}(t)$. Denote by $\widetilde{N}(t)$ the number of transitions to $\mathcal{R}$ by follow-up time t, and by $N(t)$ the corresponding observed number of transitions to $\mathcal{R}$ in $(0, t]$. Note that $\widetilde{N}(t)$ may be less than $N(t)$ because of censoring. Then one can define the increments in the cumulative intensity process, given the covariate history $\mathbf{x}(t)$, by

$$d\Lambda(t) = P\big[d\widetilde{N}(t) = 1 | \widetilde{N}(u), 0 \leq u < t, \mathbf{x}(t)\big]. \tag{3.85}$$

In the case of a continuous-time process having only unit jumps, we have $\Lambda(t) = \int_0^t \lambda(t)dt$. Independent censorship requires that the condition $C \geq t$ can be added to the conditioning event without altering (3.85). This independent censorship assumption allows the censoring rate at time t to depend on the preceding counting and covariate process histories. Let $Y(t) = 1_{\{0 < t \leq C\}}$. Independent censorship then requires that

$$E\big[dN(t) | N(u), Y(u); 0 \leq u < t, \mathbf{x}(t)\big] = Y(t)d\Lambda(t),$$

so that the failure intensity at time t for the observed point process N is equal to that (3.85) for the underlying process $\widetilde{N}$ at all $t < C$.

Various models for firms' rating transitions can be considered, depending on the questions to be addressed in practice. These models are distinguished primarily by the choice of conditioning events. For instance, under independent censorship, one could specify a Cox-type model for the intensity process of transitions to $\mathcal{R}$, $d\Lambda(t) = d\Lambda_0(t) \exp\big[\boldsymbol{\beta}'\mathbf{x}(t)\big]$, or other survival regression models in Section 3.4. Inference under these model are essentially same as those for univariate failure time data; see details in Section 3.4.

Competing risks

Firms' transitions to a specific rating category is only one of the several transition events due to the multiplicity of rating categories. Such data in statistics are commonly referred to as *competing risks data*. Suppose a firm has an underlying transition time τ that may be subject to censoring, and a basic covariate vector $\mathbf{x}$, or more generally, a covariate function $\{\mathbf{x}(u) : u \geq 0\}$. Moreover, when the transition occurs, it may be of any one of the K distinct types or causes denoted by $J \in \{1, 2, \dots, K\}$. There are three kinds of problems in the analysis of rating transitions involving competing risks. The first problem deals with how to estimate the relationship between covariates and the rate of transitions to specific rating categories. The second problem is to characterize the interrelation between transition types under a specific set of study conditions. The third problem concerns the estimation of transition intensities for certain types of transitions given the removal of some or all other transition types.

To model competing risks, we consider a type-specific hazard function or process

$$\lambda_j[t; \mathbf{x}(t)] = \lim_{h \to 0+} \frac{P(t \le \tau < t + h, J = j \mid \tau \ge t, \mathbf{x}(t))}{h}$$

for $j = 1, \ldots, K$ and $t > 0$. If only one of the transition type $\{1, 2, \ldots, K\}$ can occur at time t, then the overall hazard rate is $\lambda[t; \mathbf{x}(t)] = \sum_{j=1}^{K} \lambda_j[t; \mathbf{x}(t)]$, and the overall survivor function is

$$S(t; \mathbf{x}) = P(\tau > t | \mathbf{x}) = \exp\left[- \int_0^t \lambda(u; \mathbf{x}) du \right]. \tag{3.86}$$

The density function for the time to a type j transition is

$$f_j(t; \mathbf{x}) = \lambda_j(t; \mathbf{x}) S(t; \mathbf{x}), \qquad j = 1, \ldots, K, \tag{3.87}$$

and the density function of the time to transition is $f(t; \mathbf{x}) = \sum_{j=1}^{K} f_j(t; \mathbf{x})$.

Analogous to modeling recurrent events, one can use the survival regression models in 3.4 to model the intensity processes in competing risk. For instance, one may consider a Cox model for the cause-specific hazard functions $\lambda_j[t; \mathbf{x}(t)] = \lambda_{0j} \exp[\boldsymbol{\beta}' \mathbf{x}(t)]$, $j = 1, \ldots, K$, where $\mathbf{x}(t)$ is a vector of p derived covariates. Then, inference on the regression coefficient $\boldsymbol{\beta}$ and the unknown baseline intensity λ_{0j} can be drawn using the parametric or the semiparametric approaches in Section 3.4.

Reduced-form models for rating transitions

Combining models of recurrent events and competing risks, one can obtain reduced-form models of rating transition probabilities for heterogeneous firms. Suppose there are n heterogeneous firms in the market and a common probability space $(\Omega, \mathcal{G}, \mathbb{P})$. Let $\mathcal{G} = \{\mathcal{G}_t : t \ge 0\}$ be a complete information filtration, which contains three types of information. The first type, denoted as $\mathcal{M}_t$, consists of observed and unobserved macroeconomic variables or events. The second type, denoted as $\mathcal{B}_t$, is produced by the collection of all firms' (or borrowers') observed and unobserved covariates and events up to time t. For convenience, we assume that $\mathcal{M}_t$ and $\mathcal{B}_t$ are independent. The third type, denoted as $\mathcal{S}_t$, characterizes the time variation of market and economic environment, and summarizes the mechanism that microeconomic variables or events interact with macroeconomic variables or events. Given $\mathcal{M}_t$, $\mathcal{B}_t$, and $\mathcal{S}_t$, the complete-information filtration $\mathcal{G}_t$ is the σ-algebra generated by these three sets, $\mathcal{G}_t = \sigma\{\mathcal{M}_t \cup \mathcal{B}_t \cup \mathcal{S}_t\}$. Besides, $\mathcal{M}_t$, $\mathcal{B}_t$ and $\mathcal{S}_t$ are mutually independent.

For a firm l ($l = 1, \ldots, n$), denote by $P_l(s,t)$ firm l's rating transition probability matrix during the period (s,t), in which the ijth element of $P_l(s,t)$ represents the probability that a firm starting in state i at time s is in state j at time t. Let $A_l(t)$ be the rating category of firm l at time t, and $N^*_{ijl}(t)$ the number of transitions from rating category i to rating category j of the firm l that occur over the interval $(0,t]$ for $i, j \in \{1, \ldots, K\}, j \neq i$.

Let $\{\mathbf{x}_l(t)\}$ be a p-variate observable firm-specific covariate process during the sample period. Let $\mathcal{B}^{\mathrm{obs}}_{ijl,t}$ be the filtration generated by $\{\mathbf{x}_l(s) : e_{l,0} \leq s \leq t\}$, $\mathcal{N}_{l,t}$ the filtration generated by $\{N^*_{ijl}(s) : 1 \leq i \neq j \leq K, e_{l,0} \leq s \leq t\}$, and $\lambda^{(i,j)}_l(t)$ the intensity function of $N^*_{ijl}(t)$ associated with $\mathcal{B}^{\mathrm{obs}}_{ijl,t} \cup \mathcal{N}_{l,t}$. Note that $\mathcal{B}^{\mathrm{obs}}_t := \cup_{i,j,l} \mathcal{B}^{\mathrm{obs}}_{ijl,t}$ is only a subset of $\mathcal{B}_t$, as it does not contain firms' unobserved covariates. Assume that $\mathbf{y}(t)$ is a vector of macroeconomic variables observed at time t and $\mathcal{M}^{\mathrm{obs}}_t$ is the filtration generated by $\mathbf{y}(t)$. We denote $\mathcal{M}^{\mathrm{unobs}}_t$ the set of unobserved variables or events in $\mathcal{M}_t$, then $\mathcal{M}_t$ is the filtration generated by $\mathcal{M}^{\mathrm{obs}}_t$ and $\mathcal{M}^{\mathrm{unobs}}_t$. Let $\mathcal{F}_{l,t}$ be the information filtration generated by the observed variables $\left\{ \cup_{i,j,l} \mathcal{B}^{\mathrm{obs}}_{ijl,s}; e_{l,0} \leq s \leq \min(t, e_{l,1}) \right\} \cup \left\{ \mathcal{M}^{\mathrm{obs}}_s; 0 \leq s \leq t \right\}$. Then, the econometrician's information filtration is the union of $\mathcal{F}_{l,t}$ and firm's transition history $\mathcal{N}_{l,t}$, that is, $\mathcal{F}_{l,t} \cup \mathcal{N}_{l,t}$. Given these information, intensity processes for firm l's rating transitions can be expressed as

$$E\{dN^*_{ijl}(t) | \mathcal{F}_{l,t}, \mathcal{N}_{l,t}, \mathcal{S}\} = \lambda^{(i,j)}(t; \mathbf{x}_l(t), \mathbf{y}(t)) dt,$$

in which $dN^*_{ijl}(t)$ is the increment $N^*_{ijl}\{(t + dt)\} - N^*_{ijl}(t)$ of $N^*_{ijl}(t)$ over the small interval $[t, t + dt)$ and $\lambda^{(i,j)}(t; \mathbf{x}_l(t), \mathbf{y}(t))$ is an intensity function that is usually specified as survival regression on covariate $\mathbf{x}_l(t)$ and $\mathbf{y}_t$. Once a parametric or semi-parametric intensity function is specified for $\lambda^{(i,j)}(\cdot)$, its inference can be made by using the methods in Section 3.4.

The above model assumes that all covariates or risk factors are observable, which introduces a downward biased estimate of tail portfolio losses. To relax such restriction, Duffie et al. (2009) included unobserved covariates in $\mathbf{x}_l(t)$ and $\mathbf{y}_t$ (i.e., frailties) to the intensity process and assumes marginal intensities for firms' default (or generally, rating transitions),

$$E\{dN^*_{ijl}(t) | \mathbf{X}_l(t), \mathbf{Y}(t), \mathcal{N}_{l,t}, \mathcal{S}_t\} = \lambda^{(i,j)}(\mathbf{X}_l(t), \mathbf{Y}(t); \boldsymbol{\theta}^{(i,j)}) dt;$$

see discussion on the dynamic frailty in Section 3.8.4.

3.7   Portfolio credit risk

An important development in derivative markets since the 1990s has been the expansion of credit derivatives. These instruments enable companies to trade credit risks similar to how they trade market risks. Through the purchase and sale of credit derivative contracts, banks and financial institutions can actively oversee their credit risk portfolios and safeguard against potential losses. Retail banks typically seek protection against credit events, making them net buyers of credit protection. Conversely, other investors like hedge funds and investment banks often assume dual roles as both sellers and buyers of credit protection.

There are generally two kinds of credit derivatives, *single-name* and *multi-name*. The most popular single-name credit derivative is a credit default swap. The payoff from this instrument depends on the creditworthiness of one company or country. There are several types of multi-name credit derivatives, one popular product is *collateralized debt obligation* (CDO). In a CDO, a selection of debt instruments forms a portfolio, and a complex structure is established to distribute cash flows from this portfolio among different investor groups. This section provides a brief introduction to these two types of credit derivatives.

3.7.1 Credit default swaps

A *credit default swap* (CDS) is an agreement between a buyer and a seller. The buyer obtains the right to sell bonds issued by a specific company or country, a *single name* that is called the *reference entity*, while the seller of the CDS contract commits to purchasing these bonds at face value if a credit event occurs.

The face value of a coupon-bearing bond is the principal that the protection seller repays at maturity if there is no default. The total face value of the bonds is the notional principal of the CDS. The buyer of the CDS makes periodic payments (usually quarterly) to the seller until the expiration date of the CDS or until a credit event occurs. If the reference entity has no credit event by the expiration date, the protection buyer receives no payoff from the protection seller. However, if the reference entity has a credit event at time τ prior to the expiration, the protection seller makes a payment to the protection buyer. In this way, the protection buyer acquires financial protection against the loss on the credit event of the reference entity.

CDS contracts are usually held for protection purpose. In such a case, bond investors with a significant credit exposure to the reference entity may buy CDS protection to shield themselves from potential losses due to the default of a bond issued by the reference entity. Besides this protective function, investors also engage in CDS transactions for speculative purposes. For instance, the protection buyer may not necessarily own the bond issued by the reference entity. In such a case, the buyer speculates on the widening of the credit spread of the reference entity.

CDS spread

We next explain how CDS spread is determined. Consider a CDS contract with notional value one. Suppose that M premia are paid up to the expiration time T at $0 < t_1 < \cdots < t_M = T$. Denote by τ the default (or credit event) time of the reference entity and x the CDS spread. If $\tau > t_k$ for some $1 \leq k \leq M$, the protection buyer make premium payment $x(t_k - t_{k-1})$ at t_k and no payment after τ. When $\tau < t_M$, the protection seller pays δ to the buyer at the default time τ.

Denote by $\lambda^{\mathbb{Q}}$ the risk-neutral hazard function. Then, the *premium payment leg* of the swap (i.e., the total payments made by the protection buyer discounted to time t) is

$$V_{t,\text{buy}}(x; \lambda^{\mathbb{Q}}) = E^{\mathbb{Q}}\left(\sum_{k \,:\, t_k > t} \exp\left(-\int_t^{t_k} r_u du \right) x(t_k - t_{k-1}) 1_{\{t_k < \tau\}} \Big| \mathcal{G}_t \right)$$

$$= x \sum_{k \,:\, t_k > t} p_0(t, t_k)(t_k - t_{k-1}) P^{\mathbb{Q}}(\tau > t_k | \mathcal{G}_t),$$

in which $p_0(t, t_k)$ is the price of the default-free zero-coupon bond with maturity t_k defined by (3.26), and

$$P^{\mathbb{Q}}(\tau > t_k | \mathcal{G}_t) = 1_{\{\tau > t\}} \exp\left(-\int_t^{t_k} \lambda^{\mathbb{Q}}(\mathbf{x}_s) ds \right).$$

When the default or the credit event occurs at $\tau \in (t, t_M]$, the *default payment leg* of the swap (i.e., the payment made by the protection seller at the default or credit event time discounted to time t) is

$$V_{t,\text{sell}}(\lambda^{\mathbb{Q}}) = E^{\mathbb{Q}}\left(\exp\left(-\int_t^{\tau} r(u) du \right) \delta 1_{\{t < \tau \leq t_M\}} \Big| \mathcal{G}_t \right)$$

$$= 1_{\{t < \tau\}} \delta \int_t^{t_M} \lambda^{\mathbb{Q}}(s) \exp\left(-\int_t^{s} (r(u) + \lambda^{\mathbb{Q}}(u)) du \right) ds.$$

Then the CDS spread x should be chosen such that the value of the contract is equal to zero, i.e., $V_{t,\text{buy}}(x; \lambda^{\mathbb{Q}}) = V_{t,\text{sell}}(\lambda^{\mathbb{Q}})$. Solving this equation gives the so-called *fair CDS spread* x^*,

$$x_t^* = \frac{\delta \int_t^{t_M} \lambda^{\mathbb{Q}}(s) \exp\left(-\int_t^{s}(r + \lambda^{\mathbb{Q}})(\mathbf{x}_u) du \right) ds}{\sum_{k \,:\, t_k > t} p_0(t, t_k)(t_k - t_{k-1}) \exp\left(-\int_t^{t_k} \lambda^{\mathbb{Q}}(\mathbf{x}_s) ds \right)}.$$

CDS are typically traded over the counter and commonly associated with counterparty risk. However, CDS are also useful risk-management tools. Due to the liquidity of CDS markets, CDS are the natural underlying security for many more complex credit derivatives. Models for pricing portfolio-related credit derivatives are usually calibrated to quoted CDS spreads.

3.7.2   Collateralized debt obligations

Collateralized debt obligations (CDOs) represent a category of structured financial instrument that comprises various assets and loan products. They are created by investment banks that collect cash-flow generating assets and repackage them (through securitization) into different tranches. These tranches are differentiated based on the level of credit risk assumed by the investor and are subsequently marketed to institutional investors. The nomenclature of the structured product varies depending on the nature of the underlying assets. For example, if the asset pool consists mainly bonds the product is called *collateralized bond obligations*; similarly, if the asset pool comprises predominantly loans, it is referred to as a *collateralized loan obligation*.

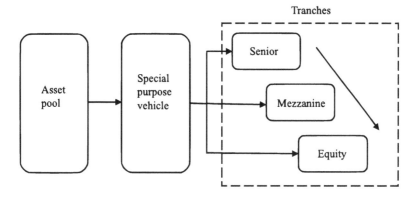

FIGURE 3.6: Schematic representation of the payments in a CDO structure.

Figure 3.6 illustrates the payment structure within a CDO. Initially, originating banks sell a portfolio of cash-flow generating assets, such as loans, to a *special purpose vehicle* (SPV). Then, the SPV issues securities comprising different tranches, typically encompassing senior, mezzanine, and equity tranches. Cash flows are distributed to tranches by rules known as a *waterfall*, which govern the allocation of principal and interest payments separately. In general, losses resulting from credit events are initially absorbed by the equity tranche, followed by the mezzanine tranche if the equity tranche is depleted, and finally by the senior tranche. Pricing tranches of CDOs typically involve models such as the one-factor Gaussian copula model, structural models, and/or conditional survival models. Some of these models are briefly introduced in the next section. Interested readers can find further details in references such as Hull (2021, Chapter 25) and McNeil et al. (2015, Chapter 12).

3.8　Mixture and copula models for static and dynamic frailty

The reduced-form credit models introduced in Sections 3.5.2 and 3.6.3 assume either no covariates or observable covariates for individual firms' credit status, and hence characterize individual firms' intensities of default or rating transitions explicitly. This usually works well for default or rating transitions in individual firms, but becomes difficult in accounting for the joint distributions of default or other credit events for a group of firms. A simple idea to overcome such difficult is to introduce frailty variables as common factors, so that the unconditional distribution of credit events among the set of firms are dependent due to the common factors, but they are conditional independent given the common factors. In this section, we introduce mixture models and copula approaches to model credit risk in which credit events are assumed to be conditionally independent given a set of unobserved common factors. The factors are usually interpreted as macroeconomic variables and can be either static variables or dynamic stochastic processes.

3.8.1　Bernoulli mixture models

Assume that there are n obligors in the portfolio and the Bernoulli random variable $Y_i = 1$ represents the ith obligor has a default. Denote by $p_i = P(Y_i = 1)$ the marginal probability that the ith obligor defaults, e_i and c_i the exposure and the loss given default of the ith obligor. Then the total loss of the portfolio is calculated as $L = \sum_{i=1}^{n} e_i c_i Y_i$. For convenience, assume that e_i and c_i are deterministic. To compute the loss distribution F_L, one can use the following Bernoulli mixture model.

Suppose that $\mathbf{X} = (X_1, \ldots X_d)'$ is a d-dimensional random vector, and $Y_1, \ldots, Y_n$ are independent Bernoulli random variables conditional on $\mathbf{X}$. In particular, there exist functions $p_i : \mathbb{R}^d \to [0, 1]$, $1 \leq i \leq n$, such that the success probability of Y_i conditional on $\mathbf{X} = \mathbf{x}$ is $p_i(\mathbf{x})$, i.e., $P(Y_i = 1 | \mathbf{X} = \mathbf{x}) = p_i(\mathbf{x})$. Then the joint probability function of $\mathbf{Y} = (Y_1, \ldots, Y_n)'$ follows a Bernoulli mixture model and is expressed as

$$P(\mathbf{Y} = \mathbf{y} | \mathbf{X} = \mathbf{x}) = \prod_{i=1}^{n} p_i(\mathbf{x})^{y_i} (1 - p_i(\mathbf{x}))^{1-y_i}, \quad \mathbf{y} = (y_1, \ldots, y_n)' \in \{0, 1\}^n.$$

The unconditional joint distribution of $\mathbf{Y}$ is obtained by integrating over the distribution of the vector $\mathbf{X}$. Besides, the unconditional probability of $Y_i = 1$ is given by $p_i = P(Y_i = 1) = E(p_i(\mathbf{X}))$.

Given the above setup, the loss distribution F_L for $L = \sum_{i=1}^{p} e_i c_i Y_i$ can be computed in two steps. First, given the conditional independence of defaults, the Laplace-Stieltjes transform of F_L can be calculated as follows,

$$E(e^{-tL}) = E_{\mathbf{X}}\left(E\left(\exp\left(-t\sum_{i=1}^{n} e_i c_i Y_i\right)\Big|\mathbf{X}\right)\right) = E_{\mathbf{X}}\left(\prod_{i=1}^{n} E(e^{-te_i c_i Y_i}|\mathbf{X})\right)$$

$$= E_{\mathbf{X}}\left(\prod_{i=1}^{n}\left[p_i(\mathbf{x})e^{-te_i c_i} + 1 - p_i(\mathbf{x})\right]\right). \tag{3.88}$$

Note that the expectation in the last equality can be calculated numerically by integrating over the distribution of the factors $\mathbf{X}$. Second, compute the unconditional distribution F_L using the inverse Laplace transformation.

In practice, the main difficulty of the above procedure comes from the valuation of the expectation in the last equality of (3.88), especially when the common factors $\mathbf{X}$ is high-dimensional.

One-factor Bernoulli mixture models

An important special case of Bernoulli-mixture models is the one-factor model in which the factor X is one dimensional. Assume that $X \in \mathbb{R}$. Then conditional on X, the default indicator $\mathbf{Y}$ is a vector of independent Bernoulli random variables with $P(Y_i = 1 | X = x) = p_i(x)$. For simplicity, we assume that all exposures e_i and loss given default c_i are identical for all firms, i.e., $e_i = e$ and $c_i = c$ for $i = 1, \ldots, p$. Then the loss of the portfolio L is equivalent to the number of defaults of obligors M in the portfolio, i.e., $M = L/(ec) = \sum_{i=1}^{n} Y_i$.

Suppose that all firms' default probabilities conditional on the factor X are identical, say $Q := p_1(X) = \cdots = p_n(X)$. Denote by F_Q the distribution function of Q. Note that conditional on $Q = q$, the number of defaults M is the sum of p independent Bernoulli variables with parameter q, hence it has a binomial distribution Binomial(n, p), i.e., $P(M = m|Q = q) = \binom{n}{m} q^n (1 - q)^{n-m}$. The unconditional distribution of M is obtained by integrating over q, that is,

$$P(M = m) = \binom{n}{m} \int_0^1 q^m (1 - q)^{n-m} dF_Q(q).$$

Let $\pi_m = P(Y_1 = 1, \ldots, Y_m = 1)$. It is easy to see that $\pi_m = E(E(Y_1 \cdots Y_m|Q)) = E(Q^m)$ for $m = 1, \ldots, n$. Moreover, for $i \neq j$, $\mathrm{Cov}(Y_i, Y_j) = \pi_2 - \pi_1^2 = \mathrm{Var}(Q) \geq 0$. This indicates the default correlation of obligor i and j is expressed as

$$\rho_Y := \rho(Y_i, Y_j) = \frac{E(Y_i Y_j) - p_i p_j}{\sqrt{(p_i - p_i^2)(p_j - p_j^2)}} = \frac{\pi_2 - \pi_1^2}{\pi_1 - \pi_1^2} > 0.$$

Then by choosing properly the mixing distribution F_Q, any value of ρ_Y in $[0, 1]$ can be obtained.

One widely used mixing distribution F_Q is the *Beta mixing distribution*. In this case, Q follows a Beta distribution $\text{Beta}(a,b)$ whose density function is given by

$$F_Q(q) = \frac{1}{B(a,b)} q^{a-1}(1-q)^{b-1}, \quad 0 < q < 1, \; a, b > 0,$$

in which $B(a,b)$ denotes the beta function. For this distribution, the joint probability $\pi_k = P(Y_1 = 1, \ldots, Y_k = 1)$ can be computed explicitly,

$$\pi_m = \frac{1}{B(a,b)} \int_0^1 q^m q^{a-1}(1-q)^{b-1} dq = \frac{B(a+m,b)}{B(a,b)} = \prod_{j=0}^{m-1} \frac{(a+j)}{(a+b+j)}.$$

In particular, this yields that

$$\pi_1 = \frac{a}{a+b}, \quad \pi_2 = \frac{\pi_1(a+1)}{(a+b+1)}, \quad \rho_Y = \frac{1}{(a+b+1)}.$$

Then the number of defaults M has a so-called beta-binomial distribution whose probability is given explicitly by

$$P(M = m) = \binom{n}{m} \frac{1}{B(a,b)} \int_0^1 q^{m+a-1}(1-q)^{n-m+b-1}$$

$$dq = \binom{n}{m} \frac{B(a+m, b+n-m)}{B(a,b)}.$$

Besides the Beta mixing distribution, one may also use the *probit-normal mixing distribution* and the *logit-normal mixing distribution*. For the probit-normal mixing distribution, we have $Q = (\mu + \sigma X)$ for $X \sim N(0,1)$, $\mu \in \mathbb{R}$ and $\sigma > 0$, where Φ is the standard normal distribution function. For the logit-normal mixing distribution, we have $Q = F(\mu + \sigma X)$ for $X \sim N(0,1)$, $\mu \in \mathbb{R}$ and $\sigma > 0$, where $F(x) = 1/(1+e^{-x})$ is the logistic function. For these two mixing distributions, π_k and $P(M = m)$ have no explicit formulas, and calculations of them usually involve numerical evaluation of the integrals.

3.8.2   Poisson mixture models

In addition to the Bernoulli mixture models, another way to obtain a tractable model for the distribution of the portfolio loss is to use the Poisson approximation and independent gamma distributions for common factors, under the assumption that obligors' default probabilities are very small.

Consider a portfolio of n obligors and n Poisson random variables Y_i ($i = 1, \ldots, n$). Assume that the event $Y_i > 0$ (or $Y_i = 0$, respectively) indicates that the ith obligor has a default (or no default, respectively). Suppose that there are functions $\lambda_i : \mathbb{R}^p \to (0, \infty)$, $1 \le i \le n$ such that, the conditional distributions of Y_i given $\mathbf{X} \in \mathbb{R}^d$ are independent and follow the Poisson distribution with mean $\lambda_i(\mathbf{x})$. This indicates that the conditional probability of no default in obligor i can be approximated by

$$P(Y_i = 0|\mathbf{X} = \mathbf{x}) = e^{-\lambda_i(\mathbf{x})} \approx 1 - \lambda_i(\mathbf{x}),$$

and hence the conditional probability of default in obligor i is given by

$$P(Y_i > 0|\mathbf{X} = \mathbf{x}) = 1 - P(Y_i = 0|\mathbf{X} = \mathbf{x}) \approx \lambda_i(\mathbf{x}).$$

This defines a Poisson mixture model with factor $\mathbf{X}$ for the random vector $\mathbf{Y} = (Y_1, \ldots, Y_n)'$.

For simplicity, the parameter $\lambda_i(\mathbf{X})$ in the conditional Poisson distribution can be assumed to take the form

$$\lambda_i(\mathbf{X}) = k_i \mathbf{w}_i'\mathbf{X}, \qquad \text{for } \mathbf{X} = (X_1, \ldots, X_d)', \qquad (3.89)$$

in which $k_i > 0$ are constant, $\mathbf{w}_i = (w_{i1}, \ldots, w_{id})'$ are non-negative factor weights satisfying $\sum_j w_{ij} = 1$, and X_j $(j = 1, \ldots, d)$ are d independent Gamma(α_j, β_j)-distributed factors with $\alpha_j = \beta_j = \sigma_j^{-2}$ and density function $f(x) = \beta_j^{\alpha_j} x^{\alpha_j - 1} \exp(-\beta_j x)/\Gamma(\alpha_j)$. Under this parameterization, we have $E(X_j) = 1$ and $\text{Var}(X_j) = \sigma_j^2$. Hence, the expected number of defaults for obligor i is expressed as

$$E(Y_i) = E(E(Y_i|\mathbf{X})) = E(\lambda_i(\mathbf{X})) = k_i E(\mathbf{w}_i'\mathbf{X}) = k_i.$$

Distribution of the number of defaults

Given the parameterization (3.89), the distribution of Y_i conditional on $\mathbf{X} = \mathbf{x}$ can be denoted by $Y_i|\mathbf{X} = \mathbf{x} \sim \text{Poisson}(k_i \mathbf{w}i'\mathbf{x})$. Consequently, the distribution of the number of defaults $M = \sum_{i=1}^{n} Y_i$ given the factor $\mathbf{X} = \mathbf{x}$ follows a Poisson distribution with mean $\sum_{i=1}^{n} k_i \mathbf{w}_i'\mathbf{x}$. This is due to the property that the sum of independent Poisson variables is itself a Poisson variable, with a rate equal to the sum of the rates of the independent variables.

To compute the unconditional distribution of M, we use a property of mixed Poisson distributions. In particular, if a random variable Λ follows a Gamma distribution, $\Lambda \sim \text{Gamma}(\alpha, \beta)$, and conditional on Λ, the random variable N follows a Poisson distribution with mean Λ, then the unconditional distribution of N follows a negative binomial distribution $\text{NB}(\alpha, p)$ with $q = \beta/(1 + \beta)$ and the probability mass function

$$P(N = k) = \binom{\alpha + k - 1}{k} q^\alpha (1 - q)^k, \quad k = 0, 1, 2, \ldots.$$

We now use conditional distribution $M|\mathbf{X} = \mathbf{x} \sim \text{Poisson}(\sum_{i=1}^{n} k_i \mathbf{w}_i'\mathbf{x})$ and the above property to find the unconditional distribution of M. First, note that

$$\sum_{i=1}^{n} k_i \mathbf{w}_i'\mathbf{x} = \sum_{i=1}^{n} k_i \sum_{j=1}^{d} w_{ij} X_j = \sum_{j=1}^{d} Z_j, \quad Z_j := \left(\sum_{i=1}^{n} k_i w_{ij}\right) X_j.$$

Since X_j is assumed to follow a Gamma(α_j, β_j) distribution, the random variable Z_j follows a Gamma$(\alpha_j, \beta_j / \sum_{i=1}^{n} k_i w_{ij})$ distribution. Due to the independence of X_j, $\{Z_1, \ldots, Z_d\}$ are also independent, hence the conditional distribution of M given $\mathbf{Z} = (Z_1, \ldots, Z_d)'$ (or equivalently, given $\mathbf{X}$) is a Poisson distribution with mean $\sum_{j=1}^{d} Z_j$. This suggests that the unconditional distribution of M is same as that of the sum of d independent random variables M_j, in which the conditional distribution $M_j | Z_j = z_j \sim$ Poisson(z_i) and $Z_j \sim$ Gamma$(\alpha_j, \beta_j / \sum_{i=1}^{n} k_i w_{ij})$. Therefore, the unconditional distribution of M is the same as that of the sum of d independent negative binomial distribution NB$(\alpha_j, \beta_j + \sum_{i=1}^{n} k_i w_{ij})$; see Exercise 3.10.

Distribution of the aggregate loss

The Poisson mixture model can also be used to compute the distribution of the aggregate loss $L = \sum_{i=1}^{p} e_i c_i Y_i$. To do that, one can assume that $e_i c_i$ $(i = 1, \ldots, n)$ are multiples of a small amount $\epsilon > 0$, then the aggregate loss L equals to $L = \epsilon \sum_{i=1}^{p} l_i Y_i$, in which l_i is a positive integer multiplier. Then the unconditional distribution of L can be computed analogously as that of the number of defaults M in the discussion above.

3.8.3   Survivor and factor copulas

The Bernoulli and Poisson mixture models describe the distribution of the number of defaults and the aggregated loss via indicator variables of default. To get a full understanding of the joint distribution of defaults, multivariate survival functions need to be considered. We next present the concept of survival copula, and then introduce frailty and factor models for correlated defaults using survival copulas.

Survival copula

Suppose that there are n obligors in the portfolio and let τ_i be the time of the ith obligor gets default. Let $\boldsymbol{\tau} = (\tau_1, \ldots, \tau_n)'$ be a n-variate vector of random time with joint distribution function $F(t_1, \ldots, t_n)$ and marginal distribution functions $F_1(t_1), \ldots, F_n(t_n)$, and denote by $S_i(t) = P(\tau_i > t) = 1 - F_i(t)$ the survivor functions of τ_i. The joint survivor function of $\boldsymbol{\tau}$ is expressed as

$$S(t_1, \ldots, t_n) = P(\tau_1 > t_1, \ldots, \tau_n > t_n) = P(S_1(\tau_1) < S_1(t_1), \ldots, S_n(\tau_n) < S_n(t_n)).$$

Then Sklar's theorem in Section 2.2.4 also applies to the above multivariate survival function, and we have the identify

$$S(t_1, \ldots, t_n) = \overline{C}(S_1(t_1), \ldots, S_n(t_n)) \qquad (3.90)$$

for a *survival copula* $\overline{C}(u_1, \ldots, u_n)$, in which $u_1, \ldots, u_n \in [0, d]$.

Note that survival copulas are different from the survival functions of copulas, since the so called survival copulas are actually not copulas. In particular, suppose that $\mathbf{U}$ has distribution function C and the survival copula of C is $\overline{C}$, then the survival function of $\mathbf{U}$ is expressed as

$$P(U_1 > u_1, \cdots, U_d > u_d) = P(1 - U_1 \leq 1 - u_1, \ldots, 1 - U_d \leq 1 - u_d)$$
$$= \overline{C}(1 - u_1, \ldots, 1 - u_d).$$

This indicates that the survival copula $\overline{C}$ can be inferred from the copula of $\boldsymbol{\tau}$. For instance, in the case of $n = 2$, we have

$$S(x_1, x_2) = 1 - F_1(x_1) - !F_2(x_2) + F(x, y) = S_1(x_1) + S_2(x_2) - 1 + C(F_1(x_1), F_2(x_2))$$
$$= S_1(x_1) + S_2(x_2) - 1 + C(1 - S_1(x_1), 1 - S_2(x_2))$$

Hence

$$\overline{C}(u, v) = u + v - 1 + C(1 - u, 1 - v).$$

Example 3.5 (Survival copula of bivariate Pareto distribution) *An extension of the univariate Pareto distribution is the bivariate Pareto distribution with survivor function given by*

$$S(t_1, t_2) = \left(\frac{t_1 + \kappa_1}{\kappa_1} + \frac{t_2 + \kappa_2}{\kappa_2} - 1\right)^{-\alpha}, \qquad t_1, t_2 \geq 0, \alpha, \kappa_1, \kappa_2 > 0.$$

The marginal survivor functions can be computed and are given by

$$S_i(t) = \left(\frac{\kappa_i}{\kappa_i + t}\right)^{\alpha}, \qquad i = 1, 2.$$

We can infer from (3.90) that the survival copula is given by

$$\overline{C}(u_1, u_2) = (u_1^{-1/\alpha} + u_2^{-1/\alpha} - 1)^{-\alpha},$$

which is the Clayton copula.   □

Copulas are widely used in pricing multi-name credit derivatives such as collateralized debt obligations and basket credit derivatives, because the joint default (or event) times can be easily modeled via copulas. Specifically, for $t > 0$, let $\lambda_i(t)$ and $\Lambda_i(t) = \int_0^t \lambda_i(s)ds$ $(i = 1, \ldots, n)$ be the nonnegative hazard and cumulative hazard functions. Define the default time τ_i by

$$\tau_i = \inf\{t \geq 0; S_i(t) \leq U_i\} = \inf\{t \geq 0; \exp[-\Lambda_i(t)] \leq U_i\}, \qquad i = 1, \ldots, m,$$

so that obligor i defaults at the first time when the marginal survivor function $S_i(t)$ crosses the random threshold U_i. Let $\boldsymbol{\tau} = (\tau_1, \ldots, \tau_n)'$ be the vector of n default times, whose joint distribution follows a copula C (or equivalently, its joint survival function follows a survival copula $\overline{C}$). Then the joint distribution function of $\boldsymbol{\tau}$ is given by

$$P(\tau_1 \leq t_1, \ldots, \tau_n \leq t_n) = C(1 - \exp(-\Lambda_1(t_1)), \ldots, 1 - \exp(-\Lambda_n(t_n))),$$

and the joint survival function of $\boldsymbol{\tau}$ is expressed as

$$S(t_1, \ldots, t_n) = \overline{C}(\exp(-\Lambda_1(t_1)), \ldots, \exp(-\Lambda_n(t_n))).$$

Consequently, if survival regression models in Section 3.4.1 are specified to model the hazard function $\lambda_i(t)$, we obtain a statistical model for joint default (or event) times.

Factor copula models

Many copula models used in credit portfolios have a factor structure, which are called *factor copula models*. These models assume that default times are independent conditional on a unobserved d-dimensional random vector $\mathbf{Z}$. Specifically, let $\lambda_i(t; \mathbf{z})$ and $\Lambda_i(t; \mathbf{z}) := \int_0^t \lambda_i(s; \mathbf{z}) ds$ be the hazard and cumulative hazard functions given that $\mathbf{Z} = \mathbf{z}$, respectively, we can express the joint survivor function of default times $\tau_1, \ldots, \tau_n$ as

$$S(t_1, \ldots, t_n) = E_{\mathbf{Z}}\Big[E\big(\mathbf{1}_{\{\tau_1 \geq t_1, \ldots, \tau_n \geq t_n\}}|\mathbf{z}\big)\Big] = E_{\mathbf{Z}}\Big[\prod_{i=1}^n P(\tau_i \geq t_i | \mathbf{z})\Big]$$

$$= E_{\mathbf{Z}}\Big[\prod_{i=1}^n \exp(-\Lambda_i(t_i; \mathbf{z}))\Big] = E_{\mathbf{X}}\Big[\prod_{i=1}^n \exp\Big(-\int_0^{t_i} \lambda_i(s; \mathbf{z}) ds\Big)\Big]. \quad (3.91)$$

A special example of the above factor copula model is the one-factor Gaussian copula model. Let

$$v_i = \rho_i Z + \sqrt{1 - \rho_i^2}\,\epsilon_i,$$

where $\rho_i \in (0, 1)$ and where the random variables $X, \epsilon_1, \ldots, \epsilon_m$ are independent and identical distributed standard normal random variables. The random vector $\mathbf{v} = (v_1, \ldots, v_n)'$ follows a n-variate Gaussian distribution with mean $\mathbf{0}$ and correlation matrix $\boldsymbol{\rho} = (\rho_i \rho_j)_{1 \leq i,j \leq m}$. Let $u_i = \Phi(v_i)$ so that $(u_1, \ldots, u_n)'$ follow a Gaussian copula $C_{\boldsymbol{\rho}}^{\mathrm{Ga}}$. Then given that the default time τ_i is defined as $\tau_i = \inf\{t \geq 0; S_i(t) \leq U_i\}$, the conditional distribution function of τ_i can be computed as follows,

$$P(\tau_i > t | z) = P(U_i \leq S_i(t) | Z = z) = P\Big(\epsilon_i \leq \frac{\Phi^{-1}(S_i(t)) - \rho_i z}{\sqrt{1 - \rho_i^2}}\Big| z\Big).$$

Hence, the joint survivor function of $\boldsymbol{\tau} = (\tau_1, \ldots, \tau_n)'$ is expressed as

$$S(t_1, \ldots, t_n) = \frac{1}{2\pi} \int_{-\infty}^{\infty} \prod_{i=1}^n \Phi\Big(\frac{\Phi^{-1}(S_i(t)) - \rho_i z}{\sqrt{1 - \rho_i^2}}\Big) e^{-z^2/2} dz,$$

which can be easily computed using one-dimensional numerical integration.

In applications of a one-factor Gauss copula model to the pricing of portfolio credit derivatives it is frequently assumed that $\rho_i = \rho$ for all i. This model is known as the *exchangeable Gauss copula model*. In this case $\rho = \mathrm{Corr}(v_i, v_j)$, so ρ is readily interpreted in terms of asset correlation. This makes the exchangeable Gauss copula model popular with practitioners. In fact, it is common practice on CDO markets to quote prices for tranches of synthetic CDOs in terms of *implied asset correlations* computed in an exchangeable Gauss copula model.

3.8.4 Dynamic frailty and correlated default

Factor copula models assume a static frailty variable $\mathbf{x}$ so that the joint default distribution of n obligors can characterized through $\mathbf{x}$. As the factor copula model is static, the idea of frailty variables can be extended to dynamic case to model dynamic correlations of n obligors' default. Duffie et al. (2009) proposed a frailty-based model of joint default risk in which firms share an unobserved common factor that changes randomly over time.

Suppose there are n heterogeneous firms in the market and a common probability space $(\Omega, \mathcal{G}, \mathbb{P})$, in which $\mathcal{G} = \{\mathcal{G}_t : t \geq 0\}$ is a complete information filtration. For a firm l ($l = 1, \ldots, n$), denote by $N_l^*(t)$ the indicator process of firm l's default over the interval $(0, t]$. Let $\mathbf{x}_l(t)$ be a p-variate observable firm-specific covariate process for firm l, $\mathbf{y}(t)$ a vector of macro-economic variables that are observable at all times, and $\mathbf{z}(t)$ a vector of unobservable frailty variables that follows a stochastic process (e.g., an Ornstein-Uhlenbeck process). Let $e_{l,0}$ be the first appearance time of firm l, denote by $\mathcal{B}_{l,t}^{\text{obs}}$, $\mathcal{M}_t^{\text{obs}}$, and $\mathcal{N}_{l,t}$ the filtrations generated by $\{\mathbf{x}_l(s) : e_{l,0} \leq s \leq t\}$, $\{\mathbf{y}(s) : 0 \leq s \leq t\}$, and $\{N_l^*(s) : e_{l,0} \leq s \leq t\}$, respectively. Then econometrician's information filtration is given by $\mathcal{F}_t = \{\cup_l \mathcal{B}_{l,t}^{\text{obs}}, \cup_l \mathcal{N}_{l,t}, \mathcal{M}_t^{\text{obs}}\}$, whereas the complete-information filtration is $\mathcal{G}_t = \mathcal{F}_t \cup \{\mathbf{z}(s) : 0 \leq s \leq t\}$.

Let $\lambda_l(t)$ be the intensity function of $N_l^*(t)$ with respect to the complete information filtration $\mathcal{G}_t$. Suppose that $\lambda_l(t)$ has the following form,

$$\lambda_l(t) = \exp\left[\beta_0 + \boldsymbol{\beta}_1' \mathbf{x}_l(t) + \boldsymbol{\beta}_2' \mathbf{y}(t) + \boldsymbol{\beta}_3' \mathbf{z}(t)\right],$$

in which the parameter vector $\boldsymbol{\beta} = (\beta_0, \boldsymbol{\beta}_1', \boldsymbol{\beta}_2', \boldsymbol{\beta}_3')'$ is common to all firms. Then the survival probability of firm i given the information $\mathcal{F}_t$ is expressed as

$$P(\tau_i > T|\mathcal{F}_t) = E[P(\tau_i > T|\mathcal{G}_t)|\mathcal{F}_t] = E\left(e^{-\int_t^T \lambda_i(s)ds}\Big|\mathcal{F}_t\right).$$

Suppose that the dynamics of firms' observable covariates $\mathbf{x}_l(t)$ follow some time series model with parameter $\boldsymbol{\gamma}$. Then the likelihood of the data with parameters $(\boldsymbol{\beta}, \boldsymbol{\gamma})$ can be obtained, and the maximum likelihood estimate of $(\boldsymbol{\beta}, \boldsymbol{\gamma})$ can be found via the expectation-maximization algorithm.

Supplements and problems

The presentation in this chapter is meant to serve as an elementary introduction to the financial derivative pricing. There is a vast of literature on this subject, and references mentioned below comprise only a small part of the literature. Hull (2021) give a comprehensive on derivatives markets and financial instruments and is widely used for market practitioners. The following references are a supplement to the brief introduction covered by this chapter.

For Ito's formula and the Girsanov Theorem used in the derivation, we refer the readers to Jarrow (2021, Chapter 1). The basic tool for derivative pricing is stochastic analysis, and there are many excellent presentations on this. For example, Karatzas and Shreve (1991) provides a comprehensive treatise on the Brownian motion and stochastic calculus. Similar treatment on the subject can also be found in Liptser and Shiryaev (2001) and Potter (2005). Shreve (2004b) introduces stochastic calculus concisely but thoroughly with the highlight of mathematical models in finance such as risk-neutral pricing, American and exotic options, and term-structure models. Besides diffusion models, jump process is also introduced in Shreve (2004b). Concerning applications of jump processes to finance, Cont and Tankov (2003) give an overview of theoretical, numerical and empirical research on the use of jump processes in financial modeling.

Introduction on the classical Black-Scholes-Merton option pricing model can be found in most surveys or overviews on derivative pricing. For problems of American option pricing and early exercise boundary, Peskir and Shiryaev (2006, Chapter 25) explain their connection with free-boundary problems. Tree and Monte Carlo methods of pricing American option can be found in Shreve (2004a, Chapter 4) and Glasserman (2003, Chapter 8). James (2003) gives a comprehensive introduction on theory and numerical methods of pricing vanilla and exotic options. Concerning volatility smile and surface, Gatheral (2006) provides both theoretical and empirical discussions on implied volatilities and introduces related stochastic volatility models.

For stochastic interest rate models, a concise presentation on the topic is given by Cairns (2004). Brigo and Mercurio (2006) provide a more thorough and mathematical treatment on short rate models, the HJM models, market models and issues related to stochastic volatility, inflation, and credit risk. Rebonato (2002) discuss thoroughly theoretical and empirical issues on the LIBOR market models. Carmona and Tehranchi (2006) cast the interest rate models as stochastic evolution equations in infinite dimensional function spaces and discuss mathematical issues that arise in modeling the interest rate term structure. In comparison to the above theoretical discussion on interest rate models, the three-volume set Andersen and Piterbarg (2010) provide a massively comprehensive treatment on modeling fixed-income producuts for practitioners.

In contrast to the brief presentation of survival analysis in Section 3.4, more detailed discussion on the subject can be found in Kalbfleisch and Prentice (2002). A through introduction on the subject from a process point of view is given by Aalen et al. (2008). For specific topics in survival analysis, Cook and Lawless (2007) introduce the statistical analysis of recurrent events, and Crowder (2012) and Beyersmann et al. (2012) explain statistical models and methods for competing risks and multistate models.

Credit risk is one of the most intensely studied topics in quantitative finance. For this subject, Duffie and Singleton (2003) give the first integrated treatment of the conceptual, practical, and empirical foundations for credit

risk pricing and risk measurement. Lando (2004) present both mathematical models and statistical techniques for estimating credit risk. Bielecki and
Rutkowski (2004) provide a comprehensive mathematical treatment on modeling, valuation, and hedging credit risk. There are many references on credit
derivatives that highlight practitioners' concerns, for example, Chaplin (2010)
introduce various credit derivatives and hedging and risk management techniques from a real world perspective. We skip other references here as there
are too many.

For the methods of using the frailty or the copula for the analysis of multivariate survival functions and correlated default, statistical treatment of the
subject can be found in Duchateau and Janssen (2008), Wienke (2010), and
Hanagal (2019). For statistical methods of analyzing multivariate event times,
Hougaard (2000) gives a comprehensive description on the subject proposed
up to that point in time, and Crowder (2012) introduces the topic with a detailed account of parametric methods for multivariate failure time data analysis. Besides, Prentice and Zhao (2019) provides an innovative discussion on
the methods for the analysis of correlated failure times by using marginal and
marginal double hazard rate estimators. For methods and models for portfolio
credit risk management and portfolio credit derivatives, McNeil et al. (2015,
Chapters 11 and 12) give a concise description on mixture and threshold models, Monte Carlo methods, and copula approaches.

1. (**Transformation of the Black-Scholes equation**). Consider the partial differential equation (3.5) and the following transformation

   $$s = \frac{1}{2}\sigma^2(T - t), \qquad x = \log S, \qquad u(s, x) = e^{\alpha s + \beta x} f(t, S),$$

   in which $\beta = (r - q - \frac{1}{2}\sigma^2)/\sigma^2$ and $\alpha = \beta^2 + (2r/\sigma^2)$. Show that the
   Black-Scholes equation (3.5) is equivalent to the heat equation

   $$\frac{\partial u}{\partial s} = \frac{\partial^2 u}{\partial x^2}.$$

2. (**Pricing of a financial claim**). Suppose that the dynamics of a non-
   dividend asset follows the geometric Brownian motion $dS_t = \mu S_t dt + \sigma S_t dW_t^{\mathbb{P}}$, where W_t is a standard Brownian motion on a physical probability space. Consider a financial claim whose payoff at maturity T is
   $g(S) = (S^3 - K)_+$ where K denote the strike price of the claim. Use the
   following steps to price this claim.

   (a) Denote by $Y_t = S_t^3$. Use Ito's lemma to show that Y_t follows the
       dynamics $dY_t = 3(\mu + \sigma^2)Y_t dt + 3\sigma Y_t dW_t^{\mathbb{P}}$.

   (b) Suppose the asset S_t^3 is tradable and infinitely divisible. Denote
       by r the constant interest rate during $[0, T]$. Use the self-financing

replication strategy to show that the price $f(t, Y)$ of the financial claim satisfies the PDE

$$\frac{\partial f}{\partial t} + rS\frac{\partial f}{\partial Y} + \frac{9}{2}\sigma^2 Y^2 \frac{\partial^2 f}{\partial Y^2} = rf \quad \text{for} \quad 0 \le t < T.$$

(c) Show the price of the financial claim at time t is $S_t^3 \Phi(d_1) - Ke^{-r(T-t)}\Phi(d_2)$, where $\Phi(\cdot)$ is the standard normal cumulative distribution function and

$$d_1 = \frac{\log(S^3/K) + (r + 9\sigma^2/2)(T - t)}{3\sigma\sqrt{T - t}}, \quad d_2 = d_1 - 3\sigma\sqrt{T - t}.$$

3. (**Option Greeks**). Consider the European call option price (3.4a).

(a) Show that
$$Se^{-q(T-t)}\phi(d_1) = Ke^{-r(T-t)}\phi(d_2).$$

(b) Make use of the above result and show that the delta, gamma, theta, vega, and rho of the European call option price are given by (3.7a), (3.7b), (3.7c), and (3.7d), respectively.

4. Show that the Kaplan-Meier estimate (3.40) reduces to $\widehat{S}(t) = P(\text{the number of observations} > t)/n$ when there is no censoring. Show that the Greenwood's formula (3.42) reduces to

$$\widehat{\text{Var}}[\widehat{S}(t)] = n^{-1}\widehat{S}(t)(1 - \widehat{S}(t)).$$

5. Let the survival time $\tau > 0$ be an integer-valued random variable with finite mean r_0 and let $r_i = E(\tau - i | \tau > i)$ be the expected residual life at time i, $i = 1, 2, \ldots$. Show that the survivor function for integer t is $S(t) = \prod_{i=1}^{t}(r_{i-1} - 1)/r_i$.

6. Suppose that $\tau_1, \ldots, \tau_n$ are independent exponential random variables with respective failure rates $\lambda_1, \ldots, \lambda_n$. Let $\gamma_1, \ldots, \gamma_m$ be the distinct elements of $\lambda_1, \ldots, \lambda_n$. Let $\tau = \sum_{i=1}^{n} \tau_i$.

(1) Show that the survivor function of τ is $S_\tau(t) = \sum_{j=1}^{m} p_j(t)e^{-\gamma_j t}$, where $p_j(t)$ are polynomials in t.

(2) Let $\lambda_{\min} = \min\{\gamma_1, \ldots, \gamma_m\}$. Show that the hazard function of τ satisfy that $\lambda_\tau(t) \le \lambda_{\min}$ for all $t > 0$ and $\lim_{t\to\infty} \lambda_\tau(t) = \lambda_{\min}$.

7. Let $\tau_1, \ldots, \tau_n$ be a random sample from a distribution with survivor function $S(t)$ and suppose that for t near 0, $S(t) = 1 - (\lambda t)^\gamma + o(t^\gamma)$, where $\lambda > 0$ and $\gamma > 0$. Show that the limiting distribution of $\tau := n^{1/\gamma}\min(\tau_1, \ldots, \tau_n)$ is Weibull with shape γ and scale λ.

8. Let W_1 and W_2 be independent with the extreme value density $\exp(w - e^w)$. Show that $V = W_1 - W_2$ has the logsitic density $f(v) = e^v(1+e^v)^{-2}$. Derive the variance and kurtosis of this distribution.

9. Let τ be a random time with absolutely continuous distribution function $F(t)$ and hazard function $\lambda(t)$. Show that $Y_t - \Lambda(\tau \wedge t)$, $t \geq 0$, is an $\mathcal{H}_t$ martingale.

10. Consider the conditional distribution of the number of defaults $M = \sum_{i=1}^n Y_i$ given the factor $\mathbf{X} = \mathbf{x}$,

$$M|\mathbf{X} = \mathbf{x} \sim \text{Poisson}\Big(\sum_{i=1}^n k_i \mathbf{w}_i' \mathbf{x}\Big),$$

in which $k_i > 0$ are constant, $\mathbf{w}_i = (w_{i1}, \ldots, w_{id})'$ are non-negative factor weights such that $\sum_j w_{ij} = 1$, $\mathbf{X} = (X_1, \ldots, X_d)'$, and X_j ($j = 1, \ldots, d$) are d independent Gamma(α_j, β_j)-distributed factors. Show that the unconditional distribution of M is the same as that of the sum of d independent negative binomial distribution NB$(\alpha_j, \beta_j + \sum_{i=1}^n k_i w_{ij})$.

Chapter 4

Insurance risk and credibility theory

The preceding chapters describe commonly used probabilistic and statistical models related to market and credit risk. Many of these concepts are originated from actuarial science and closely tied to actuarial methodology. For example, in securitization of credit derivatives, various credit assets like loans or bonds are bundled and transformed into tradable securities. Actuarial techniques are applied here to evaluate the credit risk linked to underlying assets and determine suitable pricing and risk management strategies. Utilizing actuarial methodologies enables financial institutions to better comprehend and quantify the credit risk, thus improving their capacity to manage and mitigate these risks effectively. Furthermore, actuarial expertise can inform the structuring of credit derivative products to align with risk preferences and regulatory requirements, contributing to the overall stability and efficiency of financial markets. Recognizing the increasingly important role of actuarial methodology in financial risk management, this chapter presents key concepts and models relevant to insurance contracts and actuarial methodology.

Insurance is a contract or policy under which a policyholder pays premium to the insurer in exchange for an obligation to compensate some possible losses of the policyholder. The primary objective of insurance is to mitigate financial uncertainty and provide protection to the policyholder against specific risks. An insurance contract specifies the policy term and the mechanism of compensation. When events specified in the contract take place within the policy term, an insurance claim is initiated. Conversely, if none of the events specified in the contract occurs during the contract's duration, the policyholder doesn't receive any compensation from the insurer for paid premiums.

There are two important issues involved in the insurance risk, determining the premium and assessing the likelihood of an insurer facing financial insolvency. This chapter focuses on fundamental concepts and models pertinent to these concerns, as well as associated probabilistic and statistical methods. Section 4.1 introduces two essential concepts in actuarial science. The first is the expected utility model in which a risk-averse individual makes his decision under uncertainty by comparing the expected utilities of payoffs. The second concept is the premium principle, which assigns a financial compensation to a risk. In order to understand the distribution of total claims on a portfolio of insurance contracts, Section 4.2 presents two fundamental concepts, collective risk and individual risk. Additionally, probabilistic models are presented to calculate aggregate claim distributions. The techniques outlined in this section are also applicable for computing operational risk within banks. In Section 4.3, the concept of ruin theory is introduced, which characterizes an insurer's surplus process for a portfolio of insurance policies. In contrast to the probabilistic evaluation of insurance risk discussed in Sections 4.1–4.3, Section 4.4 introduces credibility theory and regression-type models that emphasize the utilization of insurance data. As credibility theory fundamentally operates on Bayesian principles, Section 4.4 introduces Bayesian decision theory and empirical Bayes methods. Additionally, to establish a connection between insurance premiums and explanatory variables, Section 4.4 also presents linear mixed models, generalized linear models, and generalized linear mixed models. Furthermore, Section 4.5 outlines evolutionary credibility theory, which expands upon the models introduced in Section 4.4 and enables recursive assessment of premium credibility.

4.1 Utility and premium principles

This section introduces the expected utility model and premium principles in insurance. The former explains why insurers should exist due to insured individual's risk aversion, and the latter shows how to assign financial compensation to an insurance risk.

4.1.1 Utility in insurance

A *utility* function is usually assumed to satisfy the conditions

$$u'(x) > 0, \qquad u''(x) < 0, \tag{4.1}$$

indicating $u(x)$ is increasing and concave in x. The expected utility theory is based on two fundamental assumptions regarding an individual's rationale concerning financial matters. Firstly, for any monetary amount x, there exists a utility function $u(x)$ that encapsulates the individual's subjective valuation of possessing the amount x. Secondly, when confronted with decisions resulting in the possession of random monetary amounts, the individual consistently chooses the decision associated with the highest expected utility.

An individual whose utility function satisfies the condition (4.1) is said to be *risk averse*. The rationale behind the risk aversion exhibited by such individuals can be explained through the application of Jensen's inequality. This inequality provides insight into the concavity of the utility function, demonstrating that risk-averse individuals exhibit diminishing marginal utility for wealth increments. As a result, they are inclined to choose options that mitigate uncertainty and preserve their financial well-being.

Theorem 4.1 (Jensen's inequality) *If $u(x)$ is a concave function, then $E[u(X)] \leq u(E(X))$, with equality if and only if $X = a$ almost surely, where a is a non-random variable.*

Let w be the wealth of the individual. Theorem 4.1 implies that

$$E(u(w - X)) \leq u(w - E(X)).$$

This means that the individual prefers paying a predetermined fixed amount $E(X)$ over assuming the risk associated with a variable amount X. Risk aversion can be quantified by the concept of *risk aversion coefficient*, which is mathematically expressed as $r(x) = -u''(x)/u'(x)$. The risk aversion coefficient serves as a crucial parameter in the expected utility theory. It indicates how rapidly an individual's utility function declines as wealth increases (or decreases). A higher risk aversion coefficient suggests greater aversion to risk, as it implies a steeper decline in utility for incremental changes in wealth.

Utility functions can be used to explain the premium value of an insurance contract. Suppose the individual is completely insured against a loss X for a premium P^X, then his expected utility will increase when $E(u(w - X)) \leq u(w - P^X)$. Let $\overline{P}^X$ be the maximum premium to be paid. Then, $P^X \leq \overline{P}^X$ and P^X satisfies the following utility equation

$$E(u(w - X)) = u(w - \overline{P}^X).$$

By Jensen's inequality, we have $u(w - \overline{P}^X) \leq u(w - E(X))$. Since u is increasing, it follows that $\overline{P}^X \geq E(X)$.

Similar argument applies to the insurer. Suppose that the insurer has utility function $v(x)$ and wealth z. Assume that the insurer will insure the loss X for a premium P^X, then we have $E(v(z + P^X - X)) \geq v(z)$. Let $\underline{P}^X$ be the minimum premium to be asked, then $P^X \geq \underline{P}^X$ and

$$E(v(z + \underline{P}^X - X)) = v(z). \tag{4.2}$$

By Jensen's inequality, $E(v(z + \underline{P}^X - X)) \leq v(z + \underline{P}^X - E(X))$. Since $v(\cdot)$ is increasing, we have $\underline{P}^X \geq E(X)$. An insurance contract is feasible when $\overline{P}^X \geq \underline{P}^X$.

The following lists some commonly used utility functions.

(i) *Linear utility:* $u(x) = x$.

(ii) *Quadratic utility:* $u(x) = -(\alpha - x)^2$ for $x \leq \alpha$.

(iii) *Logarithmic utility:* $u(x) = \log(\alpha + x)$ for $x > -\alpha$.

(iv) *Exponential utility:* $u(x) = -\alpha e^{-\alpha x}$ $(\alpha > 0)$.

(v) *Power utility:* $u(x) = x^c$ for $x > 0$, $0 < c < 1$.

Using the utility theory, one can explain and compute the fair values that individuals are prepared to buy insurance and to pay premia.

4.1.2 Premium principles and their properties

A *premium* of an insurance contract is the payment made by a policyholder to secure either full or partial coverage against a specified risk. In assessing the value of a premium, a fundamental tool known as a *premium principle* is needed to determine the appropriate premium for a given risk.

Let X be the potential claim arising from a risk and $\pi(X)$ the premium that an insurer asks to cover the claim (or risk) X. Here, a premium principle can be represented as a function $\phi : X \to \pi(X)$, which maps the risk X to a numerical value $\pi(X)$. A premium principle usually has the following properties.

(i) *Non-negative loading*: $\pi(X) \geq E(X)$. That is, the premium should not be less than the expected claim. The difference between the premium and the expected claim, $\pi(X) - E(X)$, is called the *premium loading*.

(ii) *Additivity*: $\pi(X+Y) = \pi(X)+\pi(Y)$ for independent X and Y. This suggests that pooling independent risks does not affect the total premium needed.

(iii) *Scale invariance*: $\pi(aX) = a\pi(X)$ for any constant $a > 0$.

(iv) *Translational equivariance*: $\pi(X + c) = \pi(X) + c$ for any $c > 0$.

(v) *No ripoff*: $\pi(X) \leq \min\{p|F_X(p) = 1\}$. This means that the premium should be no more than the finite maximum claim amount, if it exists. Otherwise, the premium would be considered infinite.

There are many types of premium principles. The following shows some examples of premium principles.

(i) *Pure premium principle.* This principle sets $\pi(X) = E(X)$ and is also known as the *equivalence principle*. This premium is sufficient for a risk neutral insurer only. In practice, it is not very attractive to an insurer because the premium loading is zero.

(ii) *Expected value principle.* Under this principle, $\pi(X) = (1 + \alpha)E(X)$ for some $\alpha > 0$. The loading in this premium is $\alpha E(X)$. The expected value principle satisfies the non-negative loading property, i.e., it is additive and invariant in scale. However, it is not translational equivariance. Another drawback of the expected value principle is that the premium only involves the mean of the risk, thus fails to account for risks with identical means but different variances.

(iii) *Variance principle.* Under this principle, $\pi(X) = E(X) + \alpha \text{Var}(X)$ for some $\alpha > 0$. The loading in this premium $\alpha \text{Var}(X)$ is proportional to the variance of the claim. The variance principle has non-negative loading and possesses additivity and translational equivariance. However, it lacks scale invariance.

(iv) *Standard deviation principle.* Under this principle, $\pi(X) = E(X) + \alpha[\text{Var}(X)]^{1/2}$ for some $\alpha > 0$. This principle is additive, scale invariant and translational equivariant, but it does not satisfy the no-ripoff condition.

(v) *Zero utility premium.* Suppose that the insurer's utility function $u(x)$ satisfies condition (4.1). Denote by W the insurer's wealth, the zero utility principle assumes that the premium solves the utility equilibrium equation

$$u(W) = E[u(W + \pi(X) - X)]. \tag{4.3}$$

This premium satisfies the non-negative loading property, but it is not scale invariant. Generally, this premium depends on the value of W and is not additive. One exception is when the utility function is exponential, i.e., $u(x) = -\exp(-\alpha x)$ $(\alpha > 0)$. In such case, (4.3) yields the following *exponential principle*

$$\pi(X) = \frac{1}{\alpha} \log E[\exp(\alpha X)].$$

(vi) *Esscher principle.* Under this principle, one has

$$\pi(X) = \frac{E(Xe^{hX})}{E(e^{hX})}, \qquad h > 0. \tag{4.4}$$

Suppose that X is continuous and has density $f(x)$ on $(0, \infty)$. The Esscher principle can be considered as the pure premium for a risk $Y = Xe^{hX}/E(e^{hX})$, whose density and distribution functions are given by

$$g(x) = \frac{e^{hx} f(x)}{\int_0^\infty e^{hy} f(y)dy}, \quad G(x) = \frac{\int_0^x e^{hy} f(y)dy}{E_X(e^{hX})}.$$

Note that the random variable Y enlarges the large values of X and reduces the small values, and hence results a tail fatter than that of X. The distribution function $G(x)$ is called the Esscher transform of $F_X(x)$ with parameter h. It is easy to see that the moment generating function of Y is given by

$$E_Y(e^{tY}) = E_X(e^{(t+h)X})/E_X(e^{tX}).$$

The Esscher principle satisfies the non-negative loading and no ripoff conditions. It is additive and has translational equivariance, but it lacks scale invariance.

(vii) *Risk adjusted premium principle.* Consider the non-negative random variable X with distribution function $F(x)$. Under the risk-adjusted premium principle, one has

$$\pi(X) = \int_0^\infty [1 - F(x)]^{1/\rho}dx,$$

where $\rho \geq 1$ is the *risk index*. The risk adjusted premium can be considered as the pure premium for a risk Y with density and distribution functions

$$g_Y(y) = \frac{1}{\rho}[1 - F(y)]^{1/\rho - 1} f(y), \qquad G_Y(y) = 1 - [1 - F(y)]^{1/\rho}.$$

Similar to the Esscher principle, the risk adjusted premium enlarges large values of X and results a fatter tail than that of X.

4.2 Collective and individual risk models

Collective and individual risk models are fundamental concepts in actuarial science and insurance risk management. Both models are essential tools for insurers, actuaries, and risk managers in understanding and managing the financial risks associated with insurance policies and portfolios. The focus of a collective risk model is the aggregate losses incurred by a group of insurance policyholders or participants over a specified period. A collective risk model usually assumes that the individual losses experienced by policyholders are independent and identically distributed random variables. In contrast to collective risk models, an individual risk model focuses on the losses experienced by individual insurance policyholders. It considers the risk associated with each policyholder separately, taking into account factors such as the policyholder's characteristics, exposure, and potential losses.

Consider a general insurance risk over a short period of time. Let N be the number of claims from the risk during the period and X_i the amount of the ith claim $(i = 1, \ldots, N)$. The aggregate claim is the sum of individual claim amounts, i.e.

$$S_N = X_1 + \cdots + X_N = \sum_{i=1}^{N} X_i. \tag{4.5}$$

Note that both X_i and N are random variables, hence the total claim $S = 0$ if $N = 0$. For convenience, the following two assumptions are made through the chapter. First, the random variables $\{X_1, X_2, \ldots\}$ are independent and identically distributed with common distribution function G, $G(0) = 0$, and second, the random variables N and $\{X_1, X_2, \ldots\}$ are independent. Under these two assumptions, the sum (4.5) is called a *compound sum*. The probability mass function of N is denoted by $p_N(k) = P(N = k)$, $k = 0, 1, 2, \ldots$. The random variable N is referred to as the *compounding random variable*.

4.2.1   Individual risk models

The *individual risk model* deals with the distribution function of the total claim amount S for an insurer's portfolio. One may assume directly the random variable S follows certain distribution functions, or characterize the distribution of S for a fixed $N = n$ and specified distributions of $X_1, \ldots, X_n$. In the latter case, one usually needs to find a procedure to compute the distribution function of S. The following example illustrates such a case.

Suppose the portfolio consists of n independent policies. Assume that the number of claims arising under the ith policy is either zero with probability $1 - q_i$ or one with probability q_i, for $i = 1, \ldots, n$. The aggregate claim amount is denoted by

$$S = \sum_{i=1}^{n} X_i,$$

where X_i is the amount paid out in claims under the ith policy. Suppose further that the amount X_i of the occurred claim under the ith policy follows a distribution function $F_i(x)$ with $F_i(0) = 0$, mean μ_i and variance σ_i^2. Then we can show that X_i has a compound binomial distribution with mean $E(X_i) = q_i \mu_i$ and variance $\mathrm{Var}(X_i) = q_i \sigma_i^2 + q_i(1 - q_i)\mu_i^2$.

De Pril's recursion formula

De Pril's recursion formula provides a way to calculate the aggregate claims distribution within the individual risk model. In this context, let q_i be the mortality rate of policyholder i, μ_i denote the sum assured under policy i, and assume $\sigma_i^2 = 0$ for all i, indicating a fixed value for the benefits under the insurance contract. Suppose the sums assured in the portfolio are integers, say $\mu_i \in \{1, 2, \ldots, I\}$, and each policyholder is subject to one of J different mortality rates. Let n_{ij} denote the number of policyholders with mortality rate q_j and *sum assured i* ($i = 1, \ldots, I, j = 1, \ldots, J$), and $g_x = P(S = x)$ for nonnegative integers x. Here, the term "sum assured" refers to the fixed amount guaranteed by an insurance company to a policyholder upon the occurrence of the insured event.

Before delving into the intricacies of De Pril's recursion formula, we first introduce the concept of *probability generating function*. When X_1 is a discrete random variable distributed on the non-negative integers, its probability generating function is defined as

$$G_X(t) = \sum_{k=0}^{\infty} P(X_1 = k)t^k.$$

Consider a claim amount from a policyholder with mortality rate q_j and sum assured i, its probability generating function is given by

$$G_{ij}(t) = 1 - q_j + q_j t^i.$$

This function represents the probability distribution of the claim amount for a specific policyholder, where t is a variable representing the size of the claim, q_j is the mortality rate associated with policyholder j, and i represents the sum assured under the policy. By independence of the policyholders, the probability generating function of $S = \sum_{i=1}^{n} X_i$ is expressed as

$$G_S(t) = \prod_{i=1}^{I}\prod_{j=1}^{J}(1 - q_j + q_j t^i) = \sum_{x=0}^{\infty} g_x t^x.$$

Taking logarithm of $G_S(t)$ and differentiating it yields that

$$G'_S(t)/G_S(t) = \sum_{i=1}^{I}\sum_{j=1}^{J}\frac{n_{ij}iq_j t^{i-1}}{1 - q_j + q_j t^i} = \sum_{i=1}^{I}\sum_{j=1}^{J}n_{ij}i\frac{q_j t^i}{1 - q_j}\left(1 + \frac{q_j t^i}{1 - q_j}\right)^{-1}$$

$$= \sum_{i=1}^{I}\sum_{j=1}^{J}n_{ij}i\frac{q_j t^i}{1 - q_j}\sum_{k=1}^{\infty}(-1)^{k-1}\left(\frac{q_j t^i}{1 - q_j}\right)^{k-1}.$$

For $i = 1, \ldots, I$, define that

$$h(i, k) = i(-1)^{k-1}\sum_{j=1}^{J}n_{ij}\left(\frac{q_j}{1 - q_j}\right)^k.$$

Then we obtain that

$$tG'_S(t) = G_S(t)\sum_{i=1}^{I}\sum_{k=1}^{\infty}h(i, k)t^{ik}.$$

This indicates that

$$\sum_{x=1}^{\infty}xt^x g_x = \sum_{x=0}^{\infty}t^x g_x \sum_{i=1}^{I}\sum_{k=1}^{\infty}h(i, k)t^{ik}.$$

Denote by $[y]$ represents the largest integer value that is less than or equal to y. Comparing the coefficient of t^x on both sides yield that

$$g_x = \frac{1}{x}\sum_{i=1}^{\min x, I}\sum_{k=1}^{[x/i]}g_{x-ik}h(i, k), \quad x = 1, 2, \ldots. \tag{4.6}$$

The above equation is called *De Pril's recursion formula*, and its starting value is given by

$$g_0 = \sum_{i=1}^{I}\sum_{j=1}^{J}(1 - q_{ij})^{n_{ij}},$$

because $S = 0$ only if no claim occurs under each policy.

In practice, values of q_j are small and the recursion formula (4.6) can be computationally expensive for large portfolios. In such case, the distribution of S often needs to be approximated.

4.2.2   Aggregate claim and compound distribution

We now consider the issue of computing the distribution function of S_N, when N is a random variable. The first and second moment (or variance) of the aggregate claim S_N can be easily obtained. Note that for two random variables Y and Z,

$$E(Y) = E[E(Y|Z)], \qquad \mathrm{Var}(Y) = E[\mathrm{Var}(Y|Z)] + \mathrm{Var}[E(Y|Z)]. \qquad (4.7)$$

Let $E(X_1^k) = m_k$ for $k = 1, 2, \ldots$, the above two expressions imply that

$$E(S_N) = m_1 E(N), \qquad \mathrm{Var}(S_N) = (m_2 - m_1^2)E(N) + m_1^2 \mathrm{Var}(N); \qquad (4.8)$$

see Exercise 4.4.

The characterization of distribution of S_N is more complicated. Let $F(x)$ and $G(x)$ represent the distribution functions of individual claims X_i and aggregate claims S_N, respectively. Denote $p_n = P(N = n)$ the probability of having n claims in the portfolio. The following property can be obtained.

Proposition 4.1 *Let S_N be a compound sum defined by (4.5). Then, for all $x \geq 0$,*

$$F_{S_N}(x) = P(S_N \leq x) = \sum_{k=0}^{\infty} p_N(k) F^{(k)}(x), \qquad (4.9)$$

where $F^{(k)}(x) = P(S_k \leq x)$, the kth convolution of F. Note that $F^{(0)}(x) = 1$ for $x \geq 0$, and $F^{(0)}(x) = 0$ for $x < 0$.

For some distributions, explicit distributions for the compound random variable S_N can be obtained; see the following example.

Example 4.1 (The compound geometric-exponential distribution)
Suppose N follows a geometric distribution with success probability p, i.e., $P(N = n) = p(1-p)^{n-1}$ for $n = 1, 2, \ldots$. Suppose that $X_1, \ldots, X_n, \ldots$ follow an exponential distribution with mean 1. The MGF's of X and N are given by

$$M_X(t) = \frac{1}{1-t}, \qquad M_N(t) = \frac{p}{1 - qe^t}.$$

Then (4.10) implies that the MGF of S_N is given by

$$M_{S_N}(t) = p + (1-p)\frac{p}{q-t},$$

which is a mixture of MGF's of the constant 0 and the exponential distribution with mean $1/p$. This indicates that the distribution function of S_N can be expressed as

$$F_S(x) = p + (1-p)(1 - e^{-px}) = 1 - (1-p)e^{-px}, \qquad x \geq 0. \qquad \square$$

The calculation of (4.9) is usually difficult because the convolution powers $F^{(k)}$ of a distribution function F are generally not available in closed form. One way to characterize (4.9) is via the MGF of S_N. When X_1 is a continuous random variable, let $M_X(t)$ and $M_N(t)$ be the MGF of X and N, respectively. and then the MGF of S_N can be obtained as follows.

$$
\begin{aligned}
M_{S_N}(t) &= E_N\big[E(e^{tS_n}|N=n)\big] = E_N\big[M_X(t)^N\big] \\
&= E_N\big[\exp[N\log M_X(t)]\big] = M_N[\log M_X(t)]. \qquad (4.10)
\end{aligned}
$$

The above arguments lead to

$$
G_{S_N}(t) = G_N[G_X(t)], \qquad (4.11)
$$

where G_{S_N} and G_N are the probability generating functions of S_N and N, respectively.

The calculation of the MGF (4.10) or the PGF (4.11) can be done numerically by using the Laplace-Stieltjes transform and the Fourier inversion. Specifically, for $s \geq 0$, let $\widetilde{F}(s) = \int_0^\infty e^{-sy}dF(y)$ be the Laplace-Stieltjes transform of distribution function $F(\cdot)$. Using Proposition 4.1 and the fact that the Laplace-Stieltjes transform of a convolution is the product of the Laplace-Stieltjes transform, we have

$$
\widetilde{F}_{S_N}(s) = \sum_{k=0}^\infty p_N(k)\widetilde{F^{(k)}}(y) = \sum_{k=0}^\infty p_N(k)\big[\widetilde{F}(s)\big]^k = \Pi_N(\widetilde{F}(s)), \qquad (4.12)
$$

where $\Pi_N(s) = \sum_{k=1}^\infty p_N(s)s^k$ is the probability generating function of N. Then using the inverse Fourier transformation, the distribution function of S_N can be computed numerically.

Example 4.2 (The compound Poisson distribution) *Assume that N follows a Poisson distribution with intensity parameter $\lambda > 0$, i.e., $P(N = n) = e^{-\lambda}\lambda^n/n!$ for $n \geq 0$. Then for $s > 0$,*

$$
\Pi_N(s) = \sum_{n=0}^\infty e^{-\lambda}\frac{\lambda^n}{n!}s^n = e^{-\lambda(1-s)}.
$$

Using (4.12), we obtain the Laplace-Stieltjes transform of the distribution of S_N,

$$
\widetilde{F}_{S_N}(s) = \exp\big[-\lambda(1-\widetilde{F}(s))\big], \quad \text{for } s \geq 0.
$$

In this case, the distribution of S_N is called the compound Poisson distribution.
□

Distributions of the number of claims

When the sample mean and sample variance of the numbers of claims are approximately same, one usually assume that the number of claims N follows a Poisson(λ) distribution. If $N \sim$ Poisson(λ), then its probability mass function, mean, variance and MGF are expressed as

$$P(N = n) = e^{-\lambda} n^{\lambda} / n!, \quad E(X) = \text{Var}(X) = \lambda, \quad M_N(t) = \exp[\lambda(e^t - 1)].$$

If the data regarding the numbers of claims indicate that the sample variance exceeds the sample mean, signifying *overdispersion*, one may assume the number of claims N follows a Negative Binomial(r, p) distribution. If $N \sim$ Negative Binomial(r, p) distribution, its probability mass function, mean, variance, and mgf are given by

$$P(N = n) = \binom{r + n - 1}{n} p^r (1 - p)^n, \quad M_N(t) = \left(\frac{p}{1 - (1 - p)e^t}\right)^r,$$

$$E(N) = \frac{r(1 - p)}{p}, \quad \text{Var}(N) = \frac{E(N)}{p}.$$

We will explore two examples in the following discussion. One example involves the Poisson-Gamma mixture distribution, while the other deals with the compound Poisson-logarithmic distribution. Both compound distributions can be considered as special cases of the negative binomial distribution.

Example 4.3 (The compound Poisson-Gamma distribution) *Assume that an individual makes a Poisson(λ) distributed number of claims in one year. The parameter λ depends on the individual and follows a Gamma(α, β) distribution with density and MGF*

$$f(x) = \frac{\beta^\alpha}{\Gamma(\alpha)} x^{\alpha-1} e^{-\beta x}, \text{ for } x > 0; \quad M_\Gamma(t) = \left(\frac{\beta}{\beta - t}\right)^\alpha, \text{ for } t < \beta,$$

in which $\Gamma(\cdot)$ is the Gamma function. Let $\Lambda \sim$ Gamma(α, β). The MGF of N is given by

$$M_N(t) = E\left[E(e^{tN}|\Lambda)\right] = E\left[\exp\{\Lambda(e^t - 1)\}\right] = M_\Gamma(e^t - 1) = \left(\frac{\beta}{\beta - (e^t - 1)}\right)^\alpha.$$

Note that this is the MGF of a negative binomial(r, p) distribution with $r = \alpha$, $p = \beta/(1 + \beta)$. Hence, $N \sim$ Negative Binomial($\alpha, \beta/(1 + \beta)$) and the overdispersion is $\text{Var}(N)/E(N) = 1 + \beta^{-1}$. □

Example 4.4 (The compound Poisson-logarithmic distribution) *Suppose that the number of claims N is a Poisson(λ) random variable and the number of events X_i in the ith claim follows a logarithmic(p) distribution with probability mass function*

$$P(X_i = k) = -\frac{1}{\log(1 - c)} \frac{c^k}{k}, \quad k = 1, 2, \ldots, 0 < c < 1.$$

Note that the MGF of the logarithmic(p) distribution is $M_X(t) = \log(1 - ce^t)/\log(1-c)$ for $t < -\log c$. Then the MGF of $S_N = X_1 + \cdots + X_N$ is

$$M_{S_N}(t) = M_N\big[\log M_X(t)\big] = \exp[\lambda(M_X(t) - 1)] = \left(\frac{1-c}{1 - ce^t}\right)^{-\lambda/\log(1-c)},$$

which is the MGF of a negative binomial(r, p) distribution distribution with $r = -\lambda/\log(1-c)$ and $p = 1 - c$. $\qquad\square$

Compound Poisson distribution

Using the MGF of Poisson distribution, we obtain the following proposition.

Proposition 4.2 *Let* $S_1, \ldots, S_m$ *be independent compound Poisson random variables with parameters* λ_i *and claims distribution* F_i, $i = 1, 2, \ldots, m$. *Then* $S = S_1 + \cdots + S_m$ *is compound Poisson distributed with parameter* $\lambda = \lambda_1 + \cdots + \lambda_m$ *and distribution function* $F(x) = \sum_{i=1}^m (\lambda_i/\lambda) F_i(x)$.

The above proposition can be easily shown. Let m_i be the MGF of F_i. Then S has the MGF

$$M_S(t) = \prod_{i=1}^m \exp\big\{\lambda_i[m_i(t) - 1]\big\} = \exp\Big\{\lambda\big[\sum_{i=1}^m \frac{\lambda_i}{\lambda} m_i(t) - 1\big]\Big\}.$$

This shows that S is a compound Poisson random variable with parameter λ and distribution function $F(x)$. A special case of Proposition 4.2 is that there are N_i claims in S_i and each claim have fixed amount x_i. In such case, the total amount $S = \sum_{i=1}^m x_i N_i$ and the following proposition can be obtained.

Proposition 4.3 *Suppose that* $S = \sum_{i=1}^m x_i N_i$ *is compound Poisson distributed with parameter* λ *and with discrete distribution* $\pi_i = p(x_i) = P(X = x_i)$, $i = 1, \ldots, m$, *where* N_i *is the frequency of the claim amount* x_i ($i = 1, \ldots, m$). *Then* $N_1, \ldots, N_m$ *are independent Poisson($\lambda\pi_i$) random variables,* $i = 1, \ldots, m$.

To show this proposition, let $N = N_1 + \cdots + N_m$ and $n = n_1 + \cdots + n_m$. Then conditionally on $N = n$, we have $N_1, \ldots, N_m \sim$ Multinomial($n_1, \pi_1, \ldots, \pi_m$). Hence,

$$P(N_1 = n_1, \ldots, N_m = n_m) = P(N_1 = n_1, \ldots, N_m = n_m | N = n)P(N = n)$$

$$= \prod_{i=1}^m e^{-\lambda\pi_i} \frac{(\lambda\pi_i)^{n_i}}{n_i!}$$

By summing over all n_i $i \neq k$, we see that N_k is marginally Poisson($\lambda\pi_i$) distributed. The independence of N_i can be obtained by the joint distribution of N_i's is the product of marginal distributions of N_i's.

4.2.3   Recursive calculation of aggregate claims distribution

When individual claim amounts X_i are distributed on the non-negative integers and the distribution of the claim number N falls into the $(a, b, 0)$ class of distributions, the distribution of the aggregate claim S_N can be calculate recursively. This is referred to as the *Panjer recursion formula*.

The $(a, b, 0)$ class of distributions

The $(a, b, 0)$ class of distributions refers to a family of probability distributions commonly used in actuarial science and insurance risk modeling. In this context, a refers to the parameter governing the frequency distribution of the number of claims, b is the parameter controlling the severity distribution of individual claims, and 0 means that the distribution is truncated at zero, indicating that only non-negative values are allowed. The $(a, b, 0)$ class of distributions encompasses various models, including the Poisson, negative binomial, compound Poisson, and zero-truncated distributions, among others. It provides a flexible framework for modeling the joint distribution of the frequency and severity of insurance claims.

A counting distribution N is said to belong to the $(a, b, 0)$ class of distributions if its probability function $\{p_r := P(N = r)\}_{r=0}^{\infty}$ can be calculated recursively from the formula

$$p_r = \left(a + \frac{b}{r}\right)p_{r-1}$$

for $r = 1, 2, \ldots$, where a and b are constants. The starting value for the recursive calculation is p_0, which is assumed to be greater than 0.

Note that the probability generating function of N is given by $G_N(t) = p_0 + \sum_{n=1}^{\infty} p_n t^n$. Differentiation with respect to t yields

$$G_N'(t) = a \sum_{n=1}^{\infty} n r^{n-1} p_{n-1} + b G_N(t) = a t G_N'(t) + (a + b) G_N(t). \qquad (4.13)$$

The above equation can be solved for a and b. Actually, there are only three non-trivial distributions in the $(a, b, 0)$ class, Poisson, Binomial and negative Binomial; see Johnson and Kotz (1969) and Sundt and Jewell (1981).

Example 4.5 (Probability generating functions of Poisson, Binomial and negative Binomial distributions) *Consider the distribution of N. For $N \sim B(n, p)$ with the probability mass function $p_r = \binom{n}{r} p^r (1 - p)^{n-r}$ for $r = 0, 1, \ldots, n$, we have $p_r = (a + (b/r)) p_{r-1}$ with $a = -p/(1 - p)$ and $b = (n + 1)p/(1 - p)$. For $N \sim Poisson(\lambda)$ with the probability mass function is $p_r = e^{-\lambda} \lambda^r / r!$, we have $p_r = (a + (b/r)) p_{r-1}$ with $a = 0$ and $b = \lambda$. For $N \sim$ negative Binomial(n, p) with the probability mass function*

$$p_r = \binom{n + r - 1}{r} p^n (1 - p)^r, \qquad r \geq 0, n > 0, 0 < p < 1,$$

we have $p_r = (a + (b/r)) p_{r-1}$ with $a = 1 - p$ and $b = (n - 1)(1 - p)$.   □

Panjer's recursion

The Panjer recursion formula is one of the most important results in risk theory. It is not only useful in calculating aggregate claim distributions, but also in the ruin theory, which will be introduced in the next section. Assume that individual claim amounts X_i are distributed on the non-negative integers and $f_k = P(X_1 = k)$. Then $S_N = X_1 + \cdots + X_N$ is also distributed on the non-negative integers. Let $p_n = P(N = n)$, $s_k = P(S_N = k)$, and $G_N(y)$ the probability generating function of N. The following theorem gives a recursive formula to compute $\{s_n\}_{n \geq 0}$.

Theorem 4.2 (Panjer recursion) *Suppose that N belongs to the $(a, b, 0)$ class, then $\{s_n\}_{n \geq 0}$ are given by*

$$s_0 = p_0 + \sum_{n=1}^{\infty} p_n f_0^n, \tag{4.14}$$

$$s_n = \frac{1}{1 - af_0} \sum_{k=1}^{n} \left(a + \frac{bk}{n}\right) f_k s_{n-k}, \quad n \geq 1. \tag{4.15}$$

Proof. First, for $n = 0$,

$$s_0 = P(N = 0) + \sum_{n=1}^{\infty} P(S_n = 0 | N = n) P(N = n) = p_0 + \sum_{n=1}^{\infty} p_n f_0^n.$$

By (4.11), the probability generating function of S_N is given by $G_{S_N}(t) = G_N[G_X(t)]$. Differentiation with respect to y yields that

$$G'_{S_N}(t) = G'_N[G_X(t)]G'_X(t).$$

Then formula (4.13) implies that

$$G'_{S_N}(t) = aG_X(t)G'_S(t) + (a + b)G_S(t)G'_X(t). \tag{4.16}$$

Since $G_{S_N}(t)$ and $G_X(t)$ are probability generating functions, $G_{S_N}(t) = \sum_{j=0}^{\infty} s_j t^j$ and $G_X(t) = \sum_{k=0}^{\infty} f_k t^k$, and their differentiations with respect to y are $G'_{S_N}(t) = \sum_{j=1}^{\infty} j s_j t^{j-1}$ and $G'_X(t) = \sum_{k=1}^{\infty} k f_k t^{k-1}$. Plugging these expressions into (4.16) yields

$$\sum_{n=1}^{\infty} n s_n t^{n-1} = a \left(\sum_{k=0}^{\infty} f_k t^k\right) \left(\sum_{j=1}^{\infty} j s_j t^{j-1}\right) + (a+b) \left(\sum_{j=0}^{\infty} s_j t^j\right) \left(\sum_{k=1}^{\infty} k f_k t^{k-1}\right)$$

$$= \sum_{k=0}^{\infty} \sum_{j=1}^{\infty} \left[(aj + ak + bk) f_k s_j t^{k+j-1}\right] = \sum_{n=1}^{\infty} \left(\sum_{k=0}^{n} (an + bk) f_k s_{n-k}\right) t^{n-1}.$$

Since the coefficients of t^{n-1} on both sides of the equation are same, we obtain that

$$n s_n = \sum_{k=0}^{n} (an + bk) f_k s_{n-k} = af_0 n s_n + \sum_{k=1}^{n} (an + bk) f_k s_{n-k}.$$

Solving for s_n yields the recursion (4.15).   □

Not only the recursive formula exists for the density function s_n of S_N, but the moments of S_N can be computed recursively as well.

Theorem 4.3 *Suppose that N belongs to the $(a,b,0)$ class. For $r = 1, 2, \ldots$, the moments of S_N are given recursively by*

$$E(S_N^r) = \frac{1}{1-a}\left\{\sum_{i=0}^{r-1}\left[a\binom{r}{i} + b\binom{r-1}{i}\right]E(S_N^i)E(X_1^{r-i})\right\}. \qquad (4.17)$$

Proof. By Panjer's recursion (4.15), we have

$$E(S_N^r) = \frac{1}{1-af_0}\sum_{n=1}^{\infty}n^r\sum_{k=1}^{n}\left(a + \frac{bk}{n}\right)f_k s_{n-k}$$

$$= \frac{1}{1-af_0}\sum_{k=1}^{\infty}f_k\sum_{n=k}^{\infty}\left(an^r + bkn^{r-1}\right)s_{n-k}$$

$$= \frac{1}{1-af_0}\sum_{k=1}^{\infty}f_k\sum_{m=0}^{\infty}\left(a(m+k)^r + bk(m+k)^{r-1}\right)g_m.$$

Use the binomial expansion for $(m+k)^r$, we obtain

$$\sum_{m=0}^{\infty}(m+k)^r g_m = \sum_{m=0}^{\infty}\sum_{i=0}^{r}\binom{r}{i}m^i k^{r-i}g_m$$

$$= \sum_{i=0}^{r}\binom{r}{i}k^{r-i}\sum_{m=0}^{\infty}m^i g_m = \sum_{i=0}^{r}\binom{r}{i}k^{r-i}E(S_N^i).$$

Hence

$$E(S_N^r) = \frac{1}{1-af_0}\sum_{k=1}^{\infty}f_k\left\{a\sum_{i=0}^{r}\binom{r}{i}k^{r-i}E(S_N^i) + bk\sum_{i=0}^{r-1}\binom{r-1}{i}k^{r-1-i}E(S_N^i)\right\}$$

$$= \frac{1}{1-af_0}\left\{a\sum_{i=0}^{r}\binom{r}{i}E(S_N^i)\sum_{k=1}^{\infty}k^{r-i}f_k + b\sum_{i=0}^{r-1}\binom{r-1}{i}E(S_N^i)\sum_{k=1}^{\infty}k^{r-i}f_k\right\}$$

$$= \frac{1}{1-af_0}\left\{\sum_{i=0}^{r-1}\left[a\binom{r}{i} + b\binom{r-1}{i}\right]E(S_N^i)E(X_1^{r-i}) + aE(S_N^r)\sum_{k=1}^{\infty}f_k\right\}.$$

Since $\sum_{k=1}^{\infty}f_k = 1 - f_0$, rearranging terms in the above identity yields the result.   □

Although both the density function and the moments of S_N can be computed via recursion, recursive formulas for the distribution function of S_N generally do not exist except for some special cases. The following shows an example on this.

Example 4.6 *Assume that N follows a geometric distribution with $p_n = p(1-p)^n$ for $n = 0, 1, \ldots$. In this case, $p_n = (a + (b/n))p_{n-1}$ with $a = 1-p$ and $b = 0$. Let $F_S(y) = P(S_N \leq y)$ be the distribution function of S_N. Then (4.15) implies that*

$$s_n = \frac{1-p}{1-(1-p)f_0} \sum_{k=1}^{n} f_k s_{n-k},$$

and for $y = 1, 2, 3, \ldots$,

$$F_S(t) = \sum_{n=0}^{t} s_n = s_0 + \frac{1-p}{1-(1-p)f_0} \sum_{k=1}^{n} f_k \sum_{n=1}^{t} s_{n-k}$$

$$= s_0 + \frac{1-p}{1-(1-p)f_0} \sum_{k=1}^{n} f_k F_S(t-k). \qquad \square$$

The Panjer recursion formula can be extended to some classes of distributions, for example, the $(a, b, 1)$ class of distributions in which the probability function p_r can be calculated recursively by $p_r = (a + b/r)p_{r-1}$ for $r = 2, 3, \ldots$. This class differs from the $(a, b, 0)$ class because the recursion starts with q_1 instead of q_0. However, by modifying the mass of probability at 0 in the $(a, b, 0)$ class, one can extend Theorems 4.2 and 4.3 and calculate the probability $s_k = P(S_N = k)$. Besides the $(a, b, 1)$ class, another class of distributions can be used to calculate the distribution of S_N is the *Schröter's class of distributions* whose probability function $\{p_r\}_{r \geq 0}$ can be calculated recursively from the formula

$$p_r = (a + b/r)p_{r-1} + (c/r)p_{r-2}$$

for $r \geq 1$, where a, b and c are constants and p_{-1} is defined to be zero. We refer readers to the detailed discussion about these two classes of distributions in Dickson (2016, Chapter 4).

Recursion formula and discretization of continuous distribution

The above recursion formulas assume that individual claim amounts are distributed on the non-negative integers. In practice, however, continuous distributions such as Pareto or lognormal are used to model individual claim amounts. In order to apply the recursion formula in such case, the continuous distribution should be replaced by an appropriate discrete distribution on the non-negative integers. Such process is referred to as discretizing a distribution.

There are many ways of discretization. For a continuous distribution F with $F(0) = 0$, a simple way is to construct a discrete distribution with probability function $\{\underline{h}_j = F(j) - F(j-1); j = 1, 2, \ldots, \}$. Since for any $x > 0$, $\sum_{j < x} \underline{h}_j \leq F(x)$, the distribution function of the constructed discrete random variable is a lower bound of F. Using the same idea, we can construct a discrete distribution $\{\overline{h}_j\}_{j \geq 0}$ with $\overline{h}_0 = F(1)$ and $\overline{h}_j = F(j+1) - F(j)$ for $j = 1, 2, \dots \}$, whose distribution function provides a upper bound for F.

Besides using the lower and upper bounds of F on non-negative integers, another approach is to match moments of the discrete distribution $\{\tilde{h}_j\}_{j \geq 0}$ with those of the continuous distribution. Let Y denote the constructed discrete random variables such that

$$P(Y \leq n) = \sum_{j=0}^{n} \tilde{h}_j = \int_{n}^{n+1} F(x)dx$$

for $n = 0, 1, \dots$. Then we have

$$E(Y) = \sum_{n \geq 0}[1 - P(Y \leq n)] = \sum_{n \geq 0}\int_{n}^{n+1}(1 - F(x))dx = \int_{0}^{\infty}(1 - F(x))dx = E(X),$$

and such discretization is mean preserving.

The recursive method of calculating aggregate claims distributions also raise some computational issues. One major concern is that not all recursions are stable, which implies the procedure may produce probabilities that are outside the interval $[0, 1]$. For instance, the Panjer recursion is stable when the claim number distribution is Poisson or negative binomial, but it becomes unstable when the claim number distribution is binomial.

4.2.4 Approximation of aggregate claim distribution

The recursive method introduced above can be computationally expensive, hence approximation methods are often used to simplify the calculation. The following outlines two such approximation methods.

The first method is *Normal approximation*. Since the individual claim amount X_i are i.i.d. (assume its second moment is finite), the distribution function F_S of $S_N = X_1 + \cdots + X_N$ can be approximated by

$$F_{S_N}(x) \approx \Phi\left(\frac{x - E(N)E(X_1)}{\sqrt{\mathrm{Var}(N)[E(X_1)]^2 + E(N)\mathrm{Var}(X_1)}}\right),$$

where $\Phi(\cdot)$ is the standard normal distribution function. For compound Poisson distribution with parameter λ and distribution F, the skewness parameter satisfies

$$\frac{E[[S_N - E(S_N))^3]}{[\mathrm{Var}(S_N)]^{3/2}} = \frac{E(X_1^3)}{\sqrt{\lambda[E(X_1^2)]^3}} > 0,$$

since X_1 is non-negative. As the mean of a normal approximation is zero, one may improve the approximation by find an approximation with positive skewness.

The *translated Gamma approximation* uses the first three moments of S_N and provides nonzero skewness. The idea is to approximate the distribution of S_N by that of $Y + k$, where k is a translation parameter and $Y \sim$ follows a Gamma(α, β) distribution. The parameters (k, α, β) are found by matching the mean, the variance and the skewness of $Y + k$ and S_N, which leads to the following equations

$$E(S_N) = \frac{\alpha + k}{\beta}, \quad \text{Var}(S_N) = \frac{\alpha}{\beta^2}, \quad \text{Skewness}(S_N) = \frac{2}{\sqrt{\alpha}}.$$

4.3 Ruin theory for the surplus process

Ruin theory in insurance explores the trajectory of an insurer's financial health and the likelihood of financial distress, and plays a critical role in risk management and regulatory compliance within the insurance industry. Ruin theory is motivated by the practical issue of solvency, which refers to the question of whether an insurance company possesses adequate assets to cover its liabilities. The primary objective of ruin theory is to evaluate the sufficiency of an insurer's surplus to cover potential losses arising from a portfolio of insurance policies. While the collective risk models analyze the aggregate amount of claims paid out in a single time period, ruin theory studies the evolution of an insurance fund over time. This section presents an overview of a continuous-time risk process and investigates the probability of ruin occurring over an infinite time horizon.

4.3.1 The classical ruin process

Let $\{N(t)\}_{t \geq 0}$ be a counting process for the number of claims, so that for a fixed value $t > 0$, the random variables $N(t)$ is the number of claims that occur in the fixed time interval $[0, t]$. Suppose that $X_i \geq 0$ is the amount of the ith claim. A stochastic *surplus process* (or *risk process*) is defined as follows

$$U(t) = u + ct - S(t), \tag{4.18}$$

where $U(t)$ is the insurer's capital at time t, $u = U(0)$ is the initial capital, c is the (constant) premium income per unit of time, and $S(t) = X_1 + \cdots + X_{N(t)}$. The counting process $N(t)$ is usually assumed to be a Poisson process that is defined as follows.

Definition 4.1 *The process $N(t)$ is a Poisson process with intensity $\lambda > 0$ if the increments of the process $N(t + h) - N(t)$ is Poisson(λh) distributed for all $t > 0, h > 0$.*

The above definition implies that the increments $N(t_i + h_i) - N(t_i)$ are independent for disjoint intervals $(t_i, t_i + h_i)$, $i = 1, 2, \ldots$. Note that in infinitesimally small intervals $(t, t + dt)$, a Poisson process $N(t)$ implies that

$$P(N(t + dt) - N(t) = 0 | N(s), 0 \le s \le t) = 1 - \lambda dt,$$
$$P(N(t + dt) - N(t) = 1 | N(s), 0 \le s \le t) = \lambda dt,$$
$$P(N(t + dt) - N(t) \ge 2 | N(s), 0 \le s \le t) = 0.$$

Let T_i denote the ith event time (or the occurrence of the ith claim), and $W_i = T_i - T_{i-1}$ ($i = 1, 2, \ldots, T_0 = 0$) be the waiting times. Denote by $\mathcal{F}_t$ the information set (or the σ-algebra of events) up to time t. Since Poisson processes are memoryless,

$$P(W_i > h | \mathcal{F}_t) = P(N(t + h) - N(t) = 0 | \mathcal{F}_t) = e^{-\lambda h},$$

$\{W_1, \ldots, W_k, \ldots\}$ are independent exponential random variables with mean $1/\lambda$ and are also independent of the history of the process. The probability of ruin in infinite time (or the ultimate ruin probability), is defined as

$$\psi(u) = P(U(t) < 0 \text{ for some } t \le 0). \tag{4.19}$$

Adjustment coefficient

The *adjustment coefficient* is a parameter used to quantify the rate at which an insurer's surplus is being depleted over time. It measures the speed of the insurer's financial deterioration. The adjustment coefficient takes account of two factors in the surplus process, aggregate claims and premium income.

Definition 4.2 *In a ruin process with claims distributed as $X \ge 0$, the adjustment coefficient, denoted by R, is defined as the unique positive root of the equation*

$$\lambda + cr = \lambda M_X(r), \tag{4.20}$$

where $M_X(r)$ is the MGF of X and λ is the intensity of the Poisson process $N(t)$.

Equation (4.20) has a unique positive root. To see this, consider the function $g(r) = \lambda M_X(r) - \lambda - cr$. First,

$$g(0) = 0, \quad g'(0) = \lambda M_X'(0) - c = \lambda E(X) - c.$$

This shows that g is decreasing function at zero if $c > \lambda E(X)$. Second,

$$g''(r) = \lambda M_X''(r) = \lambda \int_0^\infty x^2 e^{rx} f(x) dx > 0.$$

This shows that function $g(r)$ attains its minimum at some point. Assume that there exists a $\gamma > 0$ (or $\gamma = \infty$), such that $M_X(r)$ is finite for all $r < \gamma$ with $\lim_{r \to \infty} M_X(r) = \infty$. Then $\lim_{r \to \infty} g(r) = \infty$. The above assumption and argument show that $g(r)$ must have a unique turning point and hence has a unique positive root.

Example 4.7 (Adjustment coefficient for an exponential distribution) *Suppose that X follows an exponential distribution with mean $1/\alpha$. Its MGF is expressed as $M_X(r) = \alpha/(\alpha - r)$, hence the adjustment coefficient solves the equation $\lambda + cR = \lambda\alpha/(\alpha - r)$. Since R is a positive root, we obtain that $R = \alpha - \lambda/c$.* □

4.3.2 Probability of ruin and Lundberg's inequality

To estimate the risk of ruin, one may use Lundberg's inequality. Lundberg's inequality, named after the Swedish actuary Thorbjörn Lundberg, provides an upper bound on the probability of ruin for an insurance company under certain assumptions. We now introduce this inequality and an identity of calculating the ruin probability using the MGF of $U(t)$.

Theorem 4.4 (Lundberg's exponential bound) *For a compound Poisson risk process with an initial capital u, a premium per unit of time c, claims X with MGF $M_X(t)$, and an adjustment coefficient R satisfying (4.20), the ruin probability satisfies the inequality*

$$\psi(u) \le e^{-Ru}. \tag{4.21}$$

Proof. Let $\psi_k(u)$ $(k = 0, 1, \dots)$ be the probability that ruin occurs at or before the kth claim. Since for $k \to \infty$, $\psi_k(u)$ increases to its limit for all u, it suffices to show that $\psi_k(u) \le e^{-Ru}$ for all k. We next use induction. For $k = 0$, since $\psi_0(u) = 1$ if $u < 0$ and $\psi_0(u) = 0$ if $u \ge 0$, the inequality holds. Assume that the induction hypothesis holds for $k - 1$, that is $\psi_{k-1}(u) \le e^{-Ru}$ for all u. We split the event "ruin at or before the kth claim" into considerations regarding both time and size of the first claim. Assume that it occurs between time t and $t + dt$, which has probability $\lambda e^{-\lambda t} dt$. Assume that it has a size between x and $x + dx$, which has a probability $f(x)dx$, where $f(x)$ is the density function of X. Then the capital right after time t equals $u + ct - x$, and

$$\psi_k(u) = \int_0^\infty \int_0^\infty \psi_{k-1}(u + ct - x)f(x)\lambda e^{-\lambda t}dxdt \le \int_0^\infty \int_0^\infty e^{-R(u+ct-x)}\lambda e^{-\lambda t}dxdt$$

$$= e^{-Ru}\int_0^\infty \lambda e^{-t(\lambda+Rc)}dt \int_0^\infty e^{Rx}f(x)dx = e^{-Ru}\frac{\lambda}{\lambda+cR}M_X(R) = e^{-Ru}.$$

By induction, (4.21) holds for all $k \ge 0$. □

Since there is a positive probability $e^{-\lambda}$ of having no claims until time 1, and $\psi(1) \le e^{-R}$ by Theorem 4.4, for any non-negative initial capital $u \ge 0$ we have $1 - \psi(u) \ge \psi(0) \ge e^{-\lambda}(1 - e^{-R})$. Hence, the probability of non-ruin (or the survival probability) $\phi(u) = 1 - \psi(u)$ is strictly positive when $R > 0$ holds. Actually, we have

$$\phi(u) = \int_0^\infty \lambda e^{-\lambda t}\int_0^{u+ct} f(x)\psi(u + ct - x)dxdt = \frac{\lambda}{c}e^{\lambda u/c}\int_u^\infty \int_0^s f(x)\phi(s - x)dxds,$$

in which the second equality is obtained by substituting $s = u + ct$. Differentiation with respect to u gives

$$\phi'(u) = \frac{\lambda}{c}\phi(u) - \frac{\lambda}{c}\int_0^u f(x)\phi(u-x)dx. \tag{4.22}$$

In some special cases, Lundberg's exponential upper bound becomes an equality.

Theorem 4.5 *Consider a Poisson ruin process with claims X, intensity λ and premium rate c. Assume that X follows an exponential distribution with mean $1/\alpha$. Then*

$$\psi(u) = \frac{\lambda}{\alpha c}e^{-(\alpha-\lambda/c)u} = \frac{\lambda}{\alpha c}e^{-Ru}.$$

Proof. Consider the survival probability $\phi(u) = 1 - \psi(u)$ and the integro-differential equation (4.22). Since $f(x) = \alpha e^{-\alpha x}$ for $x \geq 0$, we have

$$\phi'(u) = \frac{\lambda}{c}\phi(u) - \frac{\lambda}{c}\int_0^u \alpha e^{-\alpha x}\phi(u-x)dx = \frac{\lambda}{c}\phi(u) - \frac{\alpha\lambda}{c}\int_0^u e^{-\alpha(u-x)}\phi(x)dx$$

$$= \frac{\lambda}{c}\phi(u) - \frac{\alpha\lambda}{c}e^{-\alpha u}\int_0^u e^{-\alpha x}\phi(x)dx. \tag{4.23}$$

Differentiation the above equation yields

$$\phi''(u) = \frac{\lambda}{c}\phi'(u) + \frac{\alpha^2\lambda}{c}e^{-\alpha u}\int_0^u e^{\alpha x}\phi(x)dx - \frac{\alpha\lambda}{c}\phi(u). \tag{4.24}$$

Equations (4.23) and (4.24) imply a second-order differential equation $\phi''(u) + \left(\alpha - \frac{\lambda}{c}\right)\phi'(u) = 0$. The general solution of this equation is $\phi(u) = a_0 + a_1 e^{-(\alpha-\lambda/c)u}$, where a_0 and a_1 are constants. By Lundberg's inequality, we have $\lim_{u\to\infty}\phi(u) = 1$, which gives $a_0 = 1$. It then follows that $\phi(0) = 1 + a_1$ or $a_1 = -\psi(0)$. Hence

$$\psi(u) = \psi(0)e^{-(\alpha-\lambda/c)u} = \psi(0)e^{-Ru}. \tag{4.25}$$

To solve for $\psi(0)$, note that (4.22) implies that

$$\psi'(u) = \frac{\lambda}{c}\psi(u) - \frac{\lambda}{c}\int_0^u f(x)\psi(u-x)dx - \frac{\lambda}{c}[1 - F(u)].$$

Integrating this equation over $(0, \infty)$ yields that

$$-\psi(0) = \frac{\lambda}{c}\left(\int_0^\infty \psi(u)du - \int_0^\infty\int_0^u f(x)\psi(u-x)dxdu - \int_0^\infty (1 - F(u))du\right).$$

By changing the order of the integration in the double integral, the first two terms on the right-hand side of the above equation are canceled. Hence, we have

$$\psi(0) = \frac{\lambda}{c} \int_0^\infty (1 - F(u)) du = \frac{\lambda E(X)}{c} = \frac{\lambda}{\alpha c}. \qquad (4.26)$$

Combining (4.25) and (4.26) completes the proof. □

We next give an expression for the ruin probability that involves the MGF of $U(T)$, which is the capital at the moment of first ruin, conditionally given the event that ruin occurs in a finite time period.

Theorem 4.6 *Let R be the adjustment coefficient. The ruin probability for $u \geq 0$ satisfies*

$$\psi(u) = \frac{e^{-Ru}}{E[e^{-RU(T)}|T \leq \infty]}. \qquad (4.27)$$

Proof. We show the result in three steps. First, for $R > 0$ and $t > 0$, we have

$$E(e^{-RU(t)}) = E(e^{-RU(t)}|T \leq t)P(T \leq t) + E(e^{-RU(t)}1_{\{T>t\}}). \qquad (4.28)$$

Since $U(t) = u + ct - S(t)$ and $S(t)$ is a compound Poisson process with intensity λ,

$$E(e^{-RU(t)}) = E[e^{-R(u+ct-S(t))}] = e^{-Ru}\left[e^{-Rc}\exp[\lambda(M_X(R)-1)]\right]^t = e^{-Ru}.$$

Then, given $\nu \in [0,t]$, the total claims $S(t) - S(\nu)$ has a compound Poisson distribution. Moreover, $U(\nu)$ and $S(t) - S(\nu)$ are independent. These two facts imply that

$$E(e^{-RU(t)}|T = \nu) = E[e^{-R[U(\nu)+c(t-\nu)-(S(t)-S(\nu))]}|T = \nu]$$
$$= E(e^{-RU(\nu)}|T = \nu)e^{-Rc(t-\nu)}E[e^{R(S(t)-S(\nu))}|T = \nu]$$
$$= E(e^{-RU(\nu)}|T = \nu)\left[e^{-Rc}\exp[\lambda(M_X(R)-1)]\right]^{t-\nu} = E(e^{-RU(T)}|T = \nu).$$

Since the above equality holds for all $\nu < t$, we have $E(e^{-RU(t)}|T \leq \nu) = E(e^{-RU(T)}|T \leq \nu)$.

Finally, we consider the event $T > t$. It implies that $U(t) \geq 0$ or $e^{-RU(t)} \leq 1$. For some positive function $u_0(t)$, we have

$$E(e^{-RU(t)}1_{\{T>t\}}) = E(e^{-RU(t)}1_{\{T>t,0\leq U(t)\leq u_0(t)\}}) + E(e^{-RU(t)}1_{\{T>t,U(t)>u_0(t)\}})$$
$$\leq P(U(t) \leq u_0(t)) + E(e^{-Ru_0(t)}).$$

The second term vanishes if $u_0(t) \to \infty$. For the first term, note that $E(U(t)) = u + ct - \lambda t E(X)$ and $\mathrm{Var}(U(t)) = \lambda t E(X^2)$. Choose $u_0(t) = t^{2/3}$ so that $(E(U(t)) - u_0(t))/\mathrm{Var}(U(t)) \to \infty$ and apply Chebyshev's inequality, the first term vanishes. Then, letting $t \to \infty$ in (4.28) yields the identity (4.27). □

4.3.3   Maximal aggregate loss

We next show that the non-ruin probability $\phi(u) = 1 - \psi(u)$ can be written as a compound geometric distribution function with known specifications. Consider a new process $\{L(t)\}_{t \geq 0}$ with $L(t) = S(t) - ct$ for all $t \geq 0$. $L(t)$ is called the *aggregate loss* process and $U(t) = u - L(t)$. Define the *maximal aggregate loss* as follows

$$L = \max\{S(t) - ct \mid t \geq 0\}. \qquad (4.29)$$

Since $S(0) = 0$, we have $L \geq 0$. Besides, the event $L(t) > u$ is equivalent to the event $U(t) \geq 0$ for all t, hence $\phi(u) = P(L \leq u)$.

Let L_j $(j = 1, 2, \dots)$ be the difference in amounts between the jth record low and the $(j-1)$th one. Denote by M the random number of new records. We have $L = L_1 + \dots L_M$. Since a Poisson process is memoryless, the probability that a particular record low is the last one is same at every time. Hence, M follows a geometric distribution(p_M) and $L_1, L_2, \dots$ are independent and identically distributed. The success probability p_M, or the probability that the previous record is the last one, equals the probability to avoid ruin starting with initial capital 0, hence $p_M = 1 - \psi(0)$. The value of p_M and the distribution of L_1 conditional on $M \geq 1$ can be obtained from the following theorem.

Theorem 4.7 *If the initial capital u equals 0, then for all $y > 0$,*

$$P[U(T) \in (-y - dy, -y), T < \infty] = \frac{\lambda}{c}[1 - F(y)]dy. \qquad (4.30)$$

Proof. Let $G(u, y) = P(U(T) \in (-\infty, -y), T < \infty | U(0) = u)$. Note that, in the interval $(t, t + dt)$, there is either no claim and the capital increases from u to $u + cdt$, or one claim of random size X with probability λdt. The latter case has two possibilities. If the claim size is less than u, then the process continues with capital $u + cdt - X$. Otherwise ruin occurs, but the capital at ruin is only larger than y if $X > u + y$. Hence

$$G(u, y) = (1 - \lambda dt)G(u + cdt, y) + \lambda dt \left[\int_0^u G(u - x, y) dP(x) + \int_{u+y}^\infty dP(x) \right].$$

Substitute $G(u + cdt, y) = G(u, y) + cdt G'(u, y)$ into the above equality and divide by cdt, we obtain

$$G'(u, y) = \frac{\lambda}{c} \left[G(u, y) - \int_0^u G(u - x, y) dF(x) - \int_{u+y}^\infty dF(x) \right].$$

Integrate the above equality over $u \in [0, z]$ and note that

$$\int_0^z \int_0^u G(u - x, y) dF(x) du = \int_0^z \int_0^{z-v} G(v, y) dF(x) dv = \int_0^z G(v, y) P(z - v) dv,$$

$$\int_0^z \int_{u+y}^\infty dF(x)du = \int_y^{z+y} (1 - F(v))dv,$$

we obtain that

$$G(z,y) - G(0,y) = \frac{\lambda}{c}\Big[\int_0^z G(u,y)(1 - P(z-u))du - \int_y^{z+y}(1-F(u))du\Big].$$

Let $z \to \infty$ in the above equality, the first terms on both sides vanish and

$$G(0,y) = \frac{\lambda}{c}\int_y^\infty (1-F(u))du.$$

Taking derivative with respect to y yields (4.30). $\square$

Theorem 4.7 can generate interesting results, we show some of them as follows; see also Exercise 4.9.

Proposition 4.4 *(i) The ruin probability at 0 is given by*

$$\psi(0) = \frac{\lambda}{c}\int_0^\infty [1 - F(y)]dy = \lambda E(X)/c. \qquad (4.31)$$

(ii) Assume that $M \geq 1$ and L_1 has the same distribution as the amount with which ruin occurs starting from $u = 0$ (if ruin occurs). Then the density of L_1 is $f_{L_1}(y) = (1 - F(y))/E(X)$ and the MGF of L_1 is

$$M_{L_1}(r) = \frac{1}{rE(X)}\int_0^\infty (1-F(y))d(e^{ry}-1) = \frac{1}{rE(X)}[M_X(r)-1]. \quad (4.32)$$

(iii) Since $L = L_1 + \cdots + L_M$ with M following a geometric distribution with success probability $p_M = 1 - \psi(0) = 1 - \lambda E(X)/c$, identity (4.10) implies that

$$M_L(r) = \frac{p_M}{1 - (1 - p_M)M_{L_1}(r)} = \frac{r(c - \lambda E(X))}{cr - \lambda M_X(r) - 1}. \qquad (4.33)$$

(iv) The ruin probability in u can be expressed in the following recursive form,

$$\psi(u) = \frac{\lambda}{c}\Big[\int_0^u [1 - F(y)]\psi(u - y)dy + \int_u^\infty [1 - F(y)]dy\Big]. \qquad (4.34)$$

Here, the derivation of (4.34) needs to use the fact

$$P(T < \infty) = P(T < \infty|M > 0)P(M > 0) = \frac{\lambda}{c}\int_0^\infty P(T < \infty|L_1 = y)[1 - F(y)]dy.$$

4.4 Credibility theory and regression-type models

In insurance practice, determining a premium for a group of insurance contracts involves considering both the individual experience of each policyholder and the collective experience of the entire group. Credibility theory addresses this challenge by providing statistical methods to estimate premiums based on historical data. Credibility theory is essentially a Bayesian inference method, which estimate the premium using historical data. The theory focuses on developing statistical techniques that effectively balance the individual experience of policyholders with the overall experience of the group.

This section starts with an introduction to fundamental concepts of Bayesian inference and empirical Bayes methods. Bayesian inference allows us to update our beliefs about parameters based on observed data, while empirical Bayes methods combine observed data with prior information to make more accurate estimates. Subsequently, this section presents several models commonly used in credibility theory. These include the Bühlmann-Straub model, which is a traditional approach in credibility theory, as well as more modern models such as linear mixed models, generalized linear models, and generalized linear mixed models. Each model offers a different framework for incorporating individual and collective experience into premium estimation, catering to various practical scenarios encountered in insurance pricing.

4.4.1 Bayes inference and empirical Bayes

Suppose that the joint probability density function of $X_1, \ldots, X_n$ is $f_{\boldsymbol{\theta}}(x_1, \ldots, x_n)$, in which $\boldsymbol{\theta}$ is a parameter taking values in the parameter space Θ. Bayesian inference is based on a conditional distribution, called the *posterior distribution*, of $\boldsymbol{\theta}$ given the data, treating the unknown $\boldsymbol{\theta}$ as a random variable that has a *prior distribution* with density function π. The joint density function of $(\boldsymbol{\theta}, X_1, \ldots, X_n)$ is $\pi(\boldsymbol{\theta}) f_{\boldsymbol{\theta}}(x_1, \ldots, x_n)$. Hence, the posterior density function $\pi(\boldsymbol{\theta}|X_1, \ldots, X_n)$ of $\boldsymbol{\theta}$ given $X_1, \ldots, X_n$ is proportional to $\pi(\boldsymbol{\theta}) f_{\boldsymbol{\theta}}(X_1, \ldots, X_n)$, with the constant of proportionality chosen so that the posterior density function integrates to 1:

$$\pi(\boldsymbol{\theta}|X_1, \ldots, X_n) = \pi(\boldsymbol{\theta}) f_{\boldsymbol{\theta}}(X_1, \ldots, X_n)/g(X_1, \ldots, X_n), \qquad (4.35)$$

where $g(X_1, \ldots, X_n) = \int \pi(\boldsymbol{\theta}) f_{\boldsymbol{\theta}}(X_1, \ldots, X_n) d\boldsymbol{\theta}$ (with the integral replaced by a sum when the X_i are discrete) is the marginal density of $X_1, \ldots, X_n$ in the Bayesian model. Often one can ignore the constant of proportionality $1/g(X_1, \ldots, X_n)$ in identifying the posterior distribution with density function (4.35), as illustrated by the following example.

Example 4.8 *Let $X_1, \ldots, X_n$ be i.i.d. $N(\mu, \sigma^2)$, where σ is assumed to be known and μ is the unknown parameter. Suppose μ has a prior $N(\mu_0, v_0)$*

distribution. Letting $\bar{X} = n^{-1}\sum_{i=1}^{n} X_i$, the posterior density function of μ given $X_1, \ldots, X_n$ is proportional to

$$e^{-(\mu-\mu_0)^2/(2v_0)} \prod_{i=1}^{n} e^{-(X_i-\mu)^2/(2\sigma^2)} \propto \exp\left\{-\frac{\mu^2}{2}\left[\frac{1}{v_0} + \frac{n}{\sigma^2}\right] + \mu\left[\frac{\mu_0}{v_0} + \frac{n\bar{X}}{\sigma^2}\right]\right\}$$

$$\propto \exp\left\{-\frac{1}{2}\left(\frac{1}{v_0} + \frac{n}{\sigma^2}\right)\left[\mu - \left(\frac{\mu_0}{v_0} + \frac{n\bar{X}}{\sigma^2}\right)\bigg/\left(\frac{1}{v_0} + \frac{n}{\sigma^2}\right)\right]^2\right\},$$

which, as a function of μ, is proportional to a normal density function. Hence, the posterior distribution of μ given $(X_1, \ldots, X_n)$ is

$$N\left(\left(\frac{\mu_0}{v_0} + \frac{n\bar{X}}{\sigma^2}\right)\bigg/\left(\frac{1}{v_0} + \frac{n}{\sigma^2}\right), \left(\frac{1}{v_0} + \frac{n}{\sigma^2}\right)^{-1}\right). \qquad (4.36)$$

Note that we have avoided the integration to determine the constant of proportionality in the preceding argument. From (4.36), the mean of the posterior distribution is a weighted average of the sample mean $\bar{X}$ and the prior mean μ_0. Moreover, the reciprocal of the posterior variance is the sum of the reciprocal of the prior variance and n/σ^2, which is the reciprocal of the variance of $\bar{X}$. □

Bayes procedure

Bayes procedures are defined by minimizing certain functionals of the posterior distribution of θ given the data. To describe these functionals, we first give a brief introduction to statistical decision theory. In this theory, statistical decision problems such as estimation and hypothesis testing have the following ingredients:

(i) a parameter space Θ and a family of distribution $\{P_\theta, \theta \in \Theta\}$;

(ii) data $(X_1, \ldots, X_n) \in \mathcal{X}$ sampled from the distribution P_θ when θ is the true parameter, where $\mathcal{X}$ is called the "sample space";

(iii) an action space $\mathcal{A}$ consisting of all available actions to be chosen; and

(iv) a loss function $L : \Theta \times \mathcal{A} \rightarrow [0, \infty)$ representing the loss $L(\theta, a)$ when θ is the parameter value and action a is taken.

A *statistical decision rule* is a function $d : \mathcal{X} \rightarrow \mathcal{A}$ that takes action $d(\mathbf{X})$ when $\mathbf{X} = (X_1, \ldots, X_n)$ is observed. Its performance is evaluated by the *risk function*

$$R(\theta, d) = E_\theta L(\theta, d(\mathbf{X})), \qquad \theta \in \Theta.$$

A statistical decision rule d is as good as d^* if $R(\theta, d) \leq R(\theta, d^*)$ for all $\theta \in \Theta$. In this case, d is better than d^* if strict inequality also holds at some θ. A statistical decision rule is said to be *inadmissible* if there exists another rule that is better; otherwise, it is called *admissible*.

Given a prior distribution π on Θ, the *Bayes risk* of a statistical decision rule d is

$$B(d) = \int R(\theta, d) d\pi(\theta). \tag{4.37}$$

A *Bayes rule* is a statistical decision rule that minimizes the Bayes risk. It can be determined by

$$d(\mathbf{X}) = \arg \min_{a \in \mathcal{A}} E\big[L(\theta, a) | \mathbf{X}\big].$$

This follows from the key observation that (4.37) can be expressed as

$$B(d) = E\big\{E\big[L(\theta, d(\mathbf{X}) | \theta\big]\big\} = EL(\theta, d(\mathbf{X})) = E\big\{E\big[L(\theta, d(\mathbf{X})) | \mathbf{X}\big]\big\},$$

where we have used the *tower property* $E(Z) = E[E(Z|W)]$ of conditional expectations for the second and third equalities. Hence, to minimize $B(\theta, d)$ over statistical decision functions $d : \mathcal{X} \to \mathcal{A}$, it suffices to choose the action $a \in \mathcal{A}$ that minimizes for the observed data $\mathbf{X}$ the *posterior loss* $E\big[L(\theta, a) | \mathbf{X}\big]$, which is the expected value $\int L(\theta, a) d\pi(\theta|\mathbf{X})$ with respect to the posterior distribution $\pi(\cdot|\mathbf{X})$ of θ. This shows the central role of the posterior distribution in Bayes rules. Under certain weak conditions, Bayes rules can be shown to be admissible. Conversely, admissible statistical decision rules are either Bayes rules or limits of Bayes rules. This is a fundamental result in decision theory, known as the admissibility of Bayes rules.

Estimation of parameters $\boldsymbol{\theta} \in \mathbb{R}^d$ can be considered as a statistical decision problem with $d(\mathbf{X})$ being the estimator. The usual squared error loss in estimation theory corresponds to the loss function $L(\boldsymbol{\theta}, \mathbf{a}) = ||\boldsymbol{\theta} - \mathbf{a}||^2$. Hence, $\Theta = \mathcal{A} = \mathbb{R}^d$ and $E\big(||\boldsymbol{\theta} - \mathbf{a}||^2 | \mathbf{X}\big)$ is minimized by the posterior mean $\mathbf{a} = E(\boldsymbol{\theta}|\mathbf{X})$. Consider Example 4.8 in which $X_1, \ldots, X_n$ are i.i.d. $N(\mu, \sigma^2)$ with σ known and μ is unknown but has a prior $N(\mu_0, v_0)$ distribution. The posterior distribution (4.36) implies that the Bayes estimator of μ (with respect to squared error loss) is

$$E(\mu|X_1, \ldots, X_n) = \left(\frac{\mu_0}{v_0} + \frac{n\bar{X}}{\sigma^2}\right) \Big/ \left(\frac{1}{v_0} + \frac{n}{\sigma^2}\right). \tag{4.38}$$

Compound decision theory and empirical Bayes procedures

Compound decision theory, introduced by Robbins in 1950, deals with a sequence of independent statistical decision problems of the same form. The basic idea of this theory is to achieve substantial reduction of the total risk by allowing the statistical procedures for each individual problem to take into account the observations from the entire sequence. This approach enables the utilization of information across multiple decision problems, potentially leading to more efficient decision-making processes.

Let $f(x; \theta)$ be a family of probability measures with a parameter θ. Consider a sequence of independent experiments with observations $X_i \sim f(x; \theta_i)$, $i = 1, \ldots, n$, where θ_i are deterministic unknown parameters. Suppose that we are interested in making statistical decisions δ_i about θ_i with a loss function $L(a, \theta)$. Robbins (1951) formulated the compound decision problem, in which δ_i are allowed to depend on the observations $X_{(n)}$ from all n experiments, under the compound risk

$$R_n(\delta_{(n)}, \theta_{(n)}) \equiv \frac{1}{n} \sum_{i=1}^{n} E_{\theta_{(n)}} L(\theta_i, \delta_i(X_{(n)})). \tag{4.39}$$

If the ith decision is a fixed deterministic function of X_i, $\delta_i(X_{(n)}) = d(X_i)$, that is, the compound risk (4.39) is equal to the Bayes risk

$$R(d, G) := \int \left[\int L(d(x), \theta) f(x; \theta) dx \right] dG(\theta)$$

for a single decision problem under the unknown prior $G(A) = G_n(A) = n^{-1} \sum_{i=1}^{n} 1_{\{\theta_i \in A\}}$.

Empirical Bayes methodology

The *empirical Bayes* (EB) methodology was introduced by Robbins (1956) for n independent and structurally similar problems of statistical inference on unknown parameters θ_i from observed data Y_i ($i = 1, \ldots, n$), where Y_i has probability density $f(y|\theta_i)$ and θ_i and Y_i can represent vectors. The θ_i are assumed to have a common prior distribution G that has unspecified hyperparameters. Letting $d_G(y)$ be the Bayes decision rule (with respect to some loss function and assuming known hyperparameters) when $Y_i = y$ is observed, the basic principle underlying empirical Bayes is that d_G can often be consistently estimated from $Y_1, \ldots, Y_n$, leading to the empirical Bayes decision rule $d_{\widehat{G}}$. Thus, the n structurally similar problems can be pooled to provide information about unspecified hyperparameters in the prior distribution, thereby yielding $\widehat{G}$ and the decision rules $d_{\widehat{G}}(Y_i)$ for the independent problems.

In the empirical Bayes setting, the unknown parameter θ_i are assumed as independent random variables with an unknown common prior G, then the risk for the compound version (4.39) is

$$R_n(\delta_{(n)}, G) := n^{-1} \sum_{i=1}^{n} E_G L(\delta_i(X_{(n)}), \theta_i) = E_G R_n(\delta_{(n)}, \theta_{(n)}).$$

Let $R(d, G)$ be the Bayes risk (4.4.1) and let

$$d_G^*(x) := \arg\min_{d \in \mathcal{A}} R(d, G)$$

be the ideal Bayes rule, where G is the unknown prior in empirical Bayes problems and $G = G_n$ is the empirical distribution of unknown parameters in compound decision problems. Robbins' procedures can be written as

$$\delta_i(X_n) = \widetilde{d}_n(X_i),$$

where $\tilde{d}_n(X_i)$ is an estimate of $d_G^*(x)$ based on the observation $X_{(n)}$. As a general solution, he suggested using $\hat{d}_n = d_{\hat{G}}^*(x)$ with a suitable estimate $\hat{G}$ of G and formulated the problem of estimating the prior G based on $X_{(n)}$.

Linear empirical Bayes estimators

In particular, Robbins (1956) considered Poisson X_i with mean θ_i, as in the case of the number of accidents by the ith driver in a sample of size n (in a given year) from a population of drivers, with distribution G for the accident-proneness parameter θ. In this case the Bayes estimate (with respect to squared error loss) of θ_i when $X_i = x$ is given by

$$d_g(x) = (x+1)g(x+1)/g(x), \quad x = 0, 1, \ldots, \tag{4.40}$$

where $g(x) = \int_0^\infty \theta^x e^\theta dG(\theta)/(x!)$. Using $\hat{g}(k) = n^{-1}\sum_{i=1}^n I_{\{X_i=k\}}$ to replace $g(k)$ in the above estimate yields the empirical Bayes estimate $d_{\hat{g}}(x)$.

Stein (1956) and subsequently James and Stein (1961) considered the special case $X_i \sim N(\theta_i, \sigma^2)$ with known σ but unknown θ_i. Let (μ_i, X_i), $i = 1, \ldots, n$, be independent random vectors such that $X_i|\mu_i \sim N(\mu_i, \sigma^2)$ and $\mu_i \sim N(\mu_0, v_0)$. Again assuming that σ^2 is known, the Bayes estimate $E(\mu_i|X_i)$ of μ_i is given by (4.38) with $n\bar{X}$ replaced by X_i and n/σ^2 replaced by $1/\sigma^2$. Note that (4.38) requires a "subjective" choice of the "hyperparameters" μ_0 and v_0. In the present setting, we can in fact estimate them from the data since

$$E(X_i) = E\{E(X_i|\mu_i)\} = E(\mu_i) = \mu_0,$$
$$\mathrm{Var}(X_i) = E\big(\mathrm{Var}(X_i|\mu_i)\big) + \mathrm{Var}\big(E(X_i|\mu_i)\big) = \sigma^2 + v_0.$$

Since (X_i, μ_i) are i.i.d., we can use the above two formulas and the method of moments to estimate μ_0 and v_0 by

$$\hat{\mu}_0 = \bar{X}\left(= n^{-1}\sum_{i=1}^n X_i\right), \qquad \hat{v}_0 = \left(n^{-1}\sum_{i=1}^n (X_i - \bar{X})^2 - \sigma^2\right)_+.$$

Robbins (1983) gives a general description of linear empirical Bayes models. Suppose the decision rule $d(x)$ is restricted to linear functions of the form $A + Bx$ and assume that the family of density functions $f(x|\theta)$ satisfies

$$E(X|\theta) = \theta, \qquad \mathrm{Var}(X|\theta) = a + b\theta + c\theta^2 \tag{4.41}$$

for some constants a, b and $c \neq -1$. Then the Bayes rule $d_G(y)$ that minimizes the Bayes risk, with respect to squared error loss, among linear estimators of θ is given by the linear regression formula

$$d_G(x) = E\theta + \frac{\mathrm{Cov}(\theta, X)}{\mathrm{Var}(X)}(x - EX). \tag{4.42}$$

Moreover, from (4.41), it follows that $EX = E\theta$, $\mathrm{Cov}(\theta, X) = \mathrm{Var}(\theta)$, and

$$\mathrm{Var}(X) = E[\mathrm{Var}(X|\theta)] + \mathrm{Var}[E(X|\theta)] = a + bE\theta + c(E\theta)^2 + (c+1)\mathrm{Var}(\theta),$$

and therefore

$$\mathrm{Var}(\theta) = \frac{\mathrm{Var}(X) - \{a + bEX + c(EX)^2\}}{c+1}. \tag{4.43}$$

Since $(\theta_1, X_1), \ldots, (\theta_n, X_n)$ are i.i.d vectors from the distribution under which θ_i has prior distribution G and X_i has conditional density $f(\cdot|\theta_i)$ given X_i, the parameters $E\theta$, $\mathrm{Cov}(\theta, X)$, $\mathrm{Var}(X)$ and EX in the Bayes rule (4.42) can be estimated consistently by using (4.41), (4.43) and the method of moments. This yields the linear EB estimate

$$\widehat{d}(x) = \bar{X} + \left[1 - \frac{cs^2 + a + b\bar{X} + c\bar{X}^2}{(c+1)s^2}\right]_+ (x - \bar{X}),$$

where $y_+ = \max(y, 0)$, $\bar{X} = n^{-1}\sum_{i=1}^{n} X_i$ estimates the mean EX and $s^2 = \sum_{i=1}^{n}(X_i - \bar{X})^2/(n-1)$ estimates the variance $\mathrm{Var}(X)$.

In the case $X_i|\theta_i \sim N(\theta_i, \sigma^2)$ with known σ^2, (4.41) holds with $a = \sigma^2$ and $b = c = 0$, for which (4.4.1) corresponds to a variant of the James-Stein (James and Stein, 1961) estimator of $(\theta_1, \ldots, \theta_n)'$ that has a smaller mean squared error than the maximum likelihood estimator $(X_1, \ldots, X_n)'$. Moreover, if G is normal, then (4.42) minimizes the Bayes risk among all (not necessarily linear) estimators of θ.

4.4.2   The Bühlmann-Straub model

The Bühlmann-Straub model, named after Hans Bühlmann and Daniel Straub, is a popular approach in credibility theory, which is widely used in insurance pricing. This model aims to determine insurance premiums for a group of policies by combining individual experience with group experience in a systematic manner. In the Bühlmann-Straub model, premiums are calculated based on a weighted average of two components: the individual experience of each policyholder and the collective experience of the entire group. The weights assigned to these two components reflect the credibility assigned to each source of information.

Consider a sequence of observed aggregate claim amount $Y_t = S_{t,N_t}$ ($t = 1, \ldots, n$), in which S_{t,N_t} is defined in the same way as the aggregate claim amount specified in equation (4.5), i.e.,

$$S_{t,N_t} = X_{t,1} + \cdots + X_{t,N_t}, \qquad t = 1, \ldots, n.$$

Our interest now is to determine the risk premium for the aggregate claims Y_{n+1} given the observation $\mathbf{Y} = (Y_1, \ldots, Y_n)'$. In order to do this, one needs to make certain assumption on the distribution function of Y_t.

The simplest standard assumption is that $Y_1, \ldots, Y_n$ are (conditionally) independent and identically distributed with (conditional) distribution function $F_\theta(y)$, in which the parameter or the *risk profile* θ is unknown and varies from risk to risk. The correct risk premium with risk profile θ can then be expressed as $\mu(\theta) := E(X_{n+1}|\theta)$. Since θ is unknown and varies from risk to risk, one can assume that θ follows some prior distribution. Then, according to the Bayes inference procedure outlined in Section 4.4.1, the posterior mean $\widehat{\mu}(\theta) = E(\mu(\theta)|\mathbf{Y})$ is the best possible estimator within the class of all estimator functions.

However, it is usually challenging to compute $\widehat{\mu}(\theta)$ in practice for the following reasons. Firstly, $\widehat{\mu}(\theta)$ typically cannot be expressed in a closed analytical form except for some special cases, hence one can only calculate $\widehat{\mu}(\theta)$ by numerical procedures. Secondly, one may also have difficulty in finding prior distribution of θ, so it is not feasible to compute the posterior distribution or mean of $\mu(\theta)$. The basic idea of the *credibility* theory is to restrict the class of allowable estimator functions to those that are linear in the observations $(Y_1, \ldots, Y_n)'$. From this perspective, credibility estimators can be viewed as linear Bayes estimators. The following presents a simple credibility model.

Example 4.9 (A simple credibility model) *Suppose that random variables Y_t $(t = 1, \ldots, n)$ are independent with the distribution function $F_\theta(y)$ and let $\mu(\theta) = E(Y_t|\theta)$ and $\sigma^2(\theta) = Var(Y_t|\theta)$. Assume that θ follows a prior distribution $\pi(\theta)$ and let $\mu_0 = \int \mu(\theta)\pi(\theta)d\theta$. To find the linear Bayes estimator for $\mu(\theta)$, we note that the probability distribution of $Y_1, \ldots, Y_n$ is invariant under permutations of Y_t (i.e., the role of $Y_1, \ldots, Y_n$ are symmetric and exchangeable). This indicates that the linear estimator for $\mu(\theta)$ must be expressed as*

$$\widehat{\mu}(\theta) = \widehat{a} + \widehat{b}\overline{Y}$$

where $\overline{Y} = n^{-1}\sum_{t=1}^n Y_t$. Moreover, $\widehat{a}$ and $\widehat{b}$ are the solution of the minimization problem

$$\min_{a,b\in\mathbb{R}} E\left\{\left[\mu(\theta) - a - b\overline{Y}_j\right]^2\right\}.$$

To find $\widehat{a}$ and $\widehat{b}$, we take partial derivatives with respect to a and b and obtain

$$E\left[\mu(\theta) - a - b\overline{Y}\right] = 0, \qquad Cov(\overline{Y}, \mu(\theta)) = b\,Var(\overline{Y}).$$

Let $\tau^2 = Var(\mu(\theta))$, we have $Cov(\overline{Y}, \mu(\theta)) = Var(\mu(\theta)) = \tau^2$ and

$$Var(\overline{Y}) = n^{-1}E(\sigma^2(\theta)) + Var(\mu(\theta)) = \tau^2 + \sigma^2/n.$$

This shows that $\widehat{a}$ and $\widehat{b}$ are given by

$$\widehat{b} = \frac{\tau^2}{\tau^2 + \sigma^2/n}, \qquad \widehat{a} = (1 - \widehat{b})\mu_0.$$

Then the linear estimator $\widehat{\mu}(\theta)$ can be expressed as

$$\widehat{\mu}(\theta) = \alpha \overline{Y} + (1-\alpha)\mu_0, \qquad \alpha = \frac{n}{n + \sigma^2/\tau^2}.$$

Moreover, one can calculate the quadratic loss of $\widehat{\mu}(\theta)$, which is given by

$$E\left[\left(\widehat{\mu}(\theta) - \mu(\theta)\right)^2\right] = (1-\alpha)\tau^2 = \alpha \frac{\sigma^2}{n}. \qquad \square$$

In the above example, the best linear estimator is a weighted mean of $\mu_0 = E(\mu(\theta))$ and the individual observed average $\overline{Y}$. The quotient $\kappa = \sigma^2/\tau^2$ is called the *credibility coefficient*. The coefficient α is called the *credibility weight* and it increases as the number of observed samples n increases. Since κ can also be written as $\kappa = (\sigma/\mu_0)^2/(\tau/\mu_0)^2$, α also increases as the heterogeneity of the portfolio (τ/μ_0) increases or as the within risk variability (σ/μ_0) decreases.

Bühlmann and Straub (1970) developed the following credibility model that is more general than the one in Example 4.9. In particular, suppose that we have a portfolio of I risks. Denote by Y_{it} the jth claim of the ith class. Let $\mathbf{Y}_i = (Y_{i1}, \dots, Y_{i,n_i})'$. Assume that the risk i is characterized by an individual risk profile θ_i, and conditional on θ_i, the $\{Y_{it}; t = 1, \dots, n\}$ are independent with the following mean and variance,

$$E(Y_{it}|\theta_i) = \theta_i, \qquad \mathrm{Var}(Y_{it}|\theta_i) = \sigma_i^2,$$

for $1 \leq j \leq n_i, 1 \leq i \leq I$. Assuming a normal prior $N(\mu, \tau^2)$ for θ_i, the Bayes estimate of θ_i is

$$E(\theta_i|\mathbf{Y}_i) = \alpha_i \bar{Y}_i + (1-\alpha_i)\mu, \qquad (4.44)$$

where $\alpha_i = \tau^2/(\tau^2 + \sigma_i^2/n_i)$ and $\bar{Y}_i = n_i^{-1} \sum_{j=1}^{n_i} Y_{ij}$. Since $E(Y_{ij}) = E(\theta_i) = \mu$ and $\mathrm{Var}(Y_{ij}) = \mathrm{Var}(\theta_i) + E(\mathrm{Var}(Y_{ij}|\theta_i)) = \tau^2 + \sigma_i^2$, we can estimate μ, σ_i^2 and τ^2 by the method of moments when $n_i > 1$,

$$\widehat{\mu} = \frac{1}{\sum_{i=1}^{I} n_i} \sum_{i=1}^{I} \sum_{j=1}^{n_i} Y_{ij}, \quad \widehat{\sigma}_i^2 = \frac{1}{n_i - 1} \sum_{j=1}^{n_i} (Y_{ij} - \bar{Y}_i)^2, \quad \widehat{\tau}^2 = \frac{1}{\sum_{i=1}^{I} n_i} \sum_{i=1}^{I} n_i (\bar{Y}_i - \widehat{\mu})^2.$$

Replacing μ, σ_i^2 and τ^2 in (4.44) by $\widehat{\mu}$, $\widehat{\sigma}_i^2$ and $\widehat{\tau}^2$ yields the empirical Bayes estimate

$$\widehat{\theta}_i = \widehat{\alpha}_i \bar{Y}_i + (1-\widehat{\alpha}_i)\widehat{\mu}, \qquad (4.45)$$

where $\widehat{\alpha}_i = \widehat{\tau}^2/(\widehat{\tau}^2 + \widehat{\sigma}_i^2/n_i)$ is the credibility weight for the ith class.

The Bühlmann-Straub model is one of the most important credibility model for insurance practice, and it has important applications in insurance practice. In life insurance, the Bühlmann-Straub model facilitates the assessment of mortality risk and the determination of appropriate premiums based on policyholder characteristics and demographic factors. Similarly, in non-life insurance, the model aids in evaluating risks associated with property,

casualty, and liability insurance, enabling insurers to set premiums that accurately reflect the underlying risk profiles. Moreover, the Bühlmann-Straub
model finds extensive use in reinsurance, where it assists reinsurers in evaluating risks associated with large portfolios of insurance policies and determining
optimal reinsurance arrangements to manage their exposure effectively.

One of the notable advantages of the Bühlmann-Straub model is its adapt-
ability to incorporate additional factors such as covariates and random effects.
Such flexibility allows for the development of diverse extensions to the model,
catering to specific requirements and complexities encountered in real-world
insurance scenarios.

4.4.3 Linear mixed models

There are many extensions to the Bühlmann-Straub model. For example,
Hachemeister (1975) introduced the credibility regression model that relates
claim sizes to certain covariates. Frees et al. (1999) unified various credibility
models into the framework of *linear mixed models* (LMM). The LMM allow for
the incorporation of both fixed effects (such as policyholder characteristics)
and random effects (such as unobserved heterogeneity or variability across
insurance policies) into the modeling framework and are particularly useful
for modeling data with hierarchical or correlated structures encountered in
insurance settings. The LMM have the following form

$$y_{it} = \boldsymbol{\beta}'\mathbf{x}_{it} + \mathbf{b}_i'\mathbf{z}_{it} + \epsilon_{it}, \quad t = 1, \ldots, T_i, \quad i = 1, \ldots, n, \tag{4.46}$$

where $\mathbf{x}_{it}$ are p-dimensional explanatory variables, $\mathbf{z}_{it}$ are q-dimensional random effects, fixed effects $\boldsymbol{\beta} = (\beta_1, \ldots, \beta_p)'$, subject-specific random effects
$\mathbf{b}_i = (b_{i1}, \ldots, b_{iq})'$ with $E(\mathbf{b}_i) = \mathbf{0}$ and covariance matrix $\text{cov}(\mathbf{b}_i) = \sigma^2\mathbf{D}$, ϵ_{it}
are zero-mean random disturbances with variance σ^2, and ϵ_t are uncorrelated
with the random effects and the covariates $\mathbf{x}_{it}$ and $\mathbf{z}_{it}$.

Let $\mathbf{y}_i = (y_{i1}, \ldots, y_{i,T_i})'$, $\mathbf{X}_i = (\mathbf{x}_{i1}, \ldots, \mathbf{x}_{i,T_i})'$, $\mathbf{Z}_i = (\mathbf{z}_{i1}, \ldots, \mathbf{z}_{i,T_i})'$, and
$\epsilon_i = (\epsilon_{i1}, \ldots, \epsilon_{i,T_i})'$. Then (4.46) can be written as, for the ith subject,

$$\mathbf{y}_i = \mathbf{X}_i\boldsymbol{\beta} + \mathbf{Z}_i\mathbf{b}_i + \epsilon_i, \quad i = 1, \ldots, n. \tag{4.47}$$

The n equations (4.47) can be further summarized as the following model

$$\mathbf{y} = \mathbf{X}\boldsymbol{\beta} + \mathbf{Z}\mathbf{b} + \epsilon, \tag{4.48}$$

in which $\mathbf{y}$, $\mathbf{X}$, $\mathbf{Z}$ and ϵ are expressed as

$$\mathbf{y} = \begin{pmatrix} \mathbf{y}_1 \\ \vdots \\ \mathbf{y}_n \end{pmatrix}, \ \mathbf{X} = \begin{pmatrix} \mathbf{x}_1 \\ \vdots \\ \mathbf{x}_n \end{pmatrix}, \ \mathbf{b} = \begin{pmatrix} \mathbf{b}_1 \\ \vdots \\ \mathbf{b}_n \end{pmatrix}, \ \epsilon = \begin{pmatrix} \epsilon_1 \\ \vdots \\ \epsilon_n \end{pmatrix}, \ \mathbf{Z} = \begin{pmatrix} \mathbf{Z}_1 & \cdots & \mathbf{0} \\ \cdots & \cdots & \cdots \\ \mathbf{0} & \cdots & \mathbf{Z}_n \end{pmatrix}.$$

Note that, in the LLM (4.48), $\mathbf{y}$ and ϵ are $(T_1 + \cdots + T_n)$-dimensional vectors,
$\mathbf{b}$ is a nq-dimensional vector, and $\mathbf{X}$ and $\mathbf{Z}$ are $(T_1 + \cdots + T_n) \times p$ and $(T_1 + \cdots + T_n) \times q$ matrices, respectively,

Best linear unbiased estimator of fixed effect coefficients

Denote the total number of observations as $T := T_1 + \cdots + T_n$, and $\text{cov}(\mathbf{b}) = \sigma^2(\mathbf{I} \otimes \mathbf{D})$, where $\otimes$ represents the Kronecker product of two matrices. Then regression (4.48) can be rewritten with one error term as

$$\mathbf{y} = \mathbf{X}\boldsymbol{\beta} + \boldsymbol{\eta}, \tag{4.49}$$

where $\boldsymbol{\eta} = (\boldsymbol{\eta}_1, \ldots, \boldsymbol{\eta}_n)$ and $\boldsymbol{\eta}_i = \boldsymbol{\epsilon}_i + \mathbf{Z}_i\mathbf{b}_i$. It is obvious that $E(\boldsymbol{\eta}) = \mathbf{0}$ and the $T \times T$ covariance matrix of $\boldsymbol{\eta}$ has block diagonal form

$$\mathbf{V} = \text{cov}(\boldsymbol{\eta}) = \sigma^2 \cdot \text{diag}(\mathbf{V}_1, \ldots, \mathbf{V}_n),$$

where $\mathbf{V}_i = \mathbf{I}_{T_i} + \mathbf{Z}_i\mathbf{D}\mathbf{Z}_i'$. For simplicity, we further assume that $\mathbf{b}_i$ and $\boldsymbol{\epsilon}_i$ are normally distributed as

$$\boldsymbol{\epsilon}_i \sim N(\mathbf{0}, \sigma^2\mathbf{I}_{T_i}), \qquad \mathbf{b}_i \sim N(\mathbf{0}, \sigma^2\mathbf{D}).$$

Therefore, the LMM model (4.46) with normally distributed random variables can be written in marginal form as

$$\mathbf{y}_i \sim N(\mathbf{X}_i\boldsymbol{\beta}, \sigma^2\mathbf{V}_i), \qquad i = 1, \ldots, n.$$

Then dropping the constant term, the log-likelihood function for the LMM model (4.4.3) is given by

$$l(\boldsymbol{\theta}) = -\frac{1}{2\sigma^2}\sum_{i=1}^{N}(\mathbf{y}_i - \mathbf{X}_i\boldsymbol{\beta})^T\mathbf{V}_i^{-1}(\mathbf{y}_i - \mathbf{X}_i\boldsymbol{\beta}) - \frac{1}{2}\sum_{i=1}^{N}\log|\mathbf{V}_i| + \frac{N_T}{2}\log\sigma^2, \tag{4.50}$$

where the parameter $\boldsymbol{\theta} = (\boldsymbol{\beta}', \sigma^2, \text{vech}(\mathbf{D}))$ is a combined vector of unknown parameters, and $\text{vech}(\mathbf{D})$ denotes the vector of upper triangular elements in the $q \times q$ symmetric matrix $\mathbf{D}$. The maximum likelihood estimate maximizes the function $l(\boldsymbol{\theta})$ over the parameter space $\boldsymbol{\Theta} = \{\boldsymbol{\theta} : \boldsymbol{\beta} \in \mathbb{R}^p, \sigma^2 > 0, \mathbf{D} \text{ is nonnegative definite}\}$.

Given the covariance matrix $\mathbf{D}$, the log-likelihood (4.50) can be maximized by the *generalized least squares* (GLS) estimator

$$\widehat{\boldsymbol{\beta}}_{\text{GLS}} = \Big(\sum_{i=1}^{n}\mathbf{X}_i'\mathbf{V}_i^{-1}\mathbf{X}_i\Big)^{-1}\Big(\sum_{i=1}^{n}\mathbf{X}_i'\mathbf{V}_i^{-1}\mathbf{y}_i\Big) = (\mathbf{X}'\mathbf{V}^{-1}\mathbf{X})^{-1}\mathbf{X}'\mathbf{V}^{-1}\mathbf{Y},$$
$$\tag{4.51}$$

see Lai and Xing (2008a, Section 1.7). Note the GLS estimator (4.51) minimizes the variance among all linear unbiased estimators and hence is the *best linear unbiased estimator* (BLUE) of $\boldsymbol{\beta}$. Thus, $\mathbf{X}\widehat{\boldsymbol{\beta}}_{\text{GLS}}$ is the best linear unbiased estimator of $\mathbf{X}\boldsymbol{\beta}$.

Taking the derivative of (4.50) with respect to σ^2, it is easy to see that the function l is maximized at

$$\widehat{\sigma}^2 = \frac{1}{N}\sum_{i=1}^{n}(\mathbf{y}_i - \mathbf{X}_i\boldsymbol{\beta})'\mathbf{V}_i^{-1}(\mathbf{y}_i - \mathbf{X}_i\boldsymbol{\beta}). \tag{4.52}$$

If $\mathbf{V}_i$ are known, substituting $\boldsymbol{\beta}$ in (4.52) by the GLS estimator (4.51) yields that

$$\widehat{\sigma}^2_{\text{GLS}} = q/N.$$

Restricted maximum likelihood (RML)

It is known that the MLE of variances is biased for finite sample. To reduce the bias in the variance components model, it is suggested to modify the standard log-likelihood function using GLS residuals. The resulting method is referred to as *restricted maximum likelihood estimation* (RMLE).

Consider first the general linear model defined as $\mathbf{y} \sim N(\mathbf{X}\boldsymbol{\beta}, \mathbf{V})$. In RML estimation we maximize the log-likelihood function for the residual vector $\widehat{\mathbf{e}} = \mathbf{y} - \mathbf{X}\widehat{\boldsymbol{\beta}}$, where $\widehat{\boldsymbol{\beta}}$ is the GLS estimator (4.51). Note that $\widehat{\boldsymbol{\beta}}$ and $\widehat{\mathbf{e}}$ are linear functions of $\mathbf{y}$ and have normal distributions as well. Moreover, $\widehat{\boldsymbol{\beta}}$ and $\widehat{\mathbf{e}}$ are independent since $\text{Cov}(\mathbf{X}'\mathbf{V}^{-1}\mathbf{y}, \widehat{\mathbf{e}}) = \mathbf{0}$. This implies that the likelihood function for $\mathbf{y}$ is the product of the likelihood functions for $\widehat{\mathbf{e}}$ and $\widehat{\boldsymbol{\beta}}$. But $\widehat{\boldsymbol{\beta}} \sim N(\boldsymbol{\beta}, (\mathbf{X}'\mathbf{V}^{-1}\mathbf{X})^{-1})$, and therefore the log-likelihood function for the residual vector $\widehat{\mathbf{e}}$, up to a constant, is

$$l(\widehat{\mathbf{e}}, \boldsymbol{\theta}) = l(\mathbf{y}, \boldsymbol{\theta}) - l(\widehat{\boldsymbol{\beta}}, \boldsymbol{\theta}) = -\frac{1}{2}\Big(\log |\mathbf{X}'\mathbf{V}^{-1}\mathbf{X}| + \log |\mathbf{V}|$$
$$+ (\mathbf{y} - \mathbf{X}\boldsymbol{\beta})'\mathbf{V}^{-1}(\mathbf{y} - \mathbf{X}\boldsymbol{\beta}) - (\widehat{\boldsymbol{\beta}} - \boldsymbol{\beta})'\mathbf{X}'\mathbf{V}^{-1}\mathbf{X}(\widehat{\boldsymbol{\beta}} - \boldsymbol{\beta})\Big)$$
$$= -\frac{1}{2}\Big(\log |\mathbf{X}'\mathbf{V}^{-1}\mathbf{X}| + \log |\mathbf{V}| + \widehat{\mathbf{e}}'\mathbf{V}^{-1}\widehat{\mathbf{e}}\Big).$$

Clearly, the maximization of this function is equivalent to

$$l_R(\boldsymbol{\beta}, \boldsymbol{\theta}) = -\frac{1}{2}\Big(\log |\mathbf{X}'\mathbf{V}^{-1}\mathbf{X}| + \log |\mathbf{V}| + (\mathbf{y} - \mathbf{X}\boldsymbol{\beta})'\mathbf{V}^{-1}(\mathbf{y} - \mathbf{X}\boldsymbol{\beta})\Big), \quad (4.53)$$

because the maximization of l_R for $\boldsymbol{\beta}$ gives $\widehat{\mathbf{e}} = \mathbf{y} - \mathbf{X}\widehat{\boldsymbol{\beta}}$. Function (4.53) is called the *residual log-likelihood function*.

We then rewrite (4.53) using the LMM (4.47) and come to the RML function

$$l_R(\boldsymbol{\theta}) = -\frac{1}{2}\Big[(T - p)\log \sigma^2 + \log \Big| \sum_{i=1}^{n} \mathbf{X}_i'\mathbf{V}_i^{-1}\mathbf{X}_i \Big| + \sum_{i=1}^{n}\Big(\log |\mathbf{V}_i|$$
$$+ \frac{1}{\sigma^2}(\mathbf{y}_i - \mathbf{X}_i\boldsymbol{\beta})'\mathbf{V}_i^{-1}(\mathbf{y}_i - \mathbf{X}_i\boldsymbol{\beta})\Big)\Big]. \qquad (4.54)$$

Then by maximizing the RML function above, we obtain a RML estimator for σ^2

$$\widehat{\sigma}^2_R = \frac{1}{T - p}\sum_{i=1}^{n}(\mathbf{y}_i - \mathbf{X}_i\boldsymbol{\beta})'\mathbf{V}_i^{-1}(\mathbf{y}_i - \mathbf{X}_i\boldsymbol{\beta}). \qquad (4.55)$$

Best linear unbiased predictor of random effects

The subject-specific regression parameter $\mathbf{b}_i$ are called *random effects* and are assumed to be identically distributed with a common distribution G. The conditional expectation $E(\mathbf{u}|\mathbf{Y})$ minimizes the mean squared error $E||\widehat{\mathbf{u}} - \mathbf{u}||^2 = E\{(\widehat{\mathbf{u}} - \mathbf{u})'(\widehat{\mathbf{u}} - \mathbf{u})\}$ among all predictors $\widehat{\mathbf{u}} = \widehat{\mathbf{u}}(\mathbf{Y})$ of $\mathbf{u}$ based on $\mathbf{Y}$. Whereas $E(\mathbf{u}|\mathbf{Y})$ maybe a complicated function of $\mathbf{Y}$, the *best linear predictor* which minimizes $E||\widehat{\mathbf{u}} - \mathbf{u}||^2$ among all linear functions of $\mathbf{Y}$ is given explicitly by

$$\widehat{\mathbf{u}} = \mathbf{\Sigma}_{21}\mathbf{\Sigma}_{11}^{-1}(\mathbf{Y} - \mathbf{X}\boldsymbol{\beta}), \tag{4.56}$$

where

$$\mathbf{\Sigma} = \mathrm{Cov}\begin{pmatrix} \mathbf{Y} \\ \mathbf{u} \end{pmatrix} = \begin{pmatrix} \mathbf{\Sigma}_{11} & \mathbf{\Sigma}_{12} \\ \mathbf{\Sigma}_{21} & \mathbf{\Sigma}_{22} \end{pmatrix},$$

noting that $E(\mathbf{Y}) = \mathbf{X}\boldsymbol{\beta}$ and $E(\mathbf{u}) = 0$ and that $\mathbf{\Sigma}_{11} = \mathbf{V}$.

Since $\boldsymbol{\beta}$ is typically unknown in practice, we replace $\boldsymbol{\beta}$ in (4.56) by its best linear unbiased estimator $\widehat{\boldsymbol{\beta}}_{\mathrm{GLS}}$ obtained in (4.51). Then we obtain the *best linear unbiased predictor* (BLUP)

$$\widehat{\boldsymbol{\mu}}_{\mathrm{BLUP}} = \mathbf{\Sigma}_{21}\mathbf{\Sigma}_{11}^{-1}(\mathbf{Y} - \mathbf{X}\widehat{\boldsymbol{\beta}}_{\mathrm{GLS}}). \tag{4.57}$$

To show that (4.57) is optimal among all linear unbiased estimators of $\mathbf{u}$, we consider the choice of $\omega, \boldsymbol{\theta}$ and $\boldsymbol{\beta}$ to minimize the variance of the predictor error $(\omega + \boldsymbol{\theta}'\mathbf{Y}) - (\mathbf{a}'\mathbf{X}\boldsymbol{\beta} + \mathbf{c}'\mathbf{u})$ among all unbiased linear predictors $\boldsymbol{\theta}'\mathbf{Y}$, for any given $(T_1 + \cdots T_n) \times 1$ vectors $\mathbf{a}$ and $\mathbf{c}$. Since $E\mathbf{u} = 0$ and $E\mathbf{Y} = \mathbf{X}\boldsymbol{\beta}$, unbiasedness imposes the constraint $\boldsymbol{\theta}'\mathbf{X}\boldsymbol{\beta} = \mathbf{a}'\mathbf{X}\boldsymbol{\beta}$ for the minimization problem. Using Lagrange multipliers leads to the solution $\boldsymbol{\beta} = \boldsymbol{\beta}_{\mathrm{GLS}}$ and $\omega + \boldsymbol{\theta}'\mathbf{Y} = \mathbf{a}'\mathbf{X}\widehat{\boldsymbol{\beta}}_{\mathrm{GLS}} + \mathbf{c}'\mathbf{\Sigma}_{21}\mathbf{\Sigma}_{11}^{-1}(\mathbf{Y} - \mathbf{X}\widehat{\boldsymbol{\beta}}_{\mathrm{GLS}})$. This shows the optimality of (4.57).

Parameter estimation and variance components

Note that BLUE and BLUP involve matrix parameters $\mathbf{V} = \mathbf{\Sigma}_{11}$ and $\mathbf{\Sigma}_{21}$, in which $\mathbf{\Sigma}_{11}$ is the covariance matrix of $\mathbf{Y}$, or equivalently, that of $\mathbf{e}$, and $\mathbf{\Sigma}_{21}$ is the cross-variance matrix of $\mathbf{Y}$ and $\mathbf{u}$. Since

$$\mathbf{\Sigma}_{21} = E(\mathbf{u}\mathbf{Y}') = E(\mathbf{u}\mathbf{u}') = \sum_{i=1}^{n} \mathbf{Z}_i \mathrm{Cov}(\mathbf{b}_i)\mathbf{Z}_i',$$

the parameters of the LMM include the coefficient $\boldsymbol{\beta}$ and the covariance matrices $\mathrm{Cov}(\mathbf{b}_i)$ and $\mathrm{Cov}(\boldsymbol{\epsilon})$. Without further assumptions on the structures of covariance matrices, the number of parameters might be too large relative to the sample size. Hence, it is often assumed that $E(\epsilon_{it}\epsilon_{js}) = 0$ for $i \neq j$ and $E(\epsilon_{it}\epsilon_{is}) = 0$ for $t \neq s$, and that the random errors ϵ_{it} have common variance σ^2, and $\mathrm{Cov}(\mathbf{b}_i) = \sigma_i^2 \mathbf{I}_q$. This reduces the set of parameters $\{\boldsymbol{\beta}, \mathrm{Cov}(\mathbf{b}_i), \mathrm{Cov}(\boldsymbol{\epsilon})\}$ to $p+n+1$ parameters $\beta_1, \cdots, \beta_p, \sigma_1^2, \cdots, \sigma_n^2, \sigma^2$, which we denote collectively by a parameter vector $\boldsymbol{\theta}$.

Assuming the random effects $\mathbf{b}_i$ and the random errors ϵ_{it} to be normally distributed with zero means, the likelihood function $L(\boldsymbol{\theta})$ has an explicit formula and the MLE $\widehat{\boldsymbol{\theta}}$ can be computed by solving the nonlinear equation $\nabla \log L(\boldsymbol{\theta}) = 0$, where ∇ denotes the gradient vector of first partial derivatives with respect to the components of $\boldsymbol{\theta}$. The solution yields $\widehat{\sigma}^2, \widehat{\sigma}_1^2, \cdots, \widehat{\sigma}_1^2$ and $\widehat{\boldsymbol{\beta}}$ so that the MLE of $\mathbf{V}$ is

$$\widehat{\mathbf{V}} = \mathrm{diag}(\widehat{\sigma}_1^2 \mathbf{Z}_1 \mathbf{Z}_1', \cdots, \widehat{\sigma}_n^2 \mathbf{Z}_n \mathbf{Z}_n') + \widehat{\sigma}^2 \mathbf{I}_{T_1 + \cdots + T_n}. \tag{4.58}$$

An alternative to the preceding MLE of $\mathbf{V}$ is REML (restricted maximum likelihood) that maximize the likelihood function of $\mathbf{V}$ based on $\mathbf{K}'\mathbf{Y}$, where $\mathbf{K}'$ is a $(T_1 + \cdots + T_n - r) \times (T_1 + \cdots T_n)$ matrix such that $\mathbf{K}'\mathbf{X} = \mathbf{0}$ and r is the maximum number of $(T_1 + \cdots T_n) \times 1$ vectors that are orthogonal to the column vectors of $\mathbf{X}$. Since $\mathbf{K}'\mathbf{X} = \mathbf{0}$, we have $\mathbf{K}'\mathbf{Y} = \mathbf{K}'(\mathbf{u} + \boldsymbol{\epsilon})$, and therefore $\mathbf{K}'\mathbf{Y}$ only involves the variance components of random effects and random errors. REML estimates $\boldsymbol{\beta}$ by the right-hand side of (4.51) with $\mathbf{V}$ replaced by $\widehat{\mathbf{V}}_{\mathrm{REML}}$ after maximizing the restricted likelihood function based on $\mathbf{K}'\mathbf{Y}$.

The likelihood approach is based on the "working model" of normally distributed random effects and random errors. It has been shown that the performance of the variance component estimators is robust against this distributional assumption. Similar assumptions are also used in the extension of LMM to generalized linear mixed models in the next section.

The linear regression model (4.46) with $u_{it} = \mathbf{z}_{it}'\mathbf{b}_i + \epsilon_{it}$ has two sources of random variation: the random errors ϵ_{it} and the random effects $\mathbf{b}_i$. Since $\mathbf{b}_i$ and ϵ_{it} are uncorrelated and $E\mathbf{b}_i'\mathbf{b}_j = 0$ for $i \neq j$, we can decompose the covariance matrix $\mathbf{V}$ of $\mathbf{e} = \mathbf{u} + \boldsymbol{\epsilon}$ into *variance components* of the form

$$\mathbf{V} = \mathrm{diag}(\mathbf{Z}_1 \mathrm{Cov}(\mathbf{b}_1)\mathbf{Z}_1', \cdots, \mathbf{Z}_n \mathrm{Cov}(\mathbf{b}_n)\mathbf{Z}_n') + \mathrm{Cov}(\boldsymbol{\epsilon}), \tag{4.59}$$

where $\mathbf{Z}_i$ is a $T_i \times q$ matrix whose rows are $\mathbf{z}_{it}'$.

The credibility regression model

Consider the credibility regression model

$$y_i = \boldsymbol{\alpha}_i'\mathbf{x}_i + \epsilon_i,$$

in which the subject-specific regression parameters $\boldsymbol{\alpha}_i$ have distribution G. This is a special case of the LMM (4.46) with $j = 1 = n_i$, $\mathbf{z}_i = \mathbf{x}_i$, $\boldsymbol{\beta}$ being the mean of G and $\mathbf{b}_i = \boldsymbol{\alpha}_i - \boldsymbol{\beta}$. Moreover, the index j in (4.46) should be replaced by t that denotes time. Provided that insurers observe claims of risk classes over successive periods, Frees et al. (1999) consider the setting of longitudinal data in their LMM approach to credibility theory.

4.4.4 Generalized linear models

In constrat to traditional linear regression model with normal errors that assumes the response Y is normally distributed, *generalized linear models* (GLMs) are a versatile statistical framework that accommodate response variables that follow non-normal distributions or have nonlinear relationships with the predictors.

Definition 4.3 *A generalized linear model (GLM) relates the response Y to the d-dimensional predictor x by the following assumptions*

(i) Y has density function

$$f(y; \theta, \phi) = \exp\left\{\left[y\theta - b(\theta)\right]/\phi + c(y, \phi)\right\}, \qquad (4.60)$$

where θ is a function of the mean $\mu := E(Y)$, i.e., $\theta = \theta(\mu)$. The parametric family of density functions (4.60) is called an exponential family with canonical parameter θ and dispersion parameter $\phi > 0$.

(ii) For some given smooth increasing function g and unknown parameter β, $g(\theta) = \beta' \mathbf{x}$. Denote the linear combination of predictors as $\eta = \beta' \mathbf{x}$. Since $\eta = g(\theta) = g \circ \theta(\mu)$, $\eta = g \circ \theta$ is the link function to link the mean μ and the linear combination of predictors x, and one commonly used link function is the canonical link, $g(\theta) = \theta$.

(iii) The n observed predictors and responses (x_i, y_i), $1 \leq i \leq n$, are independent.

Taking derivative with respect to θ and integrate with respect to y, we obtain that
$$E(Y) = b'(\theta), \qquad \text{Var}(Y) = b''(\theta).$$
Note that if Y follow a Normal(μ, σ^2) distribution, then
$$\theta = \mu, \quad b(\theta) = \theta^2/2, \quad \phi = \sigma^2, \quad c(y, \phi) = -\{y^2/\sigma^2 + \log(2\pi\sigma^2)\}/2.$$

Poisson regression for count data response

If Y follow a Poisson distribution with density $f(y; \lambda) = e^{-\lambda}\lambda^y/y!$ for $y = 0, 1, 2, \ldots$, then $\theta = \log \lambda$, $b(\theta) = e^\theta$, and $\phi = 1$. Using the canonical link function $g(\theta) = \theta$ yields the *Poisson regression model*,

$$Y_i | \mathbf{X} = \mathbf{x}_i \sim \text{Poisson}(\exp(\beta' \mathbf{x}_i)). \qquad (4.61)$$

This regression model links the number of events with their predictor (or risk factors) and can be used in operational risk analysis. An alternative way to model the count response and predictor is to assume that the response Y follows a negative binomial distribution.

Likelihood inference

To make inference on the GLM model, we note that, given observations $\{(\mathbf{x}_i, y_i); i = 1, \ldots, n\}$, the log-likelihood function is

$$l(\theta, \phi, y) = \sum_{i=1}^{n} \left\{ \frac{y_i \theta - b(\theta)}{\phi} + c(y_i, \phi) \right\},$$

and hence the first and second derivatives of the log-likelihood are

$$\frac{\partial l}{\partial \theta} = \frac{1}{\phi} \sum_{i=1}^{n} [y_i - b'(\theta)], \qquad \frac{\partial^2 l}{\partial \theta^2} = -nb'(\theta)/\phi.$$

Note that, for a single observation (x, y), we can use the equation $E(\partial l/\partial \theta) = 0$ to conclude that $E(y) = \mu = b'(\theta)$. As an analog, from the relation $E(\partial^2 l/\partial \theta^2) + E(\partial l/\partial \theta)^2 = 0$, we have $\text{Var}(y) = \phi b''(\theta)$. This implies that b is convex and therefore b' is increasing. The first derivative of the log-likelihood with respect to the unknown parameters β is

$$\frac{\partial l}{\partial \boldsymbol{\beta}} = \frac{\partial l}{\partial \theta} \cdot \frac{d\theta}{d\mu} \cdot \frac{d\mu}{d\eta} \cdot \frac{\partial \eta}{\partial \boldsymbol{\beta}} = \sum_{i=1}^{n} \frac{y_i - \mu}{\phi} \cdot \frac{1}{b''(\theta)} \cdot \frac{d\mu}{d\eta} \cdot \mathbf{x}_i = \sum_{i=1}^{n} \left[\frac{w_i}{\phi} \cdot \frac{d\eta_i}{d\mu} \right] (y_i - \mu) \mathbf{x}_i,$$

$$(4.62)$$

where $w_i^{-1} = b''(\theta)(d\eta_i/d\mu)^2$. The second derivative is

$$\frac{\partial^2 l}{\partial \boldsymbol{\beta} \partial \boldsymbol{\beta}'} = \sum_{i=1}^{n} \left\{ \frac{\partial}{\partial \boldsymbol{\beta}'} \left[\frac{w_i}{\phi} \cdot \frac{d\eta_i}{d\mu} \right] \right\} (y - \mu) \mathbf{x}_i - \sum_{i=1}^{n} \frac{w_i}{\phi} \mathbf{x}_i \mathbf{x}_i'.$$

Since $E(y_i) = \mu$, the expectation value of the first term is zero. We have

$$E\left(\frac{\partial^2 l}{\partial \boldsymbol{\beta} \partial \boldsymbol{\beta}'} \right) = -\sum_{i=1}^{n} \frac{w_i}{\phi} \mathbf{x}_i \mathbf{x}_i'. \qquad (4.63)$$

Applying the Newton-Raphson method yields that, at the mth iteration,

$$E\left(\frac{\partial^2 l}{\partial \boldsymbol{\beta}_m \partial \boldsymbol{\beta}_m'} \right) (\boldsymbol{\beta}_{m+1} - \boldsymbol{\beta}_m) = -\left(\frac{\partial l}{\partial \boldsymbol{\beta}} \right) \Big|_{\boldsymbol{\beta} = \boldsymbol{\beta}_m}. \qquad (4.64)$$

Rewrite the left- and right-hand sides of (4.64), we have

$$-\frac{\partial l}{\partial \boldsymbol{\beta}_m} = -\sum_{i=1}^{n} \frac{w_i}{\phi} \mathbf{x}_i \left[\eta + \frac{d\eta}{d\mu} (y_i - \mu) \right] + \sum_{i=1}^{n} \frac{w_i}{\phi} \mathbf{x}_i \mu,$$

$$E\left(\frac{\partial^2 l}{\partial \boldsymbol{\beta}_m \partial \boldsymbol{\beta}_m'} \right) (\boldsymbol{\beta}_{m+1} - \boldsymbol{\beta}_m) = E\left(\frac{\partial^2 l}{\partial \boldsymbol{\beta}_m \partial \boldsymbol{\beta}_m'} \right) \boldsymbol{\beta}_{m+1} + \sum_{i=1}^{n} \frac{w_i}{\phi} \mathbf{x}_i \eta.$$

Define that $z_{m,i} = \eta + (d\eta/d\mu)(y_i - \mu)$ for $i = 1, \ldots, n$, and make use of (4.63), the Newton-Raphson iteration (4.64) becomes

$$\left(\sum_{i=1}^{n} \frac{w_i}{\phi} \mathbf{x}_i \mathbf{x}_i' \right) \boldsymbol{\beta}_{m+1} = \sum_{i=1}^{n} \frac{w_i}{\phi} \mathbf{x}_i z_{m,i}. \qquad (4.65)$$

Note that in matrix form, $\mathbf{X} = (x_1, \ldots, x_n)'$. Let $\mathbf{W}$ be a $n \times n$ diagonal matrix with ith diagonal element w_i/ϕ. Then (4.65) becomes

$$\boldsymbol{\beta}_{m+1} = (\mathbf{X}'\mathbf{W}\mathbf{X})^{-1}\mathbf{X}\mathbf{W}\mathbf{z}_m, \qquad (4.66)$$

where $\mathbf{z}_m = (z_{m,1}, \ldots, z_{m,n})'$. Note that (4.66) is a linear regression estimate with the weight $\mathbf{W}$ and dependent variable $\mathbf{z}_m$. Thus, this algorithm is referred to as *iteratively reweighted least squares* (IRLS).

In the case of a canonical link, we have $\theta = \eta$ and $b''(\eta) = d\mu/d\eta$, hence $w_i^{-1} = d\eta/d\mu$. In classical linear regression, we assume that y follows a Normal(μ, σ^2) distribution and the link function is the canonical link. Then we have $\theta = \mu = \eta$, $b(\theta) = \theta^2/2$, $w_i \equiv 1$, and $z_{m,i} = y_i$. Hence, the IRLS estimate (4.66) reduces to the *ordinary least squares* (OLS) estimate. For the case of logistic regression with binary response. Let $p_i = P(y_i = 1|\mathbf{x}_i)$, it is easy to see that, given the canonical link,

$$w_i = (d\eta/d\mu)^{-1} = (d\theta/dp)^{-1} = p_i(1 - p_i), \quad i = 1, \ldots, n,$$

$$z_{m,i} = \log\left(\frac{p_i}{1 - p_i}\right) + \frac{y_i - p_i}{p_i(1 - p_i)} = \boldsymbol{\beta}'\mathbf{x}_i + \frac{y_i - p_i}{p_i(1 - p_i)}.$$

Then we can use the IRLS (4.66) with specified $\mathbf{W}$ and $\mathbf{z}_m$ to estimate $\boldsymbol{\beta}$.

4.4.5 Generalized linear mixed models

We have pointed out in (4.46) that linear mixed models are generalizations of linear regression models by allowing subject-specific regression coefficients. The *generalized linear model* (GLM), which extends linear regression to the setting where the density function of the response variable belongs to an exponential family of distributions, has also been extended to allow subject-specific regression coefficients.

Breslow and Clayton (1993) introduced the *generalized linear mixed models* (GLMM) for longitudinal data $Y_{i,t}$ by allowing subject-specific regression parameters $\mathbf{b}_i$, i.e., the "random effects," to extend mixed effects models in linear regression to GLMM. The GLMM assumes that y_{it} are conditionally independent given the observed covariates $\mathbf{x}_{it}$ and $\mathbf{z}_{it}$ and have a conditional density of the form

$$f(y|\mathbf{b}_i, \mathbf{z}_{it}, \mathbf{x}_{it}) = \exp\{[y\theta_{it} - \psi(\theta_{it})]/\phi + c(y, \phi)\}, \qquad (4.67)$$

in which ϕ is a dispersion parameter and $\mu_{it} = d\psi/d\theta|_{\theta=\theta_{it}}$ satisfies

$$\mu_{it} = g^{-1}(\boldsymbol{\beta}'\mathbf{x}_{it} + \mathbf{b}_i'\mathbf{z}_{it}), \tag{4.68}$$

where g^{-1} is the inverse of a monotone link function g, as in the standard GLM. The random effects $\mathbf{b}_i$ can contain an intercept term a_i by augmenting the covariate vector to $(1, \mathbf{z}_{it})$ in case a_i is not included in $\mathbf{b}_i$, and $\boldsymbol{\beta}$ is a vector of fixed effects and can likewise contain an intercept term. Breslow and Clayton further assume that the $\mathbf{b}_i$ in (4.68) have a common normal distribution with mean 0 and covariance matrix $\boldsymbol{\Sigma}$, and $\boldsymbol{\Sigma}$ is dependent on an unknown parameter.

Likelihood maximization and integral approximation

The likelihood function of the GLMM defined by (4.67) and (4.68) is of the form $\prod_{i=1}^n L_i(\sigma, \boldsymbol{\alpha}, \boldsymbol{\beta})$, where

$$L_i(\sigma, \boldsymbol{\alpha}, \boldsymbol{\beta}) = \int \Big\{ \prod_{t=1}^T f(y_{it}; \theta_{it}, \phi) \Big\} \psi_{\boldsymbol{\alpha}}(\mathbf{b}) d\mathbf{b}, \tag{4.69}$$

in which $\phi_{\boldsymbol{\alpha}}$ denotes the normal density function with mean 0 and covariance matrix depending on an unknown parameter $\boldsymbol{\alpha}$. We first describe three methods to compute the likelihood function, the maximizer of which gives the MLE of σ, $\boldsymbol{\alpha}$, and $\boldsymbol{\beta}$.

Letting $e^{l_i(\mathbf{b}|\phi,\boldsymbol{\alpha},\boldsymbol{\beta})}$ be the integrand in the right-hand side of (4.69), Laplaces's asymptotic formula for integrals yields the approximation

$$\int e^{l_i(\mathbf{b}|\phi,\boldsymbol{\alpha},\boldsymbol{\beta})} d\mathbf{b} \approx (2\pi)^{q/2} \Big\{ \det[-\ddot{l}_i(\widehat{\mathbf{b}}_i|\phi, \boldsymbol{\alpha}, \boldsymbol{\beta})] \Big\}^{-1/2} \exp\Big\{ l_i(\widehat{\mathbf{b}}_i|\sigma, \boldsymbol{\alpha}, \boldsymbol{\beta}) \Big\}, \tag{4.70}$$

where q is the dimension of $\mathbf{b}_i$, $\widehat{\mathbf{b}}_i = \widehat{\mathbf{b}}_i(\phi, \boldsymbol{\alpha}, \boldsymbol{\beta})$ is the maximizer of $l_i(\mathbf{b}|\phi, \boldsymbol{\alpha}, \boldsymbol{\beta})$ and $\ddot{l}_i$ denotes the Hessian matrix consisting of second partial derivatives of l_i with respect to the components of $\mathbf{b}$. Let $\lambda_{\min}$ be the minimum eigenvalue of the Hessian matrix $-\ddot{l}_i(\widehat{\mathbf{b}}_i|\phi, \boldsymbol{\alpha}, \boldsymbol{\beta})$. Since Laplaces's asymptotic formula (4.70) is derived from the asymptotic approximation of l_i by a quadratic function of $\mathbf{b}$ in a small neighborhood of $\widehat{\mathbf{b}}_i$ as $\lambda_{\min} \to \infty$, such formula may produce significant approximation error for L_i if the corresponding $\lambda_{\min}$ is not sufficiently large. One way to reduce the possible approximation error is to compute L_i by using an adaptive Gauss-Hermite quadrature rule.

The numerical integration procedure demands a much higher computational effort and becomes computationally infeasible if n or q is large. One may use direct Monte Carlo integration, but the computational time may be too long to be of practical interest. To circumvent the issue of high dimensional numerical integration, Antonio and Beirlant (2007) proposed putting prior distributions on the unknown parameters and estimate them by using the Markov Chain Monte Carlo (MCMC) method in a Bayesian way; see Section

7.4 for details. The performance of the MCMC method, however, depends on how the prior parameters are set as well as the convergence rate of the Markov chain to its stationary distribution, which may not even exist.

Instead of relying solely on Monte Carlo, Lai and Shih (2003) proposed a hybrid approach to estimate the random effects distribution in the more general setting of nonlinear mixed effects models, and have shown that there is very low resolution in estimating the distribution. This provides a heuristic explanation for why the choice of a normal distribution, with unspecified parameters, for the random effects distribution G does not result in a worse estimate of $\boldsymbol{\beta}$ and ϕ than the semiparametric approach that estimates G nonparametrically. An advantage of assuming a normal distribution for the random effects $\mathbf{b}_i$ is that it only involves the covariance matrix of the random effects.

Model selection, prediction and residuals

As shown by Breslow and Clayton (1993), a consistent estimator of the asymptotic covariance matrix of $(\widehat{\phi}, \widehat{\boldsymbol{\alpha}}, \widehat{\boldsymbol{\beta}})$ is $\widehat{\mathbf{V}}^{-1} = \{-\sum_{i=1}^{n} \ddot{R}_i(\widehat{\phi}, \widehat{\boldsymbol{\alpha}}, \widehat{\boldsymbol{\beta}})\}^{-1}$, where $R_i = \log L_i$ and $\ddot{R}_i$ denotes the Hessian matrix of second derivatives of R_i with respect to ϕ, $\boldsymbol{\alpha}$ and the components of $\boldsymbol{\beta}$. Lai and Shih (2003) have shown that $\widehat{\mathbf{V}}^{-1/2}(\widehat{\phi} - \phi, \widehat{\boldsymbol{\alpha}} - \boldsymbol{\alpha}, \widehat{\boldsymbol{\beta}} - \boldsymbol{\beta})$ has a limiting standard multivariate normal distribution as $n \to \infty$ under certain regularity conditions, and that the sandwich estimator

$$\{-\sum_{i=1}^{n} \ddot{R}_i(\widehat{\phi}, \widehat{\boldsymbol{\alpha}}, \widehat{\boldsymbol{\beta}})\}^{-1}\{\sum_{i=1}^{n} \dot{R}_i(\widehat{\phi}, \widehat{\boldsymbol{\alpha}}, \widehat{\boldsymbol{\beta}})\dot{R}_i'(\widehat{\phi}, \widehat{\boldsymbol{\alpha}}, \widehat{\boldsymbol{\beta}})\}\{-\sum_{i=1}^{n} \ddot{R}_i(\widehat{\phi}, \widehat{\boldsymbol{\alpha}}, \widehat{\boldsymbol{\beta}})\}^{-1}$$

is more robust than $\widehat{\mathbf{V}}^{-1}$, where $\dot{R}_i$ denotes the gradient vector of partial derivatives. Since the estimation algorithm is likelihood-based, we can use *Bayesian information criterion* (BIC) (Schwarz, 1978) for model selection,

$$\text{BIC} = -2\sum_{i=1}^{n} R_i(\widehat{\phi}, \widehat{\boldsymbol{\alpha}}, \widehat{\boldsymbol{\beta}}) + (\log n)(\text{number of parameters}).$$

Moreover, the covariates in GLMM can also be selected by using a model selection procedure; see Lai et al. (2006).

Besides model selection, another goal of the inference is to predict certain characteristics of a specific subject given its covariate vector, for example, to forecast $\mu_{i,t+1} = g^{-1}(\boldsymbol{\beta}'\mathbf{x}_{i,t+1} + \mathbf{b}_i'\mathbf{z}_{i,t+1})$ at time t with observed covariate vector $(\mathbf{x}_{i,t+1}', \mathbf{z}_{i,t+1}')'$. More generally, we would like to estimate a function of the unobservable $\mathbf{b}_i$, say $h(\mathbf{b}_i)$. If σ, $\boldsymbol{\alpha}$ and $\boldsymbol{\beta}$ are known, then $h(\mathbf{b}_i)$ should be estimated by the posterior mean $E_{\sigma, \boldsymbol{\alpha}, \boldsymbol{\beta}}[h(\mathbf{b}_i)|i\text{th subject's data}]$. Otherwise, the empirical Bayes approach replaces them in the posterior mean by their estimates $(\widehat{\sigma}, \widehat{\boldsymbol{\alpha}}, \widehat{\boldsymbol{\beta}})$ so that h is estimated by

$$\widehat{h} = E_{\widehat{\sigma}, \widehat{\boldsymbol{\alpha}}, \widehat{\boldsymbol{\beta}}}[h(\mathbf{b}_i)|i\text{th subject's data}].$$

To carry out a residual diagnostics, one may consider the generalized residuals

$$\widehat{\mathbf{r}}_i = E_{\widehat{\phi},\widehat{\alpha},\widehat{\beta}}(\mathbf{b}_i|i\text{th subject's data}).$$

Substantial deviation of these residuals from an i.i.d. pattern would suggest inadequacies and possible improvements of the assumed regression model. For those subjects with sparse data, the generalized residuals $\{\widehat{\mathbf{r}}_i\}$ are better approximations to the unobservable residuals $\{\mathbf{b}_i\}$ than the computationally more convenient $\{\widehat{\mathbf{b}}_i\}$.

4.5   Evolutionary credibility

In the Bühlmann-Straub model, the best linear estimator (4.45) for the individual premium θ is a weighted average of the sample mean $\overline{Y}_i$ and the prior mean μ. While many insurance applications use the observation vector $\mathbf{Y}_i = (Y_{i1},\dots,Y_{in_i})'$ to represent the the full history of observed aggregate claim amounts, some of them only require information from the most recent period. Specifically, the premium for period $n+1$ depends solely on the premium from period n and the newly observed claims from period n. To address this scenario, Bühlmann and Gisler (2005) developed an *evolutionary credibility* theory which can evaluate the premium credibility recursively.

In particular, Bühlmann and Gisler's (2005) assume a dynamic Bayesian model for the prior means over time. Suppose that θ_{it} $(1 \le i \le n)$ have a prior distribution that has mean μ_t and variance τ^2 at period t. To characterize how μ_t changes with time, one may consider the following model

$$\mu_t = \rho\mu_{t-1} + (1-\rho)\mu + \eta_t, \tag{4.71}$$

in which the η_t are independent and identically distributed random variables with mean 0 and variance V. This Bayesian model treats the unknown μ_t as states undergoing AR(1) dynamics, yielding a linear state-space model for which the unobserved states μ_t can be estimated from the observations $\{Y_{is}, s \le t\}$ by the Kalman filter.

Let $\mathbf{Y}_t = (Y_{1t},\dots,Y_{nt})'$ and $\mathbf{1} = (1,\dots,1)'$. Denote by $\mathcal{Y}_t = (\mathbf{Y}_1,\dots,\mathbf{Y}_t)$ the set of observations up to period t, $\mu_{t|t} = E(\mu_t|\mathcal{Y}_t)$ posterior mean and $\mu_{t+1|t} = E(\mu_{t+1}|\mathcal{Y}_t)$ the prediction at period t. We can use the Kalman filter in Theorem 6.11 to derive the estimate $\mu_{t|t}$ and $\mu_{t+1|t}$. Note that the Kalman filer is the minimum-variance linear estimator of μ_t, and it is the Bayes estimator of μ_t if Y_{it} (conditional on μ_t) and η_t are normal. Then $\widehat{\mu}_{t|t}$ can be estimated by the following recursion

$$\begin{aligned}
\widehat{\mu}_{t|t} &= \widehat{\mu}_{t|t-1} + \rho^{-1}\mathbf{K}_t(\mathbf{Y}_t - \widehat{\mu}_{t|t-1}\mathbf{1}), & (4.72)\\
\widehat{\mu}_{t+1|t} &= \rho\widehat{\mu}_{t|t} + (1-\rho)\mu, & (4.73)
\end{aligned}$$

in which $\mathbf{K}_t$ is the $1 \times n$ Kalman gain matrix defined recursively in terms of the hyperparameters $V = \text{Var}(\eta_t)$, $v_t = \text{Var}(Y_{it}|\mu_t)$ and ρ by

$$
\begin{aligned}
\mathbf{K}_t &= \rho \Sigma_{t|t-1} \mathbf{1}' (\Sigma_{t|t-1} \mathbf{1}\mathbf{1}' + v_t \mathbf{I})^{-1}, \\
\Sigma_{t+1|t} &= (\rho - \mathbf{K}_t \mathbf{1})^2 \Sigma_{t|t-1} + V + v_t \mathbf{K}_t \mathbf{K}_t'.
\end{aligned}
$$

Under the assumption that $\text{Var}(\theta_{it}|\mu_t) = \tau^2$ and $\text{Var}(Y_{it}|\theta_{it}) = a + b\theta_{it} + c\theta_{it}^2$ as in (4.41), the variance of Y_{it} given θ_{it} is $v_t = \tau^2 + a + b\theta_{it} + c\theta_{it}^2$.

Instead using the Kalman filtering approach, one can also use the empirical Bayes approach in Section 4.4.1 to make inference μ_t. In particular, we can use the method of moments to estimate μ_t and v_t for every $1 \le t \le T$. Note that

$$
\bar{Y}_t = n^{-1} \sum_{i=1}^{n} Y_{it}, \qquad s_t^2 = \sum_{i=1}^{n} (Y_{it} - \bar{Y}_t)^2 / (n-1)
$$

are consistent estimates of μ_t and v_t, respectively. Moreover, assuming $E(\mu_0) = \mu$, it follows from (4.71) that $E(\mu_t) = \mu$ for all t. Therefore, μ can be consistently estimated from the observations up to time t by $\widehat{\mu}(t) = (\sum_{s=1}^{t} \bar{Y}_s)/t$, and ρ and V can be consistently estimated by

$$
\begin{aligned}
\widehat{\rho}(t) &= \sum_{s=2}^{t} (\bar{Y}_s - \widehat{\mu}(t))(\bar{Y}_{s-1} - \widehat{\mu}(t)) / \sum_{s=2}^{t} (\bar{Y}_s - \widehat{\mu}(t))^2, \\
\widehat{V}(t) &= \sum_{s=2}^{t} \{\bar{Y}_s - \rho \bar{Y}_{s-1} - (1-\rho)\widehat{\mu}(t)\}^2 / (t-2),
\end{aligned}
$$

for $t \ge 2$. Replacing the hyperparameters μ, ρ, V and v_t in the Bayes estimate $\widehat{\mu}_{t|t}$ of μ_t by their estimates $\widehat{\mu}(t)$, $\widehat{\rho}(t)$, $\widehat{V}(t)$ and s_t^2 yields the empirical Bayes estimate of μ_t. Then, under the assumption of normal priors and normal observations, the Bayes predictor for $Y_{i,t+1}$ based on observation history $\mathcal{Y}_t$ is

$$
E(Y_{i,t+1}|\mathcal{Y}_t) = E(\theta_{i,t+1}|\mathcal{Y}_t) = E(\mu_{t+1}|\mathcal{Y}_t) = \widehat{\mu}_{t+1|t},
$$

which is the Kalman predictor (4.73). This is also the best linear predictor without assuming normality. The corresponding EB predictor replaces the hyperparameters in the Kalman predictor by their method-of-moment estimates $\widehat{\mu}(t)$, $\widehat{\rho}(t)$, $\widehat{V}(t)$ and s_t^2.

In the above procedure of estimating hyperparameters, we have used the cross-sectional mean $\bar{Y}_{t-1}$ of n independent observations with mean μ_{t-1}. An alternative and more direct approach is to replace μ_{t-1} in (4.71) by $\bar{Y}_{t-1}$, leading to

$$
\mu_t = \rho \bar{Y}_{t-1} + \omega + \eta_t,
$$

where $\omega = (1 - \rho)\mu$. This equation implies the following linear mixed model for Y_{it},

$$
Y_{it} = \rho \bar{Y}_{t-1} + \omega + b_i + \epsilon_{it},
$$

in which we have absorbed η_t into ϵ_{it}. Since the above equation is in the form of a regression model, one can easily include additional covariates to increase the predictive power of the model in the LMM form

$$Y_{it} = \rho \bar{Y}_{t-1} + a_i + \boldsymbol{\beta}' \mathbf{x}_{it} + \mathbf{b}_i' \mathbf{z}_{it} + \epsilon_{it}, \qquad (4.74)$$

where a_i and $\mathbf{b}_i$ are subject-specific random effects, $\mathbf{x}_{it}$ represents a vector of subject-specific covariates that are available prior to time t (for predicting Y_{it} prior to observing it at time t), and $\mathbf{z}_{it}$ denotes a vector of additional covariates that are associated with the random effects $\mathbf{b}_i$.

One may further extend (4.74) to the form of generalized linear mixed models,

$$g(\mu_{i,t}) = \sum_{j=1}^{p} \theta_j g(\bar{Y}_{t-j}) + a_i + \boldsymbol{\beta}' \mathbf{x}_{i,t} + \mathbf{b}_i' \mathbf{z}_{i,t}, \qquad (4.75)$$

in which $\theta_1, \ldots, \theta_p$ and $\boldsymbol{\beta}$ are the fixed effects and a_i and $\mathbf{b}_i$ are subject-specific random effects. The usefulness of such GLMMs for insurance claims data has been pointed out by Hsiao et al. (1990) who note that these claims data often exhibit "excess zeros." Then using the methods described for general LMMs and GLMMs in Section 4.4, one can make inference on models (4.74) and (4.75).

Supplements and problems

There are several concise and comprehensive overviews on actuarial risk theory; see Kaas et al. (2008), Rotar (2014), Dickson (2016), and Asmussen and Steffensen (2010). Mikosch (2009) introduce basic and more advanced models and methods for non-life insurance. Asmussen and Albrecher (2010) present the risk theory with concentration on ruin probabilities and related topics. Melnikov (2011) describes methods and techniques for the analysis of risk in finance and insurance. About the credibility theory, Kaas et al. (2008, Chapter 8) provide a concise introduction on the subject. A more comprehensive treatise on the subject, including evolutionary credibility models, is given by Bühlmann and Gisler (2005). For regression-type models for insurance, Jong and Heller (2008) introduce generalized linear models and their applications to insurance data. Stroup (2012) and Demidenko (2013) give a complete overview on mixed models and their inference methods.

1. Denote by X the amount of a claim. Under a proportional reinsurance contract, the insurer pays a fixed proportion a of each claim that occurs during the period of the reinsurance contract. The remaining proportion $1 - a$ of each claim is paid by the reinsurer. Find the distribution of aX

(i.e., the part of a claim paid by the insurer) under this proportional reinsurance contract.

(a) $X \sim \text{Gamma}(\alpha, \lambda)$, i.e., $f(x) = \lambda^\alpha x^{\alpha-1} e^{-\lambda x}/\Gamma(\alpha)$.

(b) $\log X \sim N(\mu, \sigma^2)$.

2. Denote by X the amount of a claim. Under an excess of loss reinsurance arrangement, the insurer and the reinsurer share the payment of the claim X and pay the amount $Y = \min(X, M)$ and $Z = \max(0, X - M)$, respectively. The amount M is called the *retention level M*.

   (a) Let $F(x)$ and $f(x)$ be the probability distribution and density functions of X, respectively. Show that the moments of Y and Z are given by

$$E(Y^n) = \int_0^M x^n f(x)dx + M^n(1-F(M)), \quad E(Z^n) = \int_M^\infty (x-M)^n f(x)dx.$$

   (b) X follows an exponential distribution with mean $1/\lambda$, show that $E(Y) = (1 - e^{-\lambda M})/\lambda$ and $E(Z) = e^{-\lambda M}/\lambda$.

   (c) $\log X \sim N(\mu, \sigma^2)$, show that

$$E(Y^n) = \exp\left(\mu n + \frac{1}{2}\sigma^2 n^2\right)\Phi\left(\frac{\log M - \mu - \sigma^2 n}{\sigma}\right) + M^n\left(1 - \Phi\left(\frac{\log M - \mu}{\sigma}\right)\right).$$

   (d) X follows a Pareto(α, λ) distribution with density function $f(x) = \alpha\lambda^\alpha/(\lambda+x)^{\alpha+1}$ for $x > 0$. Show that

$$E(Y) = \frac{\lambda}{\alpha-1}\left(1 - \left(\frac{\lambda}{\lambda+M}\right)^{\alpha-1}\right).$$

3. (**Properties of premium principles**) Show the following properties for the given premium principles.

   (a) The zero utility premium that solves the utility equilibrium equation (4.3) satisfies the non-negative loading property.

   (b) The exponential premium principle $\pi(X) = \frac{1}{\alpha}\log E\left[\exp(\alpha X)\right]$ is additive. That is, for two independent claims (or random variables) X_1 and X_2, $\pi(X_1 + X_2) = \pi(X_1) + \pi(X_2)$.

   (c) The Esscher principle $\pi(X) = E(Xe^{hX})/E(e^{hX})$ satisfies the non-negative loading condition and is additive and translational equivariance.

   (d) Consider the risk adjusted premium principle $\pi(X) = \int_0^\infty [1 - F(x)]^{1/\rho}dx$, where $\rho \geq 1$. (i) If X follows an exponential distribution with mean $1/\lambda$, then $\pi(X) = \rho/\lambda$. (ii) If X follows a Pareto(α, λ) distribution with density function $f(x) = \alpha\lambda^\alpha/(\lambda + x)^{\alpha+1}$ for $x > 0$, then $\pi(X) = \rho\lambda/(\alpha - \rho)$ provided that $\rho < \alpha$.

4. (**First two moments of the aggregate claim**) Consider the aggregate claim $S_N = X_1 + \cdots + X_N$, in which N be the number of claims from the risk and X_i are independent claims. Show that

$$E(S_N) = E(X_1)E(N), \qquad \text{Var}(S_N) = \text{Var}(X_1)E(N) + [E(X_1)]^2 \text{Var}(N).$$

5. (**The compound negative binomial distribution**) Suppose that N has a negative binomial distribution $\text{NB}(\alpha, p)$ with $\alpha > 0$ and $0 < p < 1$. That is, $P(N = n) = \binom{n+\alpha-1}{\alpha-1}(1-p)^n p^n$ for $n \geq 0$. Consider the aggregate claim $S_N = X_1 + \cdots + X_N$, in which $X_1, \ldots, X_N$ are N independent claims. Show that for $0 < s < (1-p)^{-1}$,

$$\Pi_N(s) = \left(\frac{p}{1 - (1-p)s}\right)^\alpha, \qquad \widetilde{F}_{S_N}(s) = \left(\frac{p}{1 - (1-p)\widetilde{F}(s)}\right)^\alpha,$$

where $\Pi_N(s) = \sum_{k=1}^\infty p_N(k)s^k$ is the probability-generating function of N, $\widetilde{F}_{S_N}(s)$ and $\widetilde{F}(s)$ are the Laplace-Stieltjes transforms of distribution functions of S_N and X_1, respectively.

6. (**Zero-truncated geometric distribution for aggregate claims**) Suppose that the number of claims N follows a zero-truncated geometric distribution with probability function $p_n = \alpha p_{n-1}$ for $n \geq 2$ and the individual claim amounts X_i have probability function $\{f_x\}_{x=0}^\infty$. Let $\{g_x\}_{x=0}^\infty$ be the probability function of the aggregate claims S_N, and denote by $G(\cdot)$ the probability generating function of a random variable.

   (a) Show that

   $$G'_N(t) = 1 - \alpha + \alpha t G'_N(t) + \alpha G_N(t).$$

   (b) Using the fact $P_{S_N}(t) = P_N(P_X(t))$, show that

   $$G'_{S_N}(t) = (1-\alpha)G'_X(t) + \alpha\left[G_X(t)G'_{S_N}(t) + G_{S_N}(t)P'_X(t)\right].$$

   (c) Find a recursion formula for g_x, $x \geq 1$, and show that $g_0 = (1-\alpha)f_0/(1-\alpha f_0)$.

   (d) Show that

   $$E(S_N^r) = E(X^r) + \frac{\alpha}{1-\alpha}\sum_{j=0}^{r-1}\binom{r}{j}E(S_N^j)E(X^{r-j}).$$

7. Suppose that the probability $p_n = P(N = n)$ of the number of claims N satisfies the recursion formula

$$p_n = \sum_{i=1}^k \left(a_i + \frac{b_i}{n}\right)p_{n-i},$$

for $n \geq 0$, where $p_m = 0$ for $m < 0$ and $\{a_i\}_{i=1}^k$ and $\{b_i\}_{i=1}^k$ are constants. Denote by $\{f_x\}_{x=0}^\infty$ and $\{g_x\}_{x=0}^\infty$ the probability functions of individual claim amount X_i and aggregate claim $S_N = X_1 + \cdots + X_N$. Show that

$$G'_N(t) = \sum_{i=1}^k \left(a_i t^i G'_N(t) + (ia_i + b_i)t^{i-1}G_N(t) \right),$$

and

$$G'_{S_N}(t) = \sum_{i=1}^k a_i G_{Y_i}(t)G'_{S_N}(t) + \sum_{i=1}^k \left(a_i + \frac{b_i}{i} \right)G'_{Y_i}(t)G_{S_N}(t),$$

where $Y_i = X_1 + \cdots + X_i$ for $i = 1, \ldots, k$.

8. Let $\{N(t)\}_{t\geq 0}$ be the number of claims that occur in the fixed time interval $[0, t]$ and assume $N(t)$ is a Poisson process with intensity λ. Suppose that $X_i \geq 0$ is the amount of the ith claim and $U(t) = u + ct - S(t)$ is a stochastic surplus process, where $U(t)$ is the insurer's capital at time t, $u = U(0)$ is the initial capital, c is the (constant) premium income per unit of time, and $S(t) = X_1 + \cdots + X_{N(t)}$. Suppose that X_i are exponentially random variables with distribution function $F(x) = 1 - e^{-\alpha x}$ for $x \geq 0$. Show that the probability of ruin in infinite time is $\psi(u) = Ce^{-Ru}$, in which R is the adjustment coefficient and

$$C = \frac{c/\lambda - E(X)}{E(Xe^{RX}) - c/\lambda}.$$

9. (**Identities of the ruin probability**) Show identities (4.31)–(4.34) in Proposition 4.4.

10. Let S^2 be the sample variance based on a sample of size n from a normal distribution. We know that $(n-1)S^2/\sigma^2$ has a χ_{n-1}^2 distribution. The conjugate prior for σ^2 is the *inverted Gamma* distribution, IG(α, β), given by

$$\pi(\sigma^2) = \frac{1}{\Gamma(\alpha)\beta^\alpha} \frac{1}{(\sigma^2)^{\alpha+1}} e^{-1/(\beta\sigma^2)}, \qquad 0 < \sigma^2 < \infty,$$

where α and β are positive constants. Show that the posterior distribution of σ^2 is IG$(\alpha + (n-1)/2, (\beta^{-1} + (n-1)S^2/2)^{-1})$.

11. (**Bayesian linear regression models**) Consider the linear regression model $y_t = \boldsymbol{\beta}'\mathbf{x}_t + \epsilon_t$, in which the regressors $\mathbf{x}_t \in \mathbb{R}^d$ can depend on the past observations $y_{t-1}, \mathbf{x}_{t-1}, \ldots, y_1, \mathbf{x}_1$, and the unobservable random errors ϵ_t are independent normal with mean 0 and variance σ^2. Let $\mathbf{Y} = (y_1, \cdots, y_n)'$ and $\mathbf{X}$ be the $n \times d$ matrix whose ith row is $\mathbf{x}_i'$.

(a) Suppose that σ^2 is known and $\boldsymbol{\beta}$ follows a prior distribution $N(\mathbf{z}, \sigma^2 \mathbf{V})$. Show that the posterior distribution of $\boldsymbol{\beta}$ given $(\mathbf{Y}, \mathbf{X})$ is $N(\boldsymbol{\beta}_n, \sigma^2 \mathbf{V}_n)$, where

$$\mathbf{V}_n = (\mathbf{V}^{-1} + \mathbf{X}^T \mathbf{X})^{-1}, \qquad \boldsymbol{\beta}_n = \mathbf{V}_n (\mathbf{V}^{-1}\mathbf{z} + \mathbf{X}^T \mathbf{Y}).$$

(b) Suppose that σ^2 is unknown and $(2\sigma^2)^{-1}$ follows a Gamma(g, λ) prior distribution. Let $\tau = (2\sigma^2)^{-1}$. Assume that the prior distribution of $(\boldsymbol{\beta}, \tau)$ is given by

$$\tau \sim \text{Gamma}(g, \lambda), \qquad \boldsymbol{\beta}|\tau \sim N\big(\mathbf{z}, \mathbf{V}/(2\tau)\big).$$

Show that the posterior distribution of $(\boldsymbol{\beta}, \tau)$ given $(\mathbf{X}, \mathbf{Y})$ is also of the same form:

$$\tau \sim \text{Gamma}\Big(g + \frac{n}{2}, \frac{1}{a_n}\Big), \qquad \boldsymbol{\beta}|\tau \sim N\big(\boldsymbol{\beta}_n, \mathbf{V}_n/(2\tau)\big)$$

in which $a_n = \lambda^{-1} + \mathbf{z}'\mathbf{V}^{-1}\mathbf{z} + \mathbf{Y}'\mathbf{Y} - \boldsymbol{\beta}_n^T \mathbf{V}_n^{-1} \boldsymbol{\beta}_n$.

Part II

Advanced Data and Risk Analytics

Chapter 5

Supervised and unsupervised learning

Machine learning has been playing an important role in finance and insurance. It revolutionizes the way how financial institutions make decisions and manage risk. The applications of machine learning in finance and insurance are very broad, ranging from algorithmic trading, portfolio management, credit scoring, and risk management to predictive analytics, fraud detection, sentiment analysis, and regulatory compliance.

In general, machine learning methods are broadly categorized into three main types: *supervised learning, unsupervised learning,* and *reinforcement learning.* This chapter provides an overview of some commonly used supervised and unsupervised methods and explores their applications in finance. The subsequent chapter presents Markov decision processes and widely used reinforcement learning techniques, further expanding the understanding of machine learning's role in optimizing decision-making processes within the financial domain.

The objective of supervised learning methods is to develop a function capable of mapping observed feature vectors (or covariates) to corresponding labels (responses), based on labeled datasets. These techniques encompass various approaches such as regression, classification, support vector machines, ensemble learning, random forest models, neural networks, and more. In particular, Section 5.1 gives an overview of regression methods, a category of supervised learning techniques employed for predicting continuous outcomes. Regression models establish the relationship between a dependent variable and one or more independent variables. Section 5.2 presents another supervised learning technique, classification, which can be used for predicting categorical outcomes. In addition to the classification methods detailed in Section 5.2, Section 5.3 introduces support vector machines (SVM), a supervised learning algorithm used for classification and regression tasks. SVM works by finding the optimal hyperplane that best separates data points belonging to different classes in a high-dimensional space. It is effective for both linear and nonlinear classification tasks through the use of kernel functions.

Methods in Sections 5.1–5.3 are centered around a single learner. In contrast, ensemble learning, introduced in Section 5.4, is a powerful supervised learning technique that combines the predictions of multiple individual models (learners) to produce a stronger and more accurate prediction. The idea behind ensemble learning is based on the principle of "wisdom of the crowd," where aggregating the opinions of multiple experts often leads to better de-

DOI: 10.1201/9781315117041-5

cisions than relying on any single expert. Additionally, Section 5.5 elaborates
on another class of supervised learning models, neural networks, which are
inspired by the structure and function of the human brain. A neural network
consists of interconnected layers of neurons (nodes) organized into an input
layer, one or more hidden layers, and an output layer. Renowned for their
ability to decipher intricate patterns and relationships within data, neural networks have been found vast applications in finance and insurance. Common
types of neural networks include feedforward neural networks, convolutional
neural networks, recurrent neural networks, generative adversarial networks,
and other types of neural network models.

In contrast to supervised learning, unsupervised learning methods focus
on uncovering latent and intriguing patterns within datasets that lack labeled
information. Section 5.6 introduces clustering analysis, a technique used to
identify cohesive groups of data points within large datasets based on their
similarities. This approach enables the discovery of inherent structures and relationships in the data, fostering insights into the underlying dynamics without
relying on pre-defined labels.

5.1   Linear regression and regularization

Denote by $\mathbf{X} \in \mathbb{R}^d$ a real valued predictor vector (or input), and $Y \in \mathbb{R}$
a real valued response variable (or output) with a joint distribution $F(\mathbf{X}, Y)$.
We look for a function $h(\mathbf{X})$ to predict Y given a predictor $\mathbf{X}$. To find the best
prediction for Y, statistical decision theory requires a *loss function* $\ell(Y, h(\mathbf{X}))$
to penalize errors in prediction. This suggests us evaluating the prediction
performance by the *expected prediction error* (EPE) of h,

$$\text{EPE}(h) = E[\ell(Y, h(\mathbf{X}))] = E_X E_{Y|\mathbf{X}}[\ell(Y, h(\mathbf{X}))|\mathbf{X}], \tag{5.1}$$

in which the second equality is obtained by conditioning on $\mathbf{X}$. To minimize
(5.1), it suffices to minimize $E_{Y|\mathbf{X}}(\ell(Y, \mathbf{X})|\mathbf{X})$ for a given $\mathbf{X} = \mathbf{x}$. Hence, the
best prediction f is given by

$$f(\mathbf{x}) = \arg\min_{g} E_{Y|\mathbf{X}}[\ell(Y, h(\mathbf{X}))|\mathbf{X} = \mathbf{x}]. \tag{5.2}$$

Such prediction problem is called *regression* when the output Y is quantitative
(i.e., continuous), and *classification* when Y is qualitative or categorical (i.e.,
discrete). For notational convenience, we denote quantitative outputs by Y
and qualitative outputs by G (for group or class) from now on.

We can choose different loss functions for the prediction problem (5.1) and
(5.2). One commonly used function is the *squared error loss*, $\ell(Y, h(\mathbf{X})) = (Y - h(\mathbf{X}))^2$, when Y is a continuous response. Then it suggests that the best

prediction of Y given $\mathbf{X} = \mathbf{x}$ is the conditional mean, i.e.,

$$\widehat{Y}(\mathbf{x}) = E(Y|\mathbf{X} = \mathbf{x}). \tag{5.3}$$

In the case that Y represents categorical data, such as a binary response in a classification problem, one may use the *zero-one* loss function in (5.2). The 0-1 loss function for predictor $\mathbf{X}$ and the response G is defined as $\ell(G, h(\mathbf{X})) = 0$ if $Y = h(\mathbf{X})$, and 1 otherwise. Suppose that the collection $\mathcal{G}$ of output variables consists of K response, i.e., $\mathcal{G} = \{1, 2, \ldots, K\}$. Then the minimization problem (5.2) becomes

$$\widehat{G}(x) = \arg\min_{k \in \mathcal{G}}(1 - P(G = k|\mathbf{X} = \mathbf{x})) = \arg\max_{k \in \mathcal{G}} P(G = k|\mathbf{X} = \mathbf{x}).$$

Since $\mathcal{G}$ has only finite values, the input space $\mathcal{X} = \{\mathbf{X}\}$ can be divided into a collection of regions labeled according to the classification. Depending on the methods used in the prediction procedure, the boundaries of these regions may have different shapes. Linear methods for classification refer to a class of procedures through which the obtained *decision boundaries* are linear.

5.1.1 Linear regression and least squares

Let $\mathbf{x}_i = (x_{i1}, \ldots, x_{id})'$ be a d-dimensional vector of predictor variables. Suppose that a sample $\{(\mathbf{x}_i, y_i)\}$ is observed. A linear regression model can be written in matrix form as

$$\mathbf{Y} = \mathbf{X}\boldsymbol{\beta} + \boldsymbol{\epsilon}, \tag{5.4}$$

where $\mathbf{X} = (\mathbf{X}_1, \ldots, \mathbf{X}_p)$,

$$\mathbf{Y} = \begin{pmatrix} y_1 \\ \vdots \\ y_n \end{pmatrix}, \quad \boldsymbol{\beta} = \begin{pmatrix} \beta_1 \\ \vdots \\ \beta_p \end{pmatrix}, \quad \boldsymbol{\epsilon} = \begin{pmatrix} \epsilon_1 \\ \vdots \\ \epsilon_n \end{pmatrix}, \quad \mathbf{X}_j = \begin{pmatrix} x_{1j} \\ \vdots \\ x_{nj} \end{pmatrix},$$

and the noises ϵ_i are *independent and identically distributed* (i.i.d.) univariate normal random variables with mean 0 and variance σ^2. Sometimes, we add $\mathbf{X}_0 = (1, 1, \ldots, 1)'$ in the input data $\mathbf{X}$, i.e., $\mathbf{X} = (\mathbf{X}_0, \mathbf{X}_1, \ldots, \mathbf{X}_d)$, to include a non-zero intercept in linear regression. The least square estimate of $\boldsymbol{\beta}$ for the squared error loss function L minimizes the *residual sum of squares* (RSS)

$$\text{RSS} = (\mathbf{Y} - \mathbf{X}\boldsymbol{\beta})'(\mathbf{Y} - \mathbf{X}\boldsymbol{\beta}) = ||\mathbf{Y} - \mathbf{X}\boldsymbol{\beta}||_2^2. \tag{5.5}$$

Taking the derivative with respect to $\boldsymbol{\beta}$ and setting it to zero yield a first-order condition on $\boldsymbol{\beta}$,

$$\mathbf{X}'\mathbf{X}\boldsymbol{\beta} = \mathbf{X}'\mathbf{Y}.$$

Solving the above equation gives us the least squares estimates of $\boldsymbol{\beta}$,

$$\widehat{\boldsymbol{\beta}} = (\mathbf{X}'\mathbf{X})^{-1}\mathbf{X}'\mathbf{Y}. \tag{5.6}$$

Consequently, the fitted value of $\mathbf{Y}$ is

$$\widehat{\mathbf{Y}} = \mathbf{X}\widehat{\boldsymbol{\beta}} = \mathbf{X}(\mathbf{X}'\mathbf{X})^{-1}\mathbf{X}'\mathbf{Y}.$$

Linear regression for classification purpose

While linear regression is primarily used for predicting continuous outcomes, it can indeed be applied to classification problems as well, albeit as a crude method. Suppose that a sample of K classes $\{(\mathbf{x}_i, G_i)\}_{i=1}^{n}$ is observed, where $G_i \in \mathcal{G}$ is the corresponding class index of $\mathbf{x}_i$. Assume that $\mathcal{G}$ has K categories, the response can be represented as a vector of indicator variables. In particular, for the response $G_i = k \in \mathcal{G}$, we can define $(y_{i1}, \ldots, y_{iK})$, where $y_{ik} = 1$ and $y_{ij} = 0$ for $j \neq k$. Let $\mathbf{Y}_k = (y_{1k}, \ldots, y_{nk})'$, then the observed n samples form an $n \times K$ *indicator response matrix* $\mathbf{Y} = (\mathbf{Y}_1, \ldots, \mathbf{Y}_K)$, which is a matrix of 0's and 1's, with each row having a single 1. Then the linear regression model (5.4) can be extended as follows,

$$\mathbf{Y} = \mathbf{X}\mathbf{B} + \epsilon, \tag{5.7}$$

where $\epsilon = (\epsilon_{ij})_{1 \leq i \leq n, 1 \leq j \leq K}$ is a matrix of noises, and the noise vectors $\epsilon_i = (\epsilon_{i1}, \ldots, \epsilon_{iK})'$ are i.i.d. K-variate normal random vectors with mean $(0, \ldots 0)'$ and variance-covariance matrix $\mathbf{V}$.

Consider the RSS trace$[(\mathbf{Y} - \mathbf{X}\mathbf{B})'(\mathbf{Y} - \mathbf{X}\mathbf{B})]$ for model (5.7), in which trace$(\mathbf{A}) = \sum_{i=1}^{n} a_{ii}$ for a $n \times n$ matrix $\mathbf{A}$. Minimizing this RSS yields the least squares estimates of $\mathbf{B}$,

$$\hat{\mathbf{B}} = (\mathbf{X}'\mathbf{X})^{-1}\mathbf{X}'\mathbf{Y},$$

and hence the fitted values are given by

$$\hat{\mathbf{Y}} = \mathbf{X}\hat{\mathbf{B}} = \mathbf{X}(\mathbf{X}'\mathbf{X})^{-1}\mathbf{X}'\mathbf{Y} \quad \text{or} \quad \hat{Y}_k = \mathbf{X}(\mathbf{X}'\mathbf{X})^{-1}\mathbf{X}'Y_k.$$

Then for a new observation with input $\mathbf{x}$, the fitted K-variate output is $\hat{f}(\mathbf{x}) = [(1, \mathbf{x}')\hat{\mathbf{B}}]'$. Note that 1 here corresponds to the intercept in the regression model (5.7). Then the predicted class for the new observation can be obtained by identifying the largest component in $\hat{f}(\mathbf{x})$, i.e.,

$$\hat{G}(\mathbf{x}) = \arg\max_{k \in \mathcal{G}} \hat{f}_k(\mathbf{x}).$$

Note that the kth component Y_k of $\mathbf{Y}$ is an indicator variable if the observation belongs to the kth class. This indicates that $E(Y_k | \mathbf{X} = \mathbf{x}) = P(G = k | \mathbf{X} = \mathbf{x})$. Consequently, the above procedure essentially approximates conditional expectation $E(Y_k | \mathbf{X} = \mathbf{x})$ by a linear function of $\mathbf{x}$ and applies Bayes' rule to the approximated probability $P(G = k | \mathbf{X} = \mathbf{x})$. Since an intercept is included in the model, the sum of fitted values for all classes for any given $\mathbf{x}$ is always 1, i.e., $\sum_{k \in \mathcal{G}} \hat{f}_k(\mathbf{x}) = 1$. However, the output $\hat{f}_k(\mathbf{x})$ in this sum can be negative or greater than 1. Although this may seem to violate the constraint that the approximated probability for $P(G = k | \mathbf{X} = \mathbf{x})$ should be between 0 and 1, it actually provides results similar to those obtained using more standard linear classification methods.

5.1.2   Regularized linear regression

The minimization of RSS (5.5) can easily lead to overfitting of the data, especially when the sample size n is small, but the number of covariates d is large. To mitigate the overfitting problem, one may consider using *regularization*. Regularization is a technique used in machine learning and statistical modeling to prevent overfitting by adding a penalty term to the objective function. The penalty term imposes constraints on the model parameters during training, encouraging simpler models that demonstrate improved generalization to unobserved data. If one adds a L_2 regularization to (5.5), the optimization problem becomes

$$\min_{\boldsymbol{\beta}} ||\mathbf{Y} - \mathbf{X}\boldsymbol{\beta}||_2^2 + \lambda||\boldsymbol{\beta}||_2^2, \tag{5.8}$$

where $\lambda > 0$ is the regularization parameter. Equation (5.8) is referred to as *ridge regression* (Tikhonov and Arsenin, 1977), which is equivalent to adding a Gaussian prior on the regression coefficient $\boldsymbol{\beta}$ (see Exercise 5.1).

One can also replace the L_2 norm with the L_p norm. For example, when $p = 1$, we obtain the *least absolute shrinkage and selection operator* (LASSO) (Tibshirani, 1996),

$$\min_{\boldsymbol{\beta}} ||\mathbf{Y} - \mathbf{X}\boldsymbol{\beta}||_2^2 + \lambda||\boldsymbol{\beta}||_1. \tag{5.9}$$

The L_1 penalty encourages sparsity by driving some coefficients to exactly zero, effectively performing variable selection. In contrast, the L_2 penalty in (5.8) tends to shrink the coefficients toward zero without eliminating them entirely, leading to solutions with all non-zero components. Hence, LASSO is more likely to result in a sparse solution that has fewer non-zero components in $\boldsymbol{\beta}$. From this perspective, LASSO performs both variable selection and estimation at the same time.

In addition to L_1 and L_2 regularization, there are other regularization techniques used in machine learning and statistical modeling. For example, the *elastic net* (Zou and Hastie, 2005) combines L_1 and L_2 norm and solves the following minimization problem

$$\min_{\boldsymbol{\beta}} ||\mathbf{Y} - \mathbf{X}\boldsymbol{\beta}||_2^2 + \lambda_1||\boldsymbol{\beta}||_1 + \lambda_2||\boldsymbol{\beta}||_2^2.$$

The *group LASSO* (Yuan and Lin, 2007) partitions the d predictors into J groups, with d_j the number in the jth group. It addresses the minimization problem given by

$$\min_{\boldsymbol{\beta}} ||\mathbf{Y} - \sum_{j=1}^{J} \mathbf{X}_j\boldsymbol{\beta}_j||_2^2 + \lambda \sum_{j=1}^{J} \sqrt{p_j}||\boldsymbol{\beta}_j||_2,$$

where $\mathbf{X}_l$ and $\boldsymbol{\beta}_l$ are the predictors and corresponding coefficient vector of the lth group, respectively. This formulation applies the LASSO regularization at the group level. Specifically, if all group sizes are one, it reduces to the standard LASSO.

5.1.3   Local regression and kernel smoothing

Consider the problem of estimating $f(x)$ in the regression model (5.3) based on $(x_i, y_i), 1 \leq i \leq n$. For simplicity, we let $d = 1$ here. Let $N_k(x)$ denote the set of i's such that x_i, $i \in N_k(x)$, are the k nearest neighbors of x. We can estimate $f(x)$ by fitting the regression line $y_i = \alpha + \beta x_i + \epsilon_i$ to the data $\{(x_i, y_i) : i \in N_k(x)\}$ and using the ordinary least square estimates $\widehat{\alpha}$ and $\widehat{\beta}$ to estimate $f(x)$ by $\widehat{\alpha} + \widehat{\beta} x$. In general, one can weight the observations by some function of $x_i - x$ in the RSS, and use the generalized least square method to fit a straight line to $\{(x_i, y_i) : i \in N_k(x)\}$ by choosing α and β to minimize

$$\sum_{i \in N_k(x)} w_i \big[y_i - (\alpha + \beta x_i) \big]^2, \tag{5.10}$$

where $w_i = K(|x_i - x| / \max_{j \in N_k(x)} |x_j - x|)$ and K is a kernel function. Some popular choices include the *tri-cube kernel* $K(t) = (1 - t^3)^3 \mathbf{1}_{\{|t| \leq 1\}}$, the standard normal density $K(t) = e^{-t^2/2}/\sqrt{2\pi}$, and the *Epanechnikov kernel* $K(t) = \frac{3}{4}(1 - t^2)\mathbf{1}_{\{|t| \leq 1\}}$. Given a kernel function, one can estimate $f(x)$ by using locally weighted linear functions. This method is called the *local linear regression*, and it can be readily generalized to *local polynomial regression*, which involves choosing $\alpha, \beta_1, \ldots, \beta_p$ to minimize the RSS

$$\sum_{i \in N_k(x)} w_i \big[y_i - (\alpha + \beta_1 x_i + \cdots + \beta_p x_i^p) \big]^2.$$

A *kernel smoother* estimates $f(x)$ by the weighted average

$$\widehat{f}(x) = \sum_{i=1}^{n} y_i K\left(\frac{x - x_i}{\lambda}\right) \Big/ \sum_{i=1}^{n} K\left(\frac{x - x_i}{\lambda}\right),$$

where $\lambda > 0$ is the *bandwidth* and K is the *kernel*, which is often chosen to be a smooth even function. Note that the normal density has unbounded support, whereas the Epanechnikov and tri-cube kernels have compact support (which is needed when used with the nearest-neighbor set of size k).

The kernel smoothing method can also be applied to estimate the density function of X based on an observed sample of X. Let $x_1, \ldots, x_n$ be a random sample drawn from a distribution with density function f_X. A *kernel estimate* of f_X is of the form

$$\widehat{f}_X(x) = \frac{1}{n\lambda} \sum_{i=1}^{n} K_\lambda\left(\frac{x - x_i}{\lambda}\right),$$

where K is a probability density function (nonnegative and $\int_{-\infty}^{\infty} K(t)dt = 1$) and $\lambda > 0$ is the bandwidth.

5.1.4 Splines and basis expansions

If one partitions the domain of x into subintervals and represent (5.3) by different polynomials of the same degree in different intervals, then a piecewise polynomial function can be obtained. In particular, if there are K breakpoints (called *knots*) $\eta_1 < \cdots < \eta_K$ in the domain of x, then the piecewise polynomial $f(x)$ can be written as the linear regression function

$$f(x) = \sum_{k=1}^{K+1} (\beta_{k0} + \beta_{k1}x + \cdots + \beta_{kM}x^M) \mathbf{1}_{\{\eta_{k-1} \le x < \eta_k\}}, \qquad (5.11)$$

where $\eta_0 = -\infty$, $\eta_{K+1} = \infty$, and therefore the parameters of (5.11) can be estimated by the method of least squares.

In many applications in finance and risk management, regression functions are required to be smooth. In particular, one requires $f(x)$ to have continuous derivatives up to order $M - 1$; that is,

$$f^{(l)}(\eta_k+) = f^{(l)}(\eta_k-), \qquad k = 1, \cdots, K; l = 0, 1, \cdots, M - 1. \qquad (5.12)$$

The piecewise polynomial (5.11) that satisfies the smoothness constraint (5.12) is called a *spline* of degree M. It can be represented as a linear combination of $K + M + 1$ basis functions

$$g_j(x) = x^j, \ j = 0, \cdots, M; \quad g_{l+M}(x) = (x - \eta_l)_+^M, \ l = 1, \cdots, K. \qquad (5.13)$$

In particular, for linear splines ($M = 1$) with $K = 2$ knots η_1 and η_2, the basis functions are 1, x, $(x - \eta_1)_+$, and $(x - \eta_2)_+$. For cubic splines ($M = 3$) with knots η_1 and η_2, the basis functions are 1, x, x^2, x^3, $(x - \eta_1)_+^3$, and $(x - \eta_2)_+^3$. Splines with specified knots are also called *regression splines*, and least squares regression can be used to estimate the coefficients associated with the basis functions. In practice, one seldom uses $M > 3$, as it is difficult to identify discontinuities in derivatives of order higher than 2.

A *natural cubic spline* is a cubic spline f satisfying the additional constraint $f'' = f''' = 0$ beyond the boundary knots. Therefore, instead of extrapolation to a cubic polynomial outside the range of the data, which often results in erratic behavior of the fitted spline near the extremes of the observed predictor values, a linear extrapolation is used to alleviate this problem. Whereas a cubic spline with K knots has $K + 4$ basis functions, the additional two constraints at the boundary knots for a natural cubic spline give up 4 degrees of freedom, yielding K basis functions

$$g_1(x) = 1, \quad g_2(x) = x, \quad g_k(x) = d_k(x) - d_{K-1}(x) \text{ for } 2 \le k \le K-1, \qquad (5.14)$$

where $d_k(x) = \{(x - \eta_k)_+^3 - (x - \eta_K)_+^3\}/(\eta_K - \eta_k)$ and the knots are arranged in increasing order of magnitude, with η_1 and η_K being the boundary knots; see Exercise 5.2.

5.1.5   Smoothing cubic spline and cross-validation

Unlike the natural cubic splines, a *smoothing cubic spline* is defined implicitly via the optimization criterion that minimizes

$$\sum_{i=1}^{n}(y_i - f(x_i))^2 + \lambda \int_a^b [f''(u)]^2 du \tag{5.15}$$

over f, where $a = x_1 \leq \cdots \leq x_n = b$ and λ is called a *smoothing parameter*, which is a positive constant that penalizes the "roughness" of f. For the case $\lambda = 0$, (5.15) reduces to the residual sum of squares and f can be any function that interpolates the data. For the case $\lambda = \infty$, $f''(u)$ has to be zero everywhere so that the problem reduces to least squares linear regression. The first term in (5.15) measures the closeness of the fitted model to the data, and the second term penalizes the curvature of the function.

The criterion (5.15) is defined on an infinite-dimensional space of functions for which its second term is finite. It can be shown that (5.15) has a unique and explicit, finite-dimensional minimizer, which is a natural cubic spline with knots $\eta_1, \ldots, \eta_K$ at the K distinct values of x_i $(1 \leq i \leq n)$. Hence, the optimization criterion (5.15) can be expressed as

$$\text{RSS}(\boldsymbol{\beta}) = (\mathbf{Y} - \mathbf{G}\boldsymbol{\beta})^T(\mathbf{Y} - \mathbf{G}\boldsymbol{\beta}) + \lambda\boldsymbol{\beta}^T\boldsymbol{\Omega}\boldsymbol{\beta},$$

where $\mathbf{G} = (g_i(\eta_j))_{1 \leq i,j \leq K}$, $\boldsymbol{\Omega} = (\int g_i''(u)g_j''(u)du)_{1 \leq i,j \leq K}$, and the g_i are the natural cubic spline basis functions in (5.14). The minimizer of $\text{RSS}(\boldsymbol{\beta})$ is given by

$$\widehat{\boldsymbol{\beta}} = (\mathbf{G}^T\mathbf{G} + \lambda\boldsymbol{\Omega})^{-1}\mathbf{G}^T\mathbf{Y}.$$

The fitted smoothing spline is then $\widehat{f}(x) = \sum_{j=1} \widehat{\beta}_j g_j(x)$.

Performance measure

The dataset $\{(\mathbf{x}_i, y_i) : 1 \leq i \leq n\}$ on which an estimate $\widehat{f}$ of the regression function f is based is called a *training sample*. The bias $b(\mathbf{x})$ of $\widehat{f}(\mathbf{x})$ is $E\{\widehat{f}(\mathbf{x}) - f(\mathbf{x})\}$, and the mean squared error (MSE) has the decomposition

$$\text{MSE}(\mathbf{x}) = E[\widehat{f}(\mathbf{x}) - f(\mathbf{x})]^2 = b^2(\mathbf{x}) + \text{Var}(\widehat{f}(\mathbf{x})).$$

When the $\mathbf{x}_i$ are sampled from a population with distribution function G, the *integrated mean squared error* (IMSE) is $\int E[\widehat{f}(\mathbf{x}) - f(\mathbf{x})]^2 dG(\mathbf{x})$. Since G is unknown, replacing it by the empirical distribution function leads to the *average squared error* (ASE) as a sample analog of the IMSE:

$$\text{ASE} = n^{-1} \sum_{i=1}^{n} [\widehat{f}(\mathbf{x}_i) - f(\mathbf{x}_i)]^2.$$

The prediction performance of $\widehat{f}(\mathbf{x}_i)$ relates to how well it can predict a future observation $y_i^* = f(\mathbf{x}_i) + \epsilon_i^*$ from the regression model (5.3), where ϵ_i^* is independent of the training sample. The *prediction squared error* (PSE) is related to the ASE by

$$\text{PSE} = n^{-1} \sum_{i=1}^{n} E\left\{ \left[y_i^* - \widehat{f}(\mathbf{x}_i) \right]^2 \big| \mathbf{x}_1, \ldots, \mathbf{x}_n, y_1, \ldots, y_n \right\} = \text{ASE} + \sigma^2$$

Bias-variance trade-off and cross-validation

The above estimate $\widehat{f}$ first approximates f by a function that involves a finite number of parameters and then estimates the parameters of the approximation. The approximation inherently introduces bias, which decreases as the dimensionality of the approximation increases. On the other hand, the estimate based on a lower-dimensional approximation (that has fewer parameters to estimate) has smaller variance. Thus, There is a trade-off between bias and variance in devising an appropriate approximation to f. In particular, for linear smoothers with univariate predictors, the squared bias tends to increase while the variance decreases as the amount of smoothing increases. The smoothing parameter λ governs the level of smoothing. In kernel smoothers, λ represents the bandwidth; in smoothing splines, it's the roughness penalty; and in nearest-neighbor methods, it corresponds to the relative neighborhood size, expressed as k/n. Increasing the number of knots in regression splines results in a more accurate approximation of f, thereby reducing bias. However, this also introduces more parameters to be estimated, consequently increasing the variance of $\widehat{f}(\mathbf{x}_i)$.

To achieve a balance of bias and variance, one can use the *cross-validation* method. This technique involves replacing the predicted response y_i^* with the observed response y_i, and the estimated function $\widehat{f}(\mathbf{x}_i)$) with the nonparametric regression estimate $\widehat{f}_{(-i)}(\mathbf{x}_i)$. Here, $\widehat{f}_{(-i)}$ is obtained by excluding the ith observation $(\mathbf{x}_i, y_i)$ from the training sample. This approach, commonly referred to as the *leave-one-out* or *jackknife* method, results in the computation of the *cross-validation sum of squares* represented by the following equation, which serves as an estimate of the PSE,

$$\text{CV}(\lambda) = \frac{1}{n} \sum_{i=1}^{n} \left[y_i - \widehat{f}_{(-i)}(\mathbf{x}_i) \right]^2. \tag{5.16}$$

It is important to note that cross-validation aims to estimate the out-of-sample error. It is different from the RSS $\sum_{i=1}^{n} \left[y_i - \widehat{f}_i(\mathbf{x}_i) \right]^2$, which measures the in-sample error. The selection of the smoothing parameter λ is often guided by the behavior of $\text{CV}(\lambda)$, which depends on λ through $\widehat{f}_{(-i)}$. Typically, the optimal λ is chosen as the one that minimizes $\text{CV}(\lambda)$.

5.2   Linear and nonlinear classification

This section focuses on linear methods for classification. We first consider two special cases of generalized linear models, and then the linear discriminant method and its extensions. Two types of nonlinear classification methods based on trees and nearest neighbors are introduced afterwards.

5.2.1   Logistic and ordered probit models

Consider the generalized linear model in Definition 4.3. Suppose that the response Y is binary and y in (4.60) follow a Bernoulli distribution $f(y) = p^y(1-p)^{1-y}$ ($y = 0, 1$). Here $p = e^\theta/(1 + e^\theta)$, or equivalently $\theta = \text{logit}(p)$, where

$$\text{logit}(p) = \log[p/(1-p)], \qquad 0 < p < 1. \tag{5.17}$$

Moreover, $\phi = 1$ and $b(\theta) = -\log(1-p)$, so $b'(\theta) = (1-p)^{-1}dp/d\theta = p$. In this case, using the canonical link function $g(\theta) = \theta$ yields the *logistic regression model* $\text{logit}\big(P\{y_t = 1|\mathbf{x}_t\}\big) = \boldsymbol{\beta}'\mathbf{x}_t$, or equivalently,

$$P(y_i = 1|\mathbf{x}_i) = \frac{\exp(\boldsymbol{\beta}'\mathbf{x}_i)}{1 + \exp(\boldsymbol{\beta}'\mathbf{x}_i)}, \qquad P(y_t = 0|\mathbf{x}_t) = \frac{1}{1 + \exp(\boldsymbol{\beta}'\mathbf{x}_i)}.$$

To estimate $\boldsymbol{\beta}$ in the above probabilities, we can use the IRLS algorithm introduced in Section 4.4.4. The logistic regression with $K = 2$ is widely used to model how the probability of the occurrence of an event (e.g., default of a firm or delinquent payment of a loan in credit risk analysis) varies with explanatory variables (predictors).

The assumption of Bernoulli distributed responses can be extended to multinomial distributions. Then a logistic regression for K classes can be obtained and has the form

$$\log\frac{P(G = j|X = x)}{P(G = K|\mathbf{X} = \mathbf{x})} = \beta_{j0} + \boldsymbol{\beta}_j'\mathbf{x}, \qquad j = 1, \ldots, K - 1. \tag{5.18}$$

A simple algebra implies that

$$P(G = k|\mathbf{X} = \mathbf{x}) = \frac{\exp(\beta_{k0} + \boldsymbol{\beta}_k'\mathbf{x})}{1 + \sum_{l=1}^{K-1}\exp(\beta_{l0} + \boldsymbol{\beta}_l'\mathbf{x})}, \qquad k = 1, \ldots, K - 1,$$

$$P(G = K|\mathbf{X} = \mathbf{x}) = \frac{1}{1 + \sum_{l=1}^{K-1}\exp(\beta_{l0} + \boldsymbol{\beta}_l'\mathbf{x})}. \tag{5.19}$$

Let $\boldsymbol{\beta} = \{\beta_{10}, \boldsymbol{\beta}_1, \ldots, \beta_{K-1,0}, \boldsymbol{\beta}_{K-1}\}$. The parameter set $\boldsymbol{\beta}$ in the logistic regression (5.19) can be estimated via the likelihood method.

Instead of using the canonical link function for binary responses, another way to model the binary response and its predictor is to use the *probit* link function $g(\theta) = \Phi^{-1}(1 - p)$, where Φ is the standard normal distribution function. This yields the probit model for binary response variables $y_i \in \{0, 1\}$,

$$P(y_i = 0|\mathbf{x}_i) = \Phi(\boldsymbol{\beta}'\mathbf{x}_i), \quad P(y_i = 1|\mathbf{x}_i) = 1 - \Phi(\boldsymbol{\beta}'\mathbf{x}_i). \tag{5.20}$$

For ordinal response variables, where the outcome variable has a natural ordering or ranking but is not necessarily binary, the probit link function implies that for a response y_i and an outcome $j \in \{1, \ldots, K\}$, there exists threshold values $\eta_0, \eta_1, \ldots, \eta_K$ such that

$$y_i = j \text{ if } \eta_j < y_i^* < \eta_{j+1}, \text{ where } y_i^* = \boldsymbol{\beta}'\mathbf{x}_i + \epsilon_i, \epsilon_i \in N(0, 1).$$

This indicates that, for $j = 0, \ldots, K - 1$,

$$P(y_i = j+1|\mathbf{X} = \mathbf{x}_i) = P(\eta_j < \boldsymbol{\beta}'\mathbf{x}_i < \eta_{j+1}) = \Phi(\eta_{j+1} - \boldsymbol{\beta}'\mathbf{x}_i) - \Phi(\eta_j - \boldsymbol{\beta}'\mathbf{x}_i).$$

Example 5.1 (Logsitic and ordered probit models for bank default)
Consider the bank default data from the quarterly Call Reports filed by all the Federal Deposit Insurance Corporation (FDIC) insured commercial banks from the fourth quarter of 2002 to the third quarter of 2013. These reports include fundamental balance-sheet and income-statement data of banks. They are crucial for assessing banks' capital adequacy, asset quality, profitability, and liquidity.

Numerous studies, such as Martin (1977) and Cole and Gunther (1995), have highlighted the statistical significance of these metrics in predicting bank failures. First, banks' capital adequacy (cap.ade) is measured by the ratio of total equity capital to total asset. Bank capital can absorb unexpected losses and preserve confidence of banks, hence banks' capital adequacy is expected to be negatively correlated with the probability of bank default. Second, banks' asset quality (asset.qual) reflects potential asset quality issues and their impact on the probability of bank default. It can be assessed by the ratio of nonperforming loans (the sum of loans 90 days or more past due, nonaccrual loans, and other real estate owned) to total asset. Third, banks' profitability is measured by banks' earning ratios (earn.ratio), defined as net income divided by total assets. It is negative correlated with the probability of a bank default. Fourth, banks' liquidity can be measured by the ratio of total securities (tot.sec) and total deposit against bank's total assets (tot.depo), respectively. Securities tend to be more liquid than loans, and could help bank minimize big loss in response to unexpected demand of cash. Hence, securities contribute negatively to bank default. Bank's deposit is a liability owed by the bank to the depositor, and hence is positively correlated with the probability of bank default. Besides these five covariates, another two covariates are also considered. Tier 1 risk-based capital ratio (rbc1), defined as the ratio of tier 1 risk-based capital to total risk adjusted assets, is an important measure of a bank's financial strength from the perspective of regulators and is expected to have negative correlation with

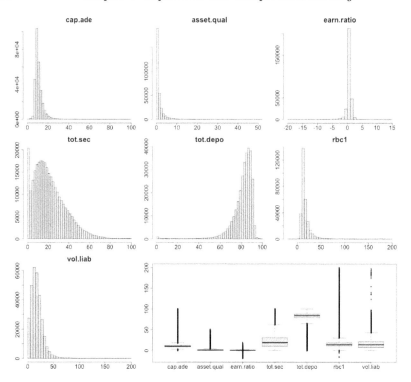

FIGURE 5.1: Histograms and a boxplot of seven covariates.

*the probability of bank default. The percentage of volatile liability (vol.liab),
defined as the ratio of volatile liability to total assets, indicates the extent of
a bank's dependence on potential volatile liabilities.*

*The data set contains 360,188 quarterly records of 9,639 U.S. banks.
Among these records, there are totally 2,467 default records from 520 defaulted
banks. All seven variables* cap.ade, asset.qual, earn.ratio, tot.sec,
tot.depo, rbc1, *and* vol.liab *are measured in percentages. Figure 5.1 shows
histograms and a boxplot of these seven covariates. We can see that all seven
variables are skewed and have heavy tails. Figure 5.2 shows the total number
of banks and the percentage of bank default in each quarter. As indicated by
Figure 5.2, banks' default rates are 0 until the end of 2007, which is the beginning of the 2007–2008 financial crisis. Then some banks began to fail from the
first quarter of 2008 and the percentage of default peaks at 1.46% in the second
quarter of 2009. After that period, the default percentage begins to drop.*

*The data are divided into two parts. The records before (and including)
the third quarter of 2011 are considered as training data, and the rest of the
records are used for testing. The training data contain 303,330 bank records,
and among these, there are 2,244 recorded failures from 496 defaulted banks.
The testing data contains 56,858 bank records, and among them, there are*

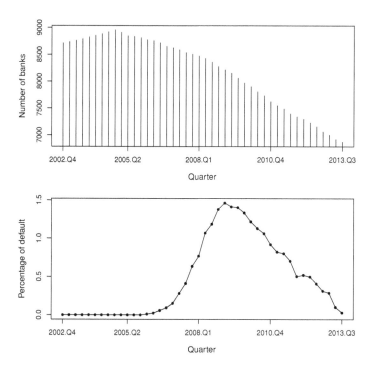

FIGURE 5.2: Total number of banks and percentage of bank default.

223 recorded failures from 70 defaulted banks. We denote "0" and "1" for nondefault and default status of banks, respectively.

We first fit a logistic regression model and an ordered probit model to the training data. The estimated regression coefficients (Coeff), their standard errors (S.E.), z-values and p-values in Table 5.1. We note that all covariates are significant in both models.

TABLE 5.1: Estimated regression coefficients.

	Logistic regression				Ordered probit model			
	Coeff.	S.E.	z-value	p-value	Coeff.	S.E.	z-value	p-value
Intercept	0.2649	0.2814	0.941	0.346	−0.4348	0.1310	−3.318	9e-4
cap.ade	−0.1282	0.0200	−6.401	2e-10	−0.0573	0.0086	−6.658	3e-11
asset.qual	0.1876	0.0049	38.172	<2e-16	0.0951	0.0023	40.069	<2e-16
earn.ratio	−0.2555	0.0141	−18.096	<2e-16	−0.1316	0.0070	−18.761	<2e-16
tot.sec	−0.0204	0.0035	−5.849	5e-9	−0.0064	0.0015	−4.384	1.2e-5
tot.depo	−0.0142	0.0027	−5.300	1.2e-7	−0.0046	0.0012	−3.750	2e-4
rbc1	−0.3875	0.0178	−21.787	<2e-16	−0.1630	0.0077	−21.085	<2e-16
vol.liab	0.0300	0.0019	16.085	2.6e-16	0.0137	0.0009	15.420	<2e-16

We then use the estimated parameters or coefficients to compute the classification for both the training and testing data. The misclassification rates of these three methods are shown in Table 5.2. Since the logistic regression model and the ordered probit model only provide predicted probabilities, a bank is considered as default if and only if the fitted or predicted probability of default is larger than a threshold η.   □

TABLE 5.2: Misclassification rates of logistic and ordered probit models.

Method		Training data			Testing data		
		Default	Nondefault	All	Default	Nondefault	All
Logistic regression	$\eta = 0.1$	31.24%	0.59%	0.81%	3.59%	1.31%	1.32%
	$\eta = 0.2$	40.73%	0.34%	0.64%	13.00%	0.86%	0.91%
	$\eta = 0.5$	58.78%	0.13%	0.57%	36.32%	0.34%	0.48%
Ordered probit	$\eta = 0.1$	28.16%	0.72%	0.93%	2.69%	1.55%	1.56%
	$\eta = 0.2$	39.04%	0.40%	0.68%	11.21%	0.96%	1.00%
	$\eta = 0.5$	61.94%	0.11%	0.57%	42.60%	0.28%	0.44%

5.2.2   Discriminant analysis

Discriminant analysis is a method of classifying observations into a number of classes. Let $f_k(\mathbf{x})$ be the class-conditional density of X in class $G = k$, and π_k the prior probability of class k such that $\sum_{k=1}^{K} \pi_k = 1$. Applying Bayes theorem implies that

$$P(G = k|\mathbf{X} = \mathbf{x}) = \frac{f_k(\mathbf{x})\pi_k}{\sum_{l=1}^{K} f_l(\mathbf{x})\pi_l}.$$

Then by the Bayes rule of 0-1 loss, the best prediction $\widehat{G}$ in (5.1) becomes

$$\widehat{G}(\mathbf{x}) = \arg\max_k P(G = k|\mathbf{X} = \mathbf{x}) = \arg\max_k \{f_k(\mathbf{x})\pi_k\} = \arg\max_k \delta_k(\mathbf{x}).$$

$$(5.21)$$

where $\delta_k(\mathbf{x}) = \log f_k(\mathbf{x}) + \log \pi_k$ is the discriminant function. Note that the classification rule (5.21) suggests that class densities $f_k(\mathbf{x})$ play an important role in classifying the new observation with predictor $\mathbf{x}$. In fact, many classification methods are based on models for $f_k(\mathbf{x})$. Here, we assume that $\mathbf{x}$ follows a d-variate Gaussian distribution, i.e., $\mathbf{x} \sim N_d(\boldsymbol{\mu}_k, \boldsymbol{\Sigma}_k)$.

Linear discriminant analysis

Linear discriminant analysis (LDA) refers to the case when we assume that all classes have a common covariance matrix $\boldsymbol{\Sigma}_k = \boldsymbol{\Sigma}$. With this assumption, the discriminant function $\delta_k(\mathbf{x})$ in (5.21) reduces to

$$\delta_k(\mathbf{x}) = \mathbf{x}'\boldsymbol{\Sigma}^{-1}\boldsymbol{\mu}_k - \frac{1}{2}\boldsymbol{\mu}_k'\boldsymbol{\Sigma}^{-1}\boldsymbol{\mu}_k + \log(\pi_k),$$

which is called the *linear discriminant function*. Then the corresponding rule with the above $\delta_k(\mathbf{x})$ (i.e., the LDA rule) suggests the decision boundary between classes k and l, which is determined by $\delta_k(\mathbf{x}) = \delta_l(\mathbf{x})$ and expressed as a linear function of $\mathbf{x}$,

$$\mathbf{x}'\mathbf{\Sigma}^{-1}(\boldsymbol{\mu}_k - \boldsymbol{\mu}_l) = \frac{1}{2}\boldsymbol{\mu}_k'\mathbf{\Sigma}^{-1}\boldsymbol{\mu}_k - \frac{1}{2}\boldsymbol{\mu}_l'\mathbf{\Sigma}^{-1}\boldsymbol{\mu}_l + \log(\pi_l) - \log(\pi_k).$$

When there are only two categories in $\mathcal{G}$, i.e., $K = 2$, the above linear boundary shows an equivalence between LDA and classification by linear regression. To implement the LDA, one can denote n_k the number of class-k observations such that $\sum_{k=1}^{K} n_k = n$ and estimate parameters π_k, μ_k, and $\mathbf{\Sigma}$ in Gaussian densities as follows

$$\widehat{\pi}_k = \frac{n_k}{n}, \quad \widehat{\boldsymbol{\mu}}_k = \frac{1}{n_k}\sum_{G_i=k}\mathbf{x}_i, \quad \widehat{\mathbf{\Sigma}} = \frac{1}{n-K}\sum_{k=1}^{K}\sum_{G_i=k}(\mathbf{x}_i - \widehat{\boldsymbol{\mu}}_k)(\mathbf{x}_i - \widehat{\boldsymbol{\mu}}_k)'.$$

$$(5.22)$$

Quadratic discriminant analysis

The LDA assumes that all $\mathbf{\Sigma}_k$ are same in the discriminant problem. When the $\mathbf{\Sigma}_k$ are not assumed to be equal, the discriminant function $\delta_k(\mathbf{x})$ in (5.21) becomes

$$\delta_k(\mathbf{x}) = \frac{1}{2}\log(|\mathbf{\Sigma}_k|) - \frac{1}{2}(\mathbf{x} - \boldsymbol{\mu}_k)'\mathbf{\Sigma}_k^{-1}(\mathbf{x} - \boldsymbol{\mu}_k) + \log(\pi_k),$$

for $k = 1, \ldots, K$. The decision boundary between classes k and l, which is determined by $\delta_k(\mathbf{x}) = \delta_l(\mathbf{x})$, is a quadratic function of $\mathbf{x}$. Hence, this is referred to as *quadratic discriminant analysis* (QDA). To implement the QDA, one can estimate parameters $\pi_k, \boldsymbol{\mu}_k$, and $\mathbf{\Sigma}_k$ in Gaussian densities. The estimates for QDA are similar to those for LDA, except that separate covariance matrices need to be estimated for each class. The following shows the estimates of $\pi_k, \boldsymbol{\mu}_k$, and $\mathbf{\Sigma}_k$ from the sample,

$$\widehat{\pi}_k = \frac{n_k}{n}, \quad \widehat{\boldsymbol{\mu}}_k = \frac{1}{n_k}\sum_{G_i=k}x_i, \quad \widehat{\mathbf{\Sigma}}_k = \frac{1}{n_k-K}\sum_{G_i=k}(\mathbf{x}_i - \widehat{\boldsymbol{\mu}}_k)(\mathbf{x}_i - \widehat{\boldsymbol{\mu}}_k)'. \quad (5.23)$$

Note that QDA fits better than LDA, but since each $\mathbf{\Sigma}_k$ is a $d \times d$ matrix, the number of parameters to be estimated in QDA is much larger than that in LDA.

Regularized discriminant analysis

To achieve a compromise between LDA and QDA, a *regularized discriminant analysis* was proposed to shrink the separate covariances of QDA toward a common covariance as in LDA. The regularized matrices have the form

$$\widehat{\mathbf{\Sigma}}_k(\alpha) = \alpha\widehat{\mathbf{\Sigma}}_k + (1-\alpha)\widehat{\mathbf{\Sigma}}, \qquad \alpha \in [0, 1],$$

where $\widehat{\boldsymbol{\Sigma}}$ is the pooled covariance matrix used in LDA. Here, α allows a continuum of models between LDA and QDA, and needs to be specified. In practice, α can be chosen based on the performance of the model on validation data, or by cross-validation. The covariance matrices $\widehat{\boldsymbol{\Sigma}}$ can also be shrunk to other covariance matrices, for instance, a scalar covariance,

$$\widehat{\boldsymbol{\Sigma}}_k(\gamma) = \gamma \widehat{\boldsymbol{\Sigma}}_k + (1-\gamma)\widehat{\sigma}^2 \mathbf{I}, \qquad \gamma \in [0,1],$$

where $\widehat{\sigma}^2$ is an estimate of diagonal elements in the pooled covariance matrix $\widehat{\boldsymbol{\Sigma}}$.

Other versions of regularized discriminant analysis can be similarly constructed. These methods are more suitable when the features or predictors of the data are high-dimensional and correlated, so that the LDA coefficients can be regularized to be smooth or sparse in the original domain of the data.

Example 5.2 (Discriminant analysis for bank default) *Consider a linear and a quadratic discriminant analysis for the FDIC-insured bank default data in Example 5.1. For the training data, the prior probabilities of banks' non-default and default class are $\widehat{\pi}_0 = 0.99261$ and $\widehat{\pi}_1 = 0.00739$. The prior means for banks' non-default and default are*

$$\widehat{\boldsymbol{\mu}}_0 = (11.74, 1.33, 0.45, 21.71, 81.02, 17.91, 17.67)'$$

and

$$\widehat{\boldsymbol{\mu}}_1 = (5.45, 12.19, -2.41, 9.98, 85.85, 6.60, 24.17)',$$

respectively. The prior covariance matrices for the LDA and the QDA can be obtained by (5.22) and (5.23). With these parameters, banks' non-default and default status in the testing data can be predicted. Table 5.3 shows the misclassification rates of the LDA and the QDA for the training and the testing data. □

TABLE 5.3: Misclassification rates of discriminant analysis.

Method	Training data			Testing data		
	Default	Nondefault	All	Default	Nondefault	All
LDA	33.47%	1.32%	1.56%	18.83%	2.75%	2.82%
QDA	26.69%	2.01%	2.19%	12.11%	3.78%	3.80%

5.2.3 Classification and regression trees

The methods introduced so far assume structured forms for the unknown regression or classification function $f(\mathbf{x})$. Now, we introduce tree-based methods, which don't assume any structures for the function $f(\mathbf{x})$, allowing them to effectively capture complex relationships within the data.

The basic idea of tree-based methods is to partition the feature space into a hierarchical structure of decision nodes, creating a tree-like model. This partitioning process involves recursively splitting the feature space based on the values of individual features, with each split creating a new decision node. At each split, the algorithm selects the feature and threshold value that best separates the data into homogeneous groups with respect to the target variable. This iterative splitting process continues until a stopping criterion is met, such as reaching a maximum tree depth or having a minimum number of data points in each leaf node.

The resulting tree structure forms a set of decision rules that collectively define regions within the feature space. Each terminal node, also known as a leaf, corresponds to a specific region and contains a prediction or classification for observations falling within that region. One of the key advantages of tree-based methods is their ability to capture nonlinear relationships and interactions between features without relying on explicit assumptions about the underlying data distribution. This makes them particularly well-suited for handling complex, high-dimensional data where traditional parametric models may have difficulty. Furthermore, the hierarchical nature of tree-based models provides much convenience for interpretation. Decision paths within the tree correspond to specific combinations of feature values, which allows users to understand and explain the reasoning behind the model's predictions or classifications.

One popular tree-based method is the *classification and regression tree* (CART). Suppose that one observes $(\mathbf{x}_i, y_i)$ for $i = 1, \ldots, n$, with $\mathbf{x}_i = (x_{i1}, \ldots, x_{id})'$. Assume first that the feature space can be partitioned into M regions $R_1, \ldots, R_M$, and the response can be represented as a constant c_m in each region:

$$f(\mathbf{x}) = \sum_{m=1}^{M} c_m I(\mathbf{x} \in R_m).$$

By minimizing the residual sum of squares $\sum_{i=1}^{n}(y_i - f(\mathbf{x}_i))^2$, one can easily see that the best $\hat{c}_m$ is the average of y_i in region R_m, $\hat{c}_m = \text{average}(y_i | \mathbf{x}_i \in R_m)$.

Note that it is generally not easy to describe regions that are obtained by an arbitrary partition, hence our attention is restricted to recursive binary partitions. Since finding the best binary partition via minimizing sum of squares is computationally expensive, a greedy algorithm can be used to simplify the searching procedure. Given a splitting variable j and split point s, we define the pair of half-planes

$$R_1(j, s) = \{X | X_j \le s\} \quad \text{and} \quad R_2(j, s) = \{X | X_j > s\}.$$

Then the best splitting variable j and the best splitting point s can be obtained by solving the optimization problem

$$\min_{j,s} \left\{ \min_{c_1} \sum_{x_i \in R_1(j,s)} (y_i - c_1)^2 + \min_{c_2} \sum_{x_i \in R_2(j,s)} (y_i - c_2)^2 \right\}.$$

After finding the best split (j, s), the data are partitioned into the two resulting regions, and one should repeat the splitting process on each of the two regions. Such process is iterated on all of the resulting regions, and should be stopped when some minimum node size is reached.

Let T_0 be the large tree grown in the above splitting process. To reduce overfitting and improve the out-of-sample performance, one needs to remove or simplify branches of the tree. This process is known as *pruning*. We next prune the large tree T_0 using *cost-complexity pruning*. Denote a subtree $T \subset T_0$ as a tree that can be obtained by pruning T_0. Let R_m be a region with terminal nodes, $m = 1, \ldots, |T|$, where $|T|$ denotes the number of terminal notes in T. Let N_m be the number of $\mathbf{x}_i$ in R_m and

$$\widehat{c}_m = \frac{1}{N_m} \sum_{\mathbf{x}_i \in R_m} y_i, \qquad Q_m(T) = \frac{1}{N_m} \sum_{\mathbf{x}_i \in R_m} (y_i - \widehat{c}_m)^2,$$

the cost complexity criterion is defined as

$$C_\alpha(T) = \sum_{m=1}^{|T|} N_m Q_m(T) + \alpha |T|.$$

The tuning parameter $\alpha \geq 0$ governs the tradeoff between tree size and its goodness of fit to the data. For each α, a unique smallest subtree T_α of T_0 can be obtained by minimizing $C_\alpha(T)$, and α can be estimated via cross-validation.

The CART algorithm described above can be extended to incorporate basis functions, offering greater flexibility in modeling complex relationships in the data. One such extension is the *multivariate adaptive regression splines* (MARS) algorithm. MARS employs expansions in piecewise linear basis functions of the form $(x - \eta)_+$ and $(x - \eta)_-$. These basis functions allow for capturing nonlinear relationships between features and the target variable by representing the data in terms of breakpoints. In contrast, the basis function used in traditional CART is piecewise constant $\mathbf{1}_x \in R_m$. It assigns a constant value to each region of the feature space determined by the tree splits, effectively creating a step-like function. By incorporating piecewise linear basis functions like those used in MARS, the CART algorithm can capture subtle relationships in the data, enabling smoother and more flexible modeling compared to traditional piecewise constant models.

The above tree algorithm can be applied to the K-class classification problem, after the criteria for splitting nodes and pruning the tree are modified. In particular, we denote the proportion of class k observations in node m as

$$\widehat{p}_{mk} = \frac{1}{N_m} \sum_{\mathbf{x}_i \in R_m} \mathbf{1}_{\{y_i = k\}},$$

and classify the observations in node m to class $k(m)$ such that $k(m) = \arg\max_k \widehat{p}_{mk}$. In the classification tree method, the squared-error node impurity measure $Q_m(T)$ can be replaced by one of the following measures:

TABLE 5.4: Misclassification rates of classification trees.

Method		Training data			Testing data		
		Default	Nondefault	All	Default	Nondefault	All
CART	Gini	49.73%	0.27%	0.64%	8.07%	1.09%	1.12%
	Entropy	33.16%	0.7%	0.94%	3.14%	2.00%	2.00%

(i) Misclassification error: $\dfrac{1}{N_m} \sum_{\mathbf{x}_i \in R_m} \mathbf{1}_{\{y_i \neq k(m)\}} = 1 - \widehat{p}_{k(m)m}$.

(ii) Gini index: $\sum_{k \neq k'} \widehat{p}_{km} \widehat{p}_{k'm} = \sum_{k=1}^{K} \widehat{p}_{km}(1 - \widehat{p}_{km})$

(iii) Cross-entropy or deviance: $-\sum_{k=1}^{K} \widehat{p}_{km} \log \widehat{p}_{km}$.

The CART methods have several advantages, making them very attractive for classification and prediction tasks. These advantages include their relative fast speed, applicability to all types of variables, and invariance under monotone transformations. This invariance ensures that they are robust against outliers and scale-irrelevant factors. Furthermore, tree methods have very few tunable parameters and provide an interpretable model representation. Consequently, CART methods have became one of the most popular procedures for classification and prediction tasks. However, despite their strength, tree methods do have certain weaknesses, such as inaccuracy, high variance, lack of smoothness and difficulty in capturing additive structure. Section 5.4 introduces techniques like bagging and boosting to address these issues.

Example 5.3 (Tree methods for bank default) *Consider classification trees for the FDIC-insured bank default data in Example 5.1. Given the training data, two pruned classification trees are obtained using the method outlined earlier, employing the Gini index and cross-entropy measures, respectively. In particular, Figure 5.3 shows the pruned classification tree using cross-entropy and Table 5.4 provides the misclassification rates of these two trees for training and testing data.* □

5.2.4 Nearest neighbor methods

In contrast to the linear methods that assume the best prediction (5.3) can be well approximated by a linear or generalized linear model, *nearest neighbor* methods assume that (5.3) can be well approximated by a locally constant function. In particular, suppose that we have n labeled data $(\mathbf{x}_i, y_i)$, where $\mathbf{x}_i \in \mathbb{R}^n$ and $y_i \in \{1, \ldots, K\}$. Assume that a distance measure $D(\mathbf{x}_i, \mathbf{x}_j)$ is

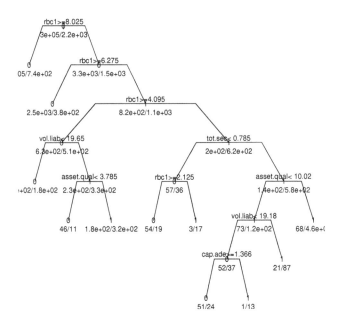

FIGURE 5.3: A pruned classification tree using cross-entropy for bank default. The numbers below each leaf are the frequencies of bank default and non-default.

given. For a new observation $\mathbf{x}$, let $r_i(\mathbf{x}) = $ rank of $\mathbf{x}_i$ sorted on $D(\mathbf{x}, \mathbf{x}_i)$. Then the nearest neighbor classifier uses majority vote from the M nearest neighbors:

$$\widehat{G}(\mathbf{x}) = \arg\max_{k \in \{1,\dots,K\}} \frac{1}{\#\{r_i(\mathbf{x}) \leq M\}} \sum_{r_i(\mathbf{x}) \leq M} \mathbf{1}_{\{y_i = k\}}, \qquad (5.24)$$

where $\#\{r_i(\mathbf{x}) \leq M\}$ is the number of elements in $\{r_i(\mathbf{x}) \leq M\}$. When the sample size n in the training set is large, the points in the neighborhood are likely to be close to x, and as M gets large the average (5.24) become more stable. Actually, under mild regularity conditions, we can show that as $n, M \to \infty$ such that $M/n \to 0$,

$$\frac{1}{M} \sum_{r_i(\mathbf{x}) \leq M} \mathbf{1}_{\{y_i = k\}} \to E(Y | X = \mathbf{x}).$$

There is a bias-variance trade-off in increasing M, the number of neighbors considered. Specifically, small M usually leads to low bias but high variance, while large M causes high bias but low variance. In practice one could use cross-validation to determine M. There are also various ways to define the metric $D(\mathbf{x}, \mathbf{x}')$. The most common choice is the weighted L_p norm:

$$D_p(\mathbf{x}, \mathbf{x}') = \left\{ \sum_{i=1}^{d} w_i d^p(\mathbf{x}_i, \mathbf{x}_i') \Big/ \sum w_i \right\}^{1/p}.$$

Although the nearest neighbor method is simple, it is often successful in many classification problems, especially the one in which the decision boundaries are very irregular.

Example 5.4 (k-nearest neighbor method for bank default) *Consider the nearest neighbor method for FDIC-insured bank default data in Example 5.1. To see how the number of neighbors affects classification, we set $M = 5, 10$ and 15 in the method. Table 5.5 lists the misclassification rate of this method for both the training and testing data.*

TABLE 5.5: Misclassification rates of the nearest neighbor method.

Method		Training data			Testing data		
		Default	Nondefault	All	Default	Nondefault	All
Nearest neighbors	$M = 5$	47.15%	0.06%	0.41%	48.43%	0.27%	0.46%
	$M = 10$	57.18%	0.06%	0.48%	46.64%	0.24%	0.42%
	$M = 15$	60.87%	0.06%	0.51%	49.78%	0.17%	0.37%

Adaptive nearest-neighbor methods

An implicit assumption in the nearest neighbor method is that the class probabilities in the neighborhood are roughly constant, and hence simple averages provide good approximations. When the nearest neighbor method is carried out in a high-dimensional feature space (i.e., d is large), the nearest neighbors of a point can be very far away and the class probabilities may vary significantly, and hence the bias of the estimate increases and the performance of the method gets worse. To overcome this difficulty, one should notice that, in high-dimensional feature space, the class probabilities may change only in a low-dimensional subspace, hence one may adapt the metric $D(\mathbf{x}, \mathbf{x}')$ so that the resulting neighborhoods can be stretched out in directions along which the class probabilities do not change much. Interested readers can find more details on these methods in Hastie et al. (2016, Section 13.4).

5.3 Support vector machine

This section introduces *support vector machines* (SVM), which extend linear methods for classification to produce nonlinear boundaries for classification in an enlarged and transformed feature space.

5.3.1   Optimal separating hyperplane and the dual problem

Consider the classification problem of $K = 2$ classes and a training set $\{(\mathbf{x}_i, y_i), y_i = G_i \in \mathcal{G} = \{-1, 1\}, i = 1, \ldots, n\}$. A hyperplane separator is a linear function that separates observations into two parts,

$$\{\mathbf{x} : f(\mathbf{x}) = \boldsymbol{\beta}'\mathbf{x} + \beta_0 = 0\},$$

where $\boldsymbol{\beta} = (\beta_1, \ldots, \beta_p)'$ is a unit vector, i.e., $\|\boldsymbol{\beta}\| = 1$. A classification rule induced by $f(\mathbf{x})$ is

$$G(\mathbf{x}) = \text{sign}\left[\boldsymbol{\beta}'\mathbf{x} + \beta_0\right]. \tag{5.25}$$

If the classes in the training sample are separable, meaning that they can be perfectly separated by a hyperplane, we can find a function $f(\mathbf{x}) = \boldsymbol{\beta}'\mathbf{x} + \beta_0$ such that $f(\mathbf{x}_i) > 0$ when $y_i = 1$ and $f(\mathbf{x}_i) < 0$ when $y_i = -1$. Additionally, we can choose the hyperplane such that $f(\mathbf{x}_i) \geq 1$ when $y_i = 1$ and $f(\mathbf{x}_i) \leq -1$ when $y_i = -1$.

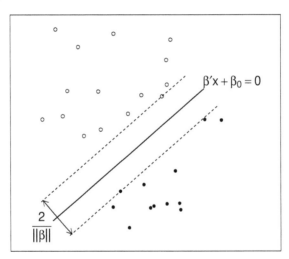

FIGURE 5.4: Support vector classifiers.

Then as illustrated in Figure 5.4, the equality holds for the sample points closest to the hyperplane, and these sample points are called *support vectors*. Note that the distance from any point $\mathbf{x}$ in the sample space to the above hyperplane is $|\boldsymbol{\beta}'\mathbf{x} + \beta_0|/\|\boldsymbol{\beta}\|$. The total distance from two support vectors of different classes to the hyperplane is $M = 2\|\boldsymbol{\beta}\|^{-1}$, which is called *margin*. This suggests that one can find a hyperplane that maximizes the *margin* $2\|\boldsymbol{\beta}\|^{-1}$ between the training points for class 1 and -1. Such maximization problem can be expressed as

$$\max_{\boldsymbol{\beta}, \beta_0, \|\boldsymbol{\beta}\|=1} \quad 2\|\boldsymbol{\beta}\|^{-1} \qquad \text{subject to } y_i(\boldsymbol{\beta}'\mathbf{x}_i + \beta_0) \geq 1, \text{ for all } i.$$

Dropping the norm constraint on $\boldsymbol{\beta}$, the maximization of $2||\boldsymbol{\beta}||^{-1}$ is equivalent to the minimization of $||\boldsymbol{\beta}||/2$. Additionally, for the convenience of using the Lagrange method, we can consider the minimization of $||\boldsymbol{\beta}||^2/2$ and write the minimization problem as

$$\min_{\boldsymbol{\beta},\beta_0} \frac{1}{2}||\boldsymbol{\beta}||^2 \quad \text{subject to } y_i(\boldsymbol{\beta}'\mathbf{x}_i + \beta_0) \geq 1, \text{ for all } i. \tag{5.26}$$

Problem (5.26) is a convex optimization problem with linear inequality constraints, which can be solved as follows. Consider the minimization problem for the Lagrange function

$$L(\boldsymbol{\beta}, \beta_0, \{\alpha_i\}) = \frac{1}{2}||\boldsymbol{\beta}||^2 - \sum_{i=1}^n \alpha_i [y_i(\boldsymbol{\beta}'\mathbf{x}_i + \beta_0) - 1]. \tag{5.27}$$

Taking derivative with respect to $\boldsymbol{\beta}$ and β_0 and setting them to zero, we obtain the first order constraint

$$\boldsymbol{\beta} = \sum_{i=1}^n \alpha_i y_i \mathbf{x}_i, \qquad 0 = \sum_{i=1}^n \alpha_i y_i.$$

Substituting these into (5.27) yields the Lagrangian (Wolfe) *dual problem*

$$\max_{\boldsymbol{\alpha}} \sum_{i=1}^n \alpha_i - \frac{1}{2} \sum_{i=1}^n \sum_{j=1}^n \alpha_i \alpha_j y_i y_j \mathbf{x}_i' \mathbf{x}_j \quad \text{subject to } \sum_{i=1}^n \alpha_i y_i = 0, \ \alpha_i \geq 0, \tag{5.28}$$

where $\boldsymbol{\alpha} = (\alpha_1, \ldots, \alpha_n)'$. The maximized objective function in (5.28) gives a lower bound on the objective function (5.26) for any feasible points. Solving the above optimization problem for α_i and we have the desired model

$$f(\mathbf{x}) = \boldsymbol{\beta}'\mathbf{x} + \beta_0 = \sum_{i=1}^n \alpha_i y_i \mathbf{x}_i' \mathbf{x} + \beta_0.$$

Note that the variable α_i solved from the dual problem (5.28) is the Lagrange multiplier in (5.27), corresponding to the training sample $(\mathbf{x}_i, y_i)$. Since (5.26) is an optimization problem with inequality constraints, it must satisfy the Karush-Kuhn-Tucker conditions

$$\alpha_i \geq 0, \quad y_i f(\mathbf{x}_i) - 1 \geq 0, \quad \alpha_i(y_i f(\mathbf{x}_i) - 1) = 0.$$

Hence, for each sample $(\mathbf{x}_i, y_i)$, we have either $\alpha_i = 0$ or $y_i f(\mathbf{x}_i) = 1$. If $\alpha_i = 0$, the sample $(\mathbf{x}_i, y_i)$ has no impact on $f(\mathbf{x})$. If $\alpha_i > 0$, then $y_i f(\mathbf{x}_i) = 1$, the sample $(\mathbf{x}_i, y_i)$ lies on the maximum-margin hyperplanes and is a support vector. This indicates an important property of support vector machine, that is, after the training is completed, most training samples are no longer needed since the final model only depends on the support vectors.

The dual problem (5.28) can be solved by quadratic programming. However, the computational cost is usually high due to the complexity caused by the number of training samples. To overcome this limitation, many efficient algorithms are proposed to exploit the structure of the optimization problem; see for example the sequential minimal optimization Platt (1998).

5.3.2 Kernel function and support vector expansion

The above discussion assumes the training samples in the feature space are linearly separable, which is often not true in practice. In such case, one can map the samples from the original feature space to a higher dimensional feature space so that the samples become linearly separable. Let $\phi(\mathbf{x})$ be the mapped feature vector of $\mathbf{x}$, then replacing $\mathbf{x}$ in the separating hyperplane $f(\mathbf{x})$, we obtain $f(\mathbf{x}) = \boldsymbol{\beta}'\phi(\mathbf{x}) + \beta_0$. Similar to (5.26), one can consider the optimization problem

$$\min_{\boldsymbol{\beta},\beta_0} \frac{1}{2}||\boldsymbol{\beta}||^2 \qquad \text{subject to } y_i(\boldsymbol{\beta}'\phi(\mathbf{x}_i) + \beta_0) \geq 1, \text{ for all } i. \qquad (5.29)$$

Its dual problem is given by

$$\max_{\boldsymbol{\alpha}} \sum_{i=1}^{n} \alpha_i - \frac{1}{2}\sum_{i=1}^{n}\sum_{j=1}^{n} \alpha_i\alpha_j y_i y_j \phi(\mathbf{x}_i)'\phi(\mathbf{x}_j) \quad \text{subject to } \sum_{i=1}^{n} \alpha_i y_i = 0, \; \alpha_i \geq 0.$$

The inner product $\phi(\mathbf{x}_i)'\phi(\mathbf{x}_j)$ in the above can be replaced by a kernel function $\kappa(\mathbf{x}_i, \mathbf{x}_j)$ such that $\kappa(\mathbf{x}_i, \mathbf{x}_j) = \phi(\mathbf{x}_i)'\phi(\mathbf{x}_j)$. Then the above dual problem can be rewritten as

$$\max_{\boldsymbol{\alpha}} \sum_{i=1}^{n} \alpha_i - \frac{1}{2}\sum_{i=1}^{n}\sum_{j=1}^{n} \alpha_i\alpha_j y_i y_j \kappa(\mathbf{x}_i, \mathbf{x}_j) \quad \text{subject to } \sum_{i=1}^{n} \alpha_i y_i = 0, \; \alpha_i \geq 0.$$
$$(5.30)$$

Solving the above problem yields that $\boldsymbol{\beta} = \sum_{i=1}^{n} \alpha_i y_i \phi(\mathbf{x}_i)$ and

$$f(\mathbf{x}) = \boldsymbol{\beta}'\phi(\mathbf{x}) + \beta_0 = \sum_{i=1}^{n} \alpha_i y_i \phi(\mathbf{x}_i)'\phi(\mathbf{x}) + \beta_0 = \sum_{i=1}^{n} \alpha_i y_i \kappa(\mathbf{x}, \mathbf{x}_i) + \beta_0.$$

In the above, the optimal solution can be expanded by training samples with the kernel functions, and this is called the *support vector expansion*.

In the above procedure, the choice of kernel is usually the uncertain step of support vector machines. A poor kernel will map the samples to a poor feature space and hence result in poor performance. Here, we list some common kernel functions.

(i) Linear kernel: $\kappa(\mathbf{x}_i, \mathbf{x}_j) = \mathbf{x}_i'\mathbf{x}_j$.

(ii) Polynomial kernel: $\kappa(\mathbf{x}_i, \mathbf{x}_j) = (\mathbf{x}_i'\mathbf{x}_j)^d$.

(iii) Gaussian kernel: $\kappa(\mathbf{x}_i, \mathbf{x}_j) = \exp\left(-||\mathbf{x}_i - \mathbf{x}_j||^2/(2\sigma^2)\right)$.

(iv) Laplacian kernel: $\kappa(\mathbf{x}_i, \mathbf{x}_j) = \exp\left(-||\mathbf{x}_i - \mathbf{x}_j||/\sigma\right)$, $\sigma > 0$.

(v) Sigmoid (hyperbolic tangent) kernel: $\kappa(\mathbf{x}_i, \mathbf{x}_j) = \tanh(\gamma\mathbf{x}_i'\mathbf{x}_j + \theta)$, where $\gamma > 0$, $\theta < 0$, and $\tanh(z) = (e^z - e^{-z})/(e^z + e^{-z})$ is the hyperbolic tangent function.

The selection of the kernel function in SVMs typically involves mapping the input space into a higher-dimensional feature space, where the data points may become linearly separable. This usually depends on the specific characteristics of the data and the problem at hand. It is often necessary to try multiple kernel functions and tune their associated hyperparameters using techniques such as cross-validation to find the best-performing model for a given dataset.

5.3.3 Nonseparable hyperplane, soft margin, and risk minimization

When the samples in the feature space are neither linearly separable nor able to find an appropriate kernel function to linearly separate them in the feature space, one strategy is to relax the *hard margin* in previous discussion by the *soft margin* and allow for some points to be on the wrong side of the margin. Of course, the number of samples on the wrong side of the margin should be minimized while maximizing the margin. Hence, the optimization objective can be written as

$$\min_{\boldsymbol{\beta},\beta_0} \frac{1}{2}||\boldsymbol{\beta}||^2 + \gamma \sum_{i=1}^{n} \ell(y_i(\boldsymbol{\beta}\mathbf{x}_i + \beta_0) - 1), \qquad (5.31)$$

where $\gamma > 0$ is a constant and $\ell(\cdot)$ is a loss function.

Consider the 0-1 loss function $\ell_{0/1}(z) = 1_{\{z<0\}}$. When $\gamma \to \infty$, (5.31) forces all samples to obey the constraint $y_i(\boldsymbol{\beta}'\mathbf{x}_i + \beta_0) \geq 1$ and (5.31) reduces to (5.26) or the hard margin. The 0-1 loss function is non-convex and discontinuous, and hence has poor mathematical properties. Therefore, one usually uses *surrogate loss* functions that are convex, continuous and are upper bound of the 0/1 loss function. The following shows three commonly used surrogate loss functions.

(i) Hinge loss: $\ell(z) = \max(0, 1 - z)$.

(ii) Exponential loss: $\ell(z) = \exp(-z)$.

(iii) Logistic loss: $\ell(z) = \log(1 + \exp(-z))$.

Take the hinge loss as an example. When it is used, problem (5.31) becomes

$$\min_{\boldsymbol{\beta},\beta_0} \frac{1}{2}||\boldsymbol{\beta}||^2 + \gamma \sum_{i=1}^{n} \max(0, 1 - y_i(\boldsymbol{\beta}'\mathbf{x}_i + \beta_0)).$$

To allow for the possibility of violating the hard margin, we introduce *slack variable* $\xi_i \geq 0$ and write the above optimization problem as

$$\min_{\boldsymbol{\beta},\beta_0} \frac{1}{2}||\boldsymbol{\beta}||^2 + C\sum_{i=1}^{n}\xi_i \quad \text{subject to } y_i(\boldsymbol{\beta}'\mathbf{x}_i + \beta_0) \geq (1-\xi_i), \tag{5.32}$$

$$\xi_i \geq 0, \text{for all } i, \text{ and } \sum_{i=1}^{n}\xi_i \leq \text{constant}.$$

This is the commonly used C-classifier of the *soft margin SVM*, where the constant $C > 0$ determines the trade-off between the training error and the margin. To solve (5.32), consider the Lagrange function

$$L(\boldsymbol{\beta},\beta_0,\{\xi_i,\alpha_i,\psi_i\}) = \frac{1}{2}||\boldsymbol{\beta}||^2 + C\sum_{i=1}^{n}\xi_i - \sum_{i=1}^{n}\alpha_i\left[y_i(\boldsymbol{\beta}'\mathbf{x}_i + \beta_0) - (1-\xi_i)\right] - \sum_{i=1}^{n}\psi_i\xi_i,$$

where $\alpha_i \geq 0$ and $\psi_i \geq 0$ are the Lagrange multipliers. Taking derivatives with respect to $\boldsymbol{\beta}$, β_0, and ξ_i and setting the respective derivatives to zero, we obtain

$$\boldsymbol{\beta} = \sum_{i=1}^{n}\alpha_i y_i \mathbf{x}_i, \qquad 0 = \sum_{i=1}^{n}\alpha_i y_i, \qquad \alpha_i = C - \psi_i, \text{ for all } i,$$

together with the positivity constraints $\alpha_i, \psi_i, \xi_i \geq 0$. Substituting them into the Lagrange function yields the Lagrangian (Wolfe) dual maximization problem

$$\max_{\boldsymbol{\alpha}} \sum_{i=1}^{n}\alpha_i - \frac{1}{2}\sum_{i=1}^{n}\sum_{j=1}^{n}\alpha_i\alpha_j y_i y_j \mathbf{x}_i'\mathbf{x}_j \quad \text{subject to } \sum_{i=1}^{n}\alpha_i y_i = 0, \ 0 \leq \alpha_i \leq C,$$

$$\tag{5.33}$$

where $\boldsymbol{\alpha} = (\alpha_1,\ldots,\alpha_n)'$. Note that the only difference between the dual problem (5.33) and the dual problem (5.28) is the constraint on dual variables, $0 \leq \alpha_i$ for hard margin and $0 \leq \alpha_i \leq C$ for soft margin. Hence, (5.33) can be solved in the same way as (5.28). By introducing kernel function, the support vector expansion can be obtained as for (5.30).

Besides the C-classifier of the SVM, another possible realization of a soft margin of the optimal hyperplane uses the ν-parameterization (Schölkopf et al., 2000); see Exercise 5.3.

Example 5.5 (SVM for bank default) *Consider the C-classification of the SVM for the bank default data in Example 5.1. To see the regularization effect of the constant C, we carry out the SVM with the Gaussian kernel for $C = 10, 50,$ and $100,$ respectively, and show the misclassification rate in Table 5.6. We see that, when C increases, the misclassification error in the training data decreases significantly, while the misclassification error in the testing data increases gradually.* □

TABLE 5.6: Misclassification rates of the SVM.

Method	Training data			Testing data		
	Default	Nondefault	All	Default	Nondefault	All
SVM ($C = 10$)	56.506%	0.066%	0.483%	47.085%	0.196%	0.380%
SVM ($C = 50$)	14.305%	0.004%	0.109%	45.740%	0.445%	0.623%
SVM ($C = 100$)	10.071%	0.002%	0.077%	46.637%	0.500%	0.681%

Empirical risk minimization

Other learning models can be obtained when other surrogate loss functions are used. Note that in the obtained models, the first term $\frac{1}{2}||\boldsymbol{\beta}||^2$ in (5.33) represents the margin size of the separating hyperplane and the second term $\sum_{i=1}^{n} \ell(f(\mathbf{x}_i), y_i)$ represents the error on the training set. This motivates us to consider a more general optimization problem

$$\min_{f} \Omega(f) + \gamma \sum_{i=1}^{n} \ell(f(\mathbf{x}_i), y_i), \tag{5.34}$$

where $\Omega(f)$ is called the *structural risk* and $\sum_{i=1}^{n} \ell(f(\mathbf{x}_i), y_i)$ is called the *empirical risk*. The structural risk represents some properties of the model f and provides a way for incorporating domain knowledge. The empirical risk accounts for how well the model matches the training sample. The constant γ is a trade-off between these two risks. From this perspective, (5.34) is a regularization problem.

5.3.4 Support vector regression

The idea of SVM can be extended from classification to regression. In traditional regression models, the loss is the difference between the model output $f(\mathbf{x})$ and the true output y and the loss is 0 if and only if $f(\mathbf{x}) = y$. In contrast, *support vector regression* (SVR) allows a margin of error between $f(\mathbf{x})$ and y, that is, a loss is incurred only if the difference between $f(\mathbf{x})$ and y exceeds certain threshold. This margin of error allows for some tolerance in fitting the data and helps prevent overfitting.

Given a training sample $\{(\mathbf{x}_i, y_i), i = 1, \ldots, n\}$. Suppose one wants to learn a regression model $f(\mathbf{x}) = \boldsymbol{\beta}'\mathbf{x} + \beta_0$, where $\boldsymbol{\beta}$ and β_0 are parameters to be learned from the training sample. The SVR problem can be written as

$$\min_{\boldsymbol{\beta}, \beta_0} \frac{1}{2}||\boldsymbol{\beta}||^2 + \gamma \sum_{i=1}^{n} \ell_\epsilon(f(\mathbf{x}_i) - y_i), \tag{5.35}$$

where γ is the regularization parameter, and ℓ_ϵ is the ϵ-insensitive loss function $\ell_\epsilon(z) = 0$ if $|z| \leq \epsilon$ and $|z| - \epsilon$ otherwise. With slack variable ξ and $\widehat{\xi}$, (5.35) can be rewritten as

$$\min_{\boldsymbol{\beta},\beta_0,\xi,\widehat{\xi}} \frac{1}{2}||\boldsymbol{\beta}||^2 + \gamma \sum_{i=1}^{n}(\xi_i + \widehat{\xi}_i) \quad \text{subject to } f(\mathbf{x}_i) - y_i \leq \epsilon + \xi_i,$$

$$y_i - f(\mathbf{x}_i) - y_i \geq \epsilon + \widehat{\xi}, \ \xi_i \geq 0, \ \widehat{\xi}_i \geq 0, i = 1,\dots,n.$$

Consider the Lagrange function

$$L(\boldsymbol{\beta},\beta_0,\alpha_i,\{\widehat{\alpha}_i,\xi_i,\widehat{\xi}_i,\phi_i,\widehat{\phi}_i\}) = \frac{1}{2}||\boldsymbol{\beta}||^2 + \gamma \sum_{i=1}^{n}(\xi_i + \widehat{\xi}_i) - \sum_{i=1}^{n}(\phi_i\xi_i + \widehat{\phi}_i\widehat{\xi}_i)$$

$$+ \sum_{i=1}^{n}\alpha_i(f(\mathbf{x}_i) - y_i - \epsilon - \xi_i) + \sum_{i=1}^{n}\widehat{\alpha}_i(y_i - f(\mathbf{x}_i) - \epsilon - \widehat{\xi}_i).$$

The first order conditions yield that

$$\boldsymbol{\beta} = \sum_{i=1}^{n}(\widehat{\alpha}_i - \alpha_i)\mathbf{x}_i, \quad \sum_{i=1}^{n}(\widehat{\alpha}_i - \alpha_i) = 0, \quad \gamma = \alpha_i + \phi_i, \quad \gamma = \widehat{\alpha}_i + \widehat{\phi}_i.$$

Substituting the above into the Lagrange function, we obtain the dual problem of SVR

$$\max_{\boldsymbol{\alpha},\widehat{\boldsymbol{\alpha}}} \sum_{i=1}^{n} y_i(\widehat{\alpha}_i - \alpha_i) - \epsilon(\widehat{\alpha}_i + \alpha_i) - \frac{1}{2}\sum_{i,j=1}^{n}(\widehat{\alpha}_i - \alpha_i)(\widehat{\alpha}_j - \alpha_j)\mathbf{x}_i'\mathbf{x}_j$$

$$\text{subject to } \sum_{i=1}^{n}(\widehat{\alpha}_i - \alpha_i) = 0, 0 \leq \alpha_i, \widehat{\alpha}_i \leq \gamma. \tag{5.36}$$

Note that the Karush-Kuhn-Tucker conditions for SVR are

$$\alpha_i(f(\mathbf{x}_i) - y_i - \epsilon - \xi_i) = 0, \quad \widehat{\alpha}_i(y_i - f(\mathbf{x}_i) - \epsilon - \widehat{\xi}_i) = 0,$$

$$\alpha_i\widehat{\alpha}_i = 0, \quad \xi_i\widehat{\xi}_i = 0, \quad (\gamma - \alpha_i)\xi_i = 0, \quad (\gamma - \widehat{\alpha}_i)\widehat{\xi}_i = 0.$$

This implies that α_i and $\widehat{\alpha}_i$ take non-zero values if and only if the sample $(\mathbf{x}_i, y_i)$ falls outside of the ϵ-insensitive region, and α_i and $\widehat{\alpha}_i$ must be zero. With these constraint, the solution of SVR is given by

$$f(\mathbf{x}) = \sum_{i=1}^{n}(\widehat{\alpha}_i - \alpha_i)\mathbf{x}_i'\mathbf{x} + \beta_0.$$

In the above solution for $\alpha_i \neq \widehat{\alpha}_i$, $(i = 1,\dots,n)$, the samples $(\mathbf{x}_i, y_i)$ are the support vectors of SVR and fall outside the ϵ-insensitive region. In addition, after mapping the sample to its feature space, we can argue similarly and obtain the SVR with kernel function,

$$f(\mathbf{x}) = \sum_{i=1}^{n}(\widehat{\alpha}_i - \alpha_i)\kappa(\mathbf{x}, \mathbf{x}_i) + \beta_0, \quad \kappa(\mathbf{x}, \mathbf{x}_i) = \phi(\mathbf{x}_i)'\phi(\mathbf{x}_j).$$

5.4 Ensemble learning

The learning methods introduced so far rely on individual learners. Actually, multiple learners can be combined to solve a learning problem, and this is called *ensemble learning*. This section introduces three types of ensemble learning methods: generalized additive models, boosting, and bagging algorithms.

5.4.1 Generalized additive models

In the regression setting, a *generalized additive model* (GAM) has the form

$$E(Y|X_1,\ldots,X_d) = \alpha + \sum_{j=1}^{d} f_j(X_j), \qquad (5.37)$$

where X_j represents the jth predictor, Y is the response, and the f_j is an unspecified smooth function for the jth predictor. Given observations $\{(x_i, y_i); i = 1,\ldots,n\}$, model (5.37) can be estimated via the *penalized residual sum of squares* (PRSS),

$$PRSS(\alpha, f_1,\ldots,f_p) = \sum_{i=1}^{n} \left\{ y_i - \alpha - \sum_{j=1}^{p} f_j(x_{ij}) \right\}^2 + \sum_{j=1}^{p} \lambda_j \int f_j''(t_j)^2 dt_j, \qquad (5.38)$$

where the $\lambda_j \geq 0$ are tuning parameters. We can show that the minimizer of (5.38) is an additive cubic spline model, that is, each of the function f_j is a cubic spline in the component X_j, with knots at each of the unique values of x_{ij}, $i = 1,\ldots,n$. Note that the constant α is not identifiable, hence the standard convention is to assume that $\hat{\alpha} = n^{-1}\sum_{i=1}^{p} y_i$. When the input matrix $\mathbf{X} = (X_1,\ldots,X_d)$ is nonsingular, (5.38) is a strictly convex criterion and the minimizer is unique. Otherwise, the linear part of the components f_j cannot be uniquely determined.

For two-class classification, the mean of the binary response $p = P(Y = 1|X)$ is related to the predictors via a linear regression model and the logit link function

$$\log \frac{p}{1-p} = \alpha + \sum_{j=1}^{p} \beta_j X_j.$$

In the spirit of GAM, the additive logistic regression model has a more general functional form

$$\log \frac{p}{1-p} = \alpha + \sum_{j=1}^{p} \beta_j f_j(X_j), \qquad (5.39)$$

where f_j is an unspecified smooth function. The functions $f_1, \ldots, f_p$ in the above model can be estimated by a backfitting algorithm within a Newton-Raphson procedure; see details in Hastie et al. (2016, Section 9.1.2). In general, generalized linear models can be similarly extended to become additive by relating the conditional mean $\mu(X)$ of a response Y to an additive function of the predictors via a link function g,

$$g[\mu(X)] = \alpha + \sum_{j=1}^{p} \beta_j f_j(X_j),$$

and functions $f_1, \ldots, f_p$ in the above model can be estimated by a backfitting algorithm within a Newton-Raphson procedure.

Example 5.6 (Additive logistic regression for bank default) *Consider an additive logistic regression model* (5.39) *for the FDIC-insured bank default data in Example 5.1. Table 5.7 shows the misclassification rates of the model for the training and testing data with threshold $\eta = 0.1, 0.2$, and 0.5. Given the fitted or predicted probabilities, a bank is considered as default if and only if the fitted or predicted probability of default is larger than a threshold η.* □

TABLE 5.7: Misclassification rates of the additive logistic regression.

Method		Training data			Testing data		
		Default	Nondefault	All	Default	Nondefault	All
Additive	$\eta = 0.1$	24.29%	0.82%	0.99%	2.24%	1.43%	1.43%
logistic	$\eta = 0.2$	34.09%	0.44%	0.69%	12.56%	0.94%	0.98%
regression	$\eta = 0.5$	55.48%	0.13%	0.54%	35.43%	0.23%	0.37%

5.4.2 Boosting algorithm

Boosting is a powerful ensemble learning technique designed to enhance the predictive performance of machine learning models by combining multiple weak learners to create a stronger, more accurate model. Boosting algorithms start with a base learner and then adjust the distribution of the training samples according to the result of the base learner such that incorrectly classified samples will receive more attention by subsequent base learners. This process repeats until the number of base learners attains a predetermined level, and then, all base learners are weighted and combined.

The most well-known boosting algorithm is AdaBoost (Freund and Schapire, 1997). Consider a two-class classification problem in which the output variables are labeled as $G \in \{-1, 1\}$. Given a vector of predictor variables $\mathbf{X}$, a classifier $G(\mathbf{X})$ produces a prediction taking values in $\{-1, 1\}$. The error rate on the training sample is

$$\overline{\text{err}} = \frac{1}{N} \sum_{i=1}^{N} 1_{\{y_i \neq G(\mathbf{x}_i)\}},$$

and the expected error rate on future predictions is $E_{\mathbf{XY}} I_{Y \neq G(\mathbf{X})}$. A weak classifier is one whose error rate is only slightly better than random guessing. The purpose of boosting is to sequentially apply the weak classification algorithm to repeatedly modified versions of the data, thereby producing a sequence of weak classifiers $G_m(\mathbf{x})$, $m = 1, \ldots, M$. The predictions from all of them are then combined through a weighted majority vote to produce the final prediction

$$G(\mathbf{x}) = \text{sign}\left(\sum_{m=1}^{M} \alpha_m G_m(\mathbf{x}) \right).$$

Here, the weight $\alpha_1, \ldots, \alpha_M$ are computed by the following boosting algorithm.

Algorithm 5.1 (AdaBoost) *Initialize the weights $w_i = 1/n$, $i = 1, \ldots, n$.*

1. *For $m = 1$ to M, fit a classifier $G_m(\mathbf{x})$ to the training data using the weights w_i and compute the error rate*

$$err_m = \frac{\sum_{i=1}^{n} w_i 1_{\{y_i \neq G_m(\mathbf{x}_i)\}}}{\sum_{i=1}^{n} w_i},$$

and $\alpha_m = \log((1 - err_m)/err_m)$. Then set $w_i \leftarrow w_i \cdot \exp(\alpha_m \cdot I(y_i \neq G_m(x_i)))$, $i = 1, \ldots, n$.

2. *Output $G(\mathbf{x}) = sign\left[\sum_{m=1}^{M} \alpha_m G_m(\mathbf{x}) \right]$.*

There are several ways to derive the AdaBoost algorithm. One way is to use the additive model $G(\mathbf{x}) = \sum_{m=1}^{M} \alpha_m G_m(\mathbf{x})$ to minimize the *exponential loss function* $\ell(f(\mathbf{x}, y)) = \exp(-yf(\mathbf{x}))$. Besides, the base classifier $G_m(\mathbf{x})$ in the AdaBoost algorithm returns a discrete class label. For problems with real-valued predictions, the algorithm can be modified appropriately; see Friedman et al. (2000).

5.4.3 Bagging and random forest

Bagging, short for *bootstrap aggregating*, is a popular ensemble learning technique used to improve the stability and accuracy of machine learning models, particularly in the context of decision trees. It was introduced by Breiman (1996). The key idea behind bagging is to create multiple subsets of the training data by randomly sampling with replacement (bootstrap sampling). Each bootstrap sample is used to train a base model independently.

These base models can be of any type, but decision trees are commonly used due to their high variance and tendency to overfit. After training the base models on different subsets of the data, bagging combines their predictions by averaging (for regression) or voting (for classification). Such ensembling of models tends to have better generalization performance than any single base model because it reduces variance and minimizes the risk of overfitting.

To further explain the idea, we consider a regression problem. Suppose that one observes $\mathbf{Z} = \{(\mathbf{x}_1, y_1), \ldots, (\mathbf{x}_n, y_n)\}$ and wants to obtain a prediction $\widehat{f}(\mathbf{x})$ at input $\mathbf{x}$. Bagging uses the average of predictions over a collection of bootstrap sample as the prediction, and hence reduces the variance of the prediction. In particular, for each bootstrap sample $\mathbf{Z}^{*b}$, $b = 1, \ldots, B$, one fits a model to $\mathbf{Z}^{*b}$ and gives prediction $\widehat{f}^{*b}(\mathbf{x})$. Then the bagging estimate is expressed as

$$\widehat{f}_{\text{bag}}(\mathbf{x}) = \frac{1}{B} \sum_{b=1}^{B} \widehat{f}^{*b}(\mathbf{x}). \tag{5.40}$$

For classification problems, suppose that $\widehat{G}(\mathbf{x})$ is the classifier for a K-class response, $\widehat{f}(\mathbf{x})$ is an underlying indicator-vector function with value a single one and $K - 1$ zeros, and $\widehat{G}(\mathbf{x}) = \arg\max_k \widehat{f}(\mathbf{x})$. Then the bagged estimate (5.40) is a K-vector, in which the kth element is the proportion of classifier predicting class k at x, and the predicted class is the one with the majority vote from the B predictions, i.e.,

$$\widehat{G}_{\text{bag}}(\mathbf{x}) = \arg\max_k \widehat{f}_{\text{bag}}(\mathbf{x}).$$

As discussed earlier, tree methods can capture complex interaction structure in the data and usually have relatively low bias. Hence, trees are natural candidates for bagging. Specifically, for $b = 1, \ldots, B$, one first draw a bootstrap sample $\mathbf{Z}^{*b}$ of size n from the observations $\{(\mathbf{x}_1, y_1), \ldots, (\mathbf{x}_n, y_n)\}$, and then grow a random-forest tree T_b to the bootstrap sample $\mathbf{Z}^{*b}$ by using the procedure in Section 5.2.3. Given a new point x, the *random forest* predictor for regression is expressed as

$$\widehat{f}_{\text{rf}}(x) = \frac{1}{B} \sum_{b=1}^{B} T_b(\mathbf{x}).$$

For a K-class classification problem, let $\widehat{G}_b(\mathbf{x})$ be the class prediction of the bth random-forest tree, then the random forest predictor is the class with the largest proportion from the B random-forest trees.

Example 5.7 (Random forest for bank default) *Consider the random forest method for classification of the FDIC-insured bank default data in Example 5.1. We run the random forest for the training data with $B = 100, 200$, and 300 trees, respectively. Table 5.8 shows the misclassification rates of the method for the training and testing data.* ☐

TABLE 5.8: Misclassification rates of the random forest method.

Method		Training data			Testing data		
		Default	Nondefault	All	Default	Nondefault	All
Random forest	$B = 100$	54.19%	0.10%	0.50%	41.70%	0.22%	0.38%
	$B = 200$	55.35%	0.10%	0.51%	40.81%	0.23%	0.39%
	$B = 300$	55.66%	0.10%	0.51%	40.36%	0.21%	0.36%

5.5 Artificial neural network

Neural networks, sometimes referred to as *artificial neural network* (ANN), are popular machine learning techniques that mimic the mechanism of learning in biological organisms. Principles of neuroscience have been used to design neural network architectures. The basic idea of neural networks is to approximate the prediction function $f(\mathbf{x})$ by combining many basic machine learning units and learning their weights jointly in order to minimize the prediction error. When a neural network is used in its most basic form, the learning algorithm reduces to simple machine learning models like least square regression or logistic regression. In the case that multiple units are used in the neural network, the power of the model to learn more sophisticated functions of the data increases significantly, hence can be used to handle learning problems in the "big data" that are usually difficult for traditional machine learning techniques.

An artificial neural network, often simply referred to as a neural network, can be conceptualized as a network of *neurons* that connects a set of inputs to a set of outputs in a potentially nonlinear manner. The predictors, or features, constitute the *input layer* of the neural network, while the predictions, or outcomes, constitute the *output layer*. The connections between the input and output layers are typically established via one or more *hidden layers* of neurons. Neurons in hidden layers are sometimes called *processing units* or *nodes*. Each node in the hidden layers receives inputs from the nodes in the previous layer and computes an output based on a weighted sum of these inputs, which is then passed through an activation function to introduce nonlinearity into the network. The activation function determines whether and to what extent the neuron should be activated, influencing the information flow through the network. Through the process of training, which involves adjusting the weights and biases of the connections between neurons, a neural network can learn to approximate complex mappings between inputs and outputs, making it a powerful tool for tasks such as classification, regression, and pattern recognition.

A neuron function $g : \mathbb{R}^d \to \mathbb{R}$ in the neural network is a mapping of the form

$$g(\mathbf{x}) = \phi(\boldsymbol{\alpha}'\mathbf{x} + \beta), \qquad (5.41)$$

where $\phi : \mathbb{R} \to \mathbb{R}$ is a continuous non-linear function called the *activation function*, $\boldsymbol{\alpha} \in \mathbb{R}^d$ is a vector of weights, and β is called the *bias*. One common example of an activation function is the *rectified linear unit* (ReLU), defined as

$$\text{ReLU}(x) = \begin{cases} x & \text{if } x > 0, \\ 0 & \text{otherwise}. \end{cases}$$

The ReLU function is a simple way to model a neuron in a brain that has two states, resting and active. Other examples of activation function include the sign function $\phi(x) = \text{sign}(x)$, the *sigmoid* function $\phi(x) = 1/(1 + e^{-x})$, and the hyperbolic tangent function $\phi(x) = (e^x - e^{-x}/(e^x + e^{-x})$.

The output of a neuron function is a scalar, one can combine several neurons to obtain a vector-valued function called a *layer function*. A layer function $h : \mathbb{R}^d \to \mathbb{R}^m$ is a mapping of the form

$$h(\mathbf{x}) = (g_1(\mathbf{x}), \dots, g_m(\mathbf{x}))', \qquad (5.42)$$

in which each g_i is a neuron function of the form (5.41) with its own vector of parameters $\boldsymbol{\alpha}_i = (\alpha_{i1}, \dots, \alpha_{id})'$ and biases β_i, $i = 1, \dots, m$. Then the layer function $h(\cdot)$ is determined by a $d \times m$ matrix of parameters $\mathbf{A} = [\boldsymbol{\alpha}_1; \dots; \boldsymbol{\alpha}_m] = (\alpha_{ij})_{1 \leq i \leq d, 1 \leq j \leq m}$ and a vector of biases $\boldsymbol{\beta} = (\beta_1, \dots, \beta_m)'$. Hence, (5.42) can be written as

$$h(\mathbf{x}) = \bar{g}(\mathbf{A}'\mathbf{x} + \boldsymbol{\beta}),$$

where $\bar{g} : \mathbb{R}^m \to \mathbb{R}^m$ is the vectorial activation function.

With the neuron and the layer functions defined, an artificial neural network is a function $g : \mathbb{R}^d \to \mathbb{R}^m$ of the form

$$h(\mathbf{x}) = h_K \circ h_{K-1} \circ \cdots \circ h_1(\mathbf{x}), \qquad K \geq 1, \qquad (5.43)$$

where each $h_k : \mathbb{R}^{d_{k-1}} \to \mathbb{R}^{d_k}$ is a layer function with its own matrix of parameters $\mathbf{A}_k$ and vector of biases $\boldsymbol{\beta}_k$. The input vector $\mathbf{x}$ is usually called the *input layer* and the result of the last layer function h_J is commonly referred to as the *output layer*. The layer functions between the input and output layers are called *hidden layers* which passes their output to another layer as the input. The number of neurons d_k in the kth layer is called the layer's *width*, and the total number K of layers in an ANN is the *depth* of the neural network. The numbers $d_1, \dots, d_K$ and K comprise the *architecture* of the network. ANNs with more than one hidden layer are called *deep neural networks* (DNN).

The unknown parameters in the ANN (5.43) are often called *weights*. Denote by $\boldsymbol{\theta}$ the complete set of weights that contains all parameters $\mathbf{A}_k \in \mathbb{R}^{d_{k-1} \times d_k}$ and $\boldsymbol{\beta}_k \in \mathbb{R}^{d_k}$ ($k = 1, \dots, K$). Let $\ell(h(\mathbf{x}), \mathbf{y})$ be the loss function measuring the error of the fit, in which $\mathbf{x} \in \mathbb{R}^d$ and $\mathbf{y} \in \mathbb{R}^m$. Provided there are n samples $(\mathbf{x}_i, \mathbf{y}_i)$ $(i = 1, \ldots, n)$ in the training set, the network can be trained by minimizing the averaged loss

$$L(\boldsymbol{\theta}) = \frac{1}{n} \sum_{i=1}^{n} \ell(h(\mathbf{x}_i), \mathbf{y}_i). \tag{5.44}$$

The ANN is a versatile tool that is able to approximate complex functions of input data. Numerous discussions in the literature have focused on its "universal approximation" property. For instance, a seminal study by Hornik et al. (1989) demonstrated that a neural network comprising a single hidden layer, given a sufficient number of neurons, could approximate continuous functions of any complexity up to any accuracy. This finding underscored the remarkable capability of neural networks to capture intricate relationships within data sets.

Despite the theoretical assurance provided by the universal approximation property, practical implementation of neural networks poses challenges, particularly in determining the appropriate number of neurons in the hidden layer. Unfortunately, there is a lack of established methods for determining this crucial parameter. Consequently, practitioners often resort to a trial-and-error approach, testing different configurations until satisfactory results are achieved. This empirical process can be time-consuming and resource-intensive but remains a common practice due to the absence of more systematic methodologies.

5.5.1 Feed-forward neural network and back-propagation

A commonly used type of neural networks in machine learning is the *feed-forward neural network* (FNN), also known as the *multilayer perceptron*. The term *feed-forward* indicates that information flows from the input layer to the hidden layers and then to output layer. This architecture lacks *feedback* connections, meaning that outputs of the model do not loop back into itself.

The *architecture* or the structure of a neural work is usually determined before the analysis, which involves decisions such as the number of layers, the number of neurons in each layer and which variables to choose as inputs and outputs. For example, Figure 5.5 shows a feed-forward neural network with $d = 5$, two hidden layer containing $d_1 = 4$ and $d_2 = 5$ nodes, respectively, and $m = 4$ output. There are no connections between units in the same layer and no feedback in the network.

The measure function $L(\boldsymbol{\theta})$ can be minimized effectively through the gradient descent approach, a technique commonly referred to as *back-propagation* in the context of neural networks. Back-propagation involves iteratively adjusting the parameters $\boldsymbol{\theta}$ of the neural network to minimize the measure function $L(\boldsymbol{\theta})$ with respect to the training data. The key step of the back-propagation approach is the computation of gradients in the FNN using the chain rule for differentiation. This rule allows the derivatives of the output with respect to

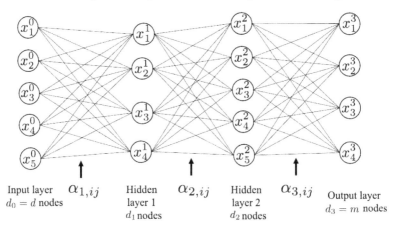

FIGURE 5.5: An artificial neural network with two hidden layers ($d = 5, d_1 = 4, d_2 = 5, m = 4$).

the parameters in each layer to be calculated by recursively applying derivatives of the subsequent layers with respect to the previous layers. This process efficiently propagates gradients backward through the network, hence the name "back-propagation."

In particular, recall that the layer function is expressed as $h_k(\mathbf{x}) = \bar{g}(\mathbf{A}_k'\mathbf{x} + \boldsymbol{\beta}_k)$, where $\mathbf{A}_k = (\alpha_{k,ij})$ is a $d_{k-1} \times d_k$ weight matrix, $\boldsymbol{\beta}_k = (\beta_{k,1}, \ldots, \beta_{k,d_k})'$ is a d_k-dimensional bias vector, and $\bar{g}$ is an activation function. Let $\mu_k = d_k(d_{k-1} + 1)$ be the total number of parameters in the kth layer, and let $\boldsymbol{\theta}_k$ be a vector containing all the parameters of $\mathbf{A}_k$ and $\boldsymbol{\beta}_k$, i.e.,

$$\boldsymbol{\theta}_k = (\alpha_{k,11}, \alpha_{k,21}, \ldots, \alpha_{k,d_{k-1}d_k}, \beta_{k,1}, \ldots, \beta_{k,d_k})'.$$

Denote by $\mu = \mu_1 + \cdots + \mu_k$ the total number of parameters of the DNN and by $\boldsymbol{\theta} \in \mathbb{R}^\mu$ the vector of all parameters of the network, $\boldsymbol{\theta} = (\boldsymbol{\theta}_1', \ldots, \boldsymbol{\theta}_K')'$. Given the loss function $L(h(\mathbf{x}), \mathbf{y})$, the gradient of the the loss function for a given $\boldsymbol{\theta} \in \mathbb{R}^\mu$ is

$$\nabla L(\boldsymbol{\theta}) = \left(\left(\frac{\partial L}{\partial \boldsymbol{\theta}_1} \right)', \ldots, \left(\frac{\partial L}{\partial \boldsymbol{\theta}_K} \right)' \right)', \tag{5.45}$$

To calculate the above gradient, we note that, for each layer function h_k ($k = 1, \ldots, K$), and for each sample $(\mathbf{x}_i, \mathbf{y}_i)$, the recursion $h_{k+1} = \bar{g}(\mathbf{A}_{k+1}'h_k + \boldsymbol{\beta}_{k+1})$ holds. Then we obtain

$$\frac{\partial \ell}{\partial \boldsymbol{\theta}_k} = \left(\frac{\partial h_k}{\partial \boldsymbol{\theta}_k} \right)\left(\frac{\partial \ell}{\partial h_k} \right) = \left(\frac{\partial h_k}{\partial \boldsymbol{\theta}_k} \right)\left(\frac{\partial h_{k+1}}{\partial h_k} \frac{\partial \ell}{\partial h_{k+1}} \right)$$

$$= \frac{\partial (\mathbf{A}_k'h_{k-1} + \boldsymbol{\beta}_k)}{\partial \boldsymbol{\theta}_k} \left(\frac{d\bar{g}}{d\mathbf{x}}\bigg|_{\mathbf{x}=\mathbf{A}_k h_{k-1}+\boldsymbol{\beta}_k} \right)\left(\mathbf{A}_{k+1} \frac{d\bar{g}}{d\mathbf{x}}\bigg|_{\mathbf{x}=\mathbf{A}_{k+1}h_k+\boldsymbol{\beta}_{k+1}} \frac{\partial \ell}{\partial h_{k+1}} \right),$$

$$\tag{5.46}$$

in which $\partial(\mathbf{A}_k h_{k-1} + \boldsymbol{\beta}_k)/\partial\boldsymbol{\theta}_k$ is a $d_k(d_{k-1}+1) \times d_k$ matrix. Summing the above derivatives over all training samples yields the derivative $\partial L/\partial\boldsymbol{\theta}_k$. Then combine all the μ_k-dimensional vector, we find the gradient $\nabla L(\boldsymbol{\theta})$ in (5.45).

The key step in the above back-propagation algorithm is the recursive formula

$$\frac{\partial\ell}{\partial h_k} = \frac{\partial h_{k+1}}{\partial h_k}\frac{\partial\ell}{\partial h_{k+1}}, \tag{5.47}$$

which starts from $k = K - 1$ and proceeds backward to $k = 1$. Due to this step, the total computational complexity of the back-propagation algorithm for one sample $(\mathbf{x}, \mathbf{y})$ is $O(\mu)$. If the derivative $\partial\ell/\partial h_k$ is not calculated by (5.47), instead, by

$$\frac{\partial\ell}{\partial h_k} = \frac{\partial\ell}{\partial h_K}\prod_{j=k}^{K-1}\frac{\partial h_{j+1}}{\partial h_j},$$

one can show that the total computational complexity based on the above equation is $O(K\mu)$, much more than the back-propagation algorithm; see Berlyand and Jabin (2023, Section 9.5).

One common concern with feed-forward neural networks is the risk of over-fitting. To mitigate this issue, two general approaches are commonly employed. The first approach is early stopping. This method involves partitioning the data into a training set, used for computing the gradient during training, and a validation set, used to estimate the error. During the training process, the model's performance on the validation set is monitored. Training is stopped when the error on the training set continues to decrease, but the error on the validation set begins to increase, indicating that the model is starting to overfit. The second approach is regularization, as introduced by (Girosi et al., 1995). The core idea behind regularization is to augment the objective function $L(\boldsymbol{\theta})$ with a penalty term. This penalty term serves to constrain the complexity of the neural network during the inference procedure. By penalizing overly complex models, regularization encourages the neural network to generalize better to unseen data, thereby reducing the risk of overfitting.

Example 5.8 (Neural network for bank default) *Consider the neural network method for classification of the FDIC-insured bank default data in Example 5.1. Five feed-forward neural networks are trained for the the bank default data. In particular, two networks have single hidden layer with nodes $M = 3$ and 5, respectively, and the other three networks have two hidden layers with the numbers of nodes $(10, 15)$, $(15, 10)$, and $(15, 15)$, respectively. Table 5.9 shows the misclassification rates of these networks for both the training and the testing data.* □

5.5.2 Radial basis function network

Radial basis function (RBF) network, proposed by Broomhead and Lowe (1988), represents a type of feed-forward neural networks characterized by a

TABLE 5.9: Misclassification rates of neural network methods.

Method		Training data			Testing data		
		Default	Nondefault	All	Default	Nondefault	All
Neural network	$M = 3$	53.34%	0.18%	0.58%	28.25%	0.45%	0.56%
	$M = 5$	54.63%	0.14%	0.54%	30.94%	0.31%	0.43%
Deep network	$(10, 15)$	20.19%	4.44%	4.56%	8.52%	4.66%	4.67%
	$(15, 10)$	21.12%	4.45%	4.57%	8.52%	4.66%	4.67%
	$(15, 15)$	16.62%	4.42%	4.51%	2.24%	4.63%	4.63%

single hidden layer. In RBF networks, the activation function for the hidden layer neurons is the radial basis function, while the output layer computes a linear combination of the outputs from these hidden layer neurons.

A radial basis function is a radially symmetric scalar function, which is defined as a monotonic function based on the Euclidean distance between a sample and a data centriod. Consider a sample $\mathbf{x}$ and a data centroid $\boldsymbol{\mu}_i$. The commonly used Gaussian radial basis function is $\rho(\mathbf{x}, \boldsymbol{\mu}) = \exp\left(-\beta||\mathbf{x}-\boldsymbol{\mu}||^2\right)$. In a RBF network with a single hidden layer, the output can be expressed as

$$\varphi(\mathbf{x}) = \sum_{m=1}^{M} w_i \rho(\mathbf{x}, \boldsymbol{\mu}_i),$$

where M is the number of hidden layer neurons, $\boldsymbol{\mu}_i$ and w_i are the center and the weight of the ith hidden layer neuron, respectively, and $\rho(\mathbf{x}, \boldsymbol{\mu}_i)$ is the radial basis function.

Park and Sangberg (1991) proved that an RBF network with a sufficient number of hidden layer neurons can approximate continuous functions of any complexity up to arbitrary accuracy. RBF networks are usually trained with two steps. The first step is to determine the neuron centers $\boldsymbol{\mu}_i$, which is often achieved through methods such as random sampling or clustering. The second step involves determining the parameters w_i and β_i using back-propagation algorithms. Through this training process, RBF networks can effectively learn complex mappings bewteen the inputs and outouts.

5.5.3   Convolutional neural network

Convolutional neural networks (CNNs) represent a specialized type of neural network architecture specifically designed to process grid-structured data. This type of data, characterized by strong spatial dependencies within localized regions of the grid, includes two-dimensional images as the most typical example. However, CNNs can also effectively handle other forms of sequential data, such as time series and sequences, which can be treated as grid-structured data. A defining characteristic of CNNs is their utilization of the convolution operation, which underpins their ability to effectively process grid-structured data. For this reason, CNNs are formally defined as networks that incorporate the convolutional operation in one or multiple layers of their architecture.

In essence, CNNs resemble traditional feed-forward neural networks, with layers that organize operations in a spatial manner. However, the distinctive feature of CNNs lies in their use of convolution and pooling layers, which are fundamental for feature extraction. These layers enable the network to automatically learn hierarchical representations of features present in the input data, capturing patterns at different levels of abstraction.

The convolutional layers within CNNs apply a set of learnable filters to the input data, effectively convolving them across the input grid. This process enables the network to extract local features and spatial patterns present in the input data. Subsequently, pooling layers are often employed to down-sample the feature maps produced by the convolutional layers, reducing their spatial dimensions while retaining important features. This downsampling operation helps to increase the network's computational efficiency and robustness to spatial translations in the input data.

Following feature extraction, the extracted features are then mapped into the output layer through one or more fully connected layers. These layers integrate the learned features from the convolutional and pooling layers, enabling the network to make predictions or classifications based on the extracted representations.

Overall, CNNs have emerged as a powerful tool for tasks involving grid-structured data, demonstrating remarkable performance in various domains such as computer vision, natural language processing, and speech recognition. Their ability to automatically learn hierarchical representations of features makes them well-suited for capturing complex patterns and relationships within input data.

5.5.4 Recurrent neural network

The neural network architectures described above is designed for data where attributes lack inherent ordering, and the input data dimensions remain fixed. However, these frameworks may not effectively accommodate sequential data, such as time-series, which inherently possess temporal dependencies and variable input dimensions. To address this issue, a specialized class of neural networks known as *recurrent neural networks* (RNNs) can be used.

Unlike traditional feed-forward neural networks, RNNs possess internal memory, allowing them to capture temporal dependencies within input sequences. This unique capability enables RNNs to excel in tasks involving time-series data, natural language processing, and sequential decision-making. At each time step, an RNN processes an input along with information stored in its internal state, updating the state and producing an output. This recurrent computation enables RNNs to dynamically adjust their behavior based on the

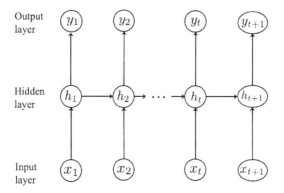

FIGURE 5.6: Time-layered representation of a recurrent neural network.

input sequence, effectively capturing patterns and dependencies across time steps.

Figure 5.6 shows the architecture of a RNN and its time-layered representation. Let $\mathbf{x}_t \in \mathbb{R}^d$, $\mathbf{h}_t \in \mathbb{R}^p$, and $\mathbf{y}_t \in \mathbb{R}^m$ be the input, the hidden state, and the output vectors at time t, respectively. At each time t, the hidden state $\mathbf{z}_t$ is represented as a function of the input vector $\mathbf{x}_x$ and the hidden state $\mathbf{z}_{t-1}$ at previous time step,

$$\mathbf{z}_t = h_1(\mathbf{z}_{t-1}, \mathbf{x}_t). \tag{5.48}$$

Then a separate function $\mathbf{y}_t = h_2(\mathbf{h}_t)$ is used to learn the output from the hidden states. Plug the hidden state equation (5.48) into the output equation recursively, we obtain that

$$\mathbf{y}_t = F_t(\mathbf{x}_1, \ldots, \mathbf{x}_t),$$

which varies with the value of t. For example, one may define a input-hidden matrix $\mathbf{W}_{xh} \in \mathbb{R}^{p \times d}$, a hidden-hidden matrix $\mathbf{W}_{hh} \in \mathbb{R}^{p \times p}$, and a hidden-output matrix $\mathbf{W}_{hy} \in \mathbb{R}^{m \times p}$, so that the hidden state and the output are written as

$$\mathbf{h}_t = \tanh(\mathbf{W}_{xh}\mathbf{x}_t + \mathbf{W}_{hh}\mathbf{h}_{t-1}), \qquad \mathbf{y}_t = \mathbf{W}_{hy}\mathbf{h}_t, \tag{5.49}$$

in which $\tanh(\mathbf{x}) := (\tanh(x_1), \ldots, \tanh(x_d))$ and $\tanh(x) = (e^x - e^{-x})/(e^x + e^{-x})$ is the hyperbolic tangent function. Parameters in this network can still be solved by the back-propagation algorithm. However, due to the time-layered representation of the RNN, one may pretend the parameters in different temporal layers are independent of each other and use the back-propagation through time algorithm to update all parameters; see Aggarwal (2023, Section 8.2).

The RNN in Figure 5.6 has only one hidden layer. One can extend it to encompass multiple hidden layers, and hence improve its representational capacity and learning capabilities. Such an extension usually involves integrating a feed-forward network with the aforementioned time-layered structure of the RNN. In a multi-layered RNN configuration, each hidden layer receives input not only from the previous time step but also from the hidden layer in the preceding time step. This enables the network to capture more sophisticated temporal dependencies within the input sequence.

5.5.5 Generative adversarial network

Generative adversarial networks (GANs) represent a class of neural networks used for unsupervised learning, whose aim is to generate new samples closely resembling those present in the training data (Goodfellow et al., 2014). GANs operate through the interaction of two adversarial networks: the *generator network* and the *discriminator network*. The objective of the generator network is to create synthetic samples that mimic the distribution of the training data, while the discriminator network's aim is to distinguish between real and synthetic samples.

During training, the generator and the discriminator networks participate in a competitive process. Specifically, the generator tries to produce increasingly realistic samples while the discriminator attempts to accurately identify whether a given sample is real or synthetic. As the training progresses, both networks iteratively refine their strategies, leading to improvements in the quality of the generated samples. The training process continues until an equilibrium is reached, characterized by the discriminator network being unable to differentiate between real and synthetic samples with high confidence. At this point, the generator has effectively learned to produce samples that closely resemble those in the training data distribution. Through this adversarial training paradigm, GANs have demonstrated remarkable capabilities in generating diverse and high-quality samples across various domains, including images, text, and music. Their ability to learn complex data distributions without the need for explicit supervision makes them a valuable tool for tasks such as image synthesis, data augmentation, and anomaly detection.

The essence of GANs can be encapsulated as follows. Let $G(\mathbf{z}; \boldsymbol{\theta}_G)$ be the multi-layer generator network that maps from the space of input noises $\mathbf{z}$ to the data space, where G is a differentiable function with parameters $\boldsymbol{\theta}_G$, and $D(\cdot; \boldsymbol{\theta}_D)$ the multi-layer discriminator network with parameters $\boldsymbol{\theta}_D$ that represents the probability that $\mathbf{x}$ are from the data rather than the generator's distribution p_g. We need to train the network D to maximize the probability of assigning the correct label to both training samples and examples from G and simultaneously train G to minimize the function $\log(1 - D(G(\mathbf{z})))$, i.e.,

$$\min_{G} \max_{D} E_{\mathbf{x} \sim p_{\text{data}}(\mathbf{x})} \big[\log D(\mathbf{x}; \boldsymbol{\theta}_D) \big] + E_{\mathbf{z} \sim p_{\mathbf{z}}(\mathbf{z})} \big[\log(1 - D(G(\mathbf{z}; \boldsymbol{\theta}_G); \boldsymbol{\theta}_D)) \big].$$

$$(5.50)$$

The above optimization problem is a two-player minimax game, and its solution is a saddle point and a Nash equilibrium of the minimax game. To solve the optimization problem (5.50), stochastic gradient ascent is applied to the parameters of the discriminator network D and stochastic gradient descent is applied to the parameters of the generator network G. We refer interested readers to Goodfellow et al. (2016) for further exploration,

5.5.6   Restricted Boltzmann machine

In contrast to feed-forward networks, the *Boltzmann machines* (Ackley et al., 1985) are a type of neural networks where the probabilistic states of neurons are learned for a given set of inputs. The Boltzmann machines characterize the joint probability distribution of observed attributes along with some hidden attributes. Boltzmann machines are typically used for unsupervised learning tasks, where the objective is to learn the underlying structure and patterns in the data without explicit labels. By modeling the joint probability distribution of observed and hidden attributes, Boltzmann machines can uncover latent factors and representations within the data. Consequently, the machines are well-suited for tasks like dimensionality reduction, feature learning, and data generation.

A typical structure of the Boltzmann machines has two types of states: the visible state and the hidden state. The visible states represent data input and output, and the hidden states represent the intrinsic structure of the data. Suppose that a Boltzmann machine contains m_v visible states and m_h hidden states. Let $q = m_v + m_h$ be the total number of neurons, $s_i \in \{0, 1\}$ the value of the ith state $(i = 1, \ldots, q)$, and w_{ij} the connection weight between neurons i and j. Let θ_i be the threshold of neuron i, then the energy of the Boltzmann machine given the state vector $\mathbf{s}$ is defined as

$$E(\mathbf{s}) = -\sum_{i=1}^{q-1} \sum_{j=i+1}^{q} w_{ij} s_i s_j - \sum_{i=1}^{q} \theta_i s_i. \tag{5.51}$$

The unconditional probability of having the state vector $\mathbf{s}$ is solely determined by its energy and the energies of all possible state vectors:

$$P(\mathbf{s}) = e^{-E(\mathbf{s})} \Big/ \sum_{\mathbf{t}} e^{-E(\mathbf{t})}. \tag{5.52}$$

Then, the training of a Boltzmann machine is to regard each training sample as a state vector and maximize its probability (5.52).

As shown in the left panel of Figure 5.7, the standard Boltzmann machines are fully connected graphs in which hidden states are connected. Since it is not convenient to use in real-world applications, one often use the *restricted Boltzmann machines* that only keeps connections between visible and hidden states and hence simplifies the network structure; see the right panel of Figure 5.7. The restricted Boltzmann machines can be trained using the contrastive divergence algorithm; see Hinton (2012).

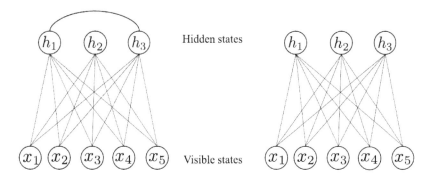

FIGURE 5.7: Boltzmann machine (left) and restricted Boltzmann machine (right).

5.6 Clustering

Clustering is a unsupervised learning method and aims to partition a data set into disjoint subsets, often referred to as *clusters*. By grouping similar data points together, clustering aims to unveil the underlying structure within the dataset. This process can serve as a critical pre-processing step for various learning tasks, as it helps identify patterns and relationships that may not be immediately apparent. Clustering techniques are widely used across various domains, including data analysis, pattern recognition, and machine learning, to gain insights into the organization and characteristics of complex datasets.

Specifically, given a data set $\mathcal{X} = \{\mathbf{x}_i, i = 1, \ldots, n\}$ containing n unlabeled samples, where each sample $\mathbf{x}_i = (x_{i1}, \ldots, x_{id})'$ is a d-dimensional vector. A clustering algorithm partitions the data set $\mathcal{X}$ into K disjoint clusters $\{C_k | k = 1, \ldots, K\}$, where $C_k \cap C_l = \emptyset$ and $\mathcal{X} = \cup_{k=1}^{K} C_k$. For each sample $\mathbf{x}_i$, we denote $\lambda_i \in \{1, 2, \ldots, K\}$ as the *cluster label* of the sample, hence the clustering result can be represented as a cluster label vector $\boldsymbol{\lambda} = (\lambda_1, \ldots, \lambda_n)'$.

5.6.1 Performance and distance measures

Intuitively, a good clustering should have high *intra-cluster similarity* and low *inter-cluster similarity*. Performance of clustering is measured by *validity indices*. There are roughly two types of clustering validity indices, one is the *external index* that compares the clustering result against a reference model, and the other is the *internal index* that evaluates the clustering result without any reference model.

Given a data set $\mathcal{X} = \{\mathbf{x}_i, i = 1, \ldots, n\}$, suppose a clustering algorithm produces the clusters $\mathcal{C} = \{C_1, \ldots, C_K\}$ and a reference model gives the clusters $\mathcal{C}^* = \{C_1^*, \ldots, C_K^*\}$. Denote by $\boldsymbol{\lambda}$ and $\boldsymbol{\lambda}^*$ by the clustering label vectors of $\mathcal{C}$ and $\mathcal{C}^*$, respectively. Then we can define the following terms for each pair of samples

$$
\begin{aligned}
SS &= \{(\mathbf{x}_i, \mathbf{x}_j) | \lambda_i = \lambda_j, \lambda_i^* = \lambda_j^*, i < j\}, \quad a = |SS|; \\
SD &= \{(\mathbf{x}_i, \mathbf{x}_j) | \lambda_i = \lambda_j, \lambda_i^* \neq \lambda_j^*, i < j\}, \quad b = |SS|; \\
DS &= \{(\mathbf{x}_i, \mathbf{x}_j) | \lambda_i \neq \lambda_j, \lambda_i^* = \lambda_j^*, i < j\}, \quad c = |DS|; \\
DD &= \{(\mathbf{x}_i, \mathbf{x}_j) | \lambda_i \neq \lambda_j, \lambda_i^* \neq \lambda_j^*, i < j\}, \quad d = |DD|.
\end{aligned}
$$

Note that SS is the set of sample pairs that both samples belong to the same cluster in $\mathcal{C}$ in $\mathcal{C}^*$, and SD, DS, DD can be interpreted similarly. Since $1 \leq i < j \leq n$, we have $a + b + c + d = n(n-1)/2$. With the above notations, some commonly used external indices can be defined as follows.

- Jaccard coefficient (JC): $JC = a/(a+b+c)$.

- Fowlkes and Mallows Index (FMI): $FMI = a/\sqrt{(a+b)(a+c)}$.

- Rand Index (RI): $RI = 2(a+d)/(n(n-1))$.

These external validity indices take values in $[0, 1]$ and a larger index value indicates better clustering quality.

Given the generated clusters $\mathcal{C} = \{C_1, \ldots, C_K\}$, consider the following definitions,

$$
\operatorname{avg}(C_k) = \frac{2}{|C_k|(|C_k| - 1)} \sum_{1 \leq i < j \leq |C_k|} \operatorname{dist}(\mathbf{x}_i, \mathbf{x}_j),
$$

$$
\operatorname{diam}(C_k) = \max_{1 \leq i < j \leq |C_k|} \operatorname{dist}(\mathbf{x}_i, \mathbf{x}_j),
$$

$$
d_{\min}(C_k, C_l) = \min_{\mathbf{x}_i \in C_k, \mathbf{x}_j \in C_l} \operatorname{dist}(\mathbf{x}_i, \mathbf{x}_j),
$$

$$
d_{\operatorname{cen}}(C_k, C_l) = \operatorname{dist}(\boldsymbol{\mu}_k, \boldsymbol{\mu}_l), \quad \boldsymbol{\mu}_k = \frac{1}{|C_k|} \sum_{1 \leq i \leq |C_k|} \mathbf{x}_i.
$$

Here, $\operatorname{avg}(C_k)$ and $\operatorname{diam}(C_k)$ are the average and the largest distances between the samples in cluster C_k, respectively. $d_{\min}(C_k, C_l)$ is the largest distance between two nearest samples in clusters C_k and C_l, and $d_{\operatorname{cen}}(C_k, C_l)$ is the distance between the centroids of clusters C_k and C_l. Then two commonly used internal validity indices can be defined as follows.

- Davies-Bouldin Index (DBI): $DBI = \dfrac{1}{K} \displaystyle\sum_{k=1}^{K} \max_{l \neq k} \left\{ \dfrac{\operatorname{avg}(C_k) + \operatorname{avg}(C_l)}{d_{\operatorname{cen}}(C_k, C_l)} \right\}.$

- Dunn Index (DI): $DI = \displaystyle\min_{1 \leq k \leq K} \left\{ \min_{l \neq k} \left(\dfrac{d_{\min}(C_k, C_l)}{\max_{1 \leq l \leq K} \operatorname{diam}(C_l)} \right) \right\}.$

Here, better clustering quality are represented by smaller DBI indices or larger DI indices.

The distance measure $\text{dist}(\cdot, \cdot)$ used in the validity indices can be of many forms, but they all need to satisfy the following axioms:

- Non-negativity: $\text{dist}(\mathbf{x}_i, \mathbf{x}_j) \geq 0$, where equality holds if and only if $\mathbf{x}_i = \mathbf{x}_j$.

- Symmetry: $\text{dist}(\mathbf{x}_i, \mathbf{x}_j) = \text{dist}(\mathbf{x}_j, \mathbf{x}_i)$.

- Subadditivity: $\text{dist}(\mathbf{x}_i, \mathbf{x}_j) \leq \text{dist}(\mathbf{x}_i, \mathbf{x}_l) + \text{dist}(\mathbf{x}_l, \mathbf{x}_j)$.

Given two samples $\mathbf{x}_i = (x_{i1}, \ldots, x_{id})'$ and $\mathbf{x}_j = (x_{j1}, \ldots, x_{jd})'$, if the attribute of the sample is ordinal, a commonly used distance measure is the *Monkowski distance*

$$\text{dist}_p(\mathbf{x}_i, \mathbf{x}_j) = \left(\sum_{l=1}^d w_l |x_{il} - x_{jl}|^p \right)^{\frac{1}{p}}, \qquad w_l \geq 0, \sum_{l=1}^d w_l = 1,$$

where w_l $(l = 1, \ldots, d)$ is the weight of the lth attribute. When all the weights are same, for $d = 1, 2$, the Monkowski distance reduces to the *Manhattan distance* and the *Euclidean distance*, respectively. Note that the Minkowski distance can only measure the distance of attributes with *ordinal information*.

For non-ordinal attributes, one can use the *value difference metric* (VDM) proposed by Stanfill and Waltz (1986). Let $m_{u,a}$ be the number of samples taking value a on the attribute u and $m_{u,a,k}$ the number of samples within the ith cluster taking value a on the attribute u. The VDM distance between two categorical values a and b of attribute u is

$$\text{VDM}_p(a, b) = \sum_{k=1}^K w_k \left| \frac{m_{u,a,k}}{m_{u,a}} - \frac{m_{u,b,k}}{m_{u,b}} \right|^p.$$

For mixed attribute types, we can combine the Minkowski distance and the VDM.

We shall note that distances used in defining similarity measures do not necessarily satisfy all axioms of distance measures. Besides, here we define the distance measures before the learning, but sometimes in practice, it is necessary to determine the distance measure based on the samples via *distance metric learning*. We refer interested readers to Xing et al. (2003) for further details.

5.6.2   Prototype clustering

A *prototype* refers to a representative data point in the sample space. *Prototype clustering* or *prototype-based clustering* is a family of clustering algorithms that assumes the clustering structure can be represented by a set of prototypes. Typically, such algorithms start with some initial prototypes, and then iteratively update and optimize the prototypes. Here, we introduce several commonly used prototype clustering algorithms.

There are several commonly used prototype clustering algorithms, and one of them is the k-means clustering. Given a data set $D = \{\mathbf{x}_i, i = 1, \ldots, n\}$ and a clustering $\mathcal{C} = \{C_1, \ldots, C_K\}$, the following squared error of clusters the closeness between the mean vector of a cluster and the samples within that cluster,

$$\text{RSS} = \sum_{k=1}^{K} \sum_{\mathbf{x} \in C_k} ||\mathbf{x} - \boldsymbol{\mu}_k||^2, \qquad \boldsymbol{\mu}_k = \frac{1}{|C_k|} \sum_{\mathbf{x} \in C_k} \mathbf{x}. \tag{5.53}$$

The following iterative descent algorithm algorithm shows how to minimize the above RSS.

Algorithm 5.2 (K-means clustering) *The iterative descent algorithm consists of two steps.*

1. *For a given cluster $\mathcal{C}$, minimize the RSS (5.53) with respect to $\{\boldsymbol{\mu}_1, \ldots, \boldsymbol{\mu}_K\}$. Then the means of the currently assigned clusters $\boldsymbol{\mu}_k = \arg\min_{\boldsymbol{\mu}} \sum_{\mathbf{x} \in C_k} ||\mathbf{x} - \boldsymbol{\mu}||^2$.*

2. *Given a current set of means $\{\boldsymbol{\mu}_1, \ldots, \boldsymbol{\mu}_K\}$, minimize (5.53) by assigning each sample to the closest (current) cluster mean. That is, $\mathbf{x} \in C_k$ if $k = \arg\min_{1 \le l \le K} ||\mathbf{x}_k - \boldsymbol{\mu}_l||^2$.*

Iterate steps 1 and 2 until the assignments do not change.

The k-means clustering is closely related to another clustering method, the mixture-of-Gaussian. Suppose that each sample in the training set $\{\mathbf{x}_1, \ldots, \mathbf{x}_n\}$ follows a mixture of Gaussian distribution with k mixture components

$$p(\mathbf{x}) = \sum_{k=1}^{K} \alpha_k p(\mathbf{x}|\boldsymbol{\mu}_k, \boldsymbol{\Sigma}_k), \tag{5.54}$$

in which $p(\mathbf{x}|\boldsymbol{\mu}_k, \boldsymbol{\Sigma}_k)$ is the density function of $\mathbf{x}$ which follows a d-dimensional Gaussian distribution with mean $\boldsymbol{\mu}_k$ and covariance matrix $\boldsymbol{\Sigma}_k$, $\alpha_1, \ldots, \alpha_K$ are mixture coefficients such that $\alpha_i > 0$ and $\sum_{i=1}^{K} \alpha_i = 1$. Here, the mixture coefficient α_k can be considered as the prior probability of a mixture component random variable $Z \in \{1, \ldots, K\}$ such that $P(Z = k) = \alpha_k$. Then the posterior distribution of Z is

$$p(Z = k|\mathbf{x}) = \frac{P(Z = k)p(\mathbf{x}|Z = k)}{p(\mathbf{x})} = \frac{\alpha_k p(\mathbf{x}|\boldsymbol{\mu}_k, \boldsymbol{\Sigma}_k)}{\sum_{j=1}^{K} \alpha_j p(\mathbf{x}|\boldsymbol{\mu}_j, \boldsymbol{\Sigma}_j)}. \tag{5.55}$$

Then when the model parameters $\{(\alpha_k, \boldsymbol{\mu}_k, \boldsymbol{\Sigma}_k)|1 \le k \le K\}$ are known, the training data $\mathbf{x}_1, \ldots, \mathbf{x}_n$ can be divided into K clusters $\mathcal{C} = \{C_1, \ldots, C_K\}$ such that $\mathbf{x} \in C_k$ if

$$p(Z = k|\mathbf{x}) = \arg\max_{l \in \{1, \ldots, K\}} p(Z = l|\mathbf{x}).$$

The model parameters $\{(\alpha_k, \boldsymbol{\mu}_k, \boldsymbol{\Sigma}_k) | 1 \leq k \leq K\}$ can be estimated by maximizing the likelihood via the *expectation-maximization* (EM) algorithm; see Exercise 4.

The data in the k-means clustering is not labeled. When the data is labeled, one can also use the *learning vector quantization* (LVQ). Suppose that training data $\{(\mathbf{x}_i, y_i); i = 1, \ldots n\}$ are given, in which $\mathbf{x}_i = (x_{i1}, \ldots, x_{id})'$ and $y_i \in \{1, \ldots, K\}$ is a class label. The goal of LVQ is to find a set of d-dimensional prototype vectors $\{\boldsymbol{\mu}_1, \ldots, \boldsymbol{\mu}_K\}$ such that each prototype vector $\boldsymbol{\mu}_i$ represents one cluster and its class label is i. The LVQ can be carried out using the following algorithm.

Algorithm 5.3 (Learning vector quantization) *Denote by $\eta \in (0, 1)$ the learning rate. Given a set of prototype vectors $\{\boldsymbol{\mu}_1, \ldots, \boldsymbol{\mu}_K\}$ and their initial labels $\{l_1, \ldots, l_K\}$, iterate the following steps.*

1. *Randomly choose a sample $(\mathbf{x}_i, y_i)$ from the training data and find the nearest prototype vector $\boldsymbol{\mu}_{i^*}$ for $\mathbf{x}_i$, i.e., $i^* = \arg\min_{k \in \{1, \ldots, K\}} ||\mathbf{x}_i - \boldsymbol{\mu}_k||_2$.*

2. *If $y_i = l_{i^*}$, let $\widetilde{\boldsymbol{\mu}} = \boldsymbol{\mu}_{i^*} + \eta(\mathbf{x}_i - \boldsymbol{\mu}_{i^*})$; otherwise, let $\widetilde{\boldsymbol{\mu}} = \boldsymbol{\mu}_{i^*} - \eta(\mathbf{x}_i - \boldsymbol{\mu}_{i^*})$. Then set $\boldsymbol{\mu}_{i^*} \leftarrow \widetilde{\boldsymbol{\mu}}$.*

Iterate steps 1 and 2 until the termination condition is satisfied.

Suppose that a set of learned prototype vectors $\{\boldsymbol{\mu}_1, \ldots, \boldsymbol{\mu}_K\}$ is provided, one can assign each sample $\mathbf{x}$ in the sample space to the cluster represented by the nearest prototype vector. In particular, let $\mathcal{R}_i$ be the region in which the distance from any sample in $\mathcal{R}_i$ to $\boldsymbol{\mu}_i$ is no larger than that to any other prototype vector $\boldsymbol{\mu}_j$ $(j \neq i)$, i.e.,

$$\mathcal{R}_i = \{\mathbf{x} \in \mathcal{X} \mid ||\mathbf{x} - \boldsymbol{\mu}_i||_2 \leq ||\mathbf{x} - \boldsymbol{\mu}_j|| \text{ for all } j \neq i\}.$$

Hence, $\{\mathcal{R}_1, \ldots, \mathcal{R}_K\}$ forms a partition of the sample space, which is called the *Voronoi tessellation*.

5.6.3 Hierarchical clustering

The clustering algorithms in the preceding section require the specification of the number of clusters to be search. In contrast, *hierarchical clustering* aims to create a tree-like clustering structure based on group dissimilarities among the observations.

There are two basic strategies for hierarchical clustering: bottom-up and top-down. The bottom-up strategy starts at the bottom and at each level recursively merge two groups with the smallest intergroup dissimilarity into a single cluster. This generates a grouping at the next higher level with one less cluster. The top-down strategy has a reversed process. It starts at the top and at each level recursively split one of clusters into two new clusters with the largest between-group dissimilarity. Suppose that n is the sample size, then both strategies generates a hierarchical clustering with $n-1$ levels and it is up to the user to decide which level represents an actual clustering.

Supplements and problems

There have been many comprehensive overviews on machine learning during the last two decades. Methods of supervised and unsupervised learning introduced in this chapter are evidently incomplete. For a complete coverage of the subject, we refer to Hastie et al. (2016) that introduce the subject from statistical learning point of view and Jo (2021) that describes learning algorithms with pseudocode. Dixon et al. (2020) provide an excellent description of supervised, unsupervised, and reinforcement learning methods and present some applications in finance such as pricing and hedging of financial products, algorithmic trading, and wealth management. References related to reinforcement learning and deep learning are given in Chapter 6.

1. (**Bayesian interpretation of ridge regression**). Consider a linear regression model $y_i = \boldsymbol{\beta}'\mathbf{x}_i + \epsilon_i$, $i = 1,\ldots,n$, in which $\mathbf{x}_i \in \mathbb{R}^d$ and ϵ_i are i.i.d. normal random variables with mean 0 and variance σ^2. Let $\mathbf{y} = (y_1,\ldots,y_n)'$, $\mathbf{X} = (\mathbf{x}_1,\ldots,\mathbf{x}_n)'$, and $\boldsymbol{\beta} = (\beta_1,\ldots,\beta_d)$. The ordinary least square method finds the coefficient $\boldsymbol{\beta}$ which minimizes the RSS $(\mathbf{y}-\mathbf{X}'\boldsymbol{\beta})'(\mathbf{y}-\mathbf{X}'\boldsymbol{\beta})$. Assume that $\boldsymbol{\beta}$ has a prior distribution Normal$(0,\tau^2\mathbf{I}_d)$. Show that

   (a) the posterior distribution of $\boldsymbol{\beta}$ has the following form

   $$f(\boldsymbol{\beta}|\mathbf{y},\mathbf{X}) \propto \exp\left[-\frac{1}{2\sigma^2}(\mathbf{y}-\mathbf{X}'\boldsymbol{\beta})'(\mathbf{y}-\mathbf{X}'\boldsymbol{\beta}) - \frac{1}{2\tau^2}||\boldsymbol{\beta}||_2^2\right].$$

   (b) The maximum posteriori estimate of the above posterior distribution yields the ridge regression that minimizes the sum of the RSS and a L_2 penalty term

   $$\widehat{\boldsymbol{\beta}}_{\text{ridge}} = \underset{\boldsymbol{\beta}}{\operatorname{argmin}}\left[(\mathbf{y}-\mathbf{X}'\boldsymbol{\beta})'(\mathbf{y}-\mathbf{X}'\boldsymbol{\beta}) + \lambda||\boldsymbol{\beta}||_2^2\right]$$

   with $\lambda = \sigma^2/\tau^2$.

2. (**Basis functions for natural cubic splines**) Consider the truncated power series representation, (5.13) with $M = 3$, for cubic splines with K interior knots. Let

   $$f(x) = \sum_{j=0}^{3}\beta_j x^j + \sum_{k=1}^{K}\beta_{3+k}(x-\eta_k)_+^3.$$

Show that, the natural boundary conditions, $f''(x) = f'''(x) = 0$ beyond the boundary knots, imply the following linear constraints on the coefficient,

$$\beta_2 = \beta_3 = 0, \qquad \sum_{k=1}^{K} \beta_{3+k} = 0, \qquad \sum_{k=1}^{K} \eta_k \beta_{3+k} = 0.$$

Hence, a natural cubic spline is characterized by the basis given in equation (5.14).

3. (ν-**support vector classifier**). One soft margin realization for the optimal hyperplane in the SVM is to use the $\nu-$parameterization. In particular, one consider

$$\min_{\beta, \beta_0} \frac{1}{2}||\beta||^2 - \nu\rho + \frac{1}{n}\sum_{i=1}^{n}\xi_i \qquad \text{subject to } y_i(\beta'\mathbf{x}_i + \beta_0) \geq (\rho - \xi_i),$$

$$\xi_i \geq 0, \text{for all } i, \text{ and } \rho > 0.$$

$$(5.56)$$

Consider the Lagrangian

$$L(\beta, \beta_0, \rho, \{\xi_i, \alpha_i, \phi_i\}, \delta) = \frac{1}{2}||\beta||^2 - \nu\rho + \frac{1}{n}\sum_{i=1}^{n}\xi_i$$

$$- \sum_{i=1}^{n}\alpha_i\left[y_i(\beta'\mathbf{x}_i + \beta_0) - \rho + \xi_i\right] + \sum_{i=1}^{n}\phi_i\xi_i - \delta\rho,$$

using multipliers $\alpha_i, \phi_i, \delta \geq 0$.

(a) Show that the first order conditions are

$$\beta = \sum_{i=1}^{n}\alpha_i y_i \mathbf{x}_i, \qquad \alpha_i + \phi_i = \frac{1}{n}, \qquad \sum_{i=1}^{n}\alpha_i y_i = 0, \qquad \sum_{i=1}^{n}\alpha_i = \delta + \nu.$$

(b) Using the first order conditions, the Lagrangian (Wolfe) dual maximimization problem for (5.56) is

$$\max_{\alpha} -\frac{1}{2}\sum_{i=1}^{n}\sum_{j=1}^{n}\alpha_i\alpha_j y_i y_j \mathbf{x}_i'\mathbf{x}_j \quad \text{subject to } \sum_{i=1}^{n}\alpha_i y_i = 0, \ 0 \leq \alpha_i \leq \frac{1}{n}, \sum_{i=1}^{n}\alpha_i \geq \nu.$$

4. (**EM algorithm for the mixture of Gaussian distribution**). Suppose the samples $\{\mathbf{x}_1, \ldots, \mathbf{x}_n\}$ follow the mixture of Gaussian distribution (5.54). The maximum likelihood estimate of $\{(\alpha_k, \boldsymbol{\mu}_k, \boldsymbol{\Sigma}_k | k = 1, \ldots, K\}$ maximizes the log-likelihood

$$L(\{\alpha_k, \boldsymbol{\mu}_k, \boldsymbol{\Sigma}_k\}) = \sum_{i=1}^{n}\log\left(\sum_{k=1}^{K}\alpha_k p(\mathbf{x}_i|\boldsymbol{\mu}_k, \boldsymbol{\Sigma}_k)\right).$$

(a) Show that, if the parameters $\{(\alpha_k, \boldsymbol{\mu}_k, \boldsymbol{\Sigma}_k | 1 \leq k \leq K\}$ maximizes the log-likelihood $L(\{\alpha_k, \boldsymbol{\mu}_k, \boldsymbol{\Sigma}_k\})$, the first order condition gives that

$$\widehat{\boldsymbol{\mu}}_k = \frac{\sum_{k=1}^{K} P(Z_i = k | \mathbf{x}_i) \mathbf{x}_i}{\sum_{k=1}^{K} P(Z_i = k | \mathbf{x}_i)}, \qquad \widehat{\alpha}_k = \frac{1}{n} \sum_{i=1}^{n} P(Z_i = k | \mathbf{x}_i),$$

$$\widehat{\boldsymbol{\Sigma}}_k = \frac{\sum_{k=1}^{K} P(Z_i = k | \mathbf{x}_i)(\mathbf{x}_i - \widehat{\boldsymbol{\mu}}_k)(\mathbf{x}_i - \widehat{\boldsymbol{\mu}}_k)'}{\sum_{k=1}^{K} P(Z_i = k | \mathbf{x}_i)}.$$

Then the optimization of L can be done by the following EM algorithm. First, given the posterior probability $P(Z = k | \mathbf{x}_i)$ for $i = 1, \ldots, n$, $k = 1, \ldots, K$, update the estimate of $\{(\alpha_k, \boldsymbol{\mu}_k, \boldsymbol{\Sigma}_k | 1 \leq k \leq K\}$. Then given the estimates of $\alpha_k, \boldsymbol{\mu}_k, \boldsymbol{\Sigma}_k$, update the posterior probability $P(Z = k | \mathbf{x}_i)$ for each sample by (5.55) and its cluster label. These two steps are iterated until the convergence.

(b) Let $\alpha_1 = 0.5$, $\alpha = 0.3$, $\alpha = 0.2$, $\boldsymbol{\mu}_1 = (-2, 0)'$, $\boldsymbol{\mu} = (0, -2)'$, $\boldsymbol{\mu}_3 = (1, 1)'$, and $\boldsymbol{\Sigma}_1 = \boldsymbol{\Sigma}_2 = \boldsymbol{\Sigma}_3 = 0.5\mathbf{I}_2$. Simulate $n = 1,000$ training samples from the mixture of Gaussian distribution (5.54). Implement the EM algorithm in (a) with the initial values $\alpha_1 = \alpha_2 = \alpha_3 = 1/3$, $\boldsymbol{\mu}_1 = (-4, 0)'$, $\boldsymbol{\mu}_2 = (-1, -3)'$, $\boldsymbol{\mu} = (0.5, 0.8)'$ and $\boldsymbol{\Sigma}_1 = \boldsymbol{\Sigma}_2 = \boldsymbol{\Sigma}_3 = \mathbf{I}_2$. Summarize the misclassification rate of the result.

Chapter 6

Bandit, Markov decision process and reinforcement learning

In contrast to supervised learning methods that learn the relationship of observed feature vectors and labels and unsupervised learning methods that explore exploratory information on unlabeled data, *reinforcement learning* (RL) enables an intelligent agent to learn optimal behavior by balancing exploration and exploitation in an environment to maximize the cumulative reward or minimize the cumulative cost. Reinforcement learning has been studied in various disciplines, such as operation research and control systems, computer science and artificial intelligence, and it has wide applications in fields such as statistics, finance, dynamic economics, neuroscience, game theory, multi-agent systems. In the operation research and control literature, reinforcement learning is also known as *approximate dynamic programming*, or *neuro-dynamic programming*, as discussed in Bertsekas (2019).

The environment of reinforcement learning is usually modeled as Markov decision processes (MDP) so that dynamic programming techniques can be used in reinforcement learning algorithms. Dynamic programming, introduced by Bellman (1957), is a collection of algorithms that can be used to find optimal policies for a fully observable MDP. Assuming that the environment (i.e., transition kernels and reward functions) of the MDP model is known, dynamic programming approaches can find exact solutions for finite MDP problems, typically with a small number of states. However, when the environment of the MDP model is unknown, or the state space is discrete and high-dimensional or continuous, dynamic programming methods become infeasible and difficult to find exact or computationally tractable solutions, which is referred to as "the curse of dimensionality" by Bellman.

Reinforcement learning methods have the same goal of finding optimal policies, as do dynamic programming approaches. However, instead of finding an exact solution in the environment of the MDP model, reinforcement learning methods use the concepts of dynamic programming and samples from a true data-generating process and aim to find a good approximation to the exact optimal solution in the unknown environment or high-dimensional state and/or action spaces. Reinforcement learning methods can be *model-free* or *model-based*. The former refers to the methods that operate directly with samples of data and rely only on samples optimizing their policy, and the latter usually builds an internal model of the environment as a part of their ultimate goal of policy optimization. Both kinds of methods use trial-and-error learning techniques and find a balance between exploration and exploitation.

This chapter introduces Markov decision processes and reinforcement learning methods. To get a flavor of the tradeoff between exploration and exploitation, Section 6.1 introduces the multi-armed bandit problem, a relatively simple reinforcement learning problem in which the agent's chosen action doesn't affect future rewards. Both Bayesian and frequentist frameworks for multi-armed bandit problems are discussed in this section. For Bayesian bandit models, the optimal policy for the bandit problems, Gittins index, and its implementation are introduced. For frequentist bandit models, approaches of comparing different strategies and their convergence to the maximum reward are covered. To provide a deeper understanding of reinforcement learning principles, Sections 6.2–6.4 introduce finite-horizon, infinite-horizon, and partially observed MDP problems, as well as their optimal control policies. Section 6.2 presents finite-horizon MDP models, Bellman's principle of optimality and backward and forward induction. Section 6.3 extends the discussion to infinite-horizon MDP problems, outlining optimal policies for scenarios involving discounted rewards. In contrast to MDP problems that assume state processes are completely observed, *partially observed MDP* or POMDP problems assume that the complete states are not observed and the reward and transition kernels of states may depend on the unobserved state process. Section 6.4 introduces these processes and explains how a POMDP can be reformulated in terms of a MDP by incorporating filtering estimates of hidden states. To better explain basic concepts of MDP and POMDP in Section 6.2–6.4, *linear-quadratic regulator* (LQR) problems and their exact solutions are also presented.

The MDP problems and the methods of solving them in Section 6.1–6.4 like value and policy iterations rely on a complete understanding of the MDP model (i.e., state transition probabilities and action probabilities) and are considered to be the *classical dynamic programming* or *classical DP* methods. However, in many real-world applications where the agent's knowledge of the MDP model is incomplete, these classical DP methods become impractical, especially when the state and/or action space is large, due to the curse of dimensionality. In such scenarios, reinforcement learning methods are often more suitable. One common way of estimating the value function, which represents the expected reward when the process starts from a given state, is through the Monte Carlo method. This technique computes an average of rewards (or costs) over multiple independent realizations initiated from the specified state, as discussed in Chapter 7. However, Monte Carlo methods can sometimes suffer from high variance in rewards (or costs) or encounter difficulties when the estimation interacts with the MDP. Consequently, their practical applications are limited, leading us to focus on reinforcement learning methods that rely on value and/or policy approximation.

Sections 6.5–6.6 present various types of reinforcement learning methods, which can be categorized into three groups: critic-only (or value-based), actor-only (or policy-based), and actor-critic methods. Section 6.5 introduces value-based reinforcement learning methods, including temporal difference learning, Q-learning, and SARSA algorithms. Section 6.6 presents policy-based reinforcement methods that include policy-gradient and actor-critic methods. In general, critic-only methods have a lower variance in the estimates of expected rewards but are usually computationally intensive, especially if the action space is continuous. Actor-only methods typically work with a parameterized family of policies over which a spectrum of continuous actions can be generated, but they can generate high variance in the estimates of gradients so the learning may be slow. The actor-critic method aims to combine the strengths of both actor-only and critic-only methods. This widely used architecture is based on the policy gradient theorem and aims to achieve more stable learning outcomes.

6.1 Multi-armed bandit

The term "multi-armed bandits" originates from a stylized gambling game in which a gambler faces several slot machines (i.e., one-armed bandits) that appear identical but yield different payoffs. The multi-armed bandit problem is a simple but powerful framework that models the decision making process over time under uncertainty. In most multi-armed bandit problems, an agent has K possible actions (i.e., *arms*) to choose over T rounds. In each round, the agent chooses an arm and receives a reward for this arm. These rewards are independently drawn from distributions that are fixed but unknown to the algorithm (or the agent). Typically, multi-armed bandits problems assume that the algorithm (or the agent) observes the reward for the chosen arm after each round, but not for the other arms that could have been chosen. Consequently, the agent must balance *exploration* and *exploitation*. On one hand, exploration is necessary to gather new information about the arms; on the other hand, exploitation involves making optimal decisions based on the available information. This inherent tradeoff between exploration and exploitation underscores the agent's need to learn which arms yield the highest rewards while not spending too much time exploring.

There are two primary approaches to formulate the bandit problems: the Bayesian and the frequentist. We begin by discussing Bayesian bandit models, which encompass techniques such as Thompson sampling, the Gittins index theorem, and their corresponding implementation algorithms. Following this, we present frequentist bandit models, exploring concepts such as lower bounds on regret and various online learning algorithms.

6.1.1 Bayesian bandit and Gittins index

Let $t = 0, 1, \ldots, T$ be discrete time. T is called the time horizon or planning horizon. Assume that $T < \infty$, but T can be infinite in some applications. A *stochastic bandit* model can be specified as follows.

Definition 6.1 *A stochastic bandit model can be represented by a 3-tuple* $(\mathcal{A}, \{P_a\}_{a \in \mathcal{A}}, \mathcal{R})$, *in which* $\mathcal{A}$ *is a set of arms (or actions) that can be chosen by the agent,* P_a *is a distribution of rewards depending on an action* $a \in \mathcal{A}$, $\mathcal{R} = \{r(P_a) : a \in \mathcal{A}\}$ *is the set of rewards, and the reward* $r(P_a) \in \mathbb{R}$ *depends on the probability distribution associated with the arm* a.

In the above model, the agent and the environment interact sequentially over T rounds. At each round $t \in \{1, \ldots, T\}$, the agent chooses an arm $a_t \in \mathcal{A}$ based on his past actions and observed rewards. The environment then generates a reward R_t from distribution P_{a_t} and reveals R_t to the agent. The interaction between the agent and the environment induces a probability measure on the sequence of outcomes $(a_1, r_1, a_2, r_2, \ldots, a_T, r_T)$. Let π_t be the map from the agent's action and reward history up to time $t - 1$ to an arm (i.e., an action). Then the sequence of functions $\pi = (\pi_1, \ldots, \pi_T)$ represents an arm-selection policy. The agent's objective is to choose an arm-selection policy (i.e., a sequence of outcomes) to maximize the expected cumulative reward over the interaction period,

$$J_T^{\pi} = E\left(\sum_{t=1}^{T} r_t\right). \tag{6.1}$$

Bayesian bandit models

Besides the total reward criterion (6.1) that assigns same weights to rewards over all rounds, one can also use the discounted reward criterion. In particular, Gittins and Jones (1974) studied the following multi-armed bandit problems.

Definition 6.2 (The Bayesian multi-armed bandit model) *Consider N arms, $1, \ldots, N$ and a single agent. Each arm i has a state space $\mathcal{X}_i$. At time $t = 1, 2, \ldots$, the agent selects one arm for activation, based on the observed states $(X_{1,t}, \ldots, X_{N,t})$ of all arms. The active arm, say arm i, generates a reward $r_i(X_{i,t})$ dependent upon its current state and then changes state according to a transition law $P_i(s, s')$ $(s, s' \in \mathcal{S}_i)$. The states of all arms except arm i remain invariant. The agent's objective is to find an arm-selection policy to maximizes the expected total discounted reward,*

$$J_T = E\left(\sum_{t=1}^{T} \beta^{t-1} r_t \Big| X_{1,0}, \ldots, X_{N,0}\right), \tag{6.2}$$

where $\beta \in (0, 1]$ is the discount factor. $\square$

The following example shows a specific multi-armed bandit model, which is the same as the first bandit problem considered by Thompson (1933).

Example 6.1 (A two-armed bandit problem and Thompson sampling) *Consider a two-armed bandit problem in which two arms generate Bernoulli random variables with unknown probability of success θ_1 and θ_2, respectively. Suppose that θ_1 and θ_2 are independent with a uniform distribution over $[0, 1]$. The agent's objective is to maximize the expected total reward (6.1) by sequentially choosing one arm to play at each time.*

At time t, let the state of arm i, $X_{i,t}$, be the posterior density of θ_i given past observations. In particular, at time 0, the initial state $X_{i,0}$ of each arm is a constant function of 1 on $[0, 1]$ since θ_i ($i = 1, 2$) at time 0 are uniformly distributed on $[0, 1]$. Given arm i's current state $X_{i,t} = f(\theta)$, the reward is the posterior mean of $X_{i,t}$, i.e.,

$$r_i(f) = E_f(\theta_i) = \int_0^1 \theta f(\theta) d\theta. \tag{6.3}$$

The state $X_{i,t+1}$ of arm i is given by

$$X_{i,t+1} = \begin{cases} \theta f(\theta)/E_f(\theta) & \text{with probability } E_f(\theta), \\ (1 - \theta)f(\theta)/[1 - E_f(\theta)] & \text{with probability } 1 - E_f(\theta). \end{cases} \tag{6.4}$$

This shows that the updated posterior distribution of θ takes two possible forms, depending on whether 1 (or respectively, 0) is observed with probability $E_f(\theta)$ (or respectively, $1 - E_f(\theta)$).

Based on the state transition given by (6.4), Thompson (1933) proposed the following heuristic randomized stationary policy. At time t, given the current states $(f_1(\theta_1), f_2(\theta_2))$ of the two arms, the probability $\pi(f_1, f_2)$ that arm 1 is better than arm 2 is expressed as

$$\pi(f_1, f_2) = P(\theta_1 > \theta_2 | f_1(\theta_1), f_2(\theta_2)) = \int_0^1 d\theta_1 \int_0^{\theta_1} d\theta_2 f_1(\theta_1) f_2(\theta_2).$$

Then a randomized action is taken that plays arm 1 (or respectively, arm 2) with probability $\pi(f_1, f_2)$ (or respectively, $1 - \pi(f_1, f_2)$). This policy was later known as Thompson sampling. Note that the above sampling is equivalent to the following policy. Draw two random values of θ_1 and θ_2 according to their posterior distribution $f_1(\theta_1)$ and $f_2(\theta_2)$, respectively, and then play arm 1 if $\theta_1 > \theta_2$ and play arm 2 otherwise. □

Gittins index

Since the bandit model can be viewed as a special class of MDP introduced in Section 6.2, it can be solved by standard methods for solving MDP directly. However, due to specific structure of the bandit model, a simpler optimal solution can be found. In particular, Gittins and Jones (1974) and Gittins

(1979) showed that an optimal index policy exists for the bandit problem and can be implemented as follows. Assign the following index (also known as *Gittins index*) to each state s of each arm i,

$$v_i(s) = \max_{\tau \geq 1} \frac{E\left[\sum_{t=1}^{\tau} \beta^{t-1} r_i(S_{i,t}) \mid S_{i,1} = s\right]}{E\left[\sum_{t=1}^{\tau} \beta^{t-1} \mid S_{i,1} = s\right]}, \tag{6.5}$$

where $\tau \geq 1$ is a stopping time. The optimal policy is simply to compare the indices of all arms at their current states and play the one with the largest index. The state of the chosen arm then evolves and its index is updated and compared with that of other arms to determine the next action.

An important feature of the Gittins index is that the index function v_i of arm i depends only on the Markov reward process of arm i so that the N arms are decoupled and treated separately in computing the optimal policy. This reduces the computational complexity from exponential to linear with N.

6.1.2 Regret for frequentist bandit models

An alternative to the Bayesian formulation of multi-armed bandits is to use the frequentist framework. A frequentist multi-armed bandit model can be specified as follows.

Definition 6.3 (The frequentist multi-armed bandit model) *Consider N arms, $1, \ldots, N$ and a single agent. Given an arm-selection policy $\pi = (\pi_1, \ldots, \pi_T)$, the agent chooses one arm to play at time $t = 1, 2, \ldots, T$. Successive play of arm $i \in \{1, \ldots, N\}$ generates rewards $r_{i,t}$ that are are independent and identically distributed random variables drawn from an unknown distribution $F_i(x)$ with an unknown mean $\mu_i = E_{F_i}(X) < \infty$. For a given configuration of the reward distributions $\mathbf{F} = (F_1, \ldots, F_N)$ and a given arm-selection policy π, define the agent's total expected reward over T plays under policy π as*

$$J^{\pi}(T; \mathbf{F}) = E\left(\sum_{t=1}^{T} r_{i,t}\right), \tag{6.6}$$

and the agent's regret or cost of learning under policy π as

$$R^{\pi}(T; \mathbf{F}) = \mu_* T - J_T^{\pi}(T; \mathbf{F}) = \sum_{i=1}^{T} \left(\mu_* - E(r_{\pi_t, t})\right), \tag{6.7}$$

where $\mu_* = \max(\mu_1, \ldots, \mu_N)$ is the maximum expected reward among the N arms. The agent's objective is to find an arm-selection policy π which minimizes the regret (6.7). □

Note that in Definition 6.3, we change the performance measure for the agent's policy from the total cumulative reward $J^{\pi}(T; \mathbf{F})$ to the difference between the agent's cumulative reward $J^{\pi}(T; \mathbf{F})$ and the best arm benchmark

μ_*T. This adjustment is made because the total cumulative reward $J^\pi(T;\mathbf{F})$ fails to adequately capture the agent's learning process regarding unknown distributions $\mathbf{F}$. For instance, consider the scenario where $\mu_1 = \mu_*$, making arm 1 the arm with the greatest expected reward. In this case, the policy that maximizes the expected total reward is to choose arm 1 all the time, hence learning is completely excluded. Furthermore, since the total expected reward depends on unknown distributions $\mathbf{F}$, an effective policy of learning $\mathbf{F}$ may still generate a low total reward due to the given $\mathbf{F}$. Therefore, in terms of evaluating a policy's effectiveness in learning the unknown $\mathbf{F}$, regret serves as a better performance measure compared to the total expected reward.

The regret $R^\pi(T;\mathbf{F})$ measures a policy's performance for a single configuration of arms' reward distributions $\mathbf{F}$. To compare policies for a set of arms' reward distributions $\mathcal{F}$, we may use the *minimax* approach or the *uniform-dominance* approach. A policy $\pi_* \in \mathcal{A}$ is minimax optimal if, for all policies $\pi \in \mathcal{A}$, we have

$$\sup_{\mathbf{F}\in\mathcal{F}} R^{\pi_*}(T;\mathbf{F}) \le \sup_{\mathbf{F}\in\mathcal{F}} R^\pi(T;\mathbf{F}),$$

in which $\sup_{\mathbf{F}\in\mathcal{F}} R^\pi(T;\mathbf{F})$ is the worst-case regret of a policy π for the set $\mathcal{F}$ of possible reward functions. A policy π_* is uniformly optimal in $\mathcal{A}$ if

$$R^{\pi_*}(T;\mathbf{F}) \le R^\pi(T;\mathbf{F}), \qquad \text{for all } \pi \in \mathcal{A}, \mathbf{F} \in \mathcal{F}.$$

A policy can be evaluated under the minimax or the uniform-dominance approach. To explore the asymptotic behavior of the regret, we need to consider strategies that generate good performance for all arm distributions in $\mathcal{F}$, and define classes of admissible policies. For a given $\alpha \in (0,1)$, a policy π is α-*consistent* if

$$\lim_{T\to\infty} \frac{1}{T^\alpha} E[R^\pi(T;\mathbf{F})] = 0, \qquad \text{for all } \mathbf{F} \in \mathcal{F}.$$

A policy that is α-consistent for all $\alpha \in (0,1)$ is called *uniformly good*.

6.1.3 Lower bounds on regret

We now explore the lower bounds on regret which can serve as benchmarks for determining the optimality of learning policies. Suppose that the family $\mathcal{F}$ of possible reward distributions is a known distribution type $F(r;\theta)$ with an unknown parameter θ belonging to a known set Θ, i.e., $\mathcal{F} = \{F(r;\theta) \mid \theta \in \Theta\}$. Assume that each $F(r;\theta)$ is differentiable and hence the corresponding density functions is $f(r;\theta)$. Then the reward generated by playing arm i at time t, $r_{i,t}$, is a random variable drawn from $F(r;\theta)$ (or $f(r;\theta)$). Let $\boldsymbol{\theta} = (\theta_1,\dots,\theta_N)'$ be the vector of unknown parameters of N arms and $\mu(\theta_i) = \int r f(r;\theta_i)dr$ the expected reward of arm i. Denote by $\mu_* = \max_{i=1,\dots,N} \mu(\theta_i)$ and μ_* is achieved by parameter θ_*, i.e., $\mu(\theta_*) = \mu_*$. To measure the difference of two reward distributions $f(r;\theta)$ and $f(r;\theta')$, we use the *Kullback-Leibler* (KL) divergence

$$D(\theta\|\theta') = E_{R\sim f(r;\theta)}\left[\log\frac{f(R;\theta)}{f(R;\theta')}\right] = \int f(r;\theta)\log\frac{f(r;\theta)}{f(r;\theta')}dr.$$

Lai and Robbins (1985) derived the following lower bound for regret.

Theorem 6.1 (Lower bound on regret) *Suppose the following regularity conditions hold on the parameter space* Θ.

(i) *Denseness of* Θ: *For all* $\delta > 0$ *and* $\theta \in \Theta$, *there exists* $\theta' \in \Theta$ *such that* $\mu(\theta') \in [\mu(\theta), \mu(\theta) + \delta]$.

(ii) *Identifiability via the mean: For all* $\theta, \theta' \in \Theta$, *if* $\mu(\theta) < \mu(\theta')$, *then* $0 < D(\theta||\theta') < \infty$.

(iii) *Continuity of* $D(\theta||\theta')$ *in* θ': *For all* $\epsilon > 0$ *and* θ, θ' *with* $\mu(\theta) < \mu(\theta')$, *there exists* $\delta = \delta(\epsilon, \theta, \theta') > 0$ *such that* $|D(\theta||\theta') - D(\theta||\widetilde{\theta})| < \epsilon$ *whenever* $\mu(\theta') < \mu(\widetilde{\theta}) < \mu(\theta') + \delta$.

Then for every α-*consistent policy* π *and every arm configuration* $\boldsymbol{\theta}$ *such that* $\mu(\theta_i)$ *are not all equal, we have*

$$\liminf_{T \to \infty} \frac{R^\pi(T; \theta)}{\log T} \geq (1 - \alpha) \sum_{i:\mu(\theta_i)<\mu_*} \frac{\mu_* - \mu(\theta_i)}{D(\theta_i||\theta_*)}. \tag{6.8}$$

For every uniformly good policy, we have

$$\liminf_{T \to \infty} \frac{R^\pi(T; \theta)}{\log T} \geq \sum_{i:\mu(\theta_i)<\mu_*} \frac{\mu_* - \mu(\theta_i)}{D(\theta_i||\theta_*)}. \tag{6.9}$$

$\square$

Theorem 6.1 implies that, in order to achieve uniformly good performance over all reward distributions, each arm with $\mu(\theta_i) < \mu_*$ needs to be explored no fewer than $(\log T)/D(\theta_i||\theta_*)$ times asymptotically. Lai and Robbins (1985) also developed a general method of constructing asymptotically optimal policies that attain the asymptotic lower bound in Theorem 6.1. In the Lai-Robbins policy, two statistics are maintained for each arm. One is the point estimate $\widehat{\mu}_{i,t}$ of the mean reward of playing arm i at time t, and the other is the upper confidence bound $U_{i,t}$. Given $\widehat{\mu}_{i,t}$ and $U_{i,t}$, the Lai-Robbins policy can be implemented as follows.

Algorithm 6.1 (Lai-Robbins policy) *Pull each arm once in the first* N *rounds. Let* $\tau_{i,t} = \sum_{s=1}^{t} 1_{\{\pi_s = i\}}$ *represent the number of times that arm* i *is played in the first* t *rounds,* $\widehat{\mu}_{i,t}$ *the point estimate of the mean of arm* i *at time* t, *and* $U_{i,t}$ *the upper confidence bound of arm* i *at time* t. *Denote by* $\delta \in (0, 1/N)$ *a predetermined constant (or parameter of the algorithm). Let* $t = N + 1, \ldots, T$.

1. *Find the leader arm* l_t *that has the largest estimated mean reward* $\widehat{\mu}_{i,t}$ *among arms that have been pulled at least* $(t - 1)\delta$ *times, i.e.,* $l_t := \arg\max_{i:\tau_{i,t-1} \geq (t-1)\delta} \widehat{\mu}_{i,t-1}$, *and the round-robin candidate* $r_t = [(t - 1) \bmod N] + 1$.

2. *If $\widehat{\mu}_{l_t,t-1} > U_{r_t,t-1}$, then play arm l_t and otherwise play r_t.*

The Lai-Robbins policy proposed by Lai and Robbins (1985) does not explicitly outline the construction of $\widehat{\mu}_{i,t}$ and $U_{i,t}$. Instead, they gave sufficient conditions on $\widehat{\mu}_{i,t}$ and $U_{i,t}$ for achieving asymptotic optimality and constructions for four types of distributions: Bernoulli, Poisson, Gaussian, and double exponential. The following example explains the Lai-Robbins policy for Bernoulli-distributed arms.

Example 6.2 (Lai-Robbins policy for Bernoulli arms) *Suppose that arm i generates independent and identically distributed Bernoulli rewards with unknown parameter θ_i ($i = 1, \ldots, N$). Then $\mu(\theta) = \theta$ and $\Theta = (0,1)$. The Kullback-Leibler divergence $D(\theta||\theta')$ is given by*

$$D(\theta||\theta') = \theta \log\left(\frac{\theta}{\theta'}\right) + (1-\theta)\log\left(\frac{1-\theta}{1-\theta'}\right).$$

At time t, the mean estimate $\widehat{\mu}_{i,t}$ is the average rewards of arm i that has been played up to time t, and the upper confidence bound $U_{i,t}$ is given by

$$U_{i,t} = \inf\left\{\theta' \in (0,1) \;:\; \theta' \geq \widehat{\mu}_{i,t} \text{ and } D(\widehat{\mu}_{i,t}||\theta') \geq \frac{\log t}{\tau_{i,t}}\right\}. \tag{6.10}$$

The lower bound sequence for $D(\widehat{\mu}_{i,t}||\theta')$ in Lai and Robbins (1985) was quite general, the sequence $(\log t)/\tau_{i,t}$ in (6.10) was suggested in Example 4.9 of Zhao (2019). □

Another lower bound on the regret uses the minimax approach. In particular, the following theorem shows that any policy suffers regret $O(\sqrt{NT})$ on certain instances of the bandit problem.

Theorem 6.2 (Minimax lower bound on regret) *Fix time horizon T and the number of arm N. Consider the bandit problem in Definition 6.3 and a policy π for the problem, we have*

$$R^\pi(T; \mathbf{F}) \geq O(\sqrt{NT}).$$

The above result traces back to Vogel (1960), and an elegant proof for this can be found in Bubeck and Cesa-Bianchi (2012) and Zhao (2019, Section 4.2.2). Since the lower bound deals with the "worst-case," it is possible to find a regret for a particular bandit problem that is lower than that of many other bandit problems. As we don't specify a particular bandit problem here, readers interested in exploring further details are encouraged to refer to the references provided in the supplements and problems of this chapter.

6.2 Finite-horizon Markov decision process

A **Markov decision process** (MDP) is a Markov process with feedback control (or closed-loop control), in which the control action taken by the decision-maker depends on the process output. The aim of the decision-maker is to choose a sequence of actions over a time horizon, aiming to minimize a cumulative cost function associated with the expected value of the trajectory of the Markov process. This section first introduces the concept of finite-state finite horizon MDP, and subsequently extends to continuous state and/or infinite horizon MDP.

6.2.1 Model specification

Let $t = 0, 1, \ldots, T$ denote discrete time. Assume that the time horizon $T < \infty$.

Definition 6.4 (Finite-horizon MDP) *A finite horizon MDP model is the 5-tuple $(\mathcal{X}, \mathcal{A}, \{P_t(x'|x, a)\}, \{r_t(x, a)\}, g_T)$ with the following interpretation:*

(i) $\mathcal{X}$ is a set of states so that each observed state $x_t \in \mathcal{X}$. The space $\mathcal{X}$ can be either discrete (e.g., finite or countable infinite) or continuous.

(ii) $\mathcal{A}$ is the set of actions or the action space that can be discrete or continuous. Let D_t be a measurable subset of $\mathcal{X} \times \mathcal{A}$ and represent the set of all state-action combinations at time t. For $x \in \mathcal{X}$, the set $D_t(x) = \{a \in \mathcal{A} | (x, a) \in D_t\}$ is the set of actions in state x at time t.

(iii) $\{P_t(x'|x, a)\}$ is the set of transition kernel from D_t to $\mathcal{X}$. That is, for any fixed pair $(x, a) \in D_{t-1}$, the transition probability of a next state x_t given a previous state x_{t-1} and an action a_{t-1} taken in this state is $P(x'|x, a) = P(x_t = x'|x_{t-1} = x, a_{t-1} = a)$.

(iv) $\{r_t(x, a)|x \in \mathcal{X}, a \in \mathcal{A}\}$ is the set of rewards. At each time t, given the state $x_s = x$ and the action $a_t = a$ taken in this state, an one-stage reward $r_t : \mathcal{X} \times \mathcal{A} \to \mathbb{R}$ is $r_t(x, a)$.

(v) At terminal time $t = T$, a terminal reward $g_T : \mathcal{X} \to \mathbb{R}$ is received for the terminal state $x \in \mathcal{X}$. □

To describe the control of a MDP, we use the following concepts of decision rules and policies.

Definition 6.5 *(i) A measurable mapping $d_t : \mathcal{X} \to \mathcal{A}$ is called a decision rule at time t. We denote by $\mathcal{D}_t$ the set of all decision rules at time t.*

(ii) A sequence of decision rules $\pi = (d_0, d_1, \ldots, d_{T-1})$ with $d_t \in \mathcal{D}_t$ is called an T-stage policy.

We now define the expected reward of a policy and the T-stage optimization problem as follows. Given a policy $\pi = (d_0, \ldots, d_{T-1})$ and let $u_t = d_t(x_t)$ for $t = 1, \ldots, T-1$, the *expected total reward* or the *state value* function at time t is defined as

$$V_t^\pi(x) = E_t^\pi \left(\sum_{k=t}^{T-1} r_k(x_k, a_k) + g_T(x_T) \middle| x_t = x, \pi \right), \qquad x \in \mathcal{X}, \qquad (6.11)$$

which only depends on the current state x and assumes that the policy π is followed starting from this state. The *state-action value* function

$$Q_t^\pi(x, a) = E_t^\pi \left(\sum_{k=t}^{T-1} r_k(x_k, a_k) + g_T(x_T) \middle| x_t = x, a_t = a, \pi \right), \qquad x \in \mathcal{X}, \quad (6.12)$$

also depends on the state x, but makes the action a chosen in this state a free variable instead of having it generated by the policy π. The state-action value function (6.12) plays an important role in the Q-learning algorithm; see Section 6.5.2. The relationship between these two definitions for the value function is given by

$$V_t^\pi(x) = E(Q_t^\pi(x, a) | a \sim \pi(x, \cdot)). \qquad (6.13)$$

Note that the terminology used in MDP and reinforcement learning literatures is slightly different. The total reward function in MDP is called the *state value function*, while the maximized total reward function in MDP is called the *optimal state value function* in RL. This is similar for the term *state-action value function* and *optimal state-action value function* in RL. See a list of terminology comparison in the two fields in Bertsekas (2019).

Let $r^+ = \max(r, 0)$ and assume the following integrability condition. For $t = 0, 1, \ldots, T$,

$$\sup_\pi E_t^\pi \left(\sum_{k=t}^{T-1} r_k^+(x_k, d_k(x_k)) + g_T^+(x_T) \right) < \infty, \qquad x \in \mathcal{X}. \qquad (6.14)$$

This integrability condition can be easily satisfied by the following boundedness assumption for reward functions and transition functions. That is, there exists a measurable function $M : \mathcal{X} \to \mathbb{R}^+$ and constant $K_r > 0$, $K_g > 0$ and $K_P > 0$ such that, for all $(x, a) \in D_t$,

$$r_t^+(x, a) \le K_r M(x), \qquad g_T^+(x) \le K_g M(x),$$

and

$$\int M(x') P_t(dx' | x, a) \le K_P M(x).$$

Then the value function $V_t(x)$ is defined as the *maximal expected total reward* at time t,

$$V_t(x) := \sup_\pi V_t^\pi(x), \qquad x \in \mathcal{X}. \tag{6.15}$$

A policy π is called *optimal* if $V_0^\pi(x) = V_0(x)$ for $x \in \mathcal{X}$. Note that under assumption (6.14),

$$V_t^\pi(x) \leq V_t(x) \leq \infty,$$

thus the performance $J_t^\pi(x)$ and the value function $V_t(x)$ are well-defined. Moreover, the following terminal condition holds

$$V_T(x) = V_T^\pi(x) = g_T(x), \qquad x \in \mathcal{X}.$$

6.2.2 The Bellman equation and backward induction

Given a policy $\pi = (d_1, \ldots, d_{T-1})$, $d_t \in \mathcal{D}_t, t = 1, \ldots, T-1$, the expected total reward can be computed recursively by the so-called *reward iteration*.

Definition 6.6 *Let $\mathcal{M}(\mathcal{X})$ be the set of measurable functions from $\mathcal{X}$ to $\mathbb{R}$. We define the following operators for $t = 0, 1, \ldots, T-1$.*

(i) For $\phi \in \mathcal{M}(\mathcal{X})$, define

$$[\mathcal{L}_t \phi](x, a) := r_t(x, a) + \int \phi(x') P_t(dx'|x, a), \qquad (x, a) \in \mathcal{D}_t.$$

(ii) Given $\phi \in \mathcal{M}(\mathcal{X})$ and $d \in \mathcal{D}_t$, define $[\mathcal{T}_t^d(\phi)](x) := [\mathcal{L}_t \phi](x, d(x))$ for $x \in \mathcal{X}$.

(iii) For $\phi \in \mathcal{M}(\mathcal{X})$, the maximal reward operator at time t is defined as $[\mathcal{T}_t \phi](x) := \sup_{d \in \mathcal{D}_t(x)} [\mathcal{L}_t \phi](x, d(x))$, $x \in \mathcal{X}$. A decision rule $d^* \in \mathcal{D}_t$ such that $\mathcal{T}_t^{d^*} \phi = \mathcal{T}_t \phi$ is called a maximizer of ϕ at time t. □

Definition 6.6 implies that $\mathcal{T}_t \phi = \sup_{d \in \mathcal{D}_t} \mathcal{L}_t^d \phi$, and for all $\phi \in \mathcal{M}(\mathcal{X})$, it holds that $\mathcal{T}_t^d \phi \in \mathcal{M}(\mathcal{X})$, but $\mathcal{T}_t \phi$ may not belong to $\mathcal{M}(\mathcal{X})$. Note that all three operators are monotone, i.e., for $\phi, \psi \in \mathcal{M}(\mathcal{X})$ with $\phi(x) \leq \psi(x)$ for all $x \in \mathcal{X}$, it holds

$$\mathcal{L}_t \phi(x, a) \leq \mathcal{L}_t \psi(x, a) \quad \text{for all } (x, a) \in \mathcal{D}_t; \tag{6.16a}$$

$$\mathcal{T}_t^d \phi(x) \leq \mathcal{T}_t^d \psi(x) \quad \text{for all } x \in \mathcal{X}, d \in \mathcal{D}_t; \tag{6.16b}$$

$$\mathcal{T}_t \phi(x) \leq \mathcal{T}_t \psi(x) \quad \text{for all } x \in \mathcal{X}; \tag{6.16c}$$

see Exercise 6.2. Using the reward iteration, these operators can be used to compute the value of a policy recursively.

Theorem 6.3 (Reward iteration) *Let $\pi = (d_0, \ldots, d_{T-1})$ be an T-stage policy. For $t = 0, 1, \ldots, T-1$, it holds that*

$$V_T^\pi = g_T, \qquad V_t^\pi = \mathcal{T}_t^{d_t} V_{t+1}^\pi = \mathcal{T}_t^{d_t} \ldots \mathcal{T}_{T-1}^{d_{T-1}} g_T. \tag{6.17}$$

Proof. For $x \in \mathcal{X}$, we have

$$V_t^\pi(x) = E_{t,x}^\pi \left(\sum_{k=t}^{T-1} r_k(x_k, d_k(x_k)) + g_T(x_T) \right)$$

$$= r_t(x, d_t(x)) + \int E_{t+1,x'}^\pi \left(\sum_{k=t+1}^{T-1} r_k(x_k, d_k(x_k)) + g_T(x_T) \right) P_t(dx'|x, d_t(x))$$

$$= r_t(x, d_t(x)) + \int V_{n+1}^\pi(x') P_t(dx'|x, d_t(x)) = \left[\mathcal{T}_t^{d_t} V_{t+1}^\pi \right](x).$$

Repeat the above argument for $t + 1, \ldots, T - 1$, one can show that (6.17) holds. □

The following theorem shows that the MDP problem can be solved by successive application of the $\mathcal{T}_t$ operators. Although it is in general not true that $\mathcal{T}_n \phi \in \mathcal{M}(\mathcal{X})$ for $\phi \in \mathcal{M}(\mathcal{X})$, one can show that V_t is measurable in $\mathcal{M}(\mathcal{X})$ and $\{V_t\}_{t=0,\ldots,T}$ satisfies the so-called *Bellman equation*.

Theorem 6.4 (Bellman equation) *Assume that there exists sets $\mathcal{M}_t \subset \mathcal{M}(\mathcal{X})$ and $\Delta_t \subset D_t$ such that for all $t = 0, 1, \ldots, T - 1$, (i) g_T is measurable on $\mathcal{M}_T$, (ii) If ϕ is measurable on $\mathcal{M}_{t+1}$, then $\mathcal{T}_t \phi$ is well-defined and measurable on $\mathcal{M}_t$, and (iii) for all measurable functions v on $\mathcal{M}_{t+1}$, there exists a maximizer d_t of ϕ such that d_t is measurable on Δ_t. Then*

(a) *V_t is measurable on $\mathcal{M}_t$, and the sequence $\{V_t\}$ satisfies the Bellman equation, i.e., for $t = 0, 1, \ldots, T - 1$, $V_T(x) = g_T(x)$ and*

$$V_t(x) = \sup_{u \in D_t(x)} E\left[r_t(x, u) + V_{t+1}(x') \right], \tag{6.18a}$$

$$Q_t(x, u) = E\left[r_t(x, u) + \sup_{u' \in D_{t+1}(x')} Q_{t+1}(x', u') \right]. \tag{6.18b}$$

(b) *$V_t = \mathcal{T}_t \mathcal{T}_{t+1} \ldots \mathcal{T}_{T-1} g_T$.*

(c) *For $t = 0, 1, \ldots, T - 1$ there exist maximizers d_t of V_{t+1} with $d_t \in \Delta_t$. Moreover, every sequence of maximizers d_t^* of V_{t+1} defines an optimal policy $(d_0^*, d_1^*, \ldots, d_{T-1}^*)$ for the T-stage Markov decision problem.*

Proof. We show the result by induction. By assumption (i), we know $V_T = g_T$ is measurable on $\mathcal{M}_T$. Suppose that the statement is true for $T-1, \ldots, t+1$. Since V_t is measurable on $\mathcal{M}_k$ for $k = T, \ldots, t+1$, the maximizers $d_t^*, \ldots, d_{T-1}^*$ exist and we obtain with the reward iteration and the induction hypothesis

$$V_t^{\pi^*} = \mathcal{T}_t^{d_n^*} V_{t+1}^{\pi^*} = \mathcal{T}_t^{d_t^*} V_{t+1} = \mathcal{T}_t V_{t+1}.$$

Hence, $V_t \geq \mathcal{T}_t V_{t+1}$. On the other hand for an arbitrary policy π,

$$V_t^\pi = \mathcal{T}_t^{d_t} V_{t+1}^\pi \leq \mathcal{T}_t^{d_t} V_{t+1} \leq \mathcal{T}_t V_{t+1}.$$

Taking the supremum over all policies yields $V_t \leq \mathcal{T}_t V_{t+1}$. Therefore, we have $V_t^{\pi^*} = \mathcal{T}_t V_{t+1} = V_t$, and according to assumption (ii), V_t is measurable on $\mathcal{M}_t$. This shows (a) and (c). Then (b) follows directly from them. □

Theorem 6.4 implies that, if assumption (i)-(iii) in the theorem are satisfied, then the following result holds for $t \leq s \leq T$,

$$V_t(x) = \sup_{\pi} E_{t,x}^{\pi} \left[\sum_{k=t}^{s-1} r_k(x_k, d_k(x_k)) + V_s(x_s) \right], \qquad x \in \mathcal{X}.$$

Theorem 6.4 also implies the following backward induction algorithm which solves the MDP problem in Theorem 6.4.

Algorithm 6.2 (Backward induction)

1. *Set $t := T$ and for $x \in \mathcal{X}$, let $V_T(x) := g_T(x)$.*

2. *Set $t \leftarrow t - 1$ and compute for all $x \in \mathcal{X}$*

$$V_t(x) = \sup_{a \in \mathcal{D}_t(x)} \left(r_t(x, a) + \int V_{t+1}(x') P_t(dx'|x, a) \right).$$

   Compute a maximizer d_t^* of V_{t+1}.

3. *If $t = 0$, then the value function is computed and the optimal policy is given by $\pi^* = (d_0^*, \ldots, d_{T-1}^*)$. Otherwise, go to step 2.*

Suppose that a solution of the Bellman equation (6.18a) exists together with a sequence of maximizers. It can be shown that this solution of the Bellman equation is also the solution of our optimization problem.

Theorem 6.5 (Verification theorem) *Let $\{\phi_t\} \in \mathcal{M}(\mathcal{X})$ be a solution of the Bellman equation (6.18a). Then*

(a) *$\phi_t \geq V_t$ for $t = 0, 1, \ldots, T$.*

(b) *If d_t^* is a maximizer of ϕ_{t+1} for $t = 0, 1, \ldots, T-1$, then $\phi_t = V_t$ and the policy $\pi^* = (d_0^*, d_1^*, \ldots, d_{T-1}^*)$ is optimal for the T-stage MDP problem.*

Proof. For $t = T$, we have $\phi_T = g_T = V_T$. Suppose $\phi_{t+1} \geq V_{t+1}$, then for all $\pi = (d_0, \ldots, d_{T-1})$

$$\phi_t = \mathcal{T}_t \phi_{t+1} \geq \mathcal{T}_t V_{t+1} \geq \mathcal{T}_t^{d_t} V_{t+1}^{\pi} = V_t^{\pi}.$$

Taking the supremum over all policies π yields $\phi_t \geq V_t$. For part (b), the result holds when $t = T$. Suppose the statement is true for $t + 1$, then

$$V_t \leq \phi_t = \mathcal{T}_t^{d_t^*} \phi_{t+1} = \mathcal{T}_t^{d_t^*} V_{t+1} \leq V_t^{\pi^*} \leq V_t.$$

Then by induction, the result holds. □

The solution method in Theorem 6.4 relies on an important concept, *principle of dynamic programming*. It essentially means that whenever there is an optimal policy π^* over a certain horizon, then for a subinterval of $[0, T]$, the corresponding policy which is obtained by restricting π^* to this subinterval is again optimal. This principle is summarized as follows.

Theorem 6.6 (Principle of dynamic programming) *Suppose assumptions (i)-(iii) in Theorem 6.4 hold. For $t \leq s \leq T$, if $(d_t^*, \ldots, d_{T-1}^*)$ is optimal for the time period $[t, T]$, then $(d_s^*, \ldots, d_{T-1}^*)$ is optimal for $[s, T]$, i.e.,*

$$V_t^{\pi^*}(x) = V_t(x) \Longrightarrow V_s^{\pi^*}(x) = V_s(x) \quad P_{t,x}^{\pi^*}\text{-almost surely.}$$

Proof. Using the reward iteration in theorem 6.3, the definition of V_s, and equation (6.18b), we obtain

$$V_t(x) = V_t^{\pi^*}(x) = \mathcal{T}_t^{d_t^*} \cdots \mathcal{T}_{s-1}^{d_{s-1}^*} V_s^{\pi^*}(x) = E_{t,x}^{\pi^*}\left[\sum_{k=t}^{s-1} r_k(x_k, d_k^*(x_k)) + V_s^{\pi^*}(x_s)\right]$$

$$\leq E_{t,x}^{\pi^*}\left[\sum_{k=t}^{s-1} r_k(x_k, d_k^*(x_k)) + V_s(x_s)\right] \leq V_t(x).$$

This implies the equality $E_{t,x}^{\pi^*}[V_s(x_s) - V_s^{\pi^*}(x_s)] = 0$. Hence, the result is proved. $\square$

In some applications, instead of running rewards r_t and the terminal reward g_T, the system generates a running cost $l_t(x_t, a_t)$ and a terminal cost $c_T(x_T)$. In such case, one wants to minimize

$$E_{n,x}^{\pi}\left[\sum_{k=t}^{T-1} l_k(x_k, d_k(x_k)) + c_T(x_T)\right], \qquad x \in \mathcal{X}$$

for $\pi = (d_0, \ldots, d_{T-1})$. This cost minimization problem is equivalent to a reward maximization problem with $r_t(x_t, a_t) = -l_t(x_t, a_t)$ and $g_T(x) = -c_T(x)$. We can still use the same notation V_t^{π} and V_t for the cost functions under policy π and the minimal cost function. The minimal cost operator $\mathcal{T}_t$ can be defined as in Definition 6.6 by replacing $r_t(x, a)$ by $l_t(x, a)$, then the value function V_t is also called *cost-to-go function*.

6.2.3 Stationary MDP and forward induction

Consider a finite-horizon MDP in Definition 6.4 such that the transition kernel and rewards functions are the form of $P_t(x'|x, a) = P(x'|x, a)$, $r_t(x, a) = \rho^t r(x, a)$ and $g_T(x) = \rho^T g(x)$ for all $t = 0, 1, \ldots, T$, $x \in \mathcal{X}$ and $a \in \mathcal{A}$. Such MDP is called *stationary* MDP and can be represented as $(\mathcal{X}, \mathcal{A}, P(x'|x, a), r(x, a), g_T, \rho)$.

Denote by $\mathcal{D}$ the set of all decision rules $d : \mathcal{X} \to \mathcal{A}$ with $d(x) \in D(x)$ for $x \in \mathcal{X}$, and $\mathcal{D}^T$ the set of all T-stage policies $\pi = (d_0, \dots, d_{T-1})$. The expected discounted reward over t stages under a policy $\pi \in \mathcal{D}^t$ is given by

$$J_t^\pi(x) := E_x^\pi \left[\sum_{k=0}^{t-1} \rho^k r(x_k, d_k(x_k)) + \rho^t g(x_t) \right], \quad x \in \mathcal{X},$$

when the system starts in state $x \in \mathcal{X}$. Assume that the following integrability condition holds,

$$\delta_T(x) := \sup_\pi E_x^\pi \left[\sum_{k=0}^{T-1} \rho^k r^+(x_k, d_k(x_k)) + \rho^T g^+(x_T) \right] < \infty, \quad x \in \mathcal{X}. \quad (6.19)$$

The maximal expected discounted reward over t stages can be defined as

$$J_0(x) := g(x), \quad J_t(x) := \sup_{\pi \in \mathcal{D}^t} J_t^\pi(x), \quad x \in \mathcal{X}, 1 \le t \le T. \quad (6.20)$$

Note that the integrability conditions (6.14) and (6.19) are equivalent when a MDP is stationary. To ensure the integrability condition (6.19) is satisfied, one usually assume the boundedness condition for the reward function and transition kernel. That is, there exists a measurable function $M : \mathcal{X} \to \mathbb{R}$ and constants $K_r > 0, K_g > 0$ and $K_P > 0$ such that, for all $x \in \mathcal{X}$ and $a \in \mathcal{A}$,

$$r(x, a) \le K_r M(x), \quad g(x) \le K_r M(x), \quad (6.21)$$

and

$$\int M(x') P(dx' | x, a) \le K_P(x). \quad (6.22)$$

It is easy to see that, under the boundedness condition, the integrability condition (6.19) is satisfied and hence the maximal expected discounted rewards (6.20) are well defined.

For a stationary MDP with the maximal expected discounted reward (6.20), when the maximal expected total reward $V_t(x)$ is defined by (6.15) as for a general MDP, the following relation between the value functions J_t and V_t hold

$$V_t(x) = \rho^t J_{T-t}(x), \quad x \in \mathcal{X}, t = 0, 1, \dots, T. \quad (6.23)$$

Moreover, every non-stationary MDP can be formulated as a stationary one by extending the state space by including the time parameter.

Analogous to Definition 6.6 and Theorems 6.3 and 6.4, one may define the following operator to derive the reward iteration and bellman equation. Let $\phi : \mathcal{X} \to \mathbb{R}$ be a measurable function, and

$$[\mathcal{L}\phi](x, a) := r(x, a) + \rho \int \phi(x') P(dx' | x, a), \quad (x, a) \in \mathcal{D}, \quad (6.24\text{a})$$

$$[\mathcal{T}^d(\phi)](x) := [\mathcal{L}\phi](x, d(x)), \quad x \in \mathcal{X}, \quad (6.24\text{b})$$

$$[\mathcal{T}\phi](x) := \sup_{d \in D(x)} [\mathcal{L}^d \phi](x, a), \quad x \in \mathcal{X}. \quad (6.24\text{c})$$

The following theorems can be shown in a similar way to Theorems 6.3 and 6.4; see Exercises 6.3 and 6.4.

Theorem 6.7 (Reward iteration) *Let* $\pi = (d_0, \ldots, d_{T-1})$ *be an T-stage policy. For* $t = 0, 1, \ldots, T - 1$, *it holds that*

$$J_t^\pi = \mathcal{T}^{d_t} \ldots \mathcal{T}^{d_{T-1}} g_T.$$

Theorem 6.8 (Bellman equation) *Suppose that there exists sets* $\mathcal{M} \subset \mathcal{M}(\mathcal{X})$ *and* $\Delta_t \subset D_t$ *such that for all* $t = 0, 1, \ldots, T - 1$, *(i)* $g(x)$ *is measurable on* $\mathcal{M}_T$, *(ii) If* ϕ *is measurable on* $\mathcal{M}$, *then* $\mathcal{T}\phi$ *is well-defined and measurable on* $\mathcal{M}$, *and (iii) for all measurable functions* v *on* $\mathcal{M}$, *there exists a maximizer* $d(x)$ *of* ϕ *such that* $\mathcal{T}^d v(x) = \mathcal{T}v(x)$ *for all* $x \in \mathcal{X}$. *Then*

(a) J_t *is measurable on* $\mathcal{M}$, *and the sequence* $\{J_t\}$ *satisfies the Bellman equation, i.e., for* $t = 0, 1, \ldots, T - 1$, $J_0(x) = g(x)$ *and*

$$J_t(x) = \sup_{a \in D_t(x)} \left(r(x, a) + \rho \int J_{t-1}(x') P(dx'|x, a) \right), \qquad x \in \mathcal{X}.$$

(b) $J_t = \mathcal{T}^t g$.

(c) *For* $t = 0, 1, \ldots, T - 1$ *there exist maximizers* d_t^* *of* J_{t-1} *with* $d_t^* \in \Delta$, *and every sequence of maximizers* d_t^* *of* J_{t-1} *defines an optimal policy* $(d_T^*, \ldots, d_1^*)$ *for the T-stage Markov decision problem.* $\square$

Analogous to the Backward induction algorithm, Theorem 6.8 implies the following forward induction algorithm.

Algorithm 6.3 (Forward induction)

1. *Set* $t := 0$ *and for* $x \in \mathcal{X}$, *let* $J_0(x) := g(x)$.

2. *Set* $t \leftarrow t + 1$ *and compute for all* $x \in \mathcal{X}$

$$J_t(x) = \sup_{a \in \mathcal{D}_t(x)} \left(r(x, a) + \int J_{t-1}(x') P(dx'|x, a) \right).$$

   Compute a maximizer d_t^* *of* J_{t-1}.

3. *If* $t = T$, *then the value function* J_T *is computed and the optimal policy is given by* $\pi^* = (d_{T-1}^*, \ldots, d_1^*)$. *Otherwise, go to step 2.*

6.2.4 Linear-quadratic regulator

An important class of MDP problems with various different applications are *linear-quadratic* (LQ) problems. The name stems from the linear state transition function and the quadratic cost function. In this section, we suppose that $\mathcal{X} := \mathbb{R}^m$ is the state space of the underlying system and $\mathcal{A} := D_t(\mathbf{x}) := \mathbb{R}^d$. The state transition functions are linear in state and action with coefficient matrices $\mathbf{A} \in \mathbb{R}^{m \times m}$, $\mathbf{B} \in \mathbb{R}^{m \times d}$. For convenience, assume that $\mathbf{A}$ and $\mathbf{B}$ are full rank. Then the system transition functions are given by

$$\mathbf{x}_{t+1} = \mathbf{A}\mathbf{x}_t + \mathbf{B}\mathbf{a}_t + \mathbf{z}_{t+1}, \tag{6.25}$$

where $\mathbf{z}_1, \ldots, \mathbf{z}_T$ are independent and identically distributed random vectors with Gaussian distribution $N_d(\mathbf{0}, \boldsymbol{\Sigma})$. The one-stage reward and the terminal reward are negative cost functions

$$r_t(\mathbf{x}, \mathbf{a}) = -(\mathbf{x}'\mathbf{Q}\mathbf{x} + 2\mathbf{x}'\mathbf{S}\mathbf{a} + \mathbf{a}'\mathbf{R}\mathbf{a}), \qquad g_T(\mathbf{x}) = -\mathbf{x}'\mathbf{Q}_T\mathbf{x},$$

where $\mathbf{Q} \in \mathbb{R}^{m \times m}$, $\mathbf{Q}_T \in \mathbb{R}^{m \times m}$, and $\mathbf{R} \in \mathbb{R}^{d \times d}$ are symmetric and positive definite matrices and determine state costs, and $\mathbf{S} \in \mathbb{R}^{m \times d}$ determines action costs. The objective of control is to take actions $\pi := \{\mathbf{a}_1, \ldots, \mathbf{a}_{T-1}\}$ to minimize the following accumulated cost function

$$\min_{\mathbf{a}_1, \ldots, \mathbf{a}_{T-1}} E^\pi \left[\sum_{t=0}^{T-1} \left(\mathbf{x}_t'\mathbf{Q}\mathbf{x}_t + 2\mathbf{x}_t'\mathbf{S}\mathbf{a}_t + \mathbf{a}_t'\mathbf{R}\mathbf{a}_t \right) + \mathbf{x}_T'\mathbf{Q}_T\mathbf{x}_T \Big| \mathbf{x}_0 = \mathbf{x} \right]. \tag{6.26}$$

Let $V_t(\mathbf{x})$ be the value function at $\mathbf{x}$ at time t. The Bellmann equation for the minimization problem (6.26) is

$$V_t(\mathbf{x}) = \min_{\mathbf{a}} E \left[\mathbf{x}'\mathbf{Q}\mathbf{x} + 2\mathbf{x}'\mathbf{S}\mathbf{a} + \mathbf{a}'\mathbf{R}\mathbf{a} + V_{t+1}(\mathbf{A}\mathbf{x} + \mathbf{B}\mathbf{a} + \mathbf{z}) \right].$$

Suppose that the value function at time $t + 1$ has a quadratic form for $\mathbf{x}$, i.e., $V_{t+1}(\mathbf{x}) = \mathbf{x}'\boldsymbol{\Lambda}_{t+1}\mathbf{x} + c_{t+1}$, where $\boldsymbol{\Lambda}_{t+1} \in \mathbb{R}^{m \times m}$ is a symmetric positive definite matrix and $c_{t+1} \in \mathbb{R}$ is a scalar. Then the Bellman equation becomes

$$V_t(\mathbf{x}) = \min_{\mathbf{a}} E \left[\mathbf{x}'\mathbf{Q}\mathbf{x} + 2\mathbf{x}'\mathbf{S}\mathbf{a} + \mathbf{a}'\mathbf{R}\mathbf{a} + (\mathbf{A}\mathbf{x} + \mathbf{B}\mathbf{a} + \mathbf{z})'\boldsymbol{\Lambda}_{t+1}(\mathbf{A}\mathbf{x} + \mathbf{B}\mathbf{a} + \mathbf{z}) + c_{t+1} \right]. \tag{6.27}$$

The first order condition suggests that

$$\mathbf{S}'\mathbf{x} + \mathbf{R}\mathbf{a} + \mathbf{B}'\boldsymbol{\Lambda}_{t+1}(\mathbf{A}\mathbf{x} + \mathbf{B}\mathbf{a}) = 0.$$

Then the optimal control at time t is

$$\mathbf{a}_t^* = -(\mathbf{R} + \mathbf{B}'\boldsymbol{\Lambda}_{t+1}\mathbf{B})^{-1}(\mathbf{B}'\boldsymbol{\Lambda}_{t+1}\mathbf{A} + \mathbf{S}')\mathbf{x}. \tag{6.28}$$

Substitute $\mathbf{a}_t^*$ back into (6.27) and note that $E(\mathbf{z}'\boldsymbol{\Lambda}_{t+1}\mathbf{z}) = \mathrm{tr}(\boldsymbol{\Lambda}_{t+1}E(\mathbf{z}\mathbf{z}')) = \mathrm{tr}(\boldsymbol{\Lambda}_{t+1}\boldsymbol{\Sigma})$, in which $\mathrm{tr}(\cdot)$ is the trace of a square matrix, we then obtain that

$$\boldsymbol{\Lambda}_t = \mathbf{Q} + \mathbf{A}'\boldsymbol{\Lambda}_{t+1}\mathbf{A} - (\mathbf{B}'\boldsymbol{\Lambda}_{t+1}\mathbf{A} + \mathbf{S}')'(\mathbf{R} + \mathbf{B}'\boldsymbol{\Lambda}_{t+1}\mathbf{B})^{-1}(\mathbf{B}'\boldsymbol{\Lambda}_{t+1}\mathbf{A} + \mathbf{S}'), \tag{6.29}$$

and

$$c_t = c_{t+1} + \mathrm{tr}(\boldsymbol{\Lambda}_{t+1}\boldsymbol{\Sigma}). \tag{6.30}$$

Equation (6.29) is known as the *Riccati difference equation*. Together with the terminal condition

$$\boldsymbol{\Lambda}_T = \mathbf{Q}_T, \qquad c_T = 0,$$

one can solve for $\{\boldsymbol{\Lambda}_t, c_t; t = T - 1, \ldots, 1\}$ backward.

6.3   Discounted infinite horizon Markov decision process

The MDPs discussed so far assume that $T < \infty$. In finite-horizon MDP problems, transition kernels and rewards are usually dependent on time, hence the state- and action-value functions depend explicitly on time. Consequently, backward recursion algorithms can be used to solve the problem of finding optimal policies. In contrast, for MDP with an infinite horizon, i.e., $T = \infty$, one usually assume that transition kernels and rewards don't vary with time, consequently their state- and action-value functions should also not depend on time. In such cases, the Bellman equation becomes a fixed point equation for the state or action-value functions and the optimal policy becomes stationary. There are two strategies to specify the reward or cost function, one is to use discounted rewards (or costs) and the other is to use average rewards (or costs). We only consider the discounted rewards (or costs) here.

6.3.1   Bellman equation and contraction mapping

In infinite-horizon MDP problems, transition kernels do not depend on time, and the terminal reward or cost degenerates due to exponentially decaying discount factors.

Definition 6.7 *An infinite horizon stationary MDP is the 5-tuple* $(\mathcal{X}, \mathcal{A}, P(x'|x, a), r(x, a), \rho)$ *in which* $\rho \in (0, 1]$ *is the discount factor,* $\mathcal{X}, \mathcal{A}, P_t(x'|x, a)$, $r_t(x, a)$ *are given in Definition 6.4 and* $P_t(x'|x, a) = P(x'|x, a)$, $r_t(x, a) = \rho^t r(x, a)$ *for all* $t = 0, 1, \dots$.

Denote by $\mathcal{D}$ the set of all decision rules $d : \mathcal{X} \to \mathcal{A}$ with $d(x) \in D(x)$ for $x \in \mathcal{X}$. Then $\mathcal{D}^\infty$ is the set of all policies $\pi = (d_0, d_1, \dots)$. The expected infinite-horizon discounted reward under a policy $\pi \in \mathcal{D}^\infty$ is given by

$$V^\pi(x) := E_x^\pi \left[\sum_{k=0}^{\infty} \rho^k r(x_k, d_k(x_k)) \Big| x_0 = x, \pi \right], \quad x \in \mathcal{X}, \qquad (6.31)$$

when the system starts in state $x \in \mathcal{X}$. The state-action value function is given by

$$Q^\pi(x, a) := E_x^\pi \left[\sum_{k=0}^{\infty} \rho^k r(x_k, d_k(x_k)) \Big| x_0 = x, d_0(x_0) = a, \pi \right], \quad x \in \mathcal{X},$$

The reward iteration in Theorem 6.3 and the stationarity of the discounted reward problems imply that

$$V^\pi(x) = E[r(x, a) + \rho V^\pi(x')],$$
$$Q^\pi(x, a) = E[r(x, a) + \rho Q^\pi(x', a')],$$

where x' is drawn from the transition kernel $P(\cdot|x,a)$ and a' is drawn from the distribution of $d(x)$. Note that equation (6.13) now becomes $V^\pi(x) = E[Q^\pi(x,a)|a \sim d(x)]$ due to the stationarity of the problem. Then the maximal expected discounted reward (or the maximal state value) function is given by

$$V(x) = \max_\pi V^\pi(x), \qquad x \in \mathcal{X}, \tag{6.32}$$

and the maximal state-action value function is

$$Q(x,a) = \max_\pi Q^\pi(x,a), \qquad x \in \mathcal{X}, a \in \mathcal{A}. \tag{6.33}$$

A policy $\pi^* \in \mathcal{D}^\infty$ is called *optimal* if $V^{\pi^*}(x) = V(x)$ for all $x \in \mathcal{X}$. To ensure that the infinite horizon optimization problems (6.32) and (6.33) are well-defined, the following integrability condition is often assumed,

$$\delta(x) := \sup_\pi E_x^\pi \left(\sum_{k=0}^\infty \rho^k r^+(x_k, d_k(x_k)) \right) < \infty, \qquad x \in \mathcal{X}.$$

To characterize the infinite horizon value function (6.32) as a solution of the Bellman equation, we proceed as follows. First we assume the following boundedness condition for the reward functions and transition kernels. That is, there exists a measurable function $M : \mathcal{X} \to \mathbb{R}$ and constants K_r and K_P, such that, for all $x \in \mathcal{X}$ and $a \in \mathcal{A}$,

$$r^+(x,a) \le K_r M(x), \quad \int M(x') P(dx'|x,a) \le K_P M(x). \tag{6.34}$$

Let $M(x)$ be a measurable function satisfying (6.21) and (6.22). For a measurable function ϕ in $\mathcal{M}(\mathcal{X})$, define its weighted supremum norm $||\cdot||_M$ by

$$||\phi||_M := \sup_{x \in \mathcal{X}} \frac{|\phi(x)|}{M(x)}.$$

Let $\mathcal{B}_M = \{\phi \in \mathcal{M}(\mathcal{X})| \, ||\phi||_M < \infty\}$. Then $(\mathcal{B}_M, ||\cdot||_M)$ is a Banach space (i.e., a complete normed vector space) and the following property can be obtained using Banach's fixed point theorem.

Proposition 6.1 *Let $\mathcal{T}^d$ and $\mathcal{T}$ be the operators defined by (6.24b) and (6.24c) and $M(x)$ be a measurable function such that condition (6.34) holds. For measurable functions $\phi \in \mathcal{B}_M$ and $\psi \in \mathcal{B}_M$ and $d \in \mathcal{D}$,*

$$||\mathcal{T}^d \phi - \mathcal{T}^d \psi||_M \le \rho K_P ||\phi - \psi||_M, \qquad ||\mathcal{T}\phi - \mathcal{T}\psi||_M \le \rho K_P ||\phi - \psi||_M.$$

If $\rho K_P < 1$, then $J^d = \lim_{n\to\infty} [\mathcal{T}^d]^n g$ for all $g \in \mathcal{B}_M$ and J^d is the unique fixed point of $\mathcal{T}^d$ in $\mathcal{B}_M$.

Proof. For $d \in \mathcal{D}$, we have

$$\mathcal{T}^d \phi(x) - \mathcal{T}^d \psi(x) \leq \rho \sup_{a \in D_x} \int \Big(\phi(x') - \psi(x') \Big) P(dx'|x,a)$$

$$\leq \rho \, ||\phi - \psi||_M \sup_{a \in D_x} \int M(x') Q(dx'|x,a).$$

Taking the norm $||\cdot||_M$ yields that $||\mathcal{T}^d \phi - \mathcal{T}^d \psi||_M \leq \rho K_P ||\phi - \psi||_M$. To show the second inequality, note that

$$|\mathcal{T}\phi(x) - \mathcal{T}\psi(x)| = \Big| \sup_{d \in D(x)} (\mathcal{T}^d \phi)(x) - \sup_{d \in D(x)} (\mathcal{T}^d \psi)(x) \Big|$$

$$\leq \sup_{d \in D(x)} \Big| (\mathcal{T}^d \phi)(x) - (\mathcal{T}^d \psi)(x) \Big|.$$

Using the first inequality and taking the norm $||\cdot||_M$, we can obtain the second inequality. To show that J^d is the unique fixed point of $\mathcal{T}^d$ in $\mathcal{B}_M$, note that the operator $\mathcal{T}^d$ is contracting on $\mathcal{B}_M$, then the result follows from Banach's fixed point theorem; see Powell (2011, Theorem 3.10.2). □

Making use of Proposition 6.1, the Bellman equation for the infinite horizon discounted MDP can be obtained as follows.

Theorem 6.9 *Let M be a measurable function satisfying condition (6.34) and $\rho K_P < 1$. Assume that there exists a closed subset $\mathcal{M} \subset \mathcal{B}_M$ and a set $\Delta \subset D$ such that (i) $0 \in \mathcal{M}$, (ii) $\mathcal{T} : \mathcal{M} \to \mathcal{M}$, and (iii) for all $\phi \in \mathcal{M}$ there exists a maximizer $d \in \Delta$ of ϕ. Then*

(a) $J_\infty \in \mathcal{M}$, $J_\infty = \mathcal{T} J_\infty$ and $J_\infty = J$, that is,

$$J(x) = \sup_{a \in D_t(x)} \Big(r(x,a) + \rho \int J(x') P(dx'|x,a) \Big), \qquad x \in \mathcal{X}.$$

(b) J_∞ is the unique fixed point of $\mathcal{T}$ in $\mathcal{M}$.

(c) J_∞ is the smallest function $\phi \in \mathcal{M}$ such that $\phi \geq \mathcal{T}\phi$.

(d) Let $g \in \mathcal{M}$. Then

$$||J_\infty - \mathcal{T}^t g||_M \leq \frac{(\rho K_P)^t}{1 - \rho K_P} ||\mathcal{T} g - g||_M.$$

(e) There exists a maximizer $d \in \Delta$ of J_∞, and every maximizer d^* of J_∞ defines an optimal stationary policy $(d^*, d^*, \dots)$.

The proof of the above theorem involves Banach's fixed point theorem, we omit the proof here and refer interested readers to Theorem 7.3.5 in Bäuerle and Rieder (2011). We want to highlight that part (d) of Theorem 6.9 suggests an important algorithm of finding optimal stationary policies for infinite-horizon MDP problems, which will be discussed in Section 6.3.2.

Example 6.3 (Linear-quadratic regulator in infinite horizon) *Consider the linear-quadratic problems in infinite horizon. Suppose that the state $\mathbf{x}_t \in \mathbb{R}^m$ follows the system transition function*

$$\mathbf{x}_{t+1} = \mathbf{A}\mathbf{x}_t + \mathbf{B}\mathbf{a}_t + \mathbf{z}_{t+1},$$

where $\mathbf{A} \in \mathbb{R}^{m \times m}$, $\mathbf{B} \in \mathbb{R}^{m \times d}$ are full rank state and action coefficient matrices, respectively, $\mathbf{a}_t \in \mathbb{R}^d$ is an action (or control) at time t, and $\mathbf{z}_1, \ldots, \mathbf{z}_t, \ldots$ are d-dimensional independent and identically distributed random vectors with Gaussian distribution $N_d(\mathbf{0}, \mathbf{\Sigma})$. The one-stage reward is negative cost function

$$r(\mathbf{x}, \mathbf{a}) = -(\mathbf{x}'\mathbf{Q}\mathbf{x} + 2\mathbf{x}'\mathbf{S}\mathbf{a} + \mathbf{a}'\mathbf{R}\mathbf{a}).$$

With the discount factor $\rho \in (0, 1]$, the expected infinite-horizon discounted cost under a policy $\pi = (\mathbf{a}_1, \ldots, \mathbf{a}_t, \ldots)$ is given by

$$V^\pi(\mathbf{x}) := E_x^\pi \left[\sum_{k=0}^{\infty} \rho^k r(\mathbf{x}_t, \mathbf{a}_t) \,\middle|\, \mathbf{x}_0 = \mathbf{x}, \pi \right].$$

The objective of the agent is to find an optimal policy π such that the above discounted cost is minimized.

Denote by $V(\mathbf{x})$ the value function at $\mathbf{x}$ and assume that $V(\mathbf{x})$ has a quadratic form for $\mathbf{x}$, i.e., $V(\mathbf{x}) = \mathbf{x}'\mathbf{\Lambda}\mathbf{x} + c$, where $\mathbf{\Lambda} \in \mathbb{R}^{m \times m}$ is a symmetric and positive definite matrix and $c \in \mathbb{R}$ is a scalar constant. Then the Bellman equation for this minimization problem is

$$V(\mathbf{x}) = \min_{\mathbf{a}} E\left[\mathbf{x}'\mathbf{Q}\mathbf{x} + 2\mathbf{x}'\mathbf{S}\mathbf{a} + \mathbf{a}'\mathbf{R}\mathbf{a} + \rho V(\mathbf{A}\mathbf{x} + \mathbf{B}\mathbf{a} + \mathbf{z})\right].$$

Follow the same procedure as in Section 6.2.4, we can show that the first order condition implies the following stationary optimal policy

$$\mathbf{a}^* = -(\mathbf{R} + \rho\mathbf{B}'\mathbf{\Lambda}\mathbf{B})^{-1}(\mathbf{S}' + \rho\mathbf{B}'\mathbf{\Lambda}\mathbf{A})\mathbf{x}. \tag{6.35}$$

Moreover, the matrix $\mathbf{\Lambda}$ satisfies the following discrete time algebraic Riccati equation,

$$\mathbf{\Lambda} = \mathbf{Q} + \rho\mathbf{A}'\mathbf{\Lambda}\mathbf{A} - (\mathbf{S}' + \rho\mathbf{B}'\mathbf{\Lambda}\mathbf{A})'(\mathbf{R} + \rho\mathbf{B}'\mathbf{\Lambda}\mathbf{B})^{-1}(\mathbf{S}' + \rho\mathbf{B}'\mathbf{\Lambda}\mathbf{A}). \tag{6.36}$$

and the constant c is given by $c = \rho(1 - \rho)^{-1} tr(\mathbf{\Lambda}\mathbf{\Sigma})$. □

6.3.2   Value and policy iteration

Value iteration is perhaps the most widely used algorithm in dynamic programming because it is the simplest to implement. It is also the most natural way of solving many infinite horizon MDP problems. It is analogous to backward dynamic programming for finite horizon MDP problems. The following algorithm is the basic version of value iteration.

Algorithm 6.4 (Value iteration for infinite horizon optimization) *Set* $V^{(0)}(x) = 0$ *for all* $x \in \mathcal{X}$ *and fix a tolerance parameter* $\epsilon > 0$. *Set* $n = 1$.

1. *For each* $x \in \mathcal{X}$ *compute*

$$V^{(n)}(x) = \max_{a \in \mathcal{A}} \left(r(x, a) + \rho \int P(dx'|x, a) V^{(n-1)}(x') \right). \qquad (6.37)$$

2. *If* $||V^{(n)}(x) - V^{(n-1)}(x)||_\infty < \epsilon(1-\rho)/(2\rho)$, *let* π^ϵ *be the resulting policy that solves (6.37), and let* $V^\epsilon = V^{(n)}$ *and stop. Otherwise set* $n \leftarrow n+1$ *and go to step 1.*

In the above algorithm, the max-norm of a function ϕ is defined by $||\phi||_\infty = \max_x |\phi(x)|$. An important property of Algorithm 6.4 is that if one's initial estimate is too low (or high), the algorithm will approach the correct value function from below (or above). Moreover, value iteration also provides a nice bound on the quality of the solution. These two properties can be formalized as follows and their rigorous proof can be found in Powell (2011, Sections 3.10).

Proposition 6.2 *Let* $V(x)$ *be the value function of the infinite horizon MDP problems. For a measurable function* $\phi : \mathcal{X} \to \mathbb{R}$,

(a) *if* ϕ *satisfies* $\phi \geq \mathcal{T}\phi$, *then* $\phi(x) \geq V(x)$;

(b) *if* ϕ *satisfies* $\phi \leq \mathcal{T}\phi$, *then* $\phi(x) \leq V(x)$;

(c) *if* ϕ *satisfies* $\phi = \mathcal{T}\phi$, *then* ϕ *is the unique solution to the Bellman equation (6.9) and* $\phi(x) = V(x)$ *for* $x \in \mathcal{X}$. □

Proposition 6.3 *Let* $V(x)$ *be the value function of the infinite horizon MDP problems. If we apply the value iteration Algorithm 6.4 with parameter* ϵ *and the algorithm terminates at iteration* n *with value function* $V^{(n+1)}$, *then*

$$||V^{(n+1)}(x) - V(x)||_\infty \leq \epsilon/2.$$

Let π^ϵ *be the policy that we terminate with, and let* V^{π^ϵ} *be the value of this policy. Then*
$$||V^{\pi^\epsilon}(x) - V(x)||_\infty \leq \epsilon.$$

□

Alternative to value iteration, one can also use policy iteration to find the optimal policy. In policy iteration, one chooses a policy and then finds the infinite horizon discounted value of the policy. The general policy iteration algorithm is given below.

Algorithm 6.5 (Policy iteration for infinite horizon optimization) *Select a policy* π^0 *and set* $n = 1$.

1. *Given a policy π^{n-1}. Compute the one-step transition kernel $P(dx'|x, \pi^{n-1}(x))$ and the one-step reward $r(x, \pi^{n-1}(x))$. Denote by $\phi = V^n$ the solution to the equation*

$$\phi(x) = r(x, \pi^{n-1}(x)) + \rho \int \phi(x') P(dx'|x, \pi^{n-1}(x)). \qquad (6.38)$$

2. *Find a policy π^n defined by*

$$d^n(x) = \arg\max_{a \in \mathcal{A}} \left(r(a) + \rho \int V^n(x') P(dx'|x, a) \right).$$

If $d^n(x) = d^{n-1}(x)$ for all states x, then set the optimal policy $d^* = d^n$. Otherwise, set $n \leftarrow n + 1$ and go to step 1.

Policy iteration algorithm works well for infinite horizon problems in which the value of a policy can be easily found. However, solving equation (6.38) is quite difficult if the number of states is large. Even if the state space is small so that the solution of (6.38) can be represented as that of a linear system, $\phi^\pi = (I - \rho P^\pi)^{-1} r^\pi$, matrix inversion can still be computationally expensive.

6.4 Partially observed Markov decision process

In many applications, the decision maker can only acquire partial information about the state process. Since the complete state remains unobserved, it is natural to assume that the admissible policies can only depend on the observed history. However, since the reward and the transition kernel may also depend on the unobserved part of the state, it becomes necessary to estimate these latent states. Provided that the *separation principle of estimation and control* holds in many cases, one can introduce an information process (filter) and reformulate the problem in terms of a MDP with complete information. In this section, we introduce a commonly used type of such processes, the *partially observed Markov decision process* (POMDP), and a filtering approach to find the optimal policy.

6.4.1 Finite-horizon POMDP

Denote by $T < \infty$ the planning horizon of the system and $t = 0, 1, \ldots, T$ the discrete time. The state space is expanded from $\mathcal{X}$ to $\mathcal{X} \times \mathcal{Y}$, in which $\mathcal{X}$ is the space of unobserved states and $\mathcal{Y}$ is the space of observable states. Then a finite horizon POMDP is defined as follows.

Definition 6.8 *A (discounted) finite horizon POMDP is the 7-tuple $(\mathcal{X}, \mathcal{Y}, \mathcal{A}, D, \{P_0, P\}, \{R, G\}, \rho)$ with the discount factor $\rho \in (0, 1]$ and the following specification:*

(i) *The state space is $\mathcal{X} \times \mathcal{Y}$, in which $\mathcal{X}$ and $\mathcal{Y}$ are finite or countable sets or Borel subsets of complete, separable metric spaces. For state $(x, y) \in \mathcal{X} \times \mathcal{Y}$, x is unobserved but y is observable.*

(ii) *The action space $\mathcal{A}$ is a finite or countable set or a Borel subset of complete, separable metric space.*

(iii) *$D \subset \mathcal{Y} \times \mathcal{A}$ is the set of possible state-action pairs. Let $D(y) = \{a \in \mathcal{A} | (y, a) \in D\}$ be the set of all state-action combinations depending only on the observable part $y \in \mathcal{Y}$.*

(iv) *P_0 is the initial distribution of X_0. $P : \mathcal{X} \times D \rightarrow \mathcal{X} \times \mathcal{Y}$ is a transition probability measure or the distribution of the new state given the current state and action.*

(v) *$R : \mathcal{X} \times \mathcal{Y} \times D \rightarrow \mathbb{R}$ and $G : \mathcal{X} \times \mathcal{Y} \rightarrow \mathbb{R}$ are measurable functions. In particular, $R(x, y, a)$ is the one-stage reward of the system in state (x, y) if action a is taken, and $G(x, y)$ gives the terminal reward of the system.* □

Let $T = \infty$, the above definition can be extended to infinite-horizon POMDP problems in which $G(x, y)$ degenerates to zero. For convenience, we will focus on finite-horizon POMDP here. In order to define policies involving unobserved states, we introduce the following definition.

Definition 6.9 *Denote the sets of observable histories by $\mathcal{H}_0 = \mathcal{Y}$ and $\mathcal{H}_t = \mathcal{H}_{t-1} \times \mathcal{A} \times \mathcal{Y}$ for $t = 1, 2, \ldots, T - 1$. Then an element $h_t = (y_0, a_0, y_1, \ldots, a_{t-1}, y_t) \in \mathcal{H}_t$ is called the observable history up to time t, and a measurable mapping $d_t : \mathcal{H}_t \rightarrow \mathcal{A}$ with the property $d_t(h_t) \in D(y_t)$ for $h_t \in \mathcal{H}_t$ is called a decision rule at stage t. Given d_t is a decision rule at stage t for all t, the sequence $\pi = (d_0, d_1, \ldots, d_{T-1})$ is called T-stage policy. The set of all T-stage policies is denoted by Π_T.* □

Note that the observed history up to stage $t + 1$ can be represented recursively, i.e., $h_{t+1} = (h_t, a_t, y_{t+1})$. This can be very convenient in deriving the filtering equation later. Given an initial state $y \in \mathcal{Y}$ and a T-stage policy $\pi = (d_0, d_1, \ldots, d_{T-1}) \in \Pi_T$, a probability measure P_y^π on $(\mathcal{X} \times \mathcal{Y})^{T+1}$ with the product σ-algebra can be defined by using the initial conditional distribution P_0 and the transition probability P. Let $\omega = (x_0, y_0, \ldots, x_T, y_T) \in (\mathcal{X} \times \mathcal{Y})^{T+1}$ be a trajectory of unobserved and observed states over all stages. One can define the random variables X_t and Y_t by the projection of ω_t onto the t-stage, i.e., $X_t(\omega) = x_t$ and $Y_t(\omega) = y_t$, and the action at the t-stage by

$$A_0 := d_0(Y_0), \quad A_t := d_t(Y_0, A_0, Y_1, \ldots, Y_t).$$

Then the T-stage optimization problem is defined as follows. Given $\pi \in \Pi_T$ and $Y_0 = y$, define the performance or the expected discounted total reward as

$$J_T^\pi(y) := \int E_{x,y}^\pi \Big[\sum_{t=0}^{T-1} \rho^t R(X_t, Y_t, A_t) + \rho^T G(X_T, Y_T) \Big] P_0(dx). \qquad (6.39)$$

Assume that for all $y \in \mathcal{Y}$, the following condition holds

$$\sup_{\pi} \int E_{x,y}^{\pi} \left[\sum_{t=0}^{T-1} \rho^t |R(X_t, Y_t, A_t)| + \rho^T |G(X_T, Y_T)| \right] P_0(dx) < \infty.$$

The objective of the agent is to find an optimal policy π so that the expected discounted total reward is maximized, i.e.,

$$J_t(y) = \sup_{\pi \in \Pi_T} J_T^{\pi}(y). \tag{6.40}$$

6.4.2 Filter equations

An important step in the analysis of POMDP is to find the conditional distribution of the unobserved state X_t given the information set of observable states and actions $\sigma(Y_0, A_0, Y_1, \ldots, A_{t-1}, Y_t)$. This conditional distribution can be computed recursively and such recursion is referred to as a *filter equation*.

In order to derive such equation, we first denote by $\mathcal{P}(\mathcal{X})$ the space of all probability measures on $\mathcal{X}$ and write the transition probability $P(x', y'|x, y, a)$ as

$$P(x', y'|x, y, a) = q(x', y'|x, y, a)\lambda_X(dx')\lambda_Y(dy'),$$

in which q is the density of the transition kernel Q with respect to some σ-finite measures λ_X and λ_Y. Then we define a *Bayes operator* $\Phi : \mathcal{Y} \times \mathcal{P}(\mathcal{X}) \times \mathcal{A} \times \mathcal{Y} \to \mathcal{P}(\mathcal{X})$ such that

$$\Phi(y, \nu, a, y')(B) := \frac{\int_B \left(\int q(x', y'|x, y, a)\nu(dx) \right) \lambda_X(dx')}{\int_{\mathcal{X}} \left(\int q(x', y'|x, y, a)\nu(dx) \right) \lambda_X(dx')}, \quad B \in \mathcal{B}(\mathcal{X}), \tag{6.41}$$

in which $\mathcal{B}(\mathcal{X})$ is a Borel subset of the Borel space $\mathcal{P}(\mathcal{X})$. Given the above operator and the recursive representation of the observed history $h_{t+1} = (h_t, a_t, y_{t+1})$, one can define the following equation for $B \in \mathcal{B}(\mathcal{X})$,

$$\mu_0 := P_0, \qquad \mu_{t+1}(B|h_{t+1}) = \Phi(y_t, \mu_t(\cdot|h_t), a_t, y_{t+1})(B). \tag{6.42}$$

The recursion (6.42) is called a *Bayes filter equation*. The following theorem shows that μ_t is actually a conditional distribution of X_t given $(Y_0, A_0, Y_1, \ldots, A_{t-1}, Y_t)$.

Theorem 6.10 (Bayes filter equation) *For all policies $\pi \in \Pi_T$, let $H_t = (Y_0, A_0, Y_1, \ldots, A_{t-1}, Y_t)$, then we have for $B \in \mathcal{B}(\mathcal{X})$,*

$$\mu_t(B|H_t) = P_y^{\pi}(X_n \in B|H_t).$$

Proof. First, for any $h_t \in \mathcal{H}_t$, $a_t \in D(y_t)$, and measurable function ν : $\mathcal{H}_t \times \mathcal{X} \times \mathcal{Y} \to \mathbb{R}$, the Bayes operator (6.41) implies that

$$\Phi(y_t, \nu, a, y_{t+1})(dx_{t+1}) \int q(x, y_{t+1}|x_t, y_t, a)\lambda_X(dx) = q(x_{t+1}, y_{t+1}|x_t, y_t, a)\lambda_X(dx_{t+1}).$$

Using this equation and Fubini's theorem, we have

$$\int \mu_t(dx_t|h_t) \int P(x_{t+1}, y_{t+1}|x_t, y_t, a_t)\nu(h_t, x_{t+1}, y_{t+1})$$

$$= \int \mu_t(dx_t|h_t) \int \lambda_Y(dy_{t+1}) \int \lambda_X(dx_{t+1})q(x_{t+1}, y_{t+1}|x_t, y_t, a_t)\nu(h_t, x_{t+1}, y_{t+1})$$

$$= \int \mu_t(dx_t|h_t) \int \lambda_Y(dy_{t+1}) \int \Big[\Phi(y_t, \nu, a, y_{t+1})(dx_{t+1})$$

$$\int q(x, y_{t+1}|x_t, y_t, a)\lambda_X(dx) \Big] \nu(h_t, x_{t+1}, y_{t+1})$$

$$= \int \mu_t(dy_t|h_t) \int P^Y(dy_{t+1}|x_t, y_t, a_t) \int \Phi(y_t, \mu_t, a_t, y_{t+1})(dx_{t+1})\nu(h_t, x_{t+1}, y_{t+1}),$$

where $P^Y(y_{t+1}|x_t, y_t, a_t)$ is the marginal distribution of Y_{t+1} conditional on (x_t, y_t, a_t).

For any measurable functions ν : $\mathcal{H}_t \times \mathcal{X} \to \mathbb{R}$, let $\nu'(h_t) := \int \nu(h_t, x_t)\mu_t(dx_t|h_t)$. We next show that

$$E_y^\pi[\nu(h_t, X_t)] = E_y^\pi[\nu'(h_t)]. \tag{6.43}$$

First for $t = 0$, we have $E_y^\pi[\nu(h_0, X_1)] = \int \nu(h_0, x)P_0(dx) = E_y^\pi[\nu'(h_0)]$. Given an observable history h_t, let $a_t = d_t(h_t)$, then

$$E_y^\pi[\nu(h_{t+1}, X_{t+1})] = \int \mu_t(dx_t|h_t) \int P(x_{t+1}, y_{t+1}|x_t, y_t, a_t)\nu(h_t, a_t, y_{t+1}, x_{t+1})$$

$$= \int \mu_t(dy_t|h_t) \int P^Y(dy_{t+1}|x_t, y_t, a_t) \int \Phi(y_t, \mu_t, a_t, y_{t+1})(dx_{t+1})\nu(h_t, a_t, y_{t+1}, x_{t+1})$$

$$= \int \mu_t(dy_t|h_t) \int P^Y(dy_{t+1}|x_t, y_t, a_t)\nu'(h_t, a_t, y_{t+1}) = E_y^\pi[\nu'(h_{t+1})],$$

in which the third equality is obtained by the definition of μ_t. Let $\nu = 1_{C \times B}$ in (6.43). We obtain

$$P_y^\pi(H_t \in C, X_t \in B) = E_y^\pi\big[1_C(H_t)\mu_t(B|H_t)\big].$$

That is, $\mu_t(\cdot|H_t)$ is a conditional P_y^π-distribution of X_t given H_t. □

Theorem 6.10 shows that, if ν is a conditional distribution of X_t, then $\Phi(y, \nu, a, y')$ defined by (6.41) is a conditional distribution of X_{t+1} given ν, the current observation y, action a and the next observation y'. We next show some examples on Bayes filters.

Example 6.4 (Finite-state hidden Markov model) *Consider a finite-state hidden Markov process $(\mathcal{X}, \mathcal{Y}, \{P_0, P\})$, in which the hidden state $x_t \in$*

$\mathcal{X} = \{1, \ldots, K\}$ *is a finite-state Markov chain. Let* $y_0 = y$ *be the initial observable state and* $\mu_0(k)$ *the initial distribution of* $x_0 = k$. *At each time* t, *the transition probabilities of* x_t *given a control* a_t *is* $p_{ij}^a = P(x_{t+1} = j | x_t = i, a_t = a)$, *and the distribution of observation* y_t *depends only the hidden state* x_t. *Denote by* $\lambda_i(y)$ *the density of* $y_t = y$ *conditional on* $x_t = i$ *and assume that the conditional distributions of* y_t *given* x_t *are independent over* t. *To find the filter equation for this problem, we note that the transition probability for the state process* $\{(x_t, y_t)\}_{t \geq 1}$ *is*

$$P(y_{t+1} \in B, x_{t+1} = j \mid y_t, x_t = i, a_t = a) = p_{ij}^a P(y_{t+1} \in B | x_{t+1} = j) = p_{ij}^a \int_B \lambda_j(y) dy.$$

Let ν *be a probability measure of* x *on* $\mathcal{X}$ *with* $P(x = i) = \nu_i$ *and* $\nu_1 + \cdots + \nu_K = 1$. *The Bayes operator for this model is given by*

$$\Phi(y, \nu, a, y')(\{k\}) = \frac{\left(\sum_{i=1}^K \nu_i p_{ik}^a \right) \lambda_k(y')}{\sum_{j=1}^K \left(\sum_{i=1}^K \nu_i p_{ij}^a \right) \lambda_j(y')}, \qquad k \in \mathcal{X}.$$

Then Theorem 6.10 implies the filter equation or the conditional distribution of x_t *given* H_t *is expressed as*

$$\mu_t(\{k\}|H_t) = P_y^\pi(X_t = k|H_t).$$

$\square$

6.4.3 Partially observable LQR and Kalman filter

Consider a controlled *linear state-space system* in which the state and observation equations are

$$\text{State equation:} \quad \mathbf{x}_{t+1} = \mathbf{A}\mathbf{x}_t + \mathbf{B}\mathbf{a}_t + \mathbf{z}_{t+1}, \qquad (6.44a)$$

$$\text{Observation equation:} \quad \mathbf{y}_{t+1} = \mathbf{C}\mathbf{x}_{t+1} + \mathbf{v}_{t+1}, \qquad (6.44b)$$

for $t \geq 0$. In the above equations, the state $\mathbf{x}_{t+1} \in \mathbb{R}^m$ are unobserved states, $\mathbf{a}_t \in \mathbb{R}^d$ are controls, and $\mathbf{y}_{t+1} \in \mathbb{R}^q$ are observations. The coefficient matrices $\mathbf{A} \in \mathbb{R}^{m \times m}$, $\mathbf{B} \in \mathbb{R}^{m \times d}$, and $\mathbf{C} \in \mathbb{R}^{q \times m}$ are full rank constant matrices. For each t, the noises $\mathbf{z}_t$ and $\mathbf{v}_t$ are independent m- and q-variate Gaussian random vectors with mean $\mathbf{0}$ and covariance matrices $\boldsymbol{\Sigma}$ and $\mathbf{V}$, respectively. The series $\{\mathbf{z}_t\}$ and $\{\mathbf{v}_t\}$ are mutually independent. Besides, we also assume that the unobserved state $\mathbf{x}_0$ follows a m-variate Gaussian distribution $N_m(\boldsymbol{\mu}_0, \mathbf{P}_0)$.

The agent's objective is to choose the control $\{\mathbf{a}_t\}$ which is a function of the observation process $\{\mathbf{y}_s; s \leq t\}$ to minimize the cost

$$\min_{\mathbf{a}_1, \ldots, \mathbf{a}_{T-1}} E^\pi \left[\sum_{t=0}^{T-1} \left(\mathbf{x}_t' \mathbf{Q}\mathbf{x}_t + 2\mathbf{x}_t' \mathbf{S}\mathbf{a}_t + \mathbf{a}_t' \mathbf{R}\mathbf{a}_t \right) + \mathbf{x}_T' \mathbf{Q}_T \mathbf{x}_T \Big| \boldsymbol{\mu}_0, \mathbf{P}_0 \right]. \quad (6.45)$$

The minimization problem (6.45) is different from problem (6.26) in that the state $\mathbf{x}_t$ here is unobserved and the optimal control is a function of observation process, instead of state $\mathbf{x}_t$ directly.

The procedure of solving (6.45) consists of two steps. The first step is to use the Kalman filter to find a filtering solution for state $\mathbf{x}_t$, and the second step is to use the separation principle to find the optimal control. The Kalman filter, denoted by $\widehat{\mathbf{x}}_{t|t}$, is the *minimum-variance linear estimate* of $\mathbf{x}_t$ based on the observations $\mathbf{y}_1, \ldots, \mathbf{y}_t$ up to stage t (Kalman, 1960). It can be represented recursively as below.

Theorem 6.11 (Kalman filter for linear state-space system) *Consider the linear state-space system (6.44a) and (6.44b). Denote by* $\mathbf{h}_t = (\mathbf{a}_0, \mathbf{y}_1, \ldots, \mathbf{a}_{t-1}, \mathbf{y}_t)$ *the control and observation processes up to time t. Suppose that the agent chooses a sequence of control* $\{\mathbf{a}_0, \ldots, \mathbf{a}_t, \ldots\}$ *such that* $\mathbf{a}_t$ *is a function of the observation process* $\{\mathbf{y}_s; s \leq t\}$, *then the linear space spanned by* $\mathbf{h}_t$ *is the same as the one spanned by* $\{\mathbf{y}_s; s \leq t\}$. *Let* $\mathbf{x}_{t|t} = E(\mathbf{x}_t|\mathbf{h}_t)$ *and* $\mathbf{x}_{t|t-1} = E(\mathbf{x}_t|\mathbf{h}_{t-1})$ *be the minimum-variance linear predictor of* $\mathbf{x}_t$ *based on* $\mathbf{h}_t$ *and* $\mathbf{h}_{t-1}$, *respectively. Let* $\mathbf{e}_{t|t} = \mathbf{x}_t - E(\mathbf{x}_t|\mathbf{h}_t)$ *and* $\mathbf{e}_{t|t-1} = \mathbf{x}_t - E(\mathbf{x}_t|\mathbf{h}_{t-1})$ *be the filter and prediction error at time t, respectively. Define the Kalman gain matrix at time t by*

$$\mathbf{K}_t = \mathbf{P}_{t|t-1}\mathbf{C}'(\mathbf{C}\mathbf{P}_{t|t-1}\mathbf{C}' + \mathbf{V})^{-1}. \tag{6.46a}$$

Then the filter and prediction of state $\mathbf{x}_t$ *at time t are recursively given by*

$$\mathbf{x}_{t|t} = \mathbf{x}_{t|t-1} + \mathbf{K}_t(\mathbf{y}_t - \mathbf{C}\mathbf{x}_{t|t-1}). \tag{6.46b}$$
$$\mathbf{x}_{t+1|t} = \mathbf{A}\mathbf{x}_{t|t} + \mathbf{B}\mathbf{a}_t, \tag{6.46c}$$

in which $\mathbf{P}_{t|t-1} = Cov(\mathbf{e}_{t|t-1}|\mathbf{h}_{t-1})$ *and* $\mathbf{P}_{t|t} = Cov(\mathbf{e}_{t|t}|\mathbf{h}_t)$ *are the covariance matrices of the filter and prediction errors at time t, respectively. They can be expressed recursively as*

$$\mathbf{P}_{t|t} = (\mathbf{I} - \mathbf{K}_t\mathbf{C})\mathbf{P}_{t|t-1}, \tag{6.46d}$$
$$\mathbf{P}_{t+1|t} = \mathbf{A}\mathbf{P}_{t|t}\mathbf{A}' + \mathbf{\Sigma}. \tag{6.46e}$$

The initial value of $\mathbf{x}_{0|0}$ *and* $\mathbf{P}_{t|t}$ *are given by* $\mathbf{x}_{0|0} = \boldsymbol{\mu}_0$ *and* $\mathbf{P}_{t|t} = \mathbf{P}_0$. $\square$

To show the filter and the prediction formulas (6.46b)-(6.46e), we need to use the following proposition, which can be easily shown; see Exercise 6.7.

Proposition 6.4 *Suppose that random vectors* $\mathbf{x}$, $\mathbf{y}$, *and* $\mathbf{z}$ *are jointly Gaussian such that* $\mathbf{y}$ *and* $\mathbf{z}$ *are independent. Denote by* $\mathbf{\Sigma}_{\mathbf{xy}} = E\{(\mathbf{x} - E\mathbf{x})(\mathbf{y} - E\mathbf{y})'\}$ *the covariance matrix of* $\mathbf{x}$ *and* $\mathbf{y}$, *and denote similarly* $\mathbf{\Sigma}_{\mathbf{xx}}$ *and* $\mathbf{\Sigma}_{\mathbf{yy}}$. *Then the conditional distribution of* $\mathbf{x}$ *given* $\mathbf{y}$ *is also Gaussian with the following mean and covariance matrix*

$$E(\mathbf{x}|\mathbf{y}) = E(\mathbf{x}) + \mathbf{\Sigma}_{\mathbf{xy}}\mathbf{\Sigma}_{\mathbf{yy}}^{-1}(\mathbf{y} - E\mathbf{y}), \qquad Cov(\mathbf{x}|\mathbf{y}) = \mathbf{\Sigma}_{\mathbf{xx}} - \mathbf{\Sigma}_{\mathbf{xy}}\mathbf{\Sigma}_{\mathbf{yy}}^{-1}\mathbf{\Sigma}_{\mathbf{yx}}.$$

Proof of Theorem 6.11. We first consider the prediction equations. By (6.44a), equation (6.46c) holds and therefore $\mathbf{e}_{t+1|t} = \mathbf{A}\widetilde{\mathbf{x}}_{t|t} + \mathbf{z}_{t+1}$, which is a sum of two independent terms. Then the covariance of $\mathbf{e}_{t+1|t}$ is given by (6.46e).

For the filter equations, we note that, under the assumption that $\mathbf{a}_{t-1}$ is a function of observation history $\mathbf{y}_1, \ldots, \mathbf{y}_{t-1}$, the information set generated by the control and observation history $\mathbf{h}_t = (\mathbf{h}_{t-1}, \mathbf{a}_{t-1}, \mathbf{y}_t)$ is the same as that generated by $(\mathbf{h}_{t-1}, \widetilde{\mathbf{y}}_{t|t-1})$, where

$$\widetilde{\mathbf{y}}_{t|t-1} = \mathbf{y}_t - E(\mathbf{y}_t|\mathbf{h}_{t-1}) = \mathbf{C}\mathbf{e}_{t|t-1} + \mathbf{v}_t.$$

This implies that $\mathrm{Cov}(\widetilde{\mathbf{y}}_{t|t-1}|\mathbf{h}_{t-1}) = \mathbf{C}\mathbf{P}_{t|t-1}\mathbf{C}' + \mathbf{V}$ and

$$\mathrm{Cov}(\mathbf{x}_t, \widetilde{\mathbf{y}}_{t|t-1}|\mathbf{h}_{t-1}) = E(\mathbf{x}_t\widetilde{\mathbf{y}}'_{t|t-1}|\mathbf{h}_{t-1}) = E(\mathbf{x}_t\mathbf{e}'_{t|t-1}|\mathbf{h}_{t-1})\mathbf{C}' = \mathbf{P}_{t|t-1}\mathbf{C}'.$$

By Proposition 6.4, we obtain that

$$\begin{aligned}
\mathbf{x}_{t|t} &= E(\mathbf{x}_t|\mathbf{h}_{t-1}, \widetilde{\mathbf{y}}_{t|t-1}) \\
&= E(\mathbf{x}_t|\mathbf{h}_{t-1}) + \mathrm{Cov}(\mathbf{x}_t, \widetilde{\mathbf{y}}_{t|t-1}|\mathbf{h}_{t-1})\left[\mathrm{Cov}(\widetilde{\mathbf{y}}_{t|t-1}|\mathbf{h}_{t-1})\right]^{-1}\widetilde{\mathbf{y}}_{t|t-1}.
\end{aligned}$$

Then $\widetilde{\mathbf{y}}_{t|t-1} = \mathbf{y}_t - \mathbf{C}\widehat{\mathbf{x}}_{t|t-1}$ implies that (6.46b) holds. Moreover, by Proposition 6.4,

$$\begin{aligned}
\mathrm{Cov}(\mathbf{x}_t|\mathbf{h}_t) = \mathrm{Cov}(\mathbf{x}_t|\mathbf{h}_{t-1}) &- \mathrm{Cov}(\mathbf{x}_t, \widetilde{\mathbf{y}}_{t|t-1}|\mathbf{h}_{t-1}) \\
&\left[\mathrm{Cov}(\widetilde{\mathbf{y}}_{t|t-1}|\mathbf{h}_{t-1})\right]^{-1}\mathrm{Cov}(\mathbf{x}_t, \widetilde{\mathbf{y}}_{t|t-1}|\mathbf{h}_{t-1})',
\end{aligned}$$

then (6.46d) follows immediately. $\square$

Provided the Kalman filter in Theorem 6.11, we can extend the method of solving the linear-quadratic regulator in Section 6.2.4 to the minimization problem (6.45). We first rewrite the state and observation equations (6.44a) and (6.44b) in the following form

$$\begin{array}{llr}
\text{State equation:} & \mathbf{x}_{t+1} = \mathbf{A}\widehat{\mathbf{x}}_{t|t} + \mathbf{B}\mathbf{a}_t + \widetilde{\mathbf{z}}_{t+1}, & (6.47\text{a}) \\
\text{Observation equation:} & \mathbf{y}_{t+1} = \mathbf{C}\widehat{\mathbf{x}}_{t+1|t} + \widetilde{\mathbf{v}}_{t+1}, & (6.47\text{b})
\end{array}$$

where $\widetilde{\mathbf{z}}_{t+1} = \mathbf{z}_{t+1} + \mathbf{A}\mathbf{e}_{t|t}$ and $\widetilde{\mathbf{v}}_{t+1} = \mathbf{v}_{t+1} + \mathbf{C}\mathbf{e}_{t+1|t}$. Since $\mathbf{z}_{t+1}$ and $\mathbf{e}_{t|t}$ are independent, we have

$$E(\widetilde{\mathbf{z}}_{t+1}|\mathbf{h}_t) = \mathbf{0}, \qquad \mathrm{Cov}(\widetilde{\mathbf{z}}_{t+1}|\mathbf{h}_t) = \mathbf{A}\mathbf{P}_{t|t}\mathbf{A}' + \mathbf{\Sigma} = \mathbf{P}_{t+1|t}.$$

Similarly, by the independence of $\mathbf{v}_{t+1}$ and $\mathbf{e}_{t+1|t}$, we have

$$E(\widetilde{\mathbf{v}}_{t+1}|\mathbf{h}_t) = \mathbf{0}, \qquad \mathrm{Cov}(\widetilde{\mathbf{v}}_{t+1}|\mathbf{h}_t) = \mathbf{C}\mathbf{P}_{t+1|t}\mathbf{C}' + \mathbf{V}.$$

Let $V_t(\mathbf{h}_t)$ be the value function at $\mathbf{h}$ at time t. The Bellman equation for the minimization problem (6.45) is

$$V_t(\mathbf{h}_t) = \min_{\mathbf{a}} E\left[\mathbf{x}'_{t|t}\mathbf{Q}\mathbf{x}_{t|t} + 2\mathbf{x}'_{t|t}\mathbf{S}\mathbf{a} + \mathbf{a}'\mathbf{R}\mathbf{a} + \widetilde{\mathbf{z}}'_{t+1}\mathbf{Q}\widetilde{\mathbf{z}}_{t+1} + V_{t+1}(\mathbf{h}_{t+1})\Big|\mathbf{h}_t\right].$$

Assume that the value function at time t has a quadratic form for the filtering estimate of $\mathbf{x}_t$, $V_t(\mathbf{h}_t) = \mathbf{x}'_{t|t}\boldsymbol{\Lambda}_t\mathbf{x}_{t|t} + c_t$, where $\boldsymbol{\Lambda}_t \in \mathbb{R}^{m \times m}$ is a symmetric and positive definite matrix and $c_t \in \mathbb{R}$ is a scalar constant. Then the Bellman equation becomes

$$V_t(\mathbf{h}_t) = \min_{\mathbf{a}} E\Big[\mathbf{x}'_{t|t}\mathbf{Q}\mathbf{x}_{t|t} + 2\mathbf{x}'_{t|t}\mathbf{S}\mathbf{a} + \mathbf{a}'\mathbf{R}\mathbf{a} + \widetilde{\mathbf{z}}'_{t+1}\mathbf{Q}\widetilde{\mathbf{z}}_{t+1}$$

$$+ (\mathbf{A}\mathbf{x}_{t|t} + \mathbf{B}\mathbf{a} + \widetilde{\mathbf{z}}_{t+1})'\boldsymbol{\Lambda}_{t+1}(\mathbf{A}\mathbf{x}_{t|t} + \mathbf{B}\mathbf{a} + \widetilde{\mathbf{z}}_{t+1}) + c_{t+1}\Big|\mathbf{h}_t\Big].$$
(6.48)

Recall that $\mathbf{a}_t$ is a function of $\{\mathbf{y}_s; s \leq t\}$ or $\mathbf{h}_t$, the first order condition on $\mathbf{a}_t$ implies that

$$\mathbf{S}'\mathbf{x}_{t|t} + \mathbf{R}\mathbf{a} + \mathbf{B}'\boldsymbol{\Lambda}_{t+1}(\mathbf{A}\mathbf{x}_{t|t} + \mathbf{B}\mathbf{a}) = 0.$$

Solving the above equation yields the optimal control at time t

$$\mathbf{a}_t^* = -(\mathbf{R} + \mathbf{B}'\boldsymbol{\Lambda}_{t+1}\mathbf{B})^{-1}(\mathbf{B}'\boldsymbol{\Lambda}_{t+1}\mathbf{A} + \mathbf{S}')\widehat{\mathbf{x}}_{t|t}.$$
(6.49)

Substitute $\mathbf{a}_t^*$ back into (6.48), we obtain the Riccati difference equation for $\mathbf{P}_t$,

$$\boldsymbol{\Lambda}_t = \mathbf{Q} + \mathbf{A}'\boldsymbol{\Lambda}_{t+1}\mathbf{A} - (\mathbf{B}'\boldsymbol{\Lambda}_{t+1}\mathbf{A} + \mathbf{S}')'(\mathbf{R} + \mathbf{B}'\boldsymbol{\Lambda}_{t+1}\mathbf{B})^{-1}(\mathbf{B}'\boldsymbol{\Lambda}_{t+1}\mathbf{A} + \mathbf{S}'),$$

which is exactly same as the Riccati difference equation (6.29). Besides, the scalar c_{t+1} satisfies the recursion

$$c_t = c_{t+1} + E(\widetilde{\mathbf{z}}'_{t+1}(\mathbf{Q} + \boldsymbol{\Lambda}_{t+1})\widetilde{\mathbf{z}}_{t+1}) = c_{t+1} + \text{tr}[(\mathbf{Q} + \boldsymbol{\Lambda}_{t+1})\mathbf{P}_{t+1|t}].$$

Together with the terminal condition

$$\boldsymbol{\Lambda}_T = \mathbf{Q}_T, \qquad c_T = 0,$$

one can solve for $\{\boldsymbol{\Lambda}_t, c_t; t = T - 1, \ldots, 1\}$ backward.

The two-step procedure above is known as the *separation* or *certainty-equivalence* principle and is one of the important results in Markov decision process. It separates a partially observed LQR problem into two stages: computation of the filter $\mathbf{x}_{t|t}$ and computation of the control value $\mathbf{a}_t$ based on $\mathbf{x}_{t|t}$. The crucial point is that these two operations are independent in the sense that the Kalman filter does not depend on the matrices $\mathbf{Q}, \mathbf{S}, \mathbf{R}$ in the cost function, whereas the control function (6.49) does not depend on parameters $\mathbf{C}, \boldsymbol{\Sigma}, \mathbf{V}$ and behaves as if $\mathbf{x}_{t|t}$ were the actual state.

6.4.4 Reformulation as a MDP

The POMDP can be reformulated as a standard MDP by adding the filtered probability distribution of X_t given H_t and enlarging the state space of the problem. Suppose that $(\mathcal{X}, \mathcal{Y}, A, D, \{P_0, P\}, R, G, \rho)$ is a POMDP in Definition 6.8. We define the following stationary T-stage MDP.

Definition 6.10 *The filtered MDP is a 6-tuple* $(\mathcal{E}, \mathcal{A}, D, \widetilde{P}, \{\widetilde{R}, \widetilde{G}\}, \rho)$ *with the discount factor $\rho \in (0, 1]$ and the following elements*

(i) $\mathcal{E} = \mathcal{Y} \times \mathcal{P}(\mathcal{X}) = \{(y, \nu) | y \in \mathcal{Y}, \nu \in \mathcal{P}(\mathcal{X})\}$ *is the state space. For $(y, \nu) \in \mathcal{E}$, y is the observable state of the POMDP and ν is the conditional distribution of the unobservable state.*

(ii) $\mathcal{A}$ *is the action space.*

(iii) $D \subset \mathcal{E} \times \mathcal{A}$ *is the set of possible state-action pairs. For $y \in \mathcal{Y}$ and $\nu \in \mathcal{P}(\mathcal{X})$, $D(y, \nu)$ is the set of actions in state (y, ν).*

(iv) $\widetilde{P}$ *is a stochastic kernel which determines the distribution of the new state defined as follows. For $(y, \nu) \in \mathcal{E}$, $a \in D(y)$ and Borel subsets $B \subset \mathcal{Y}$ and $C \in \mathcal{P}(\mathcal{X})$, we have*

$$\widetilde{P}(B \times C | y, \nu, a) := \int \int_B 1_C\big(\Phi(y, \nu, a, y')\big) P(dy' | x, y, a)\nu(dx).$$

(v) $\widetilde{R} : D \to \mathbb{R}$ *is the one-stage reward given by $\widetilde{R}(y, \nu, a) := \int R(x, y, a)\nu(dx)$, and $\widetilde{G} : \mathcal{E} \to \mathbb{R}$ is the terminal reward given by $\widetilde{G}(y, \nu) := \int G(x, y)\nu(dx)$.* □

The policies for the filtered MDP can be defined in the same way as in Section 6.2.1. Note that a policy $\pi \in \Pi_T$ now depends on the observable history and is generally not Markovian. However, if $(\tilde{d}_0, \dots, \tilde{d}_{T-1})$ is a Markov policy for the filtered MDP, then $(d_0, \dots, d_{T-1}) \in \Pi_T$ for $d_t(h_t) := \tilde{d}_t(y_t, \mu_t(\cdot | h_t))$. This shows that the set of all Markov policies for the filtered MDP is a subset of Π_T. For $(y, \nu) \in \mathcal{E}$ and $\pi \in \Pi_T$, the performance function for the filtered Markov decision process can be defined as in Section 6.2.1 and denoted by

$$\widetilde{J}_T^\pi(y, \nu) := E_{y,\nu}\Big[\sum_{t=0}^{T-1} \rho^t \widetilde{R}(y_t, \nu_t, a_t) + \widetilde{G}(y_T, \nu_T) \Big],$$

and the value function can be defined as

$$\widetilde{J}_T(y, \nu) = \sup_{\pi \in \Pi_T} \widetilde{J}_T^\pi(y, \nu).$$

In principle, when a POMDP problem is reformulated as a MDP problem, it can be solved by the theory for MDP problems in Section 6.2.1. However, when the numbers of states and actions get larger, it becomes difficulty to implement this procedure due to the curse of dimensionality.

6.5   Value function based reinforcement learning

This and next sections present some commonly used reinforcement learning methods. In this section, we focus on methods of estimating the value function $V(x)$ for some MDPs, which include temporal difference learning, Q-learning and SARSA algorithms. These algorithms are also referred to as *critic-only* methods.

6.5.1   Temporal difference learning

We first introduce *temporal difference* (TD) learning (Sutton, 1988) which is one of the most significant ideas in reinforcement learning. TD learning is a combination of Monte Carlo ideas and dynamic programming ideas. TD methods can, on the one hand, learn directly from samples without models of the environment's dynamics, and on the other hand, update estimates of the value function based in part on other learned estimates via bootstrap. To illustrate the idea, we first consider the case of small finite MDPs in which the estimates of the value function on all states can be stored in an array or table, which is called the *tabluar case* in the reinforcement learning literature. We present the most basic TD algorithm, explain how bootstrapping works and then the TD(λ) algorithm that unifies the bootstrapping and Monte Carlo techniques. The case involving large state spaces, where a tabular representation becomes infeasible, will be addressed in subsequent discussions.

Tabular TD(0)

Consider the infinite horizon stationary MDP in Definition 6.7 and the expected infinite-horizon discounted reward (6.31). The reward iteration in Theorem 6.3 and the stationarity of the discounted reward problems imply that

$$V^\pi(x) = E\big[r(x,a) + \rho V^\pi(x_{t+1})\big|x_t = x\big].$$

The TD learning involves two basic ideas in each step, one is to sample the expected values in the above equation and the other is to use the current estimate of the value function instead of the true value function.

Assume that the MDP has a finite state space. Given a realization of state $\{\mathbf{x}_t\}$ and immediate reward process $\{r_{t+1}\}$, we wish to estimate the value function $V(x)$. Let $\widehat{V}^{(t)}(x)$ be the estimate of the value function in the tth step and $\widehat{V}^{(0)}(x) \equiv 0$. In the tth step, the TD(0) or the 1-step TD learning performs the following calculation for $x \in \mathcal{X}$,

$$\widehat{V}^{(t+1)}(x_t) = \widehat{V}^{(t)}(x_t) + \alpha_t(r_{t+1} + \rho\widehat{V}^{(t)}(x_{t+1}) - \widehat{V}^{(t)}(x_t)), \qquad (6.50)$$

where the step sequence $\{\alpha_t\}_{t\geq0}$ consists of small nonnegative numbers chosen by the user. The update equation (6.50) implies that the only value changed is the one associated with x_t or the state just visited. The quantity in the bracket,

$$\delta_{t+1} := r_{t+1} + \rho \widehat{V}^{(t)}(x_{t+1}) - \widehat{V}^{(t)}(x_t), \tag{6.51}$$

is a sort of error which measures the difference between the estimated value function on x_t and the better estimate $r_{t+1} + \rho \widehat{V}^{(t)}(x_{t+1})$. δ_{t+1} is called the *TD error*. Sometimes, the 1-step TD learning can be replaced by the n-step TD learning. In such case, the TD error is represented as the difference between the n-step return prediction $r_t^{(n)}$ and the estimated value function on x_t, i.e.,

$$\delta_{t+1} := r_t^{(n)} - \widehat{V}^{(t)}(x_t), \qquad r_t^{(n)} = \sum_{k=1}^{n-1} \rho^{k-1} r_{t+k} + \rho^{n+1} \widehat{V}_t(x_{t+n}).$$

Then the value function is updated by $\widehat{V}^{(t+1)}(x_t) = \widehat{V}^{(t)}(x_t) + \alpha_t \delta_{t+1}$.

Tabluar TD(0) is a *stochastic approximation* (SA) algorithm. Define the following operator

$$\mathcal{F}\widehat{V}(x) := E\big[r_{t+1} + \rho \widehat{V}(x_{t+1}) - \widehat{V}(x_t) \big| x_t = x\big].$$

It is easy to see that $\mathcal{F}\widehat{V} = \mathcal{T}\widehat{V} - \widehat{V}$, where $\mathcal{T}$ is the Bellman operator defined before. By Theorem 6.9, $\mathcal{F}\widehat{V} = 0$ has a unique solution, which is the value function V. This indicates that, if the TD(0) algorithm converges, it must be converge to $V(x)$. Actually, if the step-size sequence $\{\alpha_t\}$ satisfies the *Robbins-Monro conditions*

$$\sum_{t=0}^{\infty} \alpha_t = \infty, \qquad \sum_{t=0}^{\infty} \alpha_t^2 < \infty,$$

one can show that the estimate $\widehat{V}(x)$ converges almost surely to $V(x)$.

TD(λ)

The convergence rate of the TD(0) algorithm sometimes becomes slower when the number of states becomes larger. In such case, one can use the so called *TD(λ)* method which averages all of the possible n-step returns into a single return. In particular, the targets of the TD(λ) update are given as the sum of the n-step return predictions $r_t^{(n)}$ with the exponential weight $(1 - \lambda)\lambda^{n-1}$,

$$R_t^{\lambda} := (1 - \lambda) \sum_{n=1}^{\infty} \lambda^{n-1} r_t^{(n)}, \qquad r_t^{(n)} = \sum_{k=1}^{n-1} \rho^{k-1} r_{t+k} + \rho^{n+1} \widehat{V}_t(x_{t+n}).$$

For $\lambda > 0$, the algorithm is a multi-step method. Consider the following *eligibility trace*,

$$z_0(x) = 0, \qquad z_{t+1}(x) = 1_{\{x = x_t\}} + \rho \lambda z_t(x),$$

which can be used as a scaling factor on the TD error. Then the update equation (6.50) is modified to

$$\widehat{V}^{(t+1)}(x) = \widehat{V}^{(t)}(x) + \alpha_t \delta_{t+1} z_{t+1}(x), \quad \delta_{t+1} = R_t^\lambda - \widehat{V}^{(t)}(x_t). \qquad (6.52)$$

The idea here is that the value of $z_t(x)$ modulates the influence of the TD error on the update of the value function on state x. Note that there are other ways to define the eligibility traces; see, for example, the replacing trace (Singh and Sutton, 1996).

In these updates, the *trace-decay parameter* λ controls the amount of bootstrapping. When $\lambda = 0$, the above algorithms are identical to TD(0) since

$$\lim_{\lambda \to 0+} (1 - \lambda) \sum_{k \geq 0} \lambda^k r_t^k = r_t^{(0)} = r_{t+1} + \rho \widehat{V}_t(x_{t+1}).$$

By tuning λ appropriately, the algorithm can converge much faster than Monte Carlo and TD(0) methods.

6.5.2   Q-learning

Q-learning is another popular algorithm for problems with small state and action spaces. It is named after the variable $Q(x, a)$ which is an estimate of the state-action value function (6.12) of being in a state x and taking action a. Let the Q-factor $\widehat{Q}^{(n)}(x, a)$ be an estimate of the true value $Q(x, a)$ after n iterations. Consider the state x_n. We choose action a_n using

$$a_n = \arg\max_{a \in \mathcal{A}} \widehat{Q}^{(n-1)}(x_n, a). \qquad (6.53)$$

Note that in the above optimization problem, we do not need to compute an expectation, nor to estimate the state in the future. After we choose an action a_n, a reward $\widehat{r}(x_n, a_n)$ and the next state x_{n+1} are observed. Then we can compute an updated estimate of the state-action value function

$$\widetilde{Q}^{(n)}(x_n, a_n) = \widehat{r}(x_n, a_n) + \gamma \max_{a' \in \mathcal{A}} \widehat{Q}^{(n-1)}(x_{n+1}, a'), \qquad (6.54)$$

and the Q-factors are updated as follows

$$\widehat{Q}^{(n)}(x_n, a_n) = (1 - \alpha_{n-1})\widehat{Q}^{(n-1)}(x_n, a_n) + \alpha_{n-1}\widetilde{Q}^{(n)}(x_n, a_n), \qquad (6.55)$$

where $\{\alpha_n\}_{n \geq 1}$ is a stepsize sequence taking values between 0 and 1.

Given a set of Q-factors, the state value function in a state x can be estimated by

$$\widehat{V}^{(n)}(x) = \max_{a \in \mathcal{A}} \widehat{Q}^{(n)}(x_n, a).$$

Note that with the above notation, (6.54) can be rewritten as

$$\widetilde{Q}^{(n)}(x_n, a_n) = \widehat{r}(x_n, a_n) + \gamma \widehat{V}^{(n-1)}(x_{n+1}). \qquad (6.56)$$

Comparing this with the update equation (6.37), we see that Q-learning depends on observations from an exogenous process and doesn't involve the transition kernel.

One concern on the optimization equation (6.53) is that the resulting algorithm may not lead to an optimal solution. The reason is that the Q-factors $\widehat{Q}^{(n)}(x, a)$ might underestimate the state-action value function at (x, a) so that the action of taking the system to state x may not be chosen. Thus, we need a rule that forces us to explore states and actions that are not visited often enough. In such case, we can use the so-called ϵ-*greedy policy* to choose an action. Using this policy, we choose an action randomly from the action space $\mathcal{A}$ with probability ϵ and choose the action based on (6.53) with probability $1 - \epsilon$, i.e.,

$$
a_n = \begin{cases} a' \in \mathcal{A} & \text{with probability } \epsilon. \\ \arg\max_{a \in \mathcal{A}} \widehat{Q}^{(n-1)}(x_n, a) & \text{with probability } 1 - \epsilon. \end{cases}
$$

The ϵ-greedy policy allows us not only exploiting our current knowledge of the state-action value function, but also exploring other states and actions as well.

6.5.3 SARSA algorithm

State-action-reward-state-action (SARSA) is an algorithm that is closely related to Q-learning, and works as follows. Suppose the system is in a state x and the agent chooses an action a. Then the agent observes a reward r, and after that, the system moves to the next state x' and then the agent chooses an action a'. The sequence x, a, r, x' and a' suggests the name SARSA to the algorithm. The difference between the Q-learning and SARSA algorithms is that SARSA chooses an action following the same current policy and updates its Q-values whereas Q-learning chooses the action that gives the maximum Q-value for the state.

Suppose the action a_n is chosen by some policy d^π, that is, $a_n = d^\pi(x_n)$. We can use the transition kernel to simulate the next state $x_{n+1} \sim P(\cdot|x_n, a_n)$ and choose the next action $a_{n+1} = d^\pi(x_{n+1})$. Then we compute a sample estimate of the state-action value as

$$
\widetilde{Q}^{(n)}(x_n, a_n) = \widehat{r}(x_n, a_n) + \gamma \widehat{Q}^{(n-1)}(x_{n+1}, a_{n+1}),
$$

and update $\widehat{Q}^{(n)}(x_n, a_n)$ as (6.55) as the Q-learning.

The SARSA algorithm allows us to learn the value of a fixed policy $d^\pi(x)$. It requires that all actions are tested infinitely often for the policy and the new state is sampled with the correct distribution determined by the policy. Instead of searching for an optimal policy, the SARSA algorithm only estimates the value of a policy. This property makes the algorithm very useful in policy iteration algorithms.

On-policy and off-policy learning

In reinforcement learning algorithms, the policy that determines which action to take so that the next state is determined is called the *behavior policy*, and the policy that determines the action that appears to be the best is often referred to as the *target policy* or the *learning policy*. The behavior policy is sometimes called the *sampling policy* because one may use this policy to control the process of sampling states. The goal of reinforcement learning is to improve the target policy, during which a sampling policy is used to ensure that states are visited often enough.

When the learning policy and the sampling policy are the same, it is called *on-policy learning*. Conversely, when the learning policy and sampling policy differ, it is referred to as *off-policy learning*. Q-learning and SARSA are good examples of off-policy and on-policy learning, respectively. In the SARSA algorithm, the same policy $d^\pi(x)$ is used to choose action a_n for the current state x_n and to choose the next action $a_{n+1} = d^\pi(x_{n+1})$ for the next state x_{n+1}. As long as all actions are chosen infinitely often, Q-factors will be guaranteed to converge to the correct value for state x and the chosen action $a = d^\pi(x)$. In contrast, Q-learning may use exploration strategies such as the ϵ-greedy policy to choose an current action and then maximize the Q-factor to choose the next action. In other words, Q-learning uses one policy to decide which action to evaluate and another policy to choose the action in the future.

When to use on-policy or off-policy learning is an important concern in reinforcement learning as both have their own issues. When the on-policy learning is used, one may not be able to ensure that the states sampled by the algorithm are enough to estimate the value functions of these states, whereas when the off-policy learning is used, the use of the sampling policy may influence the ability of learning the value of a learning policy.

6.6 Policy-based reinforcement learning

This section introduces policy-based reinforcement learning methods. We first present policy gradient methods that typically work with a parameterized family of policies over which optimization procedures can be used directly. Policy gradient algorithms are in principle *actor-only* as approximation to value functions are not involved. We then introduce *actor-critic* methods that combine the advantages of actor-only and critic-only algorithms.

6.6.1 Policy gradient methods

Policy gradient algorithms are perhaps the most popular class of *continuous action* reinforcement learning algorithms. Assume that the policy is stochastic and denoted it by $\pi_{\boldsymbol{\theta}} : \mathcal{X} \to \mathcal{P}(\mathcal{A})$, where $\mathcal{P}(\mathcal{A})$ is the set of probability measures on $\mathcal{A}$, $\boldsymbol{\theta} \in \mathbb{R}^n$ is a vector of n parameters, and $\pi_{\boldsymbol{\theta}}(a_t|x_t)$ is the conditional probability density at a_t associated with the policy. The basic idea behind policy gradient methods is to adjust the parameters $\boldsymbol{\theta}$ of the policy in the direction of the performance gradient $\nabla_{\boldsymbol{\theta}} J(\pi_{\boldsymbol{\theta}})$. A fundamental result underlying these algorithms is the following theorem (Sutton et al., 1999).

Theorem 6.12 (Policy gradient) *For any infinite-horizon MDP with discounted or average total reward $J(\pi)$, the performance gradient is given by*

$$\nabla_{\boldsymbol{\theta}} J(\pi_{\boldsymbol{\theta}}) = \int_{\mathcal{X}} d^{\pi}(x) \int_{\mathcal{A}} \nabla_{\boldsymbol{\theta}} \pi_{\boldsymbol{\theta}}(u|x) Q^{\pi}(x, u) du dx, \qquad (6.57)$$

where $d^{\pi}(x)$ is the distribution of the action given the state x and $Q^{\pi}(x, u)$ is the state-action value function.

Theorem 6.12 has important practical value, because it reduces the computation of the performance gradient to a simple expectation. Applying Theorem 6.12 and using standard optimization techniques, a locally optimal solution of the reward J can be found. The gradient $\nabla J(\pi_{\boldsymbol{\theta}})$ is estimated at each time step, and the parameters are then updated in the direction of this gradient. For example, a gradient ascent method would yield the policy gradient update equation

$$\boldsymbol{\theta}_{k+1} = \boldsymbol{\theta}_k + \alpha_{a,k} \nabla_{\boldsymbol{\theta}} J_k, \qquad (6.58)$$

where $\alpha_{a,k} > 0$ is a small enough learning rate. By (6.58), we will have the total reward $J(\boldsymbol{\theta}_{k+1}) \geq J(\boldsymbol{\theta}_k)$. Convergence is obtained if the estimated gradients are unbiased and the learning rates $\alpha_{a,k}$ satisfy the Robbins-Montro conditions in Section 6.5.1.

The gradient can be estimated by several methods, for instance, likelihood-ratio methods or infinitesimal perturbation analysis (Glynn, 1987). A drawback of the policy gradient approach is that the estimated gradient may have a large variance (Sutton et al., 1999). Besides, every gradient is computed without using any knowledge of past estimates.

Policy gradient with function approximation

In most applications, the state-action space is continuous, hence the state or the state-action value function needs to be approximated. Konda and Tsitsiklis (2003) and Sutton et al. (1999) showed that the state-action or the critic function $Q^{\pi}(x, a)$ can be approximated by a function $h_{\boldsymbol{\omega}} : X \times U \to \mathbb{R}$, where $\boldsymbol{\omega}$ is a parameter. Here, we assume that $\boldsymbol{\omega}$ doesn't affect the unbiasedness of the policy gradient estimate.

To find the closest approximation function $h_{\boldsymbol{\omega}}$, we can estimate $\boldsymbol{\omega}$ by using the least square method and minimizing the quadratic error

$$L_{\boldsymbol{\omega}}^{\pi}(x, u) = \frac{1}{2} \left[Q^{\pi}(x, u) - h_{\boldsymbol{\omega}}(x, u) \right]^2.$$

The gradient of this quadratic error with respect to $\boldsymbol{\omega}$ is

$$\nabla_{\boldsymbol{\omega}} L_{\boldsymbol{\omega}}^{\pi}(x, u) = \left[Q^{\pi}(x, u) - h_{\boldsymbol{\omega}}(x, u)\right]\nabla_{\boldsymbol{\omega}}h_{\boldsymbol{\omega}}(x, u). \tag{6.59}$$

One can use this in a gradient descent algorithm to find the optimal $\boldsymbol{\omega}$. If the estimator of $Q^{\pi}(x, u)$ is unbiased, the expected value of (6.59) is zero for the optimal $\boldsymbol{\omega}$. Then we obtain the following policy gradient theorem with function approximation (Sutton et al., 1999).

Theorem 6.13 (Policy gradient with function approximation) *Let $\pi_{\boldsymbol{\theta}}(x, u)$ be the stochastic policy parameterized by $\boldsymbol{\theta}$. Suppose that a function $h_{\boldsymbol{\omega}}$ satisfies*

$$\int_{\mathcal{X}} d^{\pi}(x) \int_{\mathcal{A}} \pi(x, u)\nabla_w L_{\boldsymbol{\omega}}^{\pi}(x, u) = 0 \tag{6.60}$$

and

$$\nabla_{\boldsymbol{\omega}} h_{\boldsymbol{\omega}}(x, u) = \nabla_{\boldsymbol{\theta}} \log \pi_{\boldsymbol{\theta}}(x, u). \tag{6.61}$$

Then

$$\nabla_{\boldsymbol{\theta}} J = \int_{\mathcal{X}} \int_{\mathcal{A}} \nabla \pi_{\boldsymbol{\theta}}(x, u) h_{\boldsymbol{\omega}}(x, u) du dx. \tag{6.62}$$

This theorem shows that, once a good parameterization for a policy has been found, a parameterization for the value function automatically follows and also guarantees convergence. When $h_{\boldsymbol{\omega}}$ is further assumed to be linear with respect to parameter $\boldsymbol{\omega}$ and features ψ, i.e., $h_{\boldsymbol{\omega}} = \boldsymbol{\omega}'\psi(x, u)$, equation (6.61) turns into

$$\psi(x, u) = \nabla_{\boldsymbol{\theta}} \log \pi_{\boldsymbol{\theta}}(x, u).$$

Features $\psi(x, u)$ satisfying the above equation are called as *compatible* features; see Konda and Tsitsiklis (2003). Then we obtain a compatible function approximation $h_{\boldsymbol{\omega}} = \boldsymbol{\omega}'\nabla_{\boldsymbol{\theta}} \log \pi_{\boldsymbol{\theta}}(x, u)$ and

$$\int_{\mathcal{U}} \pi_{\boldsymbol{\theta}}(x, u) h_{\boldsymbol{\omega}}(x, u) du = \boldsymbol{\omega}'\nabla_{\boldsymbol{\theta}} \left(\int_{\mathcal{U}} \pi_{\boldsymbol{\theta}}(x, u) du \right) = \boldsymbol{\omega}'\nabla_{\boldsymbol{\theta}}1 = 0.$$

This indicates that h_w is generally better considered as the *advantage function* (i.e., the difference between the state-action and state value functions) $A^{\pi}(x, u) = Q^{\pi}(x, u) - V^{\pi}(x)$; see Sutton et al. (1999) and Peters and Schaal (2008).

The above discussion suggests that the approximation with compatible features can only be used to represent the relative value of an action u in some state x correctly, but not the absolute value $Q(x, u)$. Hence, extra features need to be added to the gradient estimate, and such added state-dependent offset is often called an *reinforcement baseline*. For instance, denote by $b(x)$ the baseline function that can be chosen arbitrarily, equation (6.62) becomes

$$\nabla_{\boldsymbol{\theta}} J = \int_{\mathcal{X}} \int_{\mathcal{A}} \nabla \pi_{\boldsymbol{\theta}}(x, u) \left[h_{\boldsymbol{\omega}}(x, u) + b(x)\right] du dx.$$

Adding such baseline will not affect the unbiasedness of the gradient estimate, but can improve the accuracy of the approximation to the critic function. In practice, the optimal baseline should be the one that minimizes the variance in the gradient estimate for the policy π, which equals the value function $V^\pi(x)$ (Bhatnagar et al., 2009).

Natural policy gradients

The above policy-gradient algorithms use the concept of standard gradients, which is very effective for reward functions that have a single maximum and whose gradients are isotropic in any direction away from the maximum (Amari and Douglas, 1998). In many applications, the reward function may have multiple local maxima. Moreover, the performance of the policy-gradient algorithms that use standard gradients may heavily depend on the choice of the coordinate system over which the reward is defined. In such cases, using standard gradients in policy-gradient algorithms may cause problems and hence "natural gradients" may become necessary for policy-gradient algorithms.

The notion of a natural gradient is the following. Suppose that a function $J_{\boldsymbol{\theta}}(x)$ is parameterized by $\boldsymbol{\theta}$. When $\boldsymbol{\theta} \in \mathbb{R}^n$ is the Euclidean space, the squared Euclidean norm of a small increment $\Delta\boldsymbol{\theta}$ is given by

$$||\Delta\boldsymbol{\theta}||^2 = (\Delta\boldsymbol{\theta})'(\Delta\boldsymbol{\theta}).$$

When $\boldsymbol{\theta}$ is transformed to other coordinates $\boldsymbol{\vartheta} \in \mathbb{R}^n$, the squared norm of a small increment $\Delta\boldsymbol{\vartheta}$ with respect to the Riemannian space is

$$||\Delta\boldsymbol{\vartheta}||^2 = (\Delta\boldsymbol{\vartheta})'G(\boldsymbol{\vartheta})(\Delta\boldsymbol{\vartheta}),$$

in which Riemannian metric tensor $G(\boldsymbol{\vartheta})$ is an $n \times n$ positive-definite matrix and can be found from the relationship $||\Delta\boldsymbol{\theta}||^2 = ||\Delta\boldsymbol{\vartheta}||^2$. This suggests that the natural gradient $\widetilde{\nabla}_{\boldsymbol{\vartheta}} J(\boldsymbol{\vartheta})$ of the reward function is a linear transformation of the standard gradient $\nabla_{\boldsymbol{\vartheta}} J(\boldsymbol{\vartheta})$ and expressed as

$$\widetilde{\nabla}_{\boldsymbol{\vartheta}} J(\boldsymbol{\vartheta}) = G^{-1}(\boldsymbol{\vartheta})\nabla_{\boldsymbol{\vartheta}} J(\boldsymbol{\vartheta});$$

see Amari and Douglas (1998). Note that the natural and standard gradients coincide with each other only if $G(\boldsymbol{\vartheta})$ is the identity matrix. Therefore, if the parameter space of $\boldsymbol{\theta}$ is a Riemannian manifold other than the Euclidean space, natural gradients should be used in the policy gradient algorithms.

6.6.2 Actor-critic algorithms

As explained before, both actor-only and critic-only methods have their own pros and cons. Actor-only methods deal with a parameterized family of policies over which optimization procedures can be used directly, but the estimated gradient may have a large variance so that the learning may be slow. Critic-only methods that use TD learning have a lower variance in the estimates of value functions, but the computational cost might be expensive, especially if the action space is continuous.

Actor-critic methods combine the advantages of actor-only and critic-only algorithms. On the one hand, the parameterized actor (or policies) has the advantage of computing continuous actions without the need for optimization procedures on value functions, and on the other hand, the critic (or value functions) provide the actor with low-variance knowledge of the performance. Moreover, actor-critic methods usually have better convergence properties than critic-only methods (Konda and Tsitsiklis, 2003).

To illustrate the idea of actor-critic algorithms, we assume that the value or the state-action function is a linear function of basis functions $\phi(x)$, i.e.,

$$V_{\xi} = \xi' \phi(x). \tag{6.63}$$

Let $\pi_{\theta}(x)$ be the deterministic policy parameterized by θ. Suppose that the parameter θ is estimated by the policy gradient update (6.58). In order to make the critic evaluate accurately a given policy so that the best policy can be found, we need to find an approximate solution to the Bellman equation for that policy. Let $(x_k, u_k, r_{k+1}, x_{k+1})$ be a transition sample. Consider the TD(0) learning method for the value function. The TD error is expressed as

$$\delta_k = r_{k+1} + \gamma V_{\xi_k}(x_{k+1}) - V_{\xi_k}(x_k); \tag{6.64}$$

see also (6.51). To update the critic or the value function in (6.63), we use the TD error δ_k in a gradient descent update

$$\xi_{k+1} = \xi_k + \alpha_{c,k} \delta_k \nabla_{\xi} V_{\xi_k}(x_k) = \xi_k + \alpha_{c,k} \delta_k \phi(x_k), \tag{6.65}$$

where $\alpha_{c,k} > 0$ is the learning rate of the critic. Then in such case, the actor-critic template for the discounted return is given by (6.58), (6.64), and (6.65).

Besides the TD(0) learning, TD(λ) learning and other ways of determining the critic parameter ξ can be used; see Grondman et al. (2012)'s survey on actor-critic algorithms. One issue that is missing in the above discussion is how to choose right features for the actor and the critic, respectively. In many applications, this still remains a difficult problem.

Supplements and problems

The first multi-armed bandit problem was studied by Thompson (1933) for the application of clinical trial. The earlier study of this type of sequential learning problems was referred to as "sequential design of experiments," see for example, Robbins (1952) and Bellman (1956). The "multi-armed bandit" first appeared in studies in the late 1950s to early 1960s, and since then, an enormous body of work has accumulated over time. Various subsets of the topic has been summarized and reviewed. Among these presentations, Gittins et al. (2011) discuss the Gittins index for various Markovian models of multi-armed bandits and subsequent studies of various sequential resource allocation and stochastic scheduling problems. For comprehensive overviews on multi-armed bandit problem, we refer interested readers to Slivkins (2019), Zhao (2019), and Lattimore and Szepesvári (2020).

There are many excellent treatises on the subject of discrete-time optimal control and dynamic programming. The following serves as a biased selection from the literature. Bertsekas and Shreve (1978) gives a comprehensive and theoretically sound treatment of the mathematical foundations of stochastic optimal control of discrete-time systems. Bertsekas (2012) provides a complete coverage of dynamic programming methods for deterministic and stochastic control problems, in both discrete- and continuous-time. Bäuerle and Rieder (2011) presents rigorously the theory of MDP and its applications in finance and insurance. In addition to the above, Krishnamurthy (2016) describes methods for POMDP problems and its applications from engineering point of view.

Concerning methods for reinforcement learning, Powell (2011) provides a comprehensive overview to approximate dynamic programming or reinforcement learning. Bertsekas (2019) explains methods of optimal control and reinforcement learning, as well as their connections and difference. Sewak (2019) combines deep learning and reinforcement learning and explains some cutting-edge algorithms for deep reinforcement learning. Sutton and Barto (2018) give a broad and in-depth introduction to reinforcement learning, which is a standard reference for students, researchers, and practitioners in machine learning.

1. (**Thompson sampling**). Show that the reward and the state transition of an active arm in Example 6.1 are given by (6.3) and (6.4), respectively.

2. Show that the three operators in Definition 6.6 are monotone. That is, for $\phi, \psi \in \mathcal{M}(\mathcal{X})$ with $\phi(x) \leq \psi(x)$ for all $x \in \mathcal{X}$, (6.16a)-(6.16c) hold.

3. Show Theorem 6.7 by following the proof of Theorem 6.3.

4. Show Theorem 6.8 by following the proof of Theorem 6.4.

5. (**Binary MDP**) Consider a stationary MDP $(\mathcal{X}, \mathcal{A}, \{P(x'|x,a)\}, R(x,a), G(x))$, in which $\mathcal{A} = \{0,1\}$, the reward functions $R(x,a)$ and $G(x)$ are bounded. Let $R(x,1) = R_1(x)$ and $R(x,0) = r_0(x)$ for $x \in \mathcal{X}$. For a bounded and measurable function $\phi : \mathcal{X} \to \mathbb{R}$, denote $(P_a\phi)(x) := \int \phi(y)P(dy|x,a)$ for $x \in \mathcal{X}$ and J_t its value function at time t.

    (a) Let $\mathcal{L}_a J_{t-1} = r_a + \rho Q_a J_{t-1}$, $a = 0,1$. Show that the value function satisfies $J_t = \max\{r_0 + \rho Q_0 J_{t-1}, r_1 + \rho Q_1 J_{t-1}\} = \max\{\mathcal{L}_0 J_{t-1}, \mathcal{L}_0 J_{t-1}\}$.

(b) Denote $d_t(x) = \mathcal{L}_1 J_{t-1}(x) - \mathcal{L}_0 J_{t-1}(x)$ for $x \in \mathcal{X}, t = 1, \ldots, T-1$. Show that

$$d_{t+1} = \mathcal{L}_1 \mathcal{L}_0 J_{t-1} - \mathcal{L}_0 \mathcal{L}_1 J_{t-1} + \rho Q_1 d_t^+ - \rho Q_0 d_t^-,$$

where $y^+ = \max(y, 0)$ and $y^- = -\min(y, 0)$.

6. Consider the infinite horizon linear-quadratic regulator problem in Example 6.3. Show the stationary optimal policy is given by (6.35) and the matrix $\mathbf{P}$ satisfies the discrete time Riccati equation (6.36).

7. Prove Proposition 6.4.

8. (**Partially observable discounted infinite-horizon LQR**) Consider the linear state-space system (6.44a) and (6.44b). The model parameters are same as those specified in Section 6.4.3. The agent's objective is to choose $\{\mathbf{a}_t\}$ which is a function of the observation process $\{\mathbf{y}_s; s \leq t\}$ to minimize the discounted cost

$$\min_{\pi = \{\mathbf{a}_1, \ldots, \mathbf{a}_t, \ldots\}} E^{\pi} \left[\sum_{t=0}^{\infty} \rho^t \left(\mathbf{x}_t' \mathbf{Q} \mathbf{x}_t + 2\mathbf{x}_t' \mathbf{S} \mathbf{a}_t + \mathbf{a}_t' \mathbf{R} \mathbf{a}_t \right) \middle| \boldsymbol{\mu}_0, \mathbf{P}_0 \right]. \quad (6.66)$$

(a) Show that the optimal control $\mathbf{a}_t$ at time t is

$$\mathbf{a}_t^* = -(\mathbf{R} + \rho \mathbf{B}' \boldsymbol{\Lambda} \mathbf{B})^{-1}(\mathbf{S}' + \rho \mathbf{B}' \boldsymbol{\Lambda} \mathbf{A}) \mathbf{x}_{t|t}, \quad (6.67)$$

where $\mathbf{x}_{t|t}$ is the Kalman filter given in Theorem 6.11.

(b) Assume that the value function of the minimization problem (6.66) has the form $V(\mathbf{x}_{t|t}) = \mathbf{x}_{t|t}' \boldsymbol{\Lambda} \mathbf{x}_{t|t} + c$, where $\boldsymbol{\Lambda} \in \mathbb{R}^{m \times m}$ is a symmetric and positive definite matrix and c is a scalar constant. Show that $\boldsymbol{\Lambda}$ satisfies the discrete time algebraic Riccati equation (6.36).

Chapter 7

Monte Carlo methods and rare event analytics

Monte Carlo methods are basic computational tools widely used in financial engineering and risk management. They have been used to simulate various stochastic process, evaluate numerical integration, and calculate small probabilities of rare adverse events. To explain the idea, let ν be a probability measure over the Borel σ-field $\mathcal{X}$ on the sample space $\mathcal{X} \subset \mathbb{R}^d$, where $\mathbb{R}^d$ denotes the d-dimensional Euclidean space. A commonly encountered challenging problem is to evaluate integrals of the form

$$E[h(\mathbf{X})] = \int_{\mathcal{X}} h(\mathbf{x})\nu(d\mathbf{x}),$$

where $h(\mathbf{x})$ is a measurable function. Suppose that ν has a density function $f(\mathbf{x})$, then the above integral can be written as

$$E[h(\mathbf{X})] = \int_{\mathcal{X}} h(\mathbf{x})f(\mathbf{x})d\mathbf{x}. \tag{7.1}$$

Let $\mathbf{X}_1, \ldots, \mathbf{X}_n$ be a sample of size n and assume that they can be easily simulated from $f(\mathbf{x})$. Then the sample mean of $h(\mathbf{X}_1), \ldots, h(\mathbf{X}_n)$, denoted as

$$\overline{h}_n = \frac{1}{n} \sum_{i=1}^{n} h(\mathbf{X}_i),$$

can be used to approximate (7.1) because $\overline{h}$ converges to (7.1) almost surely by the Strong Law of Large Numbers. When $h(\mathbf{X})$ has a finite variance, the error of this approximation can be characterized by the central limit theorem, that is,

$$\frac{\overline{h} - E_f[h(\mathbf{X})]}{\sqrt{n\text{Var}(h(\mathbf{X}))}} \xrightarrow{d} N(0,1).$$

The variance term $\text{Var}(h(\mathbf{X}))$ can be approximated by the sample variance

$$\frac{1}{n-1} \sum_{i=1}^{n} (h(\mathbf{X}_i) - \overline{h})^2.$$

This method of approximating integrals by simulated samples is known as the

DOI: 10.1201/9781315117041-7

Monte Carlo method, which have been widely used in finance and insurance. For instance, let $h(x) = 1_{\{x \in S\}}$ for $S \in \mathcal{X}$, one can evaluate the probability $P(X \in S)$ of the event $X \in S$.

This chapter provides an overview of basic Monte Carlo methods and introduces some advancements in sequential Monte Carlo methods. In particular, Sections 7.1–7.3 cover methods for random variable generation, stochastic process generation, and variance reduction. Section 7.4 introduces Markov chain Monte Carlo (MCMC) methods and their applications to Bayesian inference. Section 7.5 discusses importance sampling methods for rare-event simulation. Section 7.6 introduces sequential importance sampling and summarizes some major advancements that have proven to be very powerful methods for rare event simulation.

7.1 Random variable generation

Monte Carlo methods reply on sampling from probability distributions. Generating a sample of i.i.d. draws on computer from the simplest continuous uniform $\mathcal{U}(0,1)$ is fundamentally important because all sampling methods depend on uniform random number generators. Many pseudo-random number generators, e.g., linear congruential generators, are available for this purpose.

To simulate a univariate random variable X with continuous distribution $f(x)$ and cumulative distribution function $F(x)$, one can in principle use the *inverse-transform* method. In particular, one can first generate a uniform random variable $U \sim \mathcal{U}(0,1)$ and then use $X = F^{-1}(U)$ to generate a random variable with distribution function F.

For discrete univariate distribution $p(x)$ with distribution function $F(x)$, one can first generate a uniform random variable U and then find X such that $F(X-1) < U \leq F(X)$. For example, to generate a Bernoulli random variable such that $P(X = 1) = p = 1 - P(X = 0)$, we can first generate U and set $X = 0$ if $U < 1 - p$, setting $X = 1$ otherwise.

7.1.1 Transformation methods

Since many distributions can be written as simple operations of certain random variables, simulation of those distributions can be obtained via some transformations. The following lists some examples of such types of distributions and simulation methods.

- *Affine transformation.* Let $\mathbf{X} = (x_1, \ldots, x_d)'$ be a random vector, $\mathbf{A}$ is an $n \times d$ matrix, and $\mathbf{b}$ is a $n \times 1$ vector. The $n \times 1$ random vector $\mathbf{Y} = \mathbf{A}\mathbf{X} + \mathbf{b}$ is said to be an *affine transformation* of $\mathbf{X}$. Then $\mathbf{Y}$ can be simulated by first generating a sample from the distribution of $\mathbf{X}$

and then using the affine transformation. A special case of this is the location-scale family; see Exercise 7.1.

- *Reciprocal transformation.* If X is a univariate random variable, then the inverse or reciprocal of X is $Z = X^{-1}$. If X has the density function f_X, Z has the density function $f_Z(z) = f_X(z^{-1})/z^2$ for $z \in \mathbb{R}$. Similarly, if $\mathbf{X}$ is an $n \times n$ invertible random matrix with the density function $f_{\mathbf{X}}$, then the inverse matrix $\mathbf{Z} = \mathbf{X}^{-1}$ has the density function $f_{\mathbf{Z}}(\mathbf{z}) = f_{\mathbf{X}}(\mathbf{Z}^{-1})/|\det(J(\mathbf{z}))|$ for $\mathbf{z} \in \mathbb{R}^{n \times n}$, where $|\det(J(\mathbf{z}))|$ is the absolute value of the determinant of the matrix of Jacobi corresponding to the transformation $\mathbf{x} \mapsto \mathbf{z} = \mathbf{x}^{-1}$. See Exercise 7.2.

- *Truncation.* Let F_A and F_B be two distribution functions on sets A and $B \subset A$, respectively. Let $\mathbf{X} \sim F_A$, then the distribution of X truncated to B, denoted by $\mathbf{Z}$, has the density function $f_{\mathbf{Z}}(\mathbf{z}) = f_{\mathbf{X}}(\mathbf{z})/\int_B f_{\mathbf{X}}(\mathbf{x})d\mathbf{x}$ for $\mathbf{z} \in B$. See Exercise 7.3.

The rest of the section introduces some methods that are based on drawing i.i.d. samples from a target distribution. Section 7.2 introduces methods for cases that generating i.i.d. samples from a target distribution is infeasible so that dependent samples shall be used instead.

7.1.2   Acceptance-rejection methods

The method above is based on transformations of random variables. In many cases, the transformation methods are inefficient. For instance, in the case where the density function is specified up to a normalizing constant, i.e., the density function is $f(x)/\int_{-\infty}^{\infty} f(t)dt$. Without computing the normalizing constant and the distribution function F, one can use the following *acceptance-rejection* method to generate $X \sim F(x)$.

Algorithm 7.1 (Acceptance-rejection) *To simulate a $X \sim F(x)$, first find an alternative probability distribution G with density function $g(x)$. Assume that the ratio $f(x)/g(x)$ is bounded by a constant $c > 0$.*

1. *Generate a random variable Y with density function g.*

2. *Generate independently $U \sim \mathcal{U}(0,1)$. If $U \leq f(Y)/(cg(Y))$, accept Y as X. Otherwise reject Y and repeat the procedure until $U \leq f(Y)/(cg(Y))$ occurs.*

Since $P(U \leq u) = u$, it follows that

$$P\left(Y \leq x \big| U \leq \frac{f(Y)}{cg(Y)}\right) = \left(\int_{-\infty}^{x} g(y)\frac{f(y)}{cg(y)}dy\right) \Big/ \int_{-\infty}^{\infty} \frac{f(y)}{cg(y)}g(y)dy$$

$$= \int_{-\infty}^{x} f(y)dy \Big/ \int_{-\infty}^{\infty} f(y)dy = F(x),$$

and therefore $X \sim F$.

Example 7.1 (A truncated Normal distribution) *Suppose we want to generate a $N(0, 1)$ distribution which is bounded away from 0 from below. Denote this random variable by X, its density function has the form of*

$$f(x) = \frac{2}{\pi} e^{-x^2/2}, \qquad x \geq 0.$$

To use the acceptance-rejection method to generate a sample from X, we can bound $f(x)$ by $cg(x)$, where $g(x) = e^{-x}$ is the probability density function of the exponential distribution with mean 1 and $c = \sqrt{2e/\pi}$.

7.1.3 Multivariate distributions and copulas

We begin by considering simulating a jointly normal $p \times 1$ random vector $\mathbf{X} \sim N(\boldsymbol{\mu}, \boldsymbol{\Sigma})$. One can rewrite $\mathbf{X}$ as $\mathbf{X} = \boldsymbol{\mu} + \mathbf{A}\mathbf{Z}$, in which $\mathbf{Z} = (Z_1, \ldots, Z_d)'$ is a vector of i.i.d. $N(0, 1)$ random variables and $\mathbf{A}\mathbf{A}' = \boldsymbol{\Sigma}$; see Section 2.2.1. The choice of a $p \times p$ lower triangular matrix for $\mathbf{A}$ corresponds to the Cholesky decomposition of $\boldsymbol{\Sigma}$ (Lai and Xing, 2008a, p. 205). Then Since $Z_1, \ldots, Z_d$ can be easily simulated, a sample of $\mathbf{X}$ can be obtained by simulating a sample of $\mathbf{Z}$. In particular,

Whereas a multivariate normal distribution is specified by its mean vector $\boldsymbol{\mu}$ and covariance $\boldsymbol{\Sigma}$, specification of non-normal joint distributions is more complicated. The Gaussian copula approach in Section 3.8.3 specifies the marginal distribution G_i of X_i and assumes $\left(\Phi^{-1}(G_1(X_1)), \ldots, \Phi^{-1}(G_p(X_d))\right)'$ to be $N(0, \boldsymbol{\Gamma})$. In this case, $(X_1, \ldots, X_d)'$ can be simulated by first generating $(Z_1, \ldots, Z_d)' \sim N(0, \boldsymbol{\Gamma})$ and then setting $X_i = \Phi^{-1}(G_i(X_i))$. For simulation of general multivariate distribution via copulas, see Section 2.2.4.

7.2 Stochastic process generation

Instead of random variable generation, many finance and risk management problems involve simulating stochastic processes. This section considers various types of stochastic processes and explains algorithms of generating those processes.

7.2.1 Gaussian and Markovian Gaussian processes

A real-valued stochastic process $\{X_t, t \in \mathcal{T}\}$ is said to be a *Gaussian process* if all its finite-dimensional distributions are Gaussian. The probability distribution of a Gaussian process is determined completely by its expectation function $\mu_t = E(X_t)$, $t \in \mathcal{T}$, and covariance function $\sigma_{s,t} = \text{Cov}(X_s, X_t)$, $s, t \in \mathcal{T}$. A zero-mean Gaussian process is one in which $\mu_t = 0$ for all $t \in \mathcal{T}$.

To generate a realization of a Gaussian process with expectation μ_t and covariance function $\Sigma_{s,t}$ at times $t_1, \ldots, t_n$, we can simply sample a multivariate normal random vector $\mathbf{X} \sim N(\boldsymbol{\mu}, \boldsymbol{\Sigma})$ with $\boldsymbol{\mu} = (\mu_{t_1}, \ldots, \mu_{t_n})'$ and $\boldsymbol{\Sigma} = (\sigma_{t_i, t_j})_{i,j=1,\ldots,n}$.

If the Gaussian process X_t also satisfies the Markov property

$$P(X_{t+s} \in A | \{X_u; u \leq t\}) = P(X_{t+s} \in A | X_t)$$

for all measurable set A and all $s, t \geq 0$, $\{X_t; t \geq 0\}$ becomes a Markovian Gaussian process. Then note that the joint normal distribution

$$\begin{pmatrix} X_{t_i} \\ X_{t_{i+1}} \end{pmatrix} \sim N\left(\begin{pmatrix} \mu_{t_i} \\ \mu_{t_{i+1}} \end{pmatrix}, \begin{pmatrix} \sigma_{t_i, t_i} & \sigma_{t_i, t_{i+1}} \\ \sigma_{t_i, t_{i+1}} & \sigma_{t_{i+1}, t_{i+1}} \end{pmatrix} \right)$$

implies the conditional distribution

$$(X_{t_{i+1}} | X_{t_i} = x) \sim N\left(\mu_{t_{i+1}} + \frac{\sigma_{t_i, t_{i+1}}}{\sigma_{t_i, t_i}} (x - \mu_{t_i}), \sigma_{t_{i+1}, t_{i+1}} - \frac{\sigma_{t_i, t_{i+1}}^2}{\sigma_{t_i, t_i}} \right).$$

A Markovian Gaussian process sampled at times $0 \leq t_1 < \cdots < t_n$ can be generated as follows. First draw $Z \sim N(0,1)$ and let $X_{t_1} = \mu_{t_1} + \sqrt{\sigma_{t_1, t_1}} Z$. Then for $i = 1, \ldots, n-1$, draw $Z \sim N(0,1)$ and let

$$X_{t_{i+1}} = \mu_{t_{i+1}} + \frac{\sigma_{t_i, t_{i+1}}}{\sigma_{t_i, t_i}} (X_{t_i} - \mu_{t_i}) + \sqrt{\sigma_{t_{i+1}, t_{i+1}} - \frac{\sigma_{t_i, t_{i+1}}^2}{\sigma_{t_i, t_i}}} Z.$$

The above definition of Markovian Gaussian process and simulation algorithms can be easily generalized to the case of d-dimensional Markovian Gaussian process; see examples for some specific Markovian Gaussian processes in Sections 7.2.4–7.2.7.

7.2.2   Discrete- and continuous-time Markov chains

A discrete-time *Markov chain* is a stochastic process $\{X_t\}_{t=1,2,\ldots}$ that satisfies the *Markov property* $P(X_{t+s} | X_u, u \leq t) = P(X_{t+s} | X_t)$. This property implies that a Markov chain $\{X_t\}_{t \geq 1}$ given X_0 can be generated sequentially from the conditional distribution of X_{t+1} given X_t for $t = 0, 1, \ldots$.

There are two common ways to specify the conditional distribution of X_{t+1} given X_t. One way is to assume that the Markov chain $\{X_0, X_1, \ldots, \}$ has a discrete state space and is time homogeneous. Then the conditional distribution of X_{t+1} given X_t is characterized by the matrix of one-step transition probabilities $P = (p_{ij})$, where $p_{ij} = P(X_{t+1} = j | X_t = i)$, $i, j \in \mathcal{K}$. Another approach to specifying the conditional distribution of X_{t+1} given X_t is to assume that the process $\{X_t; t = 0, 1, \ldots\}$ satisfies a recurrence equation $X_{t+1} = g(t, X_t, U_t)$, where U_t is an random variable whose distribution may depend on X_t and t and g is a function.

Example 7.2 *Suppose that X_t is a three-state discrete time Markov chain with transition probability matrix*

$$P = \begin{pmatrix} 0.5 & 0.3 & 0.2 \\ 0.2 & 0.7 & 0.1 \\ 0.1 & 0.1 & 0.4 \end{pmatrix}.$$

Given an initial state X_0, a sample of $\{X_t; t > 0\}$ can be generated from the transition probability matrix P. The left panel in Figure 7.1 shows a sample of such $\{X_t\}$ for $t = 1, \ldots, 100$ given $X_0 = 1$.

Continuous-time Markov chains

As introduced in Section 3.6.1, a continuous-time Markov chain (or *Markov jump*) process is a stochastic process $\{X_t, t \geq 0\}$ with a continuous index set $t \in \mathbb{R}^+$ and a discrete state space $\mathcal{K}$, which satisfies the Markov property $P(X_{t+s}|X_u, u \leq t) = P(X_{t+s}|X_t)$. A time-homogeneous Markov jump process is often defined via its generator matrix $\Lambda = (\lambda_{ij})_{i,j\in\mathcal{K}}$, where λ_{ij} is the transition rate from i to j, i.e., $\lambda_{ij} = \lim_{h\downarrow 0} P(X_{t+h} = j|X_t = i)$ for $i \neq j, i, j \in \mathcal{K}$ and $\lambda_{ii} = -\sum_{j\neq i} \lambda_{ij}$.

Note that a continuous-time Markov chain has the following property: if the process is in some state i at time t, it will remain there for an additional $\exp(-\lambda_{ii})$-distributed amount of time. When the process leaves a state i, it will move to a state j with probability $p_{ij} = -\lambda_{ij}/\lambda_{ii}$, $j \in \mathcal{K}$, independent of the history of the process. Let $X_0, X_1, \ldots$ be a Markov jump process with transition matrix $(p_{ij})_{i,j\in\mathcal{K}}$. Denote by $A_1, A_2, \ldots$ the inter-jump times (i.e., the times between jumps) and $T_1, T_2, \ldots$ by the jump times of $\{X_t\}$. They can be generated by the following algorithm.

Algorithm 7.2 (Time-homegeneous Markov chain) *Set $T_0 = 0$ and draw $X_0 = Y_0$ from its distribution. Let $n = 0$.*

1. *Draw $A_{n+1} \sim \exp(-q_{X_n, X_n})$ and set $T_{n+1} = T_n + A_{n+1}$.*

2. *Set $X_t = Y_n$ for $T_n \leq t < T_{n+1}$ and draw Y_{n+1} from the distribution $\{-\lambda_{Y_n,j}/\lambda_{Y_n,Y_n}; j \in \mathcal{K} \setminus \{Y_n\}\}$. Set $n \leftarrow n + 1$ and go to step 1.*

Example 7.3 *Suppose that X_t is a three-state continuous-time homogeneous Markov chain with generator matrix*

$$\Lambda = \begin{pmatrix} -0.2 & 0.15 & 0.05 \\ 0.15 & -0.4 & 0.25 \\ 0.35 & 0.1 & -0.45 \end{pmatrix}.$$

The right panel in Figure 7.1 shows a sample of X_t with generator matrix Λ and initial state $X_0 = 1$ using Algorithm 7.2.

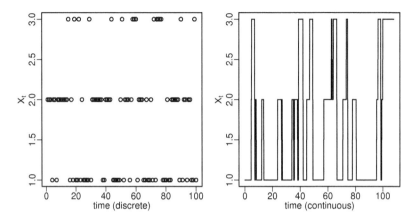

FIGURE 7.1: Realizations of two three-state Markov chains X_t (a) a discrete-time Markov chain with probability transition matrix P in Example 7.2 (left), and (b) a continuous-time Markov chain with homogeneous generator matrix Λ in Example 7.3 (right).

Algorithm 7.2 can be extended to the nonhomogeneous case, where the rates λ_{ij} depend on t. Let $0 \le \lambda_{ij}(t) < \infty$ and $\lambda_{ii}(t) = -\sum_{j\ne i} \lambda_{ij}(t)$. Suppose that at some time T_n the process jumps to state $Y_n = i$. Let A_{n+1} denote the holding time in state i. We have

$$-\lambda_{ii}(t) = \lim_{h\downarrow 0} \frac{P(t - T_n < A_{n+1} < t + h - T_n | A_{n+1} > t - T_n)}{h}$$

$$= \lim_{h\downarrow 0} \frac{F(t + h - T_n) - F(t - T_n)}{(1 - F(t - T_n))h} = \frac{f(t - T_n)}{1 - F(t - T_n)} = -\frac{d}{dt} \log(1 - F(t - T_n)),$$

where $F(t)$ denote the distribution function of A_{n+1} and $f(t)$ its density function. Then it follows that

$$F(t) = P(A_{n+1} \le t) = 1 - \exp\left(\int_{T_n}^{T_{n+1}} \lambda_{ii}(s)ds \right), \quad t \ge 0. \qquad (7.2)$$

At time $T_{n+1} = T_n + A_{n+1}$ the process jumps to state j with probability $-\lambda_{ij}(T_{n+1})/\lambda_{ii}(T_{n+1})$, $j \in \mathcal{K}$. With this result, drawing of Y_{n+1} in Step 2 of Algorithm 7.2 can be replaced by drawing Y_{n+1} from the distribution $\{-\lambda_{Y_n,j}(T_{n+1})/\lambda_{Y_n,Y_n}(T_{n+1}); j \in \mathcal{K} \setminus \{Y_n\}\}$.

7.2.3 Poisson and compound Poisson processes

Poisson processes are used to model random configurations of points in space and time. Let B be a subset of $\mathbb{R}^d$ and let $\mathcal{B}$ be the collection of Borel sets on B. For any collection of random points $\{T_n\}$ in B, there is a random counting measure N defined by $N(A) = \sum_k 1_{\{T_k \in A\}}$, $A \in \mathcal{B}$, which counts the random number of points in A. Such a random counting measure is called a *Poisson random measure* with mean measure μ if the following properties hold:

(i) $N(A) \sim \text{Poisson}(\mu(A))$ for any set $A \in \mathcal{E}$, where $\mu(A)$ is the mean measure of A.

(ii) For any disjoint sets $A_1, \ldots, A_n \in \mathcal{E}$, the random variables $N(A_1), \ldots, N(A_n)$ are independent.

In most applications, an intensity function $\lambda(\mathbf{x})$ is assumed to exist so that $\mu(A) = \int_A \lambda(\mathbf{x}) d\mathbf{x}$. Properties (i) and (ii) implies the following

(iii) Conditional on $N(A) = n$, the n points in A are independent of each other and have density function $f(\mathbf{x}) = \lambda(\mathbf{x})/\mu(\mathbf{x})$.

Property (iii) suggests us generating a Poisson process on B as follows. First, generate a Poisson random variable $N \sim \text{Poisson}(\mu(B))$, in which $\mu(B) = \int_B \lambda(\mathbf{x}) d\mathbf{x} < \infty$. Then given $N = n$, draw $\mathbf{X}_1, \ldots, \mathbf{X}_n$ i.i.d. from $f(\mathbf{x}) = \lambda(\mathbf{x})/\mu(B)$. Then samples $\mathbf{X}_1, \ldots, \mathbf{X}_n$ are the points of the Poisson process on B.

To see how this idea is applied to specific applications, we consider first one-dimensional Poisson processes on interval $[0, T]$. For simplicity, we consider a *homogeneous Poisson process* with a constant rate λ on $\mathbb{R}^+$. Denote the points of the process by $0 = T_0 < T_1, T_2, \ldots$, in which $T_1, \ldots, T_n$ are interpreted as arrival times of some events, and let $A_i = T_i - T_{i-1}$ be the i-th ($i \geq 1$) interarrival time. Then the interarrival times $\{A_i\}$ are independent and follow exponential distribution with rate λ. Hence, the points of the Poisson process on interval $[0, T]$ can be generated as follows. For each $n \geq 1$, generate $U \sim \mathcal{U}(0, 1)$ and set $A_n = -(\log U)/\lambda$ as the n-th interarrival time. Then $T_n = T_{n-1} + A_n$ is the n-th event time. The corresponding *Poisson counting process* generated this way, denoted by $\{N_t = N([0, t]), t \geq 0\}$, is a Markov jump process on the infinitely countable set $\{0, 1, 2, \ldots\}$ with $N_0 = 0$ and transition rates $\lambda_{i,i+1} = \lambda$, $i = 0, 1, 2, \ldots$ and $\lambda_{i,j} = 0$ otherwise. The process jumps at times $T_1, T_2, \ldots$ to states 1, 2, $\ldots$, staying an $\exp(\lambda)$-distributed amount of time in each state.

Similar to the above discussion, the counting process corresponding to a *nonhomogeneous* one-dimensional Poisson process on $\mathbb{R}^+$ with rate function $\lambda(t)$, $t \geq 0$ is a nonhomogeneous Markov jump process with transition rates $\lambda_{i,i+1}(t) = \lambda(t)$, $i = 0, 1, 2, \ldots$. Besides, similar to (7.2), the tail probability of the interarrival times are

$$P(A_{n+1} > t) = \exp\left(-\int_{T_n}^{T_{n+1}} \lambda(s) ds\right), \quad t \geq 0.$$

Then using property (iii), one can construct the event times $\{T_n; n \geq 0\}$ directly as follows. For $t \geq 0$, assume $\sup_{s \leq t} \lambda(s)$ exists and denote it by

$\widetilde{\lambda} = \sup_{s \le t} \lambda(s)$. Then generate the points of a two-dimensional homogeneous Poisson process $\{\widetilde{T}_m; m \ge 0\}$ on $[0, t] \times [0, \widetilde{\lambda}]$ with rate 1. Finally, project all points of $\widetilde{T}_m$ that lie below the curve of $\lambda(s)$, $s \le t$ onto the t-axis. The following algorithm implements the above procedure.

Algorithm 7.3 (One-dimensional nonhomogeneous Poisson process)
Set $T_0 = 0$, $t = 0$ and $\widetilde{\lambda} = \sup_{s \le t} \lambda(s)$.

1. *For each $n \ge 1$, generate $U \sim \mathcal{U}(0, 1)$ and assign $t \leftarrow t + (-\log U)/\widetilde{\lambda}$. If $t > T$, stop; otherwise, continue.*

2. *Generate $V \sim \mathcal{U}(0, 1)$. If $V \le \lambda(t)/\widetilde{\lambda}$, increase n by 1 and let $T_n = t$. Then go to step 1.*

Example 7.4 *Consider a time-nonhomogeneous Poisson process with rate $\lambda(t) = 1 + \sin t$ for $t \in [0, 10]$. Let $\widetilde{\lambda} = 2$ and use Algorithm 7.3, we can generate a realization of Poisson counting process. The left panel of Figure 7.2 shows how to construct the event arrival times T_n by accepting each event time with probability $\lambda(t)/\widetilde{\lambda}$, and the right panel of Figure 7.2 shows the counting process N_t which is derived from the left panel.* □

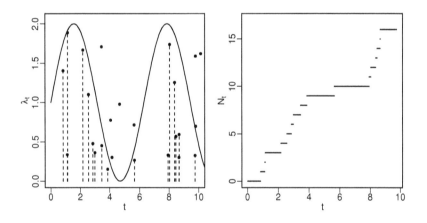

FIGURE 7.2: A realization of a nonhomogeneous Poisson counting process with rate function $\lambda(t) = 1 + \sin t$ for $t \in [0, 10]$. (a) Construction of event arrival times T_n (left). (b) Poisson counting process N_t (right).

Compound Poisson process

Let N be a Poisson random measure on $\mathbb{R}^+ \times \mathbb{R}^d$ with mean measure $\nu(d\mathbf{y})dt$. Assume that $\lambda = \nu(\mathbb{R}^d) < \infty$. The process $\{N_t\}$ with $N_t = N([0, t] \times \mathbb{R}^d)$ is a homogeneous Poisson process with rate λ. The process $\{\mathbf{X}_t\}_{t \ge 0}$ is defined as

$$\mathbf{X}_t = \int_0^t \int_{\mathbb{R}^d} \mathbf{y} N(du, d\mathbf{y}), \qquad t \geq 0,$$

is the *compound Poisson process* corresponding to the measure ν. The process can also be written as

$$\mathbf{X}_t = \sum_{i=1}^{N_t} \mathbf{Y}_i,$$

where $\mathbf{Y}_1, \mathbf{Y}_2, \ldots$ are i.i.d. random vectors with density function $\nu(d\mathbf{y})/\lambda$ and are independent of N_t. The compound Poisson process is a special case of a *Lévy process*, a stochastic process with independent and stationary increments; see Section 7.2.7. In this context, the measure ν is called the *Lévy measure*. Denote by $\{T_k\}$ the jump times of the compound Poisson process and $\{\mathbf{Y}_k\}$ the jump sizes. We can use the following algorithm to generate a compound Poisson process.

Algorithm 7.4 (Compound Poisson process I) *Set* $k = 0, T_k = 0$ *and* $\mathbf{X}_{T_k} = 0$.

1. *Generate* $A_k \sim \exp(\lambda)$ *and* $\mathbf{Y}_k \sim \nu(d\mathbf{y})/\lambda$.

2. *Set* $T_{k+1} \leftarrow T_k + A_k$, $\mathbf{X}_{T_{k+1}} = \mathbf{X}_{T_k} + \mathbf{Y}_k$, *and* $k \leftarrow k+1$, *and go to step 1.*

Consider a one-dimensional compound Poisson process $X_t = \sum_{i=1}^{N_t} Y_i$, in which N_t is a Poisson process with rate $\lambda = 2$ and Y_i are i.i.d. $N(0, 0.04)$ random variables. Figure 7.3 shows two realization of this compound Poisson process on the time interval $[0, 5]$ using Algorithm 7.4.

An alternative procedure to generate compound Poisson processes on a fixed interval $[0, T]$ is based on property (iii) of the Poisson process and summarized as follows.

Algorithm 7.5 (Compound Poisson process II) *Generate* $N \sim Poisson(\lambda T)$.

1. *Generate* $U_1, \ldots, U_N$ *i.i.d. from* $\mathcal{U}(0, T)$ *and* $\mathbf{Y}_1, \ldots, \mathbf{Y}_N$ *i.i.d. from* $\nu(d\mathbf{y})/\lambda$.

2. *Return* $\mathbf{X}_t = \sum_{i: U_i \leq t} \mathbf{Y}_i$ *for* $t \in [0, T]$.

For compound Poisson process in which the Poisson process is nonhomogeneous, Algorithms 7.4 and 7.5 can be modified in the spirit of Algorithm 7.3 to generate sample paths of the process.

7.2.4 Wiener processes and Brownian motion

A *Wiener process* is a stochastic process $W = \{W_t\}_{t \geq 0}$ with the following properties.

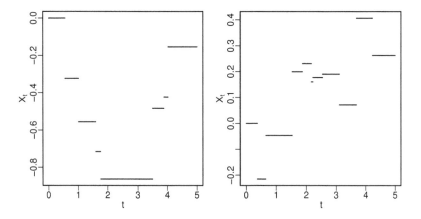

FIGURE 7.3: Two realizations of a compound Poisson process $X_t = \sum_{i=1}^{N_t} Y_i$, in which N_t is a Poisson process with rate $\lambda = 2$ and Y_i are i.i.d. $N(0, 0.04)$ random variables.

(i) W has *independent increments*, that is, for any $t_1 < t_2 \leq t_3 < t_4$, $W_{t_4} - W_{t_3}$ and $W_{t_2} - W_{t_1}$ are independent random variables.

(ii) For all $t \geq s \geq 0$, $W_t - W_s \sim N(0, t - s)$.

(iii) $\{W_t\}$ has continuous paths, with $W_0 = 0$.

The Wiener process $W = \{W_t\}_{t \geq 0}$ is a Markov process and has many interesting properties. For example, it is easy to show that W_t, $W_t^2 - t$ and $\exp(\theta W_t - (\theta^2 t)/2)$ ($\theta \in \mathbb{R}$) are martingales; besides, given W_t is a Wiener process, $tW_{1/t}$ is also a Wiener process. More properties of the Wiener process $W = \{W_t\}_{t \geq 0}$ can be found in Stroock and Varadhan (2005).

A Wiener process can be generated by using its Markovian and Gaussian properties. In particular, let $0 = t_0 < t_1 < t_2 < \cdots < t_m$ be the set of distinct times for which simulation of the process is desired. Then generate $Z_1, \ldots, Z_m$ i.i.d. from $N(0, 1)$ and let $W_{t_k} = \sum_{i=1}^{k} \sqrt{t_k - t_{k-1}} Z_i$ for $k = 1, \ldots, m$. This algorithm returns only a discrete skeleton of the true continuous process. To obtain a continuous path approximation to the exact path of the Wiener process, one could use linear interpolation on the points $W_{t_1}, \ldots, W_{t_k}$. That is,

$$\widehat{W}_s = \frac{W_{t_k}(s - t_{k-1}) + W_{t_{k-1}}(t_k - s)}{(t_k - t_{k-1})}, \qquad s \in [t_{k-1}, t_k]. \qquad (7.3)$$

Instead of using interpolation, another way is to approximate the process using the *Karhunen-Loéve expansion* (Karhunen, 1947; Loéve, 1946). In particular, the Wiener process $\{W_t; t \in \mathbb{R}^+\}$ can be viewed as a random element of $L^2(\mathbb{R}^+)$, then it has the series expansion

$$W_t = \sum_{n=0}^{\infty} Z_n \int_0^t h_n(x)dx,$$

where $Z_0, Z_1, \ldots$ are i.i.d. $N(0,1)$ random variables and $\{h_n(x)\}_{n=0}^{\infty}$ is any complete orthonormal basis of $L^2(\mathbb{R}^+)$ for which the above random series converges in L^2-norm. Typical examples for $t \in [0,1]$ are the *Harr* functions and the basis of cosine functions $h_0(x) = 1$, $h_n(x) = \sqrt{2}\cos(n\pi x)$, $n = 1,2,\ldots$. The latter gives the *sine series expansion*

$$W_t = Z_0 t + \frac{\sqrt{2}}{\pi} \sum_{n=1}^{\infty} Z_n \frac{\sin(n\pi t)}{n}, \qquad t \in [0,1].$$

When the orthonormal basis $\sqrt{2/b}\cos((1+2n)\pi x/(2b))$ for $x \in [0,b]$ and $n = 0,1,2,\ldots$ are used, we obtain the Karhunen-Loéve expansion

$$W_t = \sum_{n=0}^{\infty} Z_n \frac{2\sqrt{2b}}{(2n+1)\pi} \sin\left(\frac{(2n+1)\pi t}{2b}\right), \qquad t \in [0,b]. \qquad (7.4)$$

Using the expansion (7.4), a Wiener process W_t on the interval $[0,b]$ can be generated as follows. First, generate $Z_1, \ldots, Z_n$ i.i.d. from $N(0,1)$ for a sufficiently large n, and then output the approximation to W_t

$$\widehat{W_t} = \sum_{k=0}^{n} Z_k \frac{2\sqrt{2b}}{(2k+1)\pi} \sin\left(\frac{(2k+1)\pi t}{2b}\right). \qquad (7.5)$$

Example 7.5 (Sample paths of the Wiener process) *Consider the problem of generating sample paths of the Wiener process W_t on $[0,1]$. As discussed above, one way is to use a discrete skeleton $0 = t_0 < t_1 < \cdots < t_m = 1$ of the true continuous process with linear interpolation (7.3), and the other way is to use the approximation (7.5) to the Karhunen-Loéve expansion (7.4). Figure 7.4 shows random paths of the Wiener process on $[0,1]$ generated by these two methods. In particular, the sample path in the left panel is generated by using the first method at times $t_i = 1/m$ ($i = 1,\ldots,m$) for $m = 10^3$, and the sample path in the right panel is generated by using the approximation (7.5) with $n = 1000$.* $\square$

Multi-dimensional Wiener process and Brownian motion

The one dimensional Wiener process $\{W_t\}_{t \geq 0}$ can be generalized in several directions. One generalization is to consider a process $\{B_t\}_{t \geq 0}$ which satisfies

$$B_t = \mu t + \sigma W_t, \qquad t \geq 0, \qquad (7.6)$$

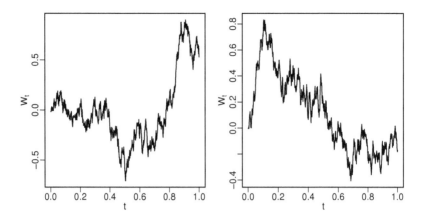

FIGURE 7.4: Realizations of a Wiener process using the discrete skeleton of the true continuous process with liner interpolation (left) and the approximation of the Karhunen-Loéve expansion (right).

where $\{W_t\}$ is a Wiener process. The stochastic process $\{B_t\}_{t\geq 0}$ is called a *Brownian motion* with *drift* μ and *volatility coefficient* σ^2. In this sense, the Wiener process is also called a standard Brownian motion, i.e. $\mu = 0, \sigma^2 = 1$. Another generalization is multi-dimensional Wiener process. Let $\{W_{t,i}\}_{t\geq 0}$, $i = 1, \ldots, d$ be independent Wiener processes and let $\mathbf{W}_t = (W_{t,1}, \ldots, W_{t,d})'$. The process $\{\mathbf{W}_t\}_{t\geq 0}$ is called a *d-dimensional Wiener process*. As an analog of (7.6), a m-dimensional Brownian motion can be defined as

$$\mathbf{B}_t = \boldsymbol{\mu} t + \mathbf{A} \mathbf{W}_t, \qquad t \geq 0, \tag{7.7}$$

in which $\boldsymbol{\mu} \in \mathbb{R}^m$, $\mathbf{A} \in \mathbb{R}^{m \times d}$, and $\mathbf{W}_t$ is a d-dimensional Wiener process.

The methods of generating sample paths for one-dimensional Wiener process can be easily extended to multi-dimensional Wiener processes and Brownian motions. Figure 7.5 shows sample paths of a two-dimensional Wiener process and a Brownian motion with drift $\mu = 0.5$ and volatility coefficient $\sigma^2 = 0.04$.

7.2.5 Stochastic differential equations and diffusion processes

A stochastic differential equation (SDE) for a stochastic process $\{X_t\}_{t\geq 0}$ is of the form

$$dX_t = a(X_t, t)dt + b(X_t, t)dW_t, \tag{7.8}$$

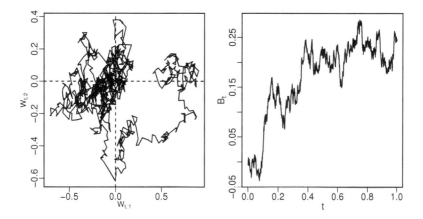

FIGURE 7.5: Realizations of a two-dimensional Wiener process (left) and a Brownian motion with $\mu = 0.5$ and volatility coefficient $\sigma^2 = 0.04$ (right).

where W_t is a Wiener process and $a(x,t)$ and $b(x,t)$ are deterministic functions. The stochastic process $\{X_t\}_{t \geq 0}$ satisfies the integral equation

$$X_t = X_0 + \int_0^t a(X_s, s)ds + \int_0^t b(X_s, s)dW_s. \tag{7.9}$$

The diffusion process $\{X_t\}$ of the form (7.8) is a Markov process with continuous sample paths. If $a(X_t, t)$ and $b(X_t, t)$ satisfy some regularity conditions, the (strong) solution to (7.9) exists and is unique on an interval $[0, T]$.

When the coefficients $a(X_t, t)$ and $b(X_t, t)$ do not depend on t, i.e., $a(x,t) = a(x)$ and $b(x,t) = b(x)$, the diffusion process X_t becomes a time-homogeneous Markov process

$$dX_t = a(X_t)dt + b(X_t)dW_t.$$

In such a case, we denote by $p_t(x,y)$ the transition density of X_t. Then under regularity conditions, the transition density satisfies the *Kolmogorov backward equations*

$$\frac{\partial}{\partial t}p_t(x,y) = a(x)\frac{\partial}{\partial x}p_t(x,y) + \frac{1}{2}b^2(x)\frac{\partial^2}{\partial x^2}p_t(x,y)$$

and the *Kolmogorov forward equations*

$$\frac{\partial}{\partial t}p_t(x,y) = -\frac{\partial}{\partial y}[a(y)p_t(x,y)] + \frac{1}{2}\frac{\partial^2}{\partial y^2}[b^2(y^2)p_t(x,y)].$$

There are several ways to approximately simulate diffusion process. Let $\{X_t\}_{t\geq 0}$ be a diffusion process defined by the SDE (7.8), where X_0 has a known distribution. A straightforward way to approximate a sample path of

$\{X_t\}$ is to use the *direct Euler method*, which replaces the SDE (7.8) with the stochastic difference equation

$$Y_{k+1} = Y_k + a(Y_k, kh)h + b(Y_k, kh)\sqrt{h}Z_k, \qquad (7.10)$$

where $Z_1, Z_2, \ldots$ are i.i.d. $N(0,1)$ random variables. For a small step size h, the time series $\{Y_k; k = 0, 1, 2, \ldots\}$ approximates the process $\{X_t; t \geq 0\}$, that is, $Y_k \approx X_{kh}$, $k = 0, 1, 2, \ldots$.

The approximate sample path generated by the direct Euler method has linear order approximation errors. To reduce the error, the *Milstein's higher order method* can be used. In particular, apply Itô's lemma to the diffusion coefficient $b(X_s, s)$, one obtains that

$$db(X_s, s) = b_x(X_s, s)[a(X_s, s)ds + b(X_s, s)dW_s] + b_s(X_s, s)ds + \frac{1}{2}b_{xx}(X_s, s)b^2(X_s, s)ds,$$

where b_x, b_t and b_{xx} are the corresponding partial derivatives of $b(x, t)$. Then the integral (7.9) can be expressed as

$$X_{t+h} - X_t = \int_t^{t+h} a(X_u, u)du + \int_t^{t+h} b(X_u, u)dW_u$$

$$= a(X_t, t)h + b(X_t, t)(W_{t+h} - W_t) + \int_t^{t+h}\int_t^u b_x(X_s, s)b(X_s, s)dW_s dW_u + O(h^{3/2}).$$

Note that the last term in the above equation can be written as $\frac{1}{2}b_x(X_t, t)b(X_t, t)[(W_{t+h} - W_t)^2 - h] + O(h^2)$. Then the SDE (7.8) can be replaced by the following difference equation

$$Y_{k+1} = Y_k + a(Y_k, kh)h + b(Y_k, kh)\sqrt{h}Z_k + b_x(Y_k, kh)b(Y_k, kh)(Z_k^2 - 1)\frac{h}{2}, \qquad (7.11)$$

where $Z_1, Z_2, \ldots$ are i.i.d. $N(0,1)$ random variables.

The direct Euler's and Milstein's methods approximate the SDE (7.8) explicitly. In contrast to this, the *implicit Euler* method replaces the Y_k in (7.10) by Y_{k+1} so that the difference equation (7.10) becomes

$$Y_{k+1} = Y_k + a(Y_{k+1}, kh)h + b(Y_k, kh)\sqrt{h}Z_k, \qquad (7.12)$$

and Y_{k+1} needs to be solved from (7.12). However, one shall note that, (7.12) may need to be solved numerically if $a(x, t)$ is nonlinear in x.

Example 7.6 (Geometric Brownian motion) *Consider the geometric Brownian motion (GBM) process defined by*

$$dX_t = \mu X_t dt + \sigma X_t dW_t,$$

in which W_t is the Wiener process. It is easy to show that the GBM X_t can be expressed as

$$X_t = X_0 \exp\left[(\mu - \frac{\sigma^2}{2})t + \sigma W_t\right], \qquad t \geq 0.$$

The above solution indicates that sample paths of X_t can be obtained by transforming a sample path of the Wiener process. In particular, let $0 = t_0 < t_1 < \cdots < t_n$ be the set of distinct times and suppose that $Z_1, \ldots, Z_n$ are i.i.d. $N(0,1)$ random variables. Then given X_0, a sample path $\{X_{t_k}; k = 1, 2, \ldots\}$ at $\{t_1, \ldots, t_n\}$ is given by

$$X_{t_k} = X_0 \exp\left(\left(\mu - \frac{\sigma^2}{2}\right)t_k + \sigma \sum_{i=1}^{k} \sqrt{t_i - t_{i-1}} Z_i\right), \qquad k = 1, \ldots, n. \quad (7.13)$$

Instead of using the above method, one can also use the Milstein's method and write the difference equation (7.11) as

$$Y_{k+1} = \left((1 + \mu h + \sigma\sqrt{h}Z_k + \frac{h}{2}\sigma^2(Z_k^2 - 1)\right)Y_k, \qquad k = 1, 2, \ldots. \quad (7.14)$$

Suppose that $\mu = 0.1, \sigma = 0.2$, and $X_0 = 1$. Let $n = 10^3$, Figure 7.6 shows two sample paths of X_t on $[0,1]$, which are generated by the equation (7.13) at $t_k = k/n$ $(1 \le k \le n)$ and the difference equation (7.14) for $k = 1, \ldots, n$.

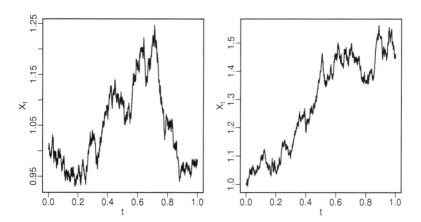

FIGURE 7.6: Two realizations of a geometric Brownian motion with $\mu = 0.1$, $\sigma = 0.2$ and $X_0 = 1$ using Milstein's method.

Multidimensional SDEs and approximation error

Multidimensional SDEs can be defined similarly as in (7.8). A SDE in $\mathbb{R}^d$ is an expression of the form

$$d\mathbf{X}_t = \mathbf{A}(\mathbf{X}_t, t)dt + \mathbf{B}(\mathbf{X}_t, t)d\mathbf{W}_t, \quad (7.15)$$

where $\{\mathbf{W}_t\}_{t \ge 0}$ is an m-dimensional Wiener process, $\mathbf{A}(\mathbf{x}, t)$ is an d-dimensional vector and $\mathbf{B}(\mathbf{x}, t)$ an $d \times m$ matrix, for each $\mathbf{x} \in \mathbb{R}^d$ and $t \ge 0$.

The $d \times d$ matrix $\mathbf{BB'}$ is called the *diffusion matrix*. The resulting diffusion process is Markov, and if $\mathbf{A}$ and $\mathbf{B}$ do not depend explicitly on t then the diffusion process is time-homogeneous. Existence and uniqueness of $\mathbf{X}_t$ can be argued similarly under the same conditions as the one-dimensional SDE case. The direct Euler method can be easily extended to generate approximate sample paths of $\{\mathbf{X}_t\}$.

Most numerical methods for solving SDEs of $\mathbf{X}_t$ are not exact. Let $\{\mathbf{X}_t\}_{t \geq 0}$ be the process of interest and $\mathbf{X}^h = \{X_t^h, t = 0, h, 2h, \dots\}$ be a discrete-time approximation. Denote by C_P^r the space of functions $g : \mathbb{R}^m \to \mathbb{R}$ such that (i) g is r times continuously differentiable and has a polynomial growth, and (ii) The partial derivatives of g up to order r also demonstrate polynomial growth. The process $\mathbf{X}^h$ is said to *converge weakly with order* $\beta > 0$ to $\mathbf{X}$ at time T as $h \downarrow 0$, if for each $g \in C_P^{2(\beta+1)}$, there exists a positive constant C, independent of h, and a finite $h_0 > 0$, such that

$$|Eg(\mathbf{X}_T) - Eg(\mathbf{X}_T^h)| \leq Ch^\beta \qquad \text{for all } h \in (0, h_0).$$

The process $\mathbf{X}^h$ is said to *converge strongly with order* $\alpha > 0$ to $\mathbf{X}$ at time T as $h \downarrow 0$, if there exists a positive constant C, independent of h, as well as some $h_0 > 0$ such that

$$E||\mathbf{X}_T - \mathbf{X}_T^h|| \leq Ch^\alpha \qquad \text{for all } h \in (0, h_0).$$

Kloeden and Platen (1999) showed that, under different regularity conditions, Euler's method converges strongly with order $\alpha = 1/2$ and weakly with order $\beta = 1$. Similarly, under different regularity conditions, they showed that Milstein's method can converge strongly with order $\alpha = 1$ and weakly with order $\beta = 1$. Finally, if we use interpolation between the points of the approximation, then under some technical conditions, a strongly convergent scheme of order α at time T is also strongly convergent uniformly over the whole path in the sense that $E \sup_{0 \leq t \leq T} ||\mathbf{X}_t - \mathbf{X}_t^h|| \leq Ch^\alpha$.

Brownian bridge

The *standard Brownian bridge* process $\{X_t, t \in [0, 1]\}$ is a stochastic process whose distribution is that of the Wiener process on $[0, 1]$ conditional on $X_1 = 0$. Let $\{W_t\}$ be a Wiener process. The standard Brownian bridge process can be viewed as a nonhomogeneous diffusion process satisfying the linear SDE

$$dX_t = -\frac{X_t}{1-t}dt + dW_t, \qquad 0 \leq t < 1, X_0 = 0; \quad X_1 = 0. \qquad (7.16)$$

The strong solution of the above SDE is given by

$$X_t = \int_0^t \frac{1-t}{1-s}dW_s, \qquad 0 \leq t \leq 1. \qquad (7.17)$$

It is easy to show that $\{X_t\}$ is a Gaussian process with mean 0 and covariance function $\mathrm{Cov}(X_s, X_t) = \min(s, t) - st$; see Exercise 7.10.

Besides using the Milstein's and the (direct and implicit) Euler's methods, sample paths of the standard Brownian bridge can be generated by using its properties. In particular, let W_t be a Wiener process, one can show that $\{W_t - tW_1; t \in [0, 1]\}$ defines a standard Brownian bridge. Then given $X_0 = X_1 = 0$, in order to generate a sample path of X_t at times $0 = t_0 < t_1 < t_2 < \cdots < t_{n+1} = 1$, one can first generate a sample path of the Wiener process W_t on $t_1, \ldots, t_n$ using the algorithm in Section 7.2.4 and then set

$$X_{t_k} = W_{t_k} - t_k W_{t_n} \tag{7.18}$$

for $k = 1, \ldots, n$. The left panel of Figure 7.7 shows a sample path that is based on the above equation for $n = 10^3$ and $t_k = k/n$ $(k = 1, \ldots, n)$.

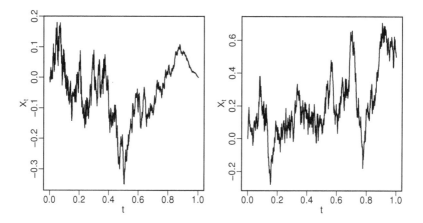

FIGURE 7.7: Realizations of a standard Brownian bridge (left) and a general Brownian bridge with $X_0 = 0$ and $X_1 = 0.5$ (right).

A *general Brownian bridge* is a stochastic process $\{X_t, t \in [t_0, t_{n+1}]\}$ whose distribution is that of the Wiener process on $[t_0, t_{n+1}]$ conditioned on $X_{t_0} = a$ and $X_{t_{n+1}} = b$. Then the distribution of X_t with $t \in [t_0, t_{n+1}]$ is Gaussian with mean $a + (b - a)(t - t_0)/(t_{n+1} - t_0)$ and variance $(t_{n+1} - t)(t - t_0)/(t_{n+1} - t_n)$. Hence, a sample path of $\{X_t\}$ at points $t_1 < \cdots < t_n$ within the interval $[t_0, t_{n+1}]$ can be generated as follows. First, generate $Z_1, \ldots, Z_n$ which are i.i.d. $N(0, 1)$ random variables, and then for $k = 1, \ldots, n$, let

$$X_{t_k} = X_{t_{k-1}} + (b - X_{t_{k-1}})\frac{t_k - t_{k-1}}{t_{n+1} - t_{k-1}} + \sqrt{\frac{(t_{n+1} - t_k)(t_k - t_{k-1})}{t_{n+1} - t_{k-1}}} Z_k. \tag{7.19}$$

Assume that $X_0 = 0, X_1 = 0.5$, and let $t_k = k/n$ $(k = 1, \ldots, n - 1)$ and $n = 10^3$. The right panel of Figure 7.7 shows a sample path of the general Brownian bridge X_t based on equation (7.19).

Ornstein-Uhlenbeck process

The *Ornstein-Uhlenbeck process* satisfies the SDE

$$dX_t = \alpha(\theta - X_t)dt + \sigma dW_t, \tag{7.20}$$

with $\sigma > 0$, $\alpha > 0$, and $\theta \in \mathbb{R}$. The strong solution of this SDE is expressed as

$$X_t = e^{-\alpha t}X_0 + \theta(1 - e^{-\alpha t}) + \sigma e^{-\alpha t}\int_0^t e^{\alpha s}dW_s. \tag{7.21}$$

It follows that $\{X_t\}$ is a Gaussian process whenever X_0 is Gaussian (or takes a fixed value), with mean function

$$E(X_t|X_0 = x_0) = e^{-\alpha t}x_0 + \theta(1 - e^{-\alpha t})$$

and covariance and variance functions

$$\text{Cov}(X_s, X_t) = \frac{\sigma^2}{2\alpha}e^{-\alpha(s+t)}\left(e^{\alpha \min(s,t)} - 1\right), \quad \text{Var}(X_t) = \frac{\sigma^2}{2\alpha}(1 - e^{-2\alpha t}).$$

This shows that X_t converges in distribution to a $N(\theta, \sigma^2/(2\alpha))$ random variable as $t \to \infty$. Moreover, when X_0 has this limiting distribution, the Markov process $\{X_t\}$ is stationary and time-reversible.

Making use of the solution (7.21), we obtain that

$$\text{Var}\left(\sigma e^{-\alpha t}\int_0^t e^{\alpha s}dW_s\right) = \frac{1 - e^{-2\alpha t}}{2\alpha}\sigma^2.$$

Hence, the distribution of X_t conditional on X_0 is a normal distribution with mean $e^{-\alpha t}X_0 + \theta(1 - e^{-\alpha t})$ and variance $\sigma^2(1 - e^{-2\alpha t})/(2\alpha)$. Then a sample path of the Ornstein-Uhlenbeck process (7.20) can be generated as follows. Let $0 = t_0 < t_1 < t_2 < \cdots < t_n$ be the set of discrete time points. Given X_0, generate $Z_1, \ldots, Z_n$ that are i.i.d. $N(0, 1)$ random variables and let

$$X_{t_k} = \theta\left(1 - e^{-\alpha(t_k - t_{k-1})}\right) + e^{-\alpha(t_k - t_{k-1})}X_{t_{k-1}} + \sigma\sqrt{\frac{1 - e^{-2\alpha(t_k - t_{k-1})}}{2\alpha}}Z_k. \tag{7.22}$$

for $k = 1, 2, \ldots, n$. Figure 7.8 shows two realizations of using equation (7.22) for $n = 10^3$, $t_k = k/n$ $(k = 1, \ldots, 2n)$.

7.2.6   Jump-diffusion processes

A jump-diffusion process is a mixture of a jump process and a diffusion process. It was first introduced by Merton (1976) to study the price of derivative securities. Merton's jump-diffusion model can be specified via the SDE

$$dS_t = S_{t-}\left(\mu dt + \sigma dW_t + dJ_t\right), \tag{7.23}$$

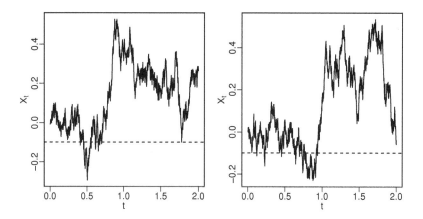

FIGURE 7.8: Realizations of the Ornstein-Uhlenbeck process (7.20) using (7.22) for $\theta = 0.1, \alpha = 1, \sigma = 0.5$ (left) and $\theta = -0.1, \alpha = 2, \sigma = 0.5$ (right).

in which μ and σ are constants, W_t is a Wiener process, J_t is a process independent of W_t with piecewise constant sample paths, and $S_{t-} = \lim_{r \uparrow t} S_r$ is the left limit of S_r when r approaches t from the left. The process J_t is given by

$$J_t = \sum_{j=1}^{N_t} (Y_j - 1), \qquad (7.24)$$

where $Y_1, Y_2, \ldots$ are random variables and N_t is a counting process. This suggests that there are random arrival times $0 < \tau_1 < \tau_2 < \ldots$ and $N_t = \sup\{n; \tau_n \le t\}$ counts the number of arrivals in $[0, t]$. The symbol dJ_t in (7.23) represents the jump at time t, and the size of this jump is $Y_j - 1$ if $t = \tau_j$ and 0 otherwise. Then the jump in S_t at τ_j is $S_{\tau_j} - S_{\tau_j^-} = S_{\tau_j^-}(Y_j - 1)$ and hence $S_{\tau_j} = S_{\tau_j^-} Y_j$. Suppose that Y_j are positive random variables. The solution of (7.23) is given by

$$S_t = S_0 \exp\left((\mu - \frac{\sigma^2}{2})t + \sigma W_t \right) \cdot \prod_{j=1}^{N_t} Y_j. \qquad (7.25)$$

To impose distributional assumptions on the jump process J_t, one may consider the compound Poisson process in Section 7.2.3. Suppose that J_t is a compound Poisson process such that the counting process N_t is a homogeneous Poisson process with rate λ. Then the interarrival times $\tau_{j+1} - \tau_j$ are independent and follow a common exponential distribution with rate λ, i.e., $P(\tau_{j+1} - \tau_j > t) = e^{-\lambda t}$ for $t \ge 0$.

In Merton's (1974) study, the jump random variables Y_j are lognormally distributed. In particular, assume that $\log Y_j \sim N(\xi, \varsigma^2)$. Since $Y_1, \ldots, Y_n, \ldots$

are independent, we have $\sum_{j=1}^{n} \log Y_j \sim N(n\xi, n\varsigma^2)$. Then conditional on $N_t = n$, $\log S_t$ follows a Gaussian distribution with mean $\log S_0 + (\mu - \sigma^2/2)t + n\xi$ and variance $\sigma^2 t + n\varsigma^2$. Since N_t has a Poisson distribution with mean λt, the unconditional distribution of S_t is expressed as a Poisson mixture of lognormal distributions, i.e.,

$$P(S_t \leq x) = \sum_{n=0}^{\infty} e^{-\lambda t} \frac{(\lambda t)^n}{n!} F_{n,t}(x),$$

in which $F_{n,t}(x)$ is the distribution function of the lognormal distribution with mean $\log S_0 + (\mu - \sigma^2/2)t + n\xi$ and variance $\sigma^2 t + n\varsigma^2$. Making use of this property, Merton (1976) express the price of an option on S_t as an infinite series, each term of which is the product of a Poisson probability and a Black-Scholes formula.

The solution of the SDE (7.23) is given by (7.25). It suggests that a sample path of X_t with the above specified J_t and Y_j can be generated as follows.

Algorithm 7.6 (Simulation of jump-diffusion processes) *Let $0 = t_0 < t_1 < t_2 < \cdots < t_n$ be the set of distinct time points.*

1. *Generate a sample path of the Wiener process $\{W_{t_k}; k = 1, \ldots, n\}$.*

2. *Generate a sample path of the compounded Poisson process with arrival times $\{\tau_j; j = 1, 2, \ldots\}$ and jumps $\{Y_j; j = 1, \ldots, N_{t_n}\}$.*

3. *Compute the sample S_{t_k} for each t_k as follows,*

$$S_{t_k} = S_0 \exp\left(\left(\mu - \frac{\sigma^2}{2}\right)t_k + \sigma W_{t_k}\right) \cdot \prod_{j=1}^{N_{t_k}} Y_j. \tag{7.26}$$

Using the above algorithm, Figure 7.9 shows a realization of the jump-diffusion process S_t with $\mu = 0.05, \sigma = 0.25, \lambda = 1.8, \xi = 0.05$, and $\varsigma = 0.3$. In particular, the left panel of Figure 7.9 shows the diffusion component (i.e., the continuous part) of the process S_t, and the discontinuous part of S_t is shown in the right panel of Figure 7.9.

The assumption of lognormal distribution for Y_j can be replaced by other distributions. For instance, Kou (2002) assumed that $\log Y_j$ follow an asymmetric double exponential distribution and found analytical approximation for finite horizon American options. Sample paths for these and more general jump-diffusion processes can be generated by modifying the above simulation procedure.

7.2.7   Purely jump and Lévy processes

A d-dimensional *Lévy process* $\{\mathbf{X}_t\}_{t \geq 0}$ is a stochastic process with stationary and independent increments such that $\mathbf{X}_t$ converges in distribution

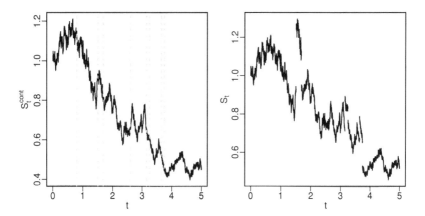

FIGURE 7.9: A realizations of the jump-diffusion process (7.25) with $\mu = 0.05, \sigma = 0.25, \lambda = 1.8, \xi = 0.05$, and $\varsigma = 0.3$. Left: The continuous part of S_t with jump times (vertical dashed lines). Right: The jump-diffusion process S_t.

to $\mathbf{X}_s$ as $t \to s$. Here, stationary increments means that $\mathbf{X}_{t+s} - \mathbf{X}_s$ has the distribution of $\mathbf{X}_t$. Every Lévy process can be represented as the sum of a deterministic drift, a Brownian motion, and a pure-jump process which are independent of the Brownian motion; see for example, Sato (1999, Chapter 4) and Tankov and Cont (2003, Chapter 3). If the number of jumps in every finite interval is almost surely finite, then the pure-jump component is a compound Poisson process. Hence, to generate a sample path of the Lévy process, the only step that goes beyond the jump-diffusion process (7.23) is to consider processes with an infinite number of jumps in finite intervals. For this reason, this section will only focus on pure-jump processes of this type, that is, Lévy processes with no Brownian component.

To simulate a pure-jump Lévy process, we consider the distribution of its increments over a fixed time interval, which involves of the concept of infinitely divisible distributions. The distribution of a random variable Y is said to be *infinitely divisible* if for each $m = 2, 3, \ldots$, there are i.i.d. random variables $Y_1^{(m)}, \ldots, Y_m^{(m)}$ such that $Y_1^{(m)} + \cdots + Y_m^{(m)}$ has the distribution of Y. Examples of infinitely divisible distributions include the normal distribution, the Gamma distribution, the Poisson distribution and the Cauchy distribution.

If X_t is Lévy process with $X_0 = 0$, then by definition of X_t, X_t can be decomposed as the sum of m i.i.d. random variables, i.e.,

$$X_t = \sum_{i=1}^{m} \left(X_{it/m} - X_{(i-1)t/m} \right).$$

This shows that X_t has an infinitely divisible distribution. Conversely, for each infinitely divisible distribution there is a Lévy process for which X_1 has that distribution. This indicates that simulating a Lévy process on a fixed time interval is equivalent to sampling from infinitely divisible distributions.

Gamma processes

As mentioned earlier, gamma distributions are infinitely divisible. This can be seen from the fact that, if $Y_1, \ldots, Y_m$ are independent with distribution Gamma$(\alpha/m, \beta)$, then $Y_1 + \cdots + Y_m$ has Gamma(α, β) distribution. Hence, given $\alpha > 0$ and $\beta > 0$, there is a Lévy process (called a *Gamma process*) such that X_1 has distribution Gamma(α, β). This indicates that, given a set of time points $t_1 < t_2 < \cdots < t_n$, the increment $X_{t_{k+1}} - X_{t_k}$ on a time grid $t_1, \ldots, t_n$ can be independently sampled from the Gamma distribution

$$X_{t_{k+1}} - X_{t_k} \sim \text{Gamma}((t_{k+1} - t_k)\alpha, \beta).$$

The left panel of Figure 7.10 shows a sample path of the Gamma process X_t with X_1 following a Gamma$(25, 10)$ distribution at times $\{t_k = k/n; k = 1, \ldots, n, n = 10^3\}$.

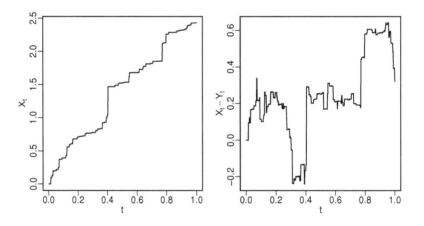

FIGURE 7.10: Left: A sample path of the Gamma process X_t with X_1 following a Gamma$(25, 10)$ distribution. Right: a sample path of the variance Gamma process $X_t - Y_t$ in which X_t and Y_t are two Gamma processes with X_1 and Y_1 following Gamma$(25, 10)$ and Gamma$(20, 15)$ distributions, respectively.

A gamma random variables takes only positive values so a gamma process is non-decreasing. To generate a sample path with ups and downs which mimics the asset price movement, one can use the *variance gamma process*, which was proposed by Madan and Seneta (1990) and is the difference of two independent gamma processes representing the up and down moves of the process. Denote by X_t and Y_t the Gamma process with X_1 and Y_1 following Gamma$(25, 10)$ and Gamma$(20, 15)$ distributions, respectively. The right panel of Figure 7.10 shows a sample path of the variance Gamma process $X_t - Y_t$ at times $\{t_k = k/n; k = 1, \ldots, n, n = 10^3\}$.

7.3 Variance reduction methods

To compute the mean $\mu = \int_{-\infty}^{\infty} x dF(x)$ of a random variable X, direct Monte Carlo involves generating n independent replicates $X_1, \ldots, X_n$ of X so that $n^{-1} \sum_{i=1}^{n} X_i$ can be used to estimate $E(X)$ with standard error $\sigma/\sqrt{n}$, where $\sigma^2 = \text{Var}(X)$. Using variance reduction methods, one can reduce the standard error of the estimates or increase the efficiency of estimates that are based on direct Monte Carlo simulations. This section introduces these methods, including antithetic variates, control variates, stratified sampling, and importance sampling.

7.3.1 Antithetic variates

A pair of real-valued random variables $(X, \widetilde{X})$ is called an *antithetic pair* if X and $\widetilde{X}$ have the same distribution and are negatively correlated. The method of *antithetic variables* aims to reduce variance by introducing negative dependence between pairs of replications. In particular, let $(X_i, \widetilde{X}_i)$, $i = 1, \ldots, n$, be n independent antithetic pairs of random variables, where each X_k and $\widetilde{X}_k$ is distributed as X. Then the antithetic estimator is given by

$$\widehat{\mu} = (2n)^{-1} \sum_{i=1}^{n} (X_i + \widetilde{X}_i).$$

Since $E(\widehat{\mu}) = E(X)$, $\widehat{\mu}$ is an unbiased estimator of $\mu = E(X)$. The variance of $\widehat{\mu}$ is given by

$$\text{Var}(\widehat{\mu}) = \frac{1}{n} \text{Var}\Big((X_1 + \widetilde{X}_1)/2\Big) = \frac{1}{2n} \{\text{Var}(X) + \text{Cov}(X, \widetilde{X})\} = \frac{\text{Var}(X)}{2n}(1 + \rho_{X,\widetilde{X}}).$$

To reduce the variance of $\widehat{\mu}$ so that it is smaller than $\text{Var}(X)/(2n)$, we need the condition for antithetic sampling, $\rho_{X,\widetilde{X}} < 0$.

In general, suppose that the random variable X has the form $X = h(\mathbf{U})$, where h is a real-valued function and $\mathbf{U} = (U_1, U_2, \ldots)$ is a random vector of i.i.d. $\mathcal{U}(0, 1)$ random variables. Let $\widetilde{\mathbf{U}}$ be another vector of i.i.d. $\mathcal{U}(0, 1)$ random variables which is dependent on $\mathbf{U}$ and for which X and $\widetilde{X} = h(\widetilde{\mathbf{U}})$ are negatively correlated. Then $(X, \widetilde{X})$ is an antithetic pair. In particular, if h

is a monotone function in each of its components, then the choice $\tilde{\mathbf{U}} = \mathbf{1} - \mathbf{U}$, where $\mathbf{1}$ is the vector of 1's, yields an antithetic pair.

7.3.2   Control variates

The method of *control variates* improves the efficiency of Monte Carlo simulation by exploiting information of estimation errors of known quantities to reduce estimation errors of an unknown quantity. The method involves simulating m i.i.d. pairs (X_i, Y_i) in which Y_i has *known* expectation $E(Y)$. Then for any fixed b, we can estimate $\mu := E(X)$ by

$$\widehat{\mu}_n(b) = \frac{1}{n} \sum_{i=1}^{n} \big[X_i - b\big(Y_i - E(Y)\big) \big]. \tag{7.27}$$

This is a control variate estimator in which the observed error $n^{-1} \sum_{i=1}^{n} Y_i - E(Y)$ and the coefficient b serve as a control and a control variable in estimating μ, respectively.

It is easy to see that $E(\widehat{\mu}_n(b)) = E(X)$, hence $\widehat{\mu}_n(b)$ is unbiased for any b. Besides, $\widehat{\mu}_n(b)$ is consistent because, with probability 1,

$$\lim_{n \to \infty} \widehat{\mu}_n(b) = \lim_{n \to \infty} \frac{1}{n} \sum_{i=1}^{n} \big[X_i - bY_i \big] + bE(Y) = E(X) = \mu.$$

Let $\sigma_X^2 = \text{Var}(X)$, $\sigma_Y^2 = \text{Var}(Y)$, and $\rho_{XY} = \text{Corr}(X, Y)$, the variance of $\widehat{\mu}_n(b)$ is given by

$$\text{Var}(\widehat{\mu}_n(b)) = \frac{1}{n} \big[\sigma_X^2 - 2b\sigma_X\sigma_Y\rho_{XY} + b^2\sigma_Y^2 \big] = \frac{1}{n} \big[(b\sigma_Y - \sigma_X\rho_{XY})^2 + (1 - \rho_{XY}^2)\sigma_X^2 \big].$$

This variance is minimized by choosing

$$b^* = \frac{\sigma_X\rho_{XY}}{\sigma_Y} = \frac{\text{Cov}(X, Y)}{\text{Var}(Y)}.$$

Then the ratio of the variance of the optimally controlled estimator to that of the uncontrolled estimator is

$$\text{Var}(\widehat{\mu}_n(b^*))/\text{Var}(\widehat{\mu}_n(0)) = (1 - \rho_{XY}^2), \tag{7.28}$$

which is smaller than 1 if $\rho_{XY} \neq 0$. Hence, the effectiveness of a control variate, measured by the variance reduction ratio (7.28), is determined by the correlation of X and Y.

The above discussion assumes $\text{Var}(Y)$ and $\text{Cov}(X, Y)$ are known. In practice, the values $\text{Var}(Y)$ and $\text{Cov}(X, Y)$ are unknown and can be estimated from the simulated samples $\{Y_1, \ldots, Y_n\}$ and $\{(X_1, Y_1), \ldots, (X_n, Y_n)\}$, then b^* should be replaced by its sample estimate. A more practical concern is how to find variable Y which is correlated to X, which usually depends on the specific problem we are dealing with and may not be easy to find a good candidate for Y; see Exercise 7.11 for an application of the method in option pricing.

7.3.3 Stratified sampling and Latin hypercube sampling

Instead of sampling from a large sample space, sampling representative values from its strata can reduce the variance of the Monte Carlo estimate of $E(X)$. Suppose that X can be generated via the composition method. Assume that there exists a random variable Z taking values in $\{1, \ldots, K\}$ with known probabilities $\{p_k, k = 1, \ldots, K\}$. The events $\{Z = k\}$, $k = 1, \ldots, K$, partition the sample space Ω into disjoint *strata*. Then the mean $E(X)$ can be expressed as

$$\mu = E(X) = E(E(X|Z)) = \sum_{k=1}^{K} p_k E(X|Z = k).$$

The above representation suggests that we can estimate μ via the following *stratified sampling estimator*

$$\widehat{\mu} = \sum_{k=1}^{K} p_k \overline{Y}_{k\cdot}, \qquad \overline{Y}_{k\cdot} := \frac{1}{N_k} \sum_{j=1}^{N_k} Y_{kj},$$

where Y_{kj} is the j-th of N_k independent observations from the conditional distribution of Y given $Z = k$, $k = 1, \ldots, K$. Then the variance of the stratified sampling estimator is expressed as

$$\mathrm{Var}(\widehat{\mu}) = \sum_{k=1}^{K} \frac{p_k^2 \sigma_k^2}{N_k}, \qquad (7.29)$$

in which $\sigma_k^2 = \mathrm{Var}(Y|Z = k)$ is the variance of Y within the k-th stratum, $k = 1, \ldots, K$. The next step is to minimize the variance (7.29) subject to $\sum_{k=1}^{K} N_k = N$. One can show that, by choosing samples sizes

$$N_k = N \frac{p_k \sigma_k}{\sum_{i=1}^{K} p_i \sigma_i}, \qquad k = 1, \ldots, K,$$

we obtain the smallest variance for $\widehat{\mu}$, which is given by $\left(\sum_{k=1}^{K} p_k \sigma_k \right)^2 / N$.

If the sample size for the k-th stratum is chosen to be proportional to p_k, that is, $N_k = p_k N$ for some overall sample size N, then

$$\mathrm{Var}(\widehat{\mu}) = \sum_{k=1}^{K} \frac{p_k^2 \sigma_k^2}{N_k} = \frac{1}{N} E(\mathrm{Var}(Y|Z)) \leq \frac{1}{N} \mathrm{Var}(Y),$$

so that the stratified estimator in this case has a variance at least as small as the variance of the conditional Mone Carlo estimator. This is called *proportional stratified sampling*.

The idea of stratified sampling can be extended for sampling in multiple dimensions, which is usually called *Latin hypercube sampling*. The difficulty of stratified sampling lies in the curse of dimensionality. For instance, consider the case of sampling from the d-dimensional hypercube $[0,1)^d$. Partitioning each coordinate into K strata produces K^d strata for the hypercube, which requires a sample size of at least K^d to ensure that each stratum is sampled. This is computationally prohibitive even for moderately large d.

7.3.4   Conditional Monte Carlo

The conditional Monte Carlo idea is sometimes referred to as *Rao-Blackwellization*. Suppose there exists a random vector $\mathbf{Z}$ that can be easily simulated and such that $E(h(\mathbf{X})|\mathbf{Z})$ has an explicit expression $g(\mathbf{Z})$. Since

$$\mu := E(h(\mathbf{X})) = E\{E[h(\mathbf{X})|\mathbf{Z}]\} = E\{g(\mathbf{Z})\}$$

and

$$\mathrm{Var}\big(h(\mathbf{X})\big) = E\{\mathrm{Var}(h(\mathbf{X})|\mathbf{Z})\} + \mathrm{Var}\big(E[h(\mathbf{X}|\mathbf{Z})]\big) \geq \mathrm{Var}\big(E[h(\mathbf{X}|\mathbf{Z})]\big) = \mathrm{Var}[g(\mathbf{Z})], \tag{7.30}$$

it is more efficient to simulate $\mathbf{Z}_1, \ldots, \mathbf{Z}_n$ and estimate μ by

$$\widehat{\mu} = \frac{1}{n} \sum_{i=1}^{n} g(\mathbf{Z}_j), \tag{7.31}$$

7.3.5   Importance sampling

One of the most important variance reduction techniques is *importance sampling*. Let $f(\mathbf{x})$ be the density of a multivariate random variable $\mathbf{X}$, or the *nominal probability density function*, and $h(\mathbf{X})$ a real-valued function of $\mathbf{X}$. Denote by $g(\mathbf{x})$ another probability density such that $h(\mathbf{x})f(\mathbf{x})$ is dominated by $g(\mathbf{x})$, i.e., $g(\mathbf{x}) = 0$ implies that $h(\mathbf{x})f(\mathbf{x}) = 0$. Then

$$E_f[h(\mathbf{X})] = \int_{\mathcal{X}} h(\mathbf{x})f(\mathbf{x})dx = \int_{\mathcal{X}} h(\mathbf{x})\frac{f(\mathbf{x})}{g(\mathbf{x})}g(\mathbf{x})dx = E_g\big[h(\mathbf{X})L(\mathbf{X})\big] = E_g\big[\widetilde{h}(\mathbf{X})\big],$$

in which $\widetilde{h}(\mathbf{x}) = h(\mathbf{x})L(\mathbf{x})$, and $L(\mathbf{x}) = f(\mathbf{x})/g(\mathbf{x})$ is the likelihood ratio of $f(\mathbf{X})$ with respect to $g(\mathbf{X})$. Instead of sampling the $\mathbf{x}_i$ from distribution $f(\mathbf{X})$, importance sampling samples i.i.d. $\mathbf{x}_i$ $(i = 1, \ldots, n)$ from distribution $g(\mathbf{X})$ and estimates $\mu := E_f[h(\mathbf{X})]$ by

$$\widehat{\mu} = \frac{1}{n} \sum_{i=1}^{n} h(\mathbf{x}_i)L(\mathbf{x}_i). \tag{7.32}$$

It is easy to see that $\widehat{\mu}$ is an unbiased estimator of μ. The density $g(\mathbf{x})$ is called the *importance sampling density* and the estimator $\widehat{\mu}$ is called the *importance sampling estimator*.

An important step in the above procedure is to choose $g(\mathbf{x})$ for both simplicity in generating Monte Carlo samples and accuracy in estimating

$E_f[h(\mathbf{X})]$ by minimizing the variance of the estimator of $E_f[h(\mathbf{X})]$ in (7.32). To find such $g(\mathbf{x})$, we note that

$$\mathrm{Var}_g[\widetilde{h}(\mathbf{X})] = E_g[\widetilde{h}^2(\mathbf{X})] - (E_g[\widetilde{h}(\mathbf{X})])^2 = \int_{\mathcal{X}} \frac{h^2(\mathbf{x})f^2(\mathbf{x})}{g(\mathbf{x})}dx - \left(E_f[h(\mathbf{X})]\right)^2$$

$$= \int_{\mathcal{X}} \frac{[h(\mathbf{x})f(\mathbf{x}) - E_f[h(\mathbf{X})]g(\mathbf{x})]^2}{g(\mathbf{x})}d\mathbf{x}.$$

Then one can choose the following density function $g^*(\mathbf{x})$ to minimize $\mathrm{Var}_g[\widetilde{h}(\mathbf{X})]$,

$$g^*(\mathbf{x}) = \frac{h(\mathbf{x})f(\mathbf{x})}{\int_{\mathcal{X}} h(\mathbf{y})f(\mathbf{y})dy} = \frac{1}{\mu}h(\mathbf{x})f(\mathbf{x}). \tag{7.33}$$

Under the above choice of $g(\mathbf{x})$, $\mathrm{Var}_{g^*}(\widehat{\mu}) = 0$ so that the estimator $\widehat{\mu}$ is constant under g^*. Obviously, it is not possible in practice to find the above optimal importance sampling density g^*, but a good importance sampling density g that is close to the minimum variance density g^* can usually be found. In general, one of the main consideration for choosing a good importance sampling probability density function is that the estimator (7.32) should have finite variance. This is equivalent to the requirement that $E_f[h^2(\mathbf{X})f(\mathbf{X})/g(\mathbf{X})] < \infty$, which implies that g should not have lighter tails than f and that the likelihood ratio f/g should be bounded. More detailed discussion on importance sampling and its applications to rare event simulation is postponed to Section 7.5.

7.4   Markov chain Monte Carlo methods

When generating i.i.d. samples from the target distribution is infeasible, dependent samples $\{\mathbf{X}_i\}$ sampling from the distribution $f(\mathbf{X})$ can be used instead, provided that the sample mean of $h(\mathbf{X}_1), \ldots, h(\mathbf{X}_n)$ converges to $E_f[h(\mathbf{X})]$. A particular class of such dependent sequences that can be simulated is the class of Markov chains. Let $P_0(d\mathbf{x})$ be the initial distribution of a Markov chain $\{\mathbf{X}_t\}$, the Markov chain $\{\mathbf{X}_t\}$ evolves according to

$$P_{t+1}(d\mathbf{y}) = \int P_t(d\mathbf{x})P_t(\mathbf{x}, d\mathbf{y}),$$

where $P_t(\mathbf{x}, d\mathbf{y})$ is the transition probability function at time t. A commonly used class of Markov chains in MCMC is the class of time-homogeneous Markov chains or stationary Markov chains where

$$P_t(\mathbf{x}, d\mathbf{y}) = P(\mathbf{x}, d\mathbf{y}), \qquad t = 1, 2, \ldots. \tag{7.34}$$

The basic idea of creating Markov chains for approximating $E_\pi(h(\mathbf{X}))$ is to construct a transition kernel $P(\mathbf{x}, d\mathbf{y})$ with $\pi(d\mathbf{x})$ as its invariant distribution, that is, $P(\mathbf{x}, d\mathbf{y})$ and $\pi(d\mathbf{x})$ satisfy the condition

$$\pi(d\mathbf{y}) = \int \pi(d\mathbf{x})P(\mathbf{x}, d\mathbf{y}). \tag{7.35}$$

When the target distribution π has the density $f(\mathbf{x})$ and the transition kernel $P(\mathbf{x}, d\mathbf{y})$ has the conditional density $p(\mathbf{y}|\mathbf{x})$, condition (7.35) can be written as

$$f(\mathbf{y}) = \int p(\mathbf{y}|\mathbf{x})f(\mathbf{x})dx.$$

The condition (7.34) or (7.35) says that, if $\mathbf{X}_t$ is a draw from the target $\pi(\mathbf{x})$ then $\mathbf{X}_{t+1}$ is also a draw, possibly dependent on $\mathbf{X}_t$, from $\pi(\mathbf{x})$. Moreover, for almost any $P_0(d\mathbf{x})$ under mild conditions $P_t(d\mathbf{x})$ converges to $\pi(d\mathbf{x})$.

Suppose that $\{\mathbf{X}_i\}$ is a simulated Markov chain, one can approximate $E_\pi(h) = \int h(\mathbf{x})\pi(d\mathbf{x})$ by

$$\bar{h}_{m,n} = \frac{1}{n}\sum_{i=1}^{n} h(\mathbf{X}_{i+m}), \tag{7.36}$$

where m is non-negative integer denoting the length of the so-called *burn-in* period.

7.4.1   The Gibbs sampler

The Gibbs sampler (Geman and Geman, 1984) is a method of generating d-dimensional random vectors, in which the underlying Markov chain is constructed from a sequence of conditional distributions. Suppose that we want to sample a random vector $\mathbf{X} = (X_1, \ldots, X_d)$ according to a target probability density function $f(\mathbf{x})$. Let $f_k(X_k|X_1, \ldots, X_{k-1}, X_{k+1}, \ldots, X_d)$ represent the conditional density function of the k-th component. The Gibbs sampler can be described as follows.

Algorithm 7.7 (Gibbs sampler) *Given an initial state $\mathbf{X}_0$, iterate the following steps for $t = 0, 1, \ldots$*

1. *For a given $\mathbf{X}_t$, generate $\mathbf{Y} = (Y_1, \ldots, Y_d)$ as follows:*

   (a) *Draw Y_1 from the conditional probability density function $f_1(x_1|X_{t,1}, \ldots, X_{t,d})$.*

   (b) *Draw Y_i from $f_i(x_i|Y_1, \ldots, Y_{i-1}, X_{t,i+1}, \ldots, X_{t,d})$ for $i = 2, \ldots, d - 1$.*

   (c) *Draw Y_d from $f_n(x_d|Y_1, \ldots, Y_{d-1})$.*

2. *Let $\mathbf{X}_{t+1} = \mathbf{Y}$.*

For illustration purpose, we consider the following example.

Example 7.7 (Normal bivariate Gibbs) *To simulate a X from $N(0,1)$ using Gibbs sampler, we consider a bivariate variable $(X, Y)'$ which follows a bivariate normal distribution*

$$\begin{pmatrix} X \\ Y \end{pmatrix} \sim N_2 \left(\begin{pmatrix} 0 \\ 0 \end{pmatrix}, \begin{pmatrix} 1 & \rho \\ \rho & 1 \end{pmatrix} \right). \qquad (7.37)$$

Algorithm 7.7 implies the following Gibbs sampler: Given x_t, generate $Y_{t+1}|x_t \sim N(\rho x_t, 1 - \rho^2)$, and then given y_{t+1}, generate $x_{t+1}|y_{t+1} \sim N(\rho y_{t+1}, 1 - \rho^2)$. This implies that the transition kernel in the chain $\{X_t\}$ is given by

$$P(x_t, x_{t+1}) = \frac{1}{2\pi(1 - \rho^2)} \int e^{-\frac{1}{2(1-\rho^2)} \left[(x_{t+1}-\rho y)^2 + (y-\rho x_t)^2 \right]} dy,$$

and the corresponding marginal Markov chain in X is an AR(1) process

$$X_{t+1} = \rho^2 X_t + \epsilon_t, \qquad \epsilon_t \sim N(0, 1 - \rho^4).$$

By the argument in Section 2.4.2, the stationary variance of X_t is 1, and the stationary distribution of the chain $\{X_t\}$ is $N(0,1)$. $\square$

In Algorithm 7.7, the transition probability density function from $\mathbf{x}$ to $\mathbf{y}$ is given by

$$\kappa_{1 \to d}(\mathbf{y}|\mathbf{x}) = \prod_{i=1}^{d} f(y_i|y_1, \dots, y_{i-1}, x_{i+1}, \dots, x_d).$$

The transition density of the reverse move is

$$\kappa_{d \to 1}(\mathbf{x}|\mathbf{y}) = \prod_{i=1}^{d} f(x_i|y_1, \dots, y_{i-1}, x_{i+1}, \dots, x_d).$$

Observing that

$$\frac{\kappa_{1 \to d}(\mathbf{y}|\mathbf{x})}{\kappa_{d \to 1}(\mathbf{x}|\mathbf{y})} = \prod_{i=1}^{d} \frac{f(y_i|y_1, \dots, y_{i-1}, x_{i+1}, \dots, x_d)}{f(x_i|y_1, \dots, y_{i-1}, x_{i+1}, \dots, x_d)} = \prod_{i=1}^{d} \frac{f(y_1, \dots, y_i, x_{i+1}, \dots, x_d)}{f(y_1, \dots, y_{i-1}, x_i, \dots, x_d)}$$

$$= \frac{f(\mathbf{y}) \prod_{i=1}^{d-1} f(y_1, \dots, y_i, x_{i+1}, \dots, x_d)}{f(\mathbf{x}) \prod_{j=2}^{d} f(y_1, \dots, y_{j-1}, x_j, \dots, x_d)} = \frac{f(\mathbf{y})}{f(\mathbf{x})}.$$

This implies the following result (Hammersley and Clifford, 1970; Besag, 1974; Gelman and Speed, 1993).

Theorem 7.1 (Hammersley-Clifford) *Let $f(x_i)$ be the i-th marginal density of the probability density function $f(\mathbf{x})$. Suppose that density $f(\mathbf{x})$ satisfies the positive condition, that is, for every $\mathbf{y} \in \{\mathbf{x} : f(x_i) > 0, i = 1, \ldots, d\}$, we have $f(\mathbf{y}) > 0$. Then,*

$$f(\mathbf{y})\kappa_{d \to 1}(\mathbf{x}|\mathbf{y}) = f(\mathbf{x})\kappa_{1 \to d}(\mathbf{y}|\mathbf{x}). \tag{7.38}$$

Integrate both sides of (7.38) with respect to $\mathbf{x}$ yields that

$$\int f(\mathbf{x})\kappa_{1 \to d}(\mathbf{y}|\mathbf{x})d\mathbf{x} = f(\mathbf{y}),$$

from which we can conclude that f is the stationary probability density function of the Markov chain with transition density $\kappa_{1 \to d}(\mathbf{y}|\mathbf{x})$. In addition, it can be shown that the positivity assumption on f implies that the Gibbs Markov chain is irreducible and that f is its limiting probability density function (Robert and Casella, 2004). In practice the positivity condition is difficult to verify. However, there are a number of weaker and more technical conditions which ensure that the limiting probability density function of the process $\{\mathbf{X}_t, t = 1, 2, \ldots\}$ generated via the Gibbs sampler is f, and that the convergence to f is geometrically fast.

Gibbs sampling for stochastic volatility models

An important application of MCMC methods in empirical finance is the inference of stochastic volatility models; see Jacquier et al. (1994, 1997); Chib et al. (2006) and reference therein. To illustrate the idea, we consider a univariate stochastic volatility model in which the mean and volatility of a series x_t are

$$x_t = \boldsymbol{\beta}'\mathbf{z}_t + u_t, \qquad u_t = e^{h_t/2}\epsilon_t, \tag{7.39}$$

$$h_t = \phi_0 + \phi_1 h_{t-1} + \cdots + \phi_p h_{t-p} + \eta_t, \tag{7.40}$$

where $\{\mathbf{z}_t = (1, z_{t,1}, \ldots, z_{t,k})'\}$ are explanatory variables available at time $t - 1$, $\boldsymbol{\beta} \in \mathbb{R}^{k+1}$ are parameters, $\{\epsilon_t\}$ is a series of i.i.d. Gaussian random variables with mean 0 and variance 1, $\{\eta_t\}$ is a series of Gaussian random variables with mean 0 and variance σ_η^2, and $\{\epsilon_t\}$ and $\{\eta_t\}$ are independent. To make sure that the log volatility process h_t is stationary, we assume that roots of $1 - \phi_1 y - \cdots - \phi_p y^p = 0$ lie outside the unit circle. This model extends the univariate stochastic volatility model in Section 2.3.3 by adding explanatory variables into the mean of the series $\{x_t\}$.

Denote the coefficient vector of the volatility equation by $\boldsymbol{\phi} = (\phi_0, \phi_1, \ldots, \phi_p)'$ and the parameter vector of the volatility equation by $\boldsymbol{\omega} = (\boldsymbol{\phi}', \sigma_\eta^2)'$. Suppose that $\mathbf{X} = (x_1, \ldots, x_n)'$, $\mathbf{Z} = (\mathbf{z}_1, \ldots, \mathbf{z}_n)$, and $\mathbf{H} = (h_1, \ldots, h_n)'$ are the collection of observed returns, explanatory variables, and unobserved volatilities, respectively. The standard maximum likelihood method for estimation of $(\boldsymbol{\beta}, \boldsymbol{\omega})$ is difficult because the likelihood function involves a n-fold integral with respect to $\mathbf{H}$ and is almost impossible to evaluate, that is,

$$f(\mathbf{X}|\mathbf{Z}, \boldsymbol{\beta}, \boldsymbol{\omega}) = \int f(\mathbf{X}|\mathbf{Z}, \boldsymbol{\beta}, \mathbf{H}) f(\mathbf{H}|\boldsymbol{\omega}) d\mathbf{H}.$$

Due to the above difficulty, we make inference on the model by the MCMC method, or more specifically, the Gibbs sampling approach. Suppose that $\boldsymbol{\beta}$, $\boldsymbol{\phi}$, and σ_η^2 have independent prior distributions

$$\boldsymbol{\beta} \sim N_{k+1}(\boldsymbol{\mu}_1, \mathbf{V}_1), \qquad \boldsymbol{\phi} \sim N_{p+1}(\boldsymbol{\mu}_2, \mathbf{V}_2), \qquad (m\lambda)/\sigma_\eta^2 \sim \chi_m^2.$$

We next explain how to use Gibbs sampling approach to draw random samples from conditional posterior distributions $f(\boldsymbol{\beta}|\mathbf{X}, \mathbf{Z}, \mathbf{H}, \boldsymbol{\omega})$, $f(\mathbf{H}|\mathbf{X}, \mathbf{Z}, \boldsymbol{\beta}, \boldsymbol{\omega})$, and $f(\boldsymbol{\omega}|\mathbf{X}, \mathbf{Z}, \mathbf{H}, \boldsymbol{\beta})$.

Consider first the posterior distribution $f(\boldsymbol{\beta}|\mathbf{X}, \mathbf{Z}, \mathbf{H}, \boldsymbol{\omega})$. Rewrite equation (7.39) as $x_t e^{-h_t/2} = \boldsymbol{\beta}'(\mathbf{z}_t e^{-h_t/2}) + \epsilon_t$, for $t = 1, \ldots, n$. Since the prior distribution of $\boldsymbol{\beta}$ is a $(k+1)$-dimensional normal $N_{k+1}(\boldsymbol{\mu}_1, \mathbf{V}_1)$, it is easy to show that the posterior distribution of $\boldsymbol{\beta}$ is a multivariate normal $N_{k+1}(\tilde{\boldsymbol{\mu}}_1, \tilde{\mathbf{V}}_1)$, where

$$\tilde{\boldsymbol{\mu}}_1 = \tilde{\mathbf{V}}_1 \left(\mathbf{V}_1^{-1} \boldsymbol{\mu}_1 + \sum_{i=1}^n e^{-h_t} \mathbf{z}_t x_t \right), \qquad \tilde{\mathbf{V}}_1 = \left(\mathbf{V}_1 + \sum_{i=1}^n e^{-h_t} \mathbf{z}_t \mathbf{z}_t' \right)^{-1}.$$

To draw $\mathbf{H}$ from the conditional distribution $f(\mathbf{H}|\mathbf{X}, \mathbf{Z}, \boldsymbol{\beta}, \boldsymbol{\omega})$, we can use the Gibbs sampling method by finding the conditional distribution $f(h_t|\mathbf{X}, \mathbf{Z}, \mathbf{H}_{-t}, \boldsymbol{\beta}, \boldsymbol{\omega})$, where $\mathbf{H}_{-t} = \mathbf{H} - \{h_t\}$. Note that

$$f(h_t|\mathbf{X}, \mathbf{Z}, \mathbf{H}_{-t}, \boldsymbol{\beta}, \boldsymbol{\omega})$$

$$\propto \quad f(u_t|h_t, x_t, \mathbf{z}_t, \boldsymbol{\beta}) f(h_t|h_{t-p}, \ldots, h_{t-1}, \boldsymbol{\omega}) \prod_{i=1}^p f(h_{t+i}|h_{t+i-1}, \ldots, h_{t+i-p}, \boldsymbol{\omega})$$

$$\propto \quad e^{-h_t/2} \exp\left[-\frac{1}{2} e^{-h_t} (x_t - \boldsymbol{\beta}' \mathbf{z}_t)^2 - \frac{1}{2\sigma_\eta^2} \left((1 + \sum_{i=1}^p \phi_i^2) h_t^2 - 2 h_t \psi_t \right) \right], \quad (7.41)$$

where $\psi_t := \phi_0(1 - \sum_{i=1}^p \phi_i) + \sum_{i=1}^p (h_{t-i} + h_{t+i}) \phi_i$. Since the AR($p$) model of h_t involves p initial values of h_t, i.e., $h_1, \ldots, h_p$, we can either assume they are fixed or use their backward prediction before applying formula (7.41). Besides, (7.41) implies the Gibbs sampler for h_t, but h_t can also be drawn by other methods, see for example, the Metropolis algorithm used by Jacquier et al. (1994).

The volatility parameter $\boldsymbol{\omega} = (\boldsymbol{\phi}', \sigma_\eta^2)'$ can be drawn in two steps due to the decomposition

$$f(\boldsymbol{\omega}|\mathbf{X}, \mathbf{Z}, \mathbf{H}, \boldsymbol{\beta}) = f(\boldsymbol{\phi}|\mathbf{X}, \mathbf{Z}, \mathbf{H}, \boldsymbol{\beta}, \sigma_\eta^2) f(\sigma_\eta^2|\mathbf{X}, \mathbf{Z}, \mathbf{H}, \boldsymbol{\beta}) = f(\boldsymbol{\phi}|\mathbf{H}, \sigma_\eta^2) f(\sigma_\eta^2|\mathbf{H}, \boldsymbol{\phi}).$$

First, given that h_t follows an AR(p) model and the AR coefficient ϕ has a $(p+1)$-dimensional $N_{p+1}(\boldsymbol{\mu}_2, \mathbf{V}_2)$ prior, the conditional distribution $f(\phi|\mathbf{H}, \sigma_\eta^2)$ is a multivariate normal $N_{p+1}(\widetilde{\boldsymbol{\mu}}_2, \widetilde{\mathbf{V}}_2)$, where

$$\widetilde{\boldsymbol{\mu}}_2 = \widetilde{\mathbf{V}}_2 \Big(\mathbf{V}_2^{-1}\boldsymbol{\mu}_2 + \frac{1}{\sigma_\eta^2} \sum_{t=p}^{n} \mathbf{y}_t h_t \Big), \qquad \widetilde{\mathbf{V}}_2 = \Big(\mathbf{V}_2^{-1} + \frac{1}{\sigma_\eta^2} \sum_{t=p}^{n} \mathbf{y}_t \mathbf{y}_t' \Big)^{-1},$$

and $\mathbf{y}_t = (1, h_{t-1}, \ldots, h_{t-p})'$. Then by the assumption that $(m\lambda/\sigma_\eta^2)$ follows a chi-squared distribution χ_m^2 with m degrees of freedom, the posterior distribution of σ_η^2 given $\mathbf{H}$ and ϕ follows an inverted chi-square distribution with $m + n - 1$ degrees of freedom, i.e.,

$$\frac{m\lambda + \sum_{t=p}^{n} v_t^2}{\sigma_\eta^2} \sim \chi_{m+n-1}^2,$$

where $v_t := h_t - \phi_0 - \sum_{i=1}^{p} \phi_i h_{t-i}$.

7.4.2   The Metropolis-Hastings sampler

Suppose that we want to generate samples from an arbitrary multidimensional probability density function $f(\mathbf{x})$. Let $q(\mathbf{y}|\mathbf{x})$ be an instrumental density, i.e., a Markov transition density describing how to go from state $\mathbf{x}$ to $\mathbf{y}$. The *Metropolis-Hastings* algorithm is based on the following "trial-and-error" strategy.

Algorithm 7.8 (Metropolis-Hastings algorithm) *To sample from a density $f(\mathbf{x})$, initialize with some $\mathbf{X}_0$ for which $f(\mathbf{X}_0) > 0$. Then, for each $t = 0, 1, \ldots, T - 1$ execute the following steps:*

1. *Given the current state $\mathbf{X}_t$, generate $\mathbf{Y} \sim q(\mathbf{y}|\mathbf{X}_t)$ and compute the acceptance ratio*

$$\alpha(\mathbf{X}_t, \mathbf{Y}) = \min \Big\{ 1, \frac{f(\mathbf{Y})q(\mathbf{X}_t|\mathbf{Y})}{f(\mathbf{X}_t)q(\mathbf{Y}|\mathbf{X}_t)} \Big\}. \tag{7.42}$$

2. *Generate $U \sim \mathcal{U}(0,1)$ and let $\mathbf{X}_{t+1} = \mathbf{Y}$ if $U \leq \alpha(\mathbf{X}_t, \mathbf{Y})$ and $\mathbf{X}_{t+1} = \mathbf{X}_t$ otherwise.*

The samples $\mathbf{X}_0, \mathbf{X}_1, \ldots, \mathbf{X}_T$ generated by the above algorithm is called *Metropolis-Hastings Markov chain*, in which $\mathbf{X}_T$ approximately distributed according to $f(\mathbf{x})$ for large T. In each iteration, the algorithm generates a sample from the transition density $P(\mathbf{x}_{t+1}|\mathbf{x}_t)$, which is given by

$$P(\mathbf{y}|\mathbf{x}) = \begin{cases} \alpha(\mathbf{x}, \mathbf{y})q(\mathbf{y}|\mathbf{x}) & \text{if } \mathbf{y} \neq \mathbf{x}, \\ \int \alpha(\mathbf{x}, \mathbf{y})q(\mathbf{y}|\mathbf{x})d\mathbf{y} & \text{if } \mathbf{y} = \mathbf{x}. \end{cases}$$

It is easy to check that the transition density satisfies the equation $f(\mathbf{x})P(\mathbf{y}|\mathbf{x}) = f(\mathbf{y})P(\mathbf{x}|\mathbf{y})$. This shows that $f(\mathbf{x})$ is the stationary distribution density function of the chain $\{\mathbf{X}_t\}$. Moreover, if the transition density $q(\mathbf{y}|\mathbf{x})$ satisfies the condition $P(\alpha(\mathbf{X}_t, \mathbf{Y}) < 1|\mathbf{X}_t) > 0$, i.e., the event $\{\mathbf{X}_{t+1} = \mathbf{X}_t\}$ has positive probability, and $q(\mathbf{y}|\mathbf{x}) > 0$ for all $\mathbf{x}, \mathbf{y} \in \mathcal{X}$, then $f(\mathbf{x})$ is the limiting probability density function of the chain. In such a case, to estimate an expectation $E(h(\mathbf{X}))$, with $\mathbf{X} \sim f(\mathbf{x})$, we can use the *ergodic* estimator (7.36).

Sometimes, the proposal function $q(\mathbf{y}|\mathbf{x})$ can be chosen to simply the algorithm. For instance, if the proposal function $q(\mathbf{y}|\mathbf{x})$ does not depend on $\mathbf{x}$, i.e., $q(\mathbf{y}|\mathbf{x}) = g(\mathbf{y})$ for some probability density function $g(\mathbf{y})$, then the acceptance probability is $\alpha(\mathbf{x}, \mathbf{y}) = \min\{f(\mathbf{y})g(\mathbf{x})/[f(\mathbf{x})g(\mathbf{y})], 1\}$ and Algorithm 7.8 is referred to as the *independence sampler*. If the proposal is symmetric, that is $q(\mathbf{y}|\mathbf{x}) = q(\mathbf{x}|\mathbf{y})$, then the acceptance probability is $\alpha(\mathbf{x}, \mathbf{y}) = \min\{f(\mathbf{y})/f(\mathbf{x}), 1\}$. Algorithm 7.8 is referred to as the *random walk sampler*. The following example shows how to use random walk sampler to perform a Bayesian analysis for logistic regressions.

Example 7.8 (Bayesian analysis of logistic regressions) *Consider the Bayesian analysis of logistic regression. Let $\mathbf{X} \in \mathbb{R}^d$ be an explanatory variable for the binary response Y, and $\boldsymbol{\beta} \in \mathbb{R}^d$ the regression coefficient with multivariate normal prior $N(\boldsymbol{\beta}_0, \mathbf{V}_0)$. Conditional on $\mathbf{X} = \mathbf{x}_i$, $i = 1, \ldots, n$, the binary observations y_i are independent with Bernoulli distributions with the probability of success*

$$p_i = P(y_i = 1|\mathbf{X} = \mathbf{x}_i) = \frac{1}{1 + e^{-\mathbf{x}_i'\boldsymbol{\beta}}}, \qquad i = 1, \ldots, n.$$

Let $\mathbf{y} = (y_1, \ldots, y_n)'$ and denote by $f(\boldsymbol{\beta}|\mathbf{y})$ the posterior density function of $\boldsymbol{\beta}$ which is expressed as

$$\log f(\boldsymbol{\beta}|\mathbf{y}) = -\frac{1}{2}(\boldsymbol{\beta} - \boldsymbol{\beta}_0)'\mathbf{V}_0^{-1}(\boldsymbol{\beta} - \boldsymbol{\beta}_0) - \sum_{i=1}^{n}\left[(\mathbf{x}_i'\boldsymbol{\beta})y_i - \log(1 + e^{\mathbf{x}_i'\boldsymbol{\beta}})\right]. \quad (7.43)$$

To obtain posterior estimates of mean and covariance matrix of $\boldsymbol{\beta}$, we use the random walk sampler with a multivariate $\mathbf{t}_\nu(\boldsymbol{\mu}, \boldsymbol{\Sigma})$ proposal distribution which matches the shape of the posterior distribution of $\boldsymbol{\beta}$. Let $\boldsymbol{\mu}$ be the mode of the posterior of $\boldsymbol{\beta}$, i.e., $\boldsymbol{\mu} = \arg\max_{\boldsymbol{\beta}} \log f(\boldsymbol{\beta}|\mathbf{y})$. $\boldsymbol{\mu}$ can be found approximately via a Newton-Raphson procedure with gradient

$$\nabla \log f(\boldsymbol{\beta}|\mathbf{y}) = -\mathbf{V}_0^{-1}(\boldsymbol{\beta} - \boldsymbol{\beta}_0) + \sum_{i=1}^{n}\left(y_i - \frac{e^{\mathbf{x}_i'\boldsymbol{\beta}}}{1 + e^{\mathbf{x}_i'\boldsymbol{\beta}}}\right)\mathbf{x}_i,$$

and Hessian matrix

$$\nabla^2 \log f(\boldsymbol{\beta}|\mathbf{y}) = -\mathbf{V}_0^{-1} - \sum_{i=1}^{n}\frac{e^{\mathbf{x}_i'\boldsymbol{\beta}}}{(1 + e^{\mathbf{x}_i'\boldsymbol{\beta}})^2}\mathbf{x}_i\mathbf{x}_i'.$$

The scale matrix Σ of the proposal distribution $\mathbf{t}_\nu(\boldsymbol{\mu}, \Sigma)$ is chosen as the inverse of the observed Fisher information matrix, i.e., $\Sigma = -(\nabla^2 \log f(\boldsymbol{\beta}|\mathbf{y}))^{-1}$, and the degrees of freedom ν is arbitrarily chosen as $\nu = 5$. The random walk sampler is initialized at $\boldsymbol{\mu}$. Denote by $\boldsymbol{\beta}_{\text{old}}$ and $\boldsymbol{\beta}_{\text{new}}$ the current value and the newly generated values, respectively. The acceptance ratio in Algorithm 7.8 is given by

$$\alpha(\boldsymbol{\beta}_{\text{old}}, \boldsymbol{\beta}_{\text{new}}) = \min\left\{\frac{f(\boldsymbol{\beta}_{\text{new}}, \mathbf{y})}{f(\boldsymbol{\beta}_{\text{old}}, \mathbf{y})}, 1\right\} = \min\left\{\frac{f(\boldsymbol{\beta}_{\text{new}}|\mathbf{y})}{f(\boldsymbol{\beta}_{\text{old}}|\mathbf{y})}, 1\right\}.$$

Implement Algorithm 7.8 with the above setting and appropriately chosen m and n, say $m = 10^3$ and $n = 10^3$, we obtain an estimate of $E(\boldsymbol{\beta}|\mathbf{y})$ and $Cov(\boldsymbol{\beta}|\mathbf{y})$. □

The original Metropolis algorithm (Metropolis et al., 1953) is suggested for symmetric proposal functions; that is, for $q(\mathbf{y}|\mathbf{x}) = q(\mathbf{x}|\mathbf{y})$. Hastings (1970) modified the original MCMC algorithm to allow nonsymmetric proposal functions, hence the name Metropolis-Hastings algorithm. The Metropolis-Hastings algorithm has proven to be fundamental and plays a central role in Monte Carlo computation. However, as pointed out by many researchers, it suffers from two difficulties. The first is the local-trap problem which means that the sampler gets trapped in a local energy minimum indefinitely and hence simulation becomes ineffective or useless. The second difficulty is the inability to sample from distributions with intractable integrals. Suppose that the target distribution of interest has density of the form $f(x) \propto c(x)\psi(x)$, where $c(x)$ denotes an intractable integral. Clearly, the Metropolis-Hastings algorithm cannot be applied to sample from $f(x)$, as the acceptance probability involves an unknown ratio $c(y)/c(x)$, where y is the proposed value. This difficulty naturally arises in Bayesian inference for many statistical models. To alleviate these two difficulties, various advanced MCMC methods need to be used, such as auxiliary variable-based methods, population-based methods, importance weight-based methods, stochastic approximation-based methods, and so on; see details in Liang et al. (2010).

7.4.3   Reversible jump sampler

The Gibbs and Metropolis-Hastings samplers assume that the dimension of random variables $\mathbf{X}$ remains the same across iterations. *Reversible jump samplers* (Green, 1995) relaxes this restriction and allows to sample from target spaces that contain vectors of different dimensions. This often arises in Bayesian inference when different models for a given dataset (e.g., change-points identification) are considered.

Suppose that the model we are interested in consists of a countable set of possible sub-models $\{1, 2, \ldots, M\}$. The prior distribution of these M sub-models is $\pi(m)$, $m = 1, \ldots, M$, and each sub-model is characterized by a parameter vector $\boldsymbol{\beta}^{(m)} = (\beta_1, \ldots, \beta_m)$. The prior distribution of $\boldsymbol{\beta}^{(m)}$ given the sub-model m is $g(\boldsymbol{\beta}^{(m)}|m)$. Given observations $\mathbf{x}$ generated from the model, the likelihood of the data given the sub-model m with parameter $\boldsymbol{\beta}^{(m)}$ is $f(\mathbf{x}|\boldsymbol{\beta}^{(m)}, m)$. Then the posterior density of $(\boldsymbol{\beta}^{(m)}, m)$ can be expressed as follows

$$f(\boldsymbol{\beta}^{(m)}, m|\mathbf{x}) \propto f(\mathbf{x}|\boldsymbol{\beta}^{(m)}, m)g(\boldsymbol{\beta}^{(m)}|m)\pi(m).$$

The reversible jump sampler jumps between spaces of different dimensions according to a set of pre-defined jumps. If we allow only jumps between vectors that differ in dimension at most 1, then all possible jumps are

$$\mathbf{x}_1 \to \mathbf{x}_1', \quad \mathbf{x}_1 \to (\mathbf{x}_1', \mathbf{x}_2'), \quad (\mathbf{x}_1, \mathbf{x}_2) \to \mathbf{x}_1'.$$

The reversible jump sampler may be viewed as a generalization of the Metropolis-Hastings sampler in which a move is proposed from the density $q(n, \mathbf{y}|\mathbf{x}) = q(n|\mathbf{x})q(\mathbf{y}|n, \mathbf{x})$. Thus, given the current state $\mathbf{x}$, a new dimension n is sampled from $q(n|\mathbf{x})$, where typically $q(n|\mathbf{x}) = q(n|m)$. Given the new dimension n, a new state $\mathbf{y}$ with $\dim(\mathbf{y}) = n$ is selected according to transition function $q(\mathbf{y}|\mathbf{x}, n)$. This gives the following algorithm.

Algorithm 7.9 (Reversible jump sampler) *Given the current state* $\mathbf{X}_t$ *with* $\dim(\mathbf{X}_t) = n$, *iterate the following steps.*

1. *Generate* $n \sim q(n|m)$, $\mathbf{Y} \sim q(\mathbf{y}|\mathbf{X}_t, n)$ *with* $\dim(\mathbf{Y}) = n$.

2. *Generate* $U \sim \mathcal{U}(0, 1)$ *and let* $\mathbf{X}_{t+1} = \mathbf{Y}$ *if* $U \leq \alpha(\mathbf{X}_t, \mathbf{Y})$ *and* $\mathbf{X}_{t+1} = \mathbf{X}_t$ *otherwise, where*

$$\alpha(\mathbf{x}, \mathbf{y}) = \min\left\{\frac{f(\mathbf{y}, n)q(m|n)q(\mathbf{x}|\mathbf{y}, m)}{f(\mathbf{x}, m)q(n|m)q(\mathbf{y}|\mathbf{x}, n)}, 1\right\}.$$

7.5   Importance sampling for rare-event simulation

Many problems of evaluating financial risk involve rare-event simulation. To simulate rare events efficiently, importance sampling methods are often used. To fix the idea, denote by A an event and its probability by $p = P(A) = E(1_A)$. When p is small, say $p < 10^{-3}$, the event A is called a *rare event* and p is called a rare-event probability. Suppose that A can be represented as $A = \{g(\mathbf{X}) \geq \gamma\}$ for some function $g : \mathbb{R}^n \to \mathbb{R}$, vector $\mathbf{X} = (X_1, \ldots, X_n)'$, and threshold parameter γ. Let $Z = 1_A$. If $Z_1, \ldots, Z_n$ are independent realizations of some random variable Z for which $E(Z) = p$, then the unbiased estimator

$$\widehat{p} = \frac{1}{n}\sum_{i=1}^{n} Z_i, \tag{7.44}$$

is the crude Monte Carlo estimator of p. The accuracy of the estimator $\widehat{p}$ can be measured in terms of its relative error, $\sqrt{\mathrm{Var}(\widehat{p})}/p = \sigma/(p\sqrt{n})$, where $\sigma = \sqrt{\mathrm{Var}(p)}$. Given a set of i.i.d. samples $Z_1, \ldots, Z_n$, we can estimate the relative error $\sigma/(p\sqrt{n})$ by $\widehat{\sigma}/(\widehat{p}\sqrt{n})$, where $\widehat{\sigma} = \sqrt{\sum_{i=1}^{n}(Z_i - \widehat{p})^2/n}$.

There are two interesting issues in the above procedure. First, by definition, the probability $p = p(\gamma) = P(g(\mathbf{X}) \geq \gamma)$ is a function of parameter γ. Clearly, as $\gamma \to \infty$, $p(\gamma) \to 0$. Thus, it is interesting to study the asymptotic properties of $\widehat{p}(\gamma)$ as a function of γ. Second, the relative error satisfies

$$\kappa = \frac{\sqrt{\mathrm{Var}(\widehat{p})}}{E\widehat{p}} = \sqrt{\frac{1-p}{np}} \approx \frac{1}{\sqrt{np}}.$$

This shows that a large number of simulations may be needed in order to make the relative error small. For example, for $p = 10^{-6}$, in order to estimate p accurately with relative error of 1%, one needs to choose a sample size $n \approx (\kappa^2 p)^{-2} = 10^{10}$. To circumvent the difficulty of generating a very large number of events in direct Monte Carlo computation of $P(A)$, one can use importance sampling instead of direct Monte Carlo. In the follows, we first introduce importance sampler for light and heavy tails distributions, and then state-dependent importance sampler.

7.5.1   Light tail estimation

We first consider methods of finding importance sampling densities when X is light tailed. A random variable X with distribution function F is said to have a (right) *light-tailed* distribution if its moment generating function is finite for some $t > 0$, that is, $E(e^{tX}) \leq c < \infty$. Otherwise, that is, when $E(e^{tX}) = \infty$ for all $t > 0$, X is said to have a *heavy-tailed* distribution.

Since for every x,

$$E(e^{tX}) \geq E(e^{tX}1_{\{X \geq x\}}) \geq e^{tx}P(X > x),$$

it follows that for any $t > 0$ and c satisfying $E(e^{tX}) \leq c < \infty$, we have

$$P(X > x) \leq ce^{-tx}.$$

That is, if X has a light tail, then $1 - F(x)$ decays at an exponential rate or faster. Similarly, heavy-tailedness is equivalent to $\lim_{x \to \infty} e^{tx}(1 - F(x)) = \infty$ for all $t > 0$. Any distribution with bounded support is light-tailed. Examples of light-tailed distributions with unbounded support includes Poisson, geometric, exponential, Gamma, Gumbel, Laplace, logistic, normal, Weilbull distributions.

For $t \in \mathbb{R}$, denote by $M_X(t) = E(e^{tX}) = \int e^{tx} f(x)dx$ the moment-generating function of X and $\zeta(t) = \log M_X(t)$ the cumulant function of X. If $M_X(t)$ is finite, we can define an exponential family of probability density functions $\{f_\theta, \theta \in \Theta\}$ via an *exponential twist* θ,

$$f_\theta(x) = \frac{e^{\theta x}}{M_X(\theta)} f(x) = \exp[\theta x - \log M_X(\theta)] f(x) = \exp[\theta x - \zeta(\theta)] f(x).$$

The likelihood ratio is $r_\theta(x) = f(x)/f_\theta(x) = \exp[\zeta(\theta) - \theta x]$. Define μ_θ as the mean of X with respect to the density f_θ, i.e.,

$$\mu_\theta := E_\theta(X) = E[X \exp(\theta X)]/M_X(\theta). \tag{7.45}$$

To choose θ optimally for a particular importance sampling problem, we consider the case of small tail probability

$$E(r(X); X \geq c) = E[1_{\{X \geq c\}} M_X(\theta) e^{-\theta X}]. \tag{7.46}$$

Since $e^{-\theta x} \leq e^{-\theta c}$ for $x \geq c$ and $t \geq 0$, we have

$$E[1_{\{X \geq c\}} M_X(\theta) e^{-\theta X}] \leq M_X(\theta) e^{-\theta c}.$$

Instead of solving the problem of minimizing (7.46) over θ, we choose θ so that the bound $M_X(\theta) e^{-\theta c}$ is minimized. This is equivalent to find θ to minimize $\log M_X(\theta) - \theta c$. Using (7.45) we obtain the first order condition

$$0 = \frac{d}{dt}\left(\log M_X(\theta) - \theta c\right) = \frac{E(X \exp(\theta X))}{M_X(\theta)} - c = \mu_\theta - c,$$

which suggests choosing $\theta^* = \theta^*(c)$ as the solution of the equation $\mu_\theta = c$, so that the rare event $\{X \geq c\}$ becomes a normal event if we compute probabilities using the density $f_{\theta^*(c)}$. Since a unique solution of the equation $\mu_\theta = c$ exists for all relevant values of c, many rare event probabilities can be computed with this method.

Stopping time probabilities

Let $X_1, X_2, \ldots, X_n$ be n i.i.d. light-tailed random variables with density function $f(X)$, distribution function F, and finite expectation $\mu = E(X)$. Denote P_θ by the probability measure under which $X_1, \ldots, X_n$ are i.i.d. random variables with density $f_\theta(X)$. Let $S_n = \sum_{i=1}^n X_i$ and define the stopping time $\tau = \inf\{n : S_n \geq \gamma\}$. Note that the likelihood ratio is given by

$$R(\mathbf{X}; \theta) = \frac{\prod_{i=1}^n f(X_i)}{\prod_{i=1}^n f_\theta(X_i)} = e^{-S_n \theta + n\zeta(\theta)}, \quad \mathbf{X} = (X_1, X_2, \ldots, X_n).$$

Consider the problem of estimating stopping time probability $p = P(\tau < \infty)$. Assume that $\mu < 0$ so that p becomes small as $\gamma \to \infty$. Suppose that there exists a $\theta^* > 0$ such that $M_X(\theta^*) = 1$. Since $M_X(t)$ is a convex function with $M_X(0) = 1$ and $M_X'(0) = \mu < 0$, such a θ^* exists if $\lim_{t \to T} M_X(t) = \infty$ for some $0 < T \leq \infty$. Siegmund (1976) proposed the following algorithm to estimate p.

Algorithm 7.10 (Importance sampler for stopping time probabilities) *Compute the root $\theta^* > 0$ of $M_X(\theta) = 1$ (or equivalently $\zeta(\theta) = 0$).*

1. *Let $S_0 = 0$. Until $S_n \geq \gamma$, set $S_{n+1} = S_n + X_{n+1}$, $n = 0, 1, 2, \ldots$, where $X_1, X_2, \ldots$ are i.i.d. random variables with density function f_{θ^*}.*

2. *Let τ be the smallest n for which $S_n \geq \gamma$ and set $Z = R(\mathbf{X}; \theta^*) = \exp(-S_\tau \theta^*)$.*

3. *Repeat Steps 1 and 2 to obtain m independent replications $Z_1, \ldots, Z_m$ of Z and return $\widehat{p} = m^{-1} \sum_{i=1}^{m} Z_i$.*

Asmussen and Glynn (2007, Section VI.2) shows that the Siegmund's estimator in Algorithm 7.10 has bounded relative error, i.e.,

$$\limsup_{\gamma \to \infty} \frac{\mathrm{Var}(Z(\gamma))}{p^2(\gamma)} \leq K < \infty,$$

where K is a constant that does not depend on γ. Define the *overshoot* at time τ as $\xi(\gamma) = S_\tau - \gamma$, then

$$p = e^{-\gamma \theta^*} E(e^{-\xi(\gamma)\theta^*}).$$

Then for some constant $0 < C < \infty$, $E(e^{-\xi(\gamma)\theta^*}) \to C$ as $\gamma \to \infty$. It follows that

$$p = P(\tau < \infty) \approx C e^{-\gamma \theta^*} \qquad \text{as } \gamma \to \infty,$$

a celebrated result which is commonly referred to as the *Cramér-Lundberg approximation*. Another property of the Siegmund's estimator is its uniqueness of the change of measure. In particular, among all importance sampling estimators whose change of measure is of the form $X_1, \ldots, X_n$ i.i.d. g, the Siegmund's estimator in Algorithm 7.10 is the only *logarithmically efficient* estimator, i.e.,

$$\limsup_{\gamma \to \infty} \frac{\log E(Z(\gamma))}{\log p(\gamma)} = 1.$$

Overflow probabilities

Consider the efficient estimation of $p_n = P(S_n \geq nb)$ for large n and fixed $b > \mu$. Note that $\mu = E(X)$ is not necessarily negative. Consider the following algorithm.

Algorithm 7.11 (Importance sampler for overflow probabilities) *Compute the root θ^* of $\zeta'(\theta) = E_\theta(X) = b$.*

1. *Generate $X_1, \ldots, X_n$ i.i.d. from $f_{\theta^*}(X)$ and compute the likelihood ratio $R_n = \exp(-\theta^* S_n + n\zeta(\theta^*))$.*

2. *Use m independent replications of S_n and W_n to compute the estimator $\widehat{p}_n = m^{-1} \sum_{k=1}^{m} W_n^{(k)} 1_{\{S_n^{(k)} \geq nb\}}$.*

It has been shown that Algorithm 7.11 has the following efficiency property (Petrov, 1965; Bucklew et al., 1990; Jensen, 1995). Among all importance sampling estimators whose change of measure is of the form $X_1, \ldots, X_n$ i.i.d. g, the estimator with $g = f_{\theta^*}$ in Algorithm 7.11 is the only one that is logarithmically efficient. Moreover, under some smoothness conditions on $M_X(\theta)$, the first-order asymptotic behavior of p_n is given by

$$p_n = \frac{e^{-n(\theta^* b - \zeta(\theta^*))}}{\theta^* |\zeta''(\theta^*)| \sqrt{2\pi n}} (1 + o(1)) \qquad \text{as } n \to \infty.$$

Compound Poisson sums

Simulation of compound Poisson sums frequently arises in insurance problems; see Chapter 4 for more details. We now consider the problem of estimating the distribution of the compound Poisson sums, i.e.,

$$p(\gamma) = P(S_N \geq \gamma) = P(X_1 + \cdots + X_N \geq \gamma),$$

where $X_1, X_2, \ldots$ are i.i.d. positive random variables with density function $f(X)$ and $N \sim \text{Poisson}(\lambda)$ is independent of $X_1, X_2, \ldots$. Assume that $f(X)$ is light-tailed and satisfies condition (i) or (ii).

(i) $f(x)$ decays like the density of a Gamma(α, λ) distribution, $f(x) = cx^{\alpha-1}e^{-\lambda x}(1 + o(1))$ as $x \to \infty$, where $c > 0$ is a constant. Examples of such include the probability density functions of the exponential and inverse-Gaussian distributions.

(ii) For any $\alpha \in (1, 2)$, $\int_0^\infty f^\alpha(x)dx < \infty$, and f can be written as $f(x) = q(x)e^{-h(x)}$, where $0 < q(x) < \infty$ and $h(x)$ is convex on $[a, b]$ for some $a < b = \sup\{x : f(x) > 0\}$. Examples of such include densities with finite support and the probability density function of the Weibull distribution $W(\alpha, \lambda)$ for $\alpha > 1$.

Let $M_X(t)$ be the moment generating function of X and $\zeta(t)$ the cumulant function of S_N. Note that

$$\zeta(t) = \log E e^{t S_N} = \log e^{\lambda(M_X(t)-1)} = \lambda(M_X(t) - 1).$$

Using an exponential tilting of S_N with parameter θ, we have

$$\zeta_\theta(t) = \log E_\theta e^{t S_N} = \log E[e^{t S_N} e^{\theta S_N - \zeta(\theta)}] = \zeta(t + \theta) - \zeta(\theta)$$
$$= \lambda[M_X(t + \theta) - M_X(\theta)] = \lambda_\theta(M_\theta(t) - 1),$$

where $\lambda_\theta = \lambda M_X(\theta)$ and $M_\theta(t) = M_X(t + \theta)/M_X(\theta)$. In particular, $E_\theta S_N = \zeta_\theta'(0) = \lambda M_X'(\theta)$. We see that, to obtain realizations of S_N under θ, one may simulate compound Poisson sums with rate λ_θ and increments X with moment generating function $M_\theta(t)$. Then, one needs to find a good twisting parameter θ. A simple choice is to take θ such that $E_\theta S_N = \lambda M_X'(\theta)$, which leads to the following algorithm.

Algorithm 7.12 (Importance sampler for compound Poisson sum)
Compute the root θ^* of the equation $\lambda M_X'(\theta) = \gamma$ and simulate $N \sim$ Poisson$(\lambda M_X(\theta^*))$.

1. *Generate N i.i.d. random variables $X_1, \ldots, X_N$ with moment generating function $M_{\theta^*}(t) = M_X(t + \theta^*)/M_X(\theta^*)$ and set $S_N = X_1 + \ldots X_N$.*

2. *Generate m independent replications of S_N and return the estimator*

$$\widehat{p}(\gamma) = \frac{1}{m} \sum_{k=1}^{m} 1_{\{S_N^{(k)} \geq \gamma\}} \exp\left[-\theta^* S_N^{(k)} + \lambda(M_X(\theta^*) - 1)\right]. \quad (7.47)$$

It can be shown that the estimator (7.47) is logarithmically efficient (Jensen, 1995) and

$$p(\gamma) = P(S_N \geq \gamma) = \frac{\exp\left[-\theta^* \gamma + \lambda(M_X(\theta^*) - 1)\right]}{\theta^* \sqrt{2\pi \lambda M''(\theta^*)}}(1 + o(1)) \quad \text{as } \gamma \to \infty.$$

Bernoulli-mixture models

Consider a portfolio loss of the form $L = \sum_{i=1}^{m} e_i Y_i$, where e_i are deterministic, positive exposures and that Y_i are default indicators with default probabilities $\bar{p}_i$. Assume that $\mathbf{Y}$ follows a Bernoulli mixture model with factor vector $\mathbf{x}$ and conditional default probabilities $p_i(\mathbf{x})$. Glasserman and Li (2005) considered the problem of estimating exceedance probabilities $\theta = P(L \geq c)$ for c substantially larger than $E(L)$ using importance sampling.

Suppose that the default indicators $Y_1, \ldots, Y_m$ are independent. Note that the state space of $\mathbf{Y}$ is $\{0,1\}^m$ and the probability distribution of $(Y_1, \ldots, Y_m)$ is

$$P(y_1, \ldots, y_m) = \prod_{i=1}^{m} \bar{p}_i^{y_i}(1 - \bar{p}_i)^{1-y_i}, \quad \mathbf{y} \in \{0,1\}^m. \quad (7.48)$$

The moment-generating function of the portfolio loss L can be calculated as follows,

$$M_L(t) = E\left(\exp\left(t \sum_{i=1}^{m} e_i Y_i\right)\right) = \prod_{i=1}^{m} E(e^{t e_i Y_i}) = \prod_{i=1}^{m}(e^{t e_i}\bar{p}_i + 1 - \bar{p}_i). \quad (7.49)$$

Consider the following measure Q

$$Q_t(\{\mathbf{y}\}) = \frac{e^{t \sum_{i=1}^{m} e_i y_i}}{M_L(t)} P(\{\mathbf{y}\}) = \prod_{i=1}^{m} \frac{e^{t e_i y_i}}{e^{t e_i}\bar{p}_i + 1 - \bar{p}_i}\bar{p}_i^{y_i}(1 - \bar{p}_i)^{1-y_i}. \quad (7.50)$$

Define new default probabilities

$$\bar{q}_{t,i} = \frac{e^{t e_i y_i}}{e^{t e_i}\bar{p}_i + 1 - \bar{p}_i}.$$

It follows that $Q_t(\{\mathbf{y}\}) = \prod_{i=1}^m \bar{q}_{t,i}^{y_i}(1-\bar{q}_{t,i})^{1-y_i}$, so that after exponential tilting the default indicators remain independent but with new default probability $\bar{q}_{t,i}$. Note that $\bar{q}_{t,i}$ goes to one for $t \to \infty$ and to zero for $t \to -\infty$, this means that we can shift the mean of L to any point in $(0, \sum_{i=1}^m e_i)$.

In analog with previous discussions, for importance sampling purposes, the optimal value of t is chosen such that $E^{Q_t}(L) = c$, leading to the equation $\sum_{i=1}^m e_i \bar{q}_{t,i} = c$. The extension to the case of conditionally independent default indicators is easy. Given a realization $\mathbf{x}$ of the economic factors, the conditional exceedance probability $\theta(\mathbf{x}) := P(L \geq c | \mathbf{X} = \mathbf{x})$ can be estimated using the approach for independent default indicators described above.

7.5.2 State-dependent importance sampling

The importance sampling densities in the preceding section are independent of the state, we now extend the importance sampling method to the state-dependent case that yields estimators with bounded relative error. In particular, we assume that the state X_t follows a time-homogeneous Markov chain and present an importance sampling method for hitting probabilities. Procedures of applying importance sampling to general dynamic processes, such as Markov chains, Markov jump processes, and generalized semi-Markov processes, can be found in Glynn and Iglehart (1989).

Let $\mathcal{X}$ be the state space. Suppose that $\{X_t \in \mathcal{X}, t = 0, 1, \dots\}$ is a time-homogeneous Markov chain that is characterized by the transition density $p(y|x)$. Denote by P^x the probability measure under which $\{X_t\}$ starts at $X_0 = x$. Suppose that $\mathcal{R} \subset \mathcal{X}$ is a region in $\mathcal{X}$. Denote by $\tau = \inf\{t \geq 0; X_t \in \mathcal{R}\}$ the first hitting time of the process $\{X_t\}$ entering the region $\mathcal{R}$. We are interested in estimating the probability that the process starts at $x \in \mathcal{R}^c := \mathcal{X} - \mathcal{R}$ and enters a region $\mathcal{A} \subset \mathcal{R}$ through $\mathcal{R}$, that is,

$$h(x) = P^x(X_\tau \in \mathcal{A}, \tau < \infty) = \sum_{t=0}^\infty P^x(X_t \in \mathcal{A}, \tau = t).$$

Condition on $X_1 = x_1$, $h(x)$ can be expressed as

$$h(x) = \sum_{t=1}^\infty \sum_{x_1 \in \mathcal{X}} P^x(X_t \in \mathcal{A}, \tau = t | X_1 = x_1) p(x_1|x)$$

$$= \sum_{x_1 \in \mathcal{X}} p(x_1|x) \sum_{t=1}^\infty P^{x_1}(x_{t-1} \in \mathcal{A}, \tau = t-1) = \sum_{x_1 \in \mathcal{X}} p(x_1|x) h(x_1).$$

Let $q(y|x) = h(y)p(y|x)/h(x)$ for $y \in \mathcal{X}$ and $x \in \mathcal{R}^c$. It is easy to see that $q(y|x)$ is a transition density on $\mathcal{X}$. Moreover, denote by Q^x the probability measure under which $\{X_t\}$ is a Markov process starting at $x \in \mathcal{X}$ with transition density $q(y|x)$. It can be shown that the distribution of $\{X_t\}$ under Q^x is the same as the distribution under P^x given the rare event $\{X_\tau \in \mathcal{A}, \tau < \infty\}$.

In practice, since $h(x)$ is unknown, we usually assume there exists a good approximation to h and use it to construct the transition density q.

Algorithm 7.13 (Importance sampler for hitting probability) *Suppose that $\widetilde{h}(x)$ is a good approximation to the probability $h(x)$. Define the transition density $\widetilde{q}$ as*

$$\widetilde{q}(x|y) = p(x|y)\frac{\widetilde{h}(x)}{c(y)}, \quad where \ c(y) = \int p(x|y)\widetilde{h}(x)dz.$$

1. *Simulate a trajectory $X_1, X_2, \ldots$ of the Markov chain $\{X_n\}$ with the transition density $\widetilde{q}(z|y)$ until the stopping time $\tau = \inf\{t : Z_t \in \mathcal{R}\}$.*

2. *Compute the likelihood ratio*

$$L(x_1, \ldots, x_\tau) = \prod_{t=1}^{\tau} \frac{p(x_t|x_{t-1})}{\widetilde{q}(x_t|x_{t-1})} = \prod_{t=1}^{\tau} \frac{c(x_{t-1})}{\widetilde{h}(x_t)}.$$

3. *Repeat steps 1 and 2 and generate m instances of $Y = L(X_1, \ldots, X_\tau)$ $1_{\{X_\tau \in \mathcal{A}\}}$ and output the estimator $\widehat{h}(x) = m^{-1}\sum_{i=1}^{m} Y_i$.*

It is easy to see that the key of the above algorithm is to find an approximated probability $\widetilde{h}(x)$. In practice, such $\widetilde{h}(x)$ may not be easy to find. Hence, although the state-dependent importance sampling is more efficient and reliable than the state-independent sampling procedure, it is more difficult to implement it.

7.6   Sequential Monte Carlo methods

Sequential Monte Carlo (SMC) methods are a general class of Monte Carlo methods that sample sequentially from a sequence of target probability densities. Some literature present SMC and particle filters as the same thing, but actually SMC encompasses a broader range of algorithm (Doucet and Johansen, 2011). This section presents first a generic form of sequential importance sampling with resampling methods and then their applications to simulating dynamic portfolio risks.

7.6.1   Sequential importance sampling with resampling

Suppose that $\{X_t; t = 1, 2, \ldots\}$ is a sequence of random variables such that the joint density of $\mathbf{x}_{1:t} = (x_1, \ldots x_t)$ is represented by $f(\mathbf{x}_{1:t})$ for each

$t = 1, \ldots, n$. We are interested in sampling from the density $f(\mathbf{x})$ using an importance sampling density

$$g(\mathbf{x}_{1:n}) = g_1(x_1)g_2(x_2|x_1)\ldots g_n(x_n|x_{n-1}).$$

Here, we assume that it is easy to sample from $g_t(x_t|\mathbf{x}_{1:t-1})$. Each element in the set $\{\mathbf{x}_{1:t}\}$ is referred to as a *particle*. Suppose that $f_1, f_2, \ldots f_n$ be a sequence of *auxiliary* probability density functions that can be easily evaluated and each $f_t(\mathbf{x}_{1:t})$ is a good approximation to $f(\mathbf{x}_{1:t})$. Then the likelihood ratio $f(\mathbf{x}_{1:n})/g(\mathbf{x}_{1:n})$ can be expressed as

$$\frac{f_1(x_1)}{g_1(x_1)} \frac{f_2(\mathbf{x}_{1:2})}{f_1(x_1)g_2(x_2|x_1)} \frac{f_3(\mathbf{x}_{1:3})}{f_2(\mathbf{x}_{1:2})g_3(x_3|\mathbf{x}_{1:2})} \cdots \frac{f_n(\mathbf{x}_{1:n})}{f_{n-1}(\mathbf{x}_{1:n-1})g_n(x_n|\mathbf{x}_{1:n-1})}.$$

It is easy to see that this likelihood ratio can be computed recursively via $w_t = u_t w_{t-1}$, $t = 1, \ldots, n$, with $w_0 = 1$ and

$$u_t = \frac{f_t(\mathbf{x}_{1:t})}{f_{t-1}(\mathbf{x}_{1:t-1})g_t(x_t|\mathbf{x}_{1:t-1})}, \quad f_0(\mathbf{x}_{1:0}) = 1.$$

Making use of this fact, the following algorithm can be used to generate a sample from $f(\mathbf{x}_{1:n})$.

Algorithm 7.14 (Sequential Monte Carlo sampler) *Let m be the number of particles in the sampling procedure.*

1. *Set $t = 1$ and sample independently $x_1^j \sim g_1(x_1)$ for each $j = 1, \ldots, m$.*

2. *For each $j = 1, \ldots, m$, sample $y_t^j \sim g_t(y_t|\mathbf{z}_{1:t-1}^j)$, and compute the importance weights*

$$u_{t,j} = \frac{f_t(\mathbf{z}_{1:t}^j)}{f_{t-1}(\mathbf{x}_{1:t-1}^j)g_t(y_t^j|\mathbf{x}_{1:t-1}^j)}, \quad \mathbf{z}_{1:t}^j = (\mathbf{x}_{1:t-1}^j, y_t^j).$$

   Renormalize the weights so that $\sum_{j=1}^m u_{t,j} = 1$.

3. *Given the population of particles $\mathbf{z}_{1:t}^1, \ldots, \mathbf{z}_{1:t}^m$, generate the new population $\mathbf{x}_{1:t}^1, \ldots, \mathbf{x}_{1:t}^m$ by sampling independently n times from the mixture distribution $\sum_{j=1}^m u_{t,j} 1_{\{\mathbf{x}_{1:t} = \mathbf{z}_{1:t}^j\}}$.*

4. *If $t = n$, exit and output the population of particles $\mathbf{x}_{1:t}^1, \ldots, \mathbf{x}_{1:t}^m$; otherwise, set $t \leftarrow t + 1$ and repeat from step 2.*

Step 3 in the above algorithm, which is also known as *bootstrap resampling*, *bootstrap filter*, or *sample importance resampling*, is a key ingredient of SMC methods. Since the importance sampling step provides estimates whose variance may increases, typically exponentially, with t, resampling can partially solve this problem in some important scenarios. Convergence results can be obtained for SMC algorithms under different scenarios, and we skip these results and refer the interested reader to Crisan and Doucet (2002) and Del Moral (2004).

7.6.2   Applications to simulating dynamic portfolio risk

Monte Carlo methods are widely used to simulate distribution of financial loss due to defaults in a portfolio of credit-sensitive assets such as loans and corporate bonds. Deng et al. (2012) have shown how the sequential importance sampling with resampling method can be applied to overcome some of the difficulties in the existing Monte Carlo methods for these simulation problems.

Consider a portfolio of n firms that are subject to default risk. The random default times of these firms are modeled by distinct stopping times $\tau^i > 0$. In risk management applications, we consider the real world probability measure P. Associated to the τ^i are indicator processes N^i given by $N_t^i = \mathbf{1}_{\{\tau^i \leq t\}}$. For each i, there is a strictly positive process λ^i such that the random variables

$$N_t^i - \int_0^t \lambda_s^i (1 - N_s^i) ds \tag{7.51}$$

form a martingale; see Section 3.5. The process $\lambda_s^i (1 - N_s^i)$ represents the conditional default rate, or intensity of firm i. The λ^i are correlated stochastic processes that we take as given in the Monte Carlo simulations. The credit risk associated with the portfolio is described by the distribution of the portfolio loss $L = l \cdot N$, where $N = (N^1, \ldots, N^n)$ is the vector of default indicators and $l = (l^1, \ldots, l^n)$ is the vector of position losses. The tail of the loss distribution represents the probability of atypically large default losses. These *rare-event probabilities* are at the center of portfolio risk management and other applications. For example, they are used to estimate portfolio risk measures such as VaR.

The problem of computing the distribution of N_T and L_T at a fixed horizon $T > 0$ can be cast as a Markov chain problem. Proposition 3.1 of Giesecke et al. (2010) states that there exists a continuous-time Markov chain $M = (M^1, \ldots, M^n) \in \mathcal{S} = \{0, 1\}^n$ such that $P(N_t = B) = P(M_t = B)$ for fixed t and all $B \in \mathcal{S}$. The mimicking chain M has no joint transitions in any of its components, and a component M^i starts at 0 and has transition intensity $\pi^i(\cdot, M)$, where

$$\pi^i(t, B) = E(\lambda_t^i (1 - \lambda_t^i) | N_t = B), \quad B \in \mathcal{S}. \tag{7.52}$$

The expectation (7.52) can be computed for many standard models of λ; see Giesecke et al. (2010, Section 5). The existence of M reduces the problem of computing the distribution of N_T to that of computing the distribution of M_T. Similarly, under the assumption that each l^i is drawn, independently of N, from a fixed distribution, it reduces the problem of computing the distribution of L_T to that of computing the distribution of J_T, where $J = l \cdot M$. This reduction allows us to analyze a Markov chain model M rather than a potentially complex point process model N, as noted by Deng et al. (2012). Although the jump process J is not itself a Markov chain because its jump times have an intensity of the form $\sum_{i=1}^n \pi^i(t, M_t)$, we can estimate the tail distribution of J_T by simulating M.

Let $Y_k = (T_k, U_k)$, where T_k is the kth arrival time of the jump process J, and $U_k = M_{T_k}$. Moreover, let $Y_0 = (0, (0, \ldots, 0))$ and $K \leq n$. The sequence $Y = (Y_k)_{0 \leq k \leq K}$ is a discrete-time Markov chain on $\mathbb{R}_+ \times \mathcal{S}$. We denote its P-transition probabilities by $p_k(x, y)$. Suppose the rare event of interest takes the form $\{\mathbf{Y}_K \in \Gamma\}$ for some suitable set Γ. Here and in the sequel, we let $\mathbf{Y}_k = (Y_0, \ldots, Y_k)$ for $0 \leq k \leq K$ and then the SISR algorithm requires the selection of a sampling measure $\widetilde{P}$ and the weight functions w_k. Consider an event $\{\mathbf{Y}_K \in \Gamma\}$ of the form $\{J_T = x\} = \{T_x \leq T < T_{x+1}\}$, setting $K = \min(x + 1, n)$, which is the event that the portfolio loss at T is equal to $x \in \{0, 1, \ldots, n\}$. Deng et al. (2012) used resampling weight functions $w_k = w_k(\mathbf{Y}_k)$ of the form

$$
w_k = \begin{cases} \dfrac{Z_k}{\pi_{k-1} Z_{k-1}} \left(\dfrac{\pi_k}{\pi_{k-1}}\right)^{x-k} \exp\left\{ \left(\pi_k - \dfrac{x}{T}\right)(T_k - T_{k-1}) + (\pi_{k-1} - \pi_k)T \right\} & \text{if } T_k < T, \\ 0 & \text{otherwise,} \end{cases}
$$

where $\pi_k = \sum_{i=1}^n \pi^i(T_k, U_k)$ is the intensity of J at the kth arrival time T_k for $1 \leq k \leq x$, and Z_k is the likelihood ratio

$$
Z_k(\widetilde{\mathbf{Y}}_k) = \frac{dP}{d\widetilde{P}}(\widetilde{\mathbf{Y}}_k) = \prod_{i=1}^k \frac{p_i(Y_{i-1}, \widetilde{Y}_i)}{\widetilde{p}_i(Y_{i-1}, \widetilde{Y}_i)}.
$$

Besides the convenient choice $P = \widetilde{P}$, one can also choose other $\widetilde{P}$ to improve the efficiency of the SISR scheme.

With the aforementioned resampling mechanism, Deng et al. (2012) propose a sequential Monte Carlo method for efficiently and unbiasedly estimating probabilities of large losses resulting from defaults in widely used dynamic point process models of portfolio credit risk.

Supplements and problems

There are several excellent textbooks on Monte Carlo statistical methods. Robert and Casella (2004) and Robert and Casella (2010) give a comprehensive introduction on Monte Carlo methods. The handbook Kroese et al. (2003) present the subject with a mix of theory, algorithms, and applications in finance and engineering. Asmussen and Glynn (2007) and Graham and Talay (2013) describe the basic ideas and algorithms of various Monte Carlo methods.

In terms of applications of Monte Carlo methods in finance, Glasserman (2003) provide an excellent review to bridge financial engineering and Monte Carlo methods. Dagpunar (2007) introduce basic Monte Carlo methods with some applications in finance, and Korn et al. (2010) give a systematic treatment on Monte Carlo methods and their application in finance and insurance.

Overviews and description on specific Monte Carlo methods can be found in the following reference. Gamerman and Lopes (2006) give a concise description on MCMC methods and provide implementable codes in R. Liang et al. (2010) provide a unified and up-to-date treatment of advanced MCMC algorithms and their variants. Chib and Greenberg (1996) review some applications of MCMC methods in econometrics. Doucet et al. (2001) present some advanced topics on sequential Monte Carlo methods. Chopin and Papaspiliopoulos (2020) provide a comprehensive description on the underlying theory and implementation of sequential Monte Carlo methods.

1. A *location-scale* family of continuous distributions has the probability density function $\{f(x; \mu, \sigma); \mu \in \mathbb{R}, \sigma > 0\}$ of the form

$$f(x; \mu, \sigma) = \frac{1}{\sigma} \tilde{f}\left(\frac{x - \mu}{\sigma}\right), \qquad x \in \mathbb{R}.$$

Parameter μ is called the location, σ is called the scale. To simulate a random variable $Z \sim f(z; \mu, \sigma)$, one can first simulate $X \sim f(x; 0, 1)$ and apply the transformation $Z = \mu + \sigma X$. Using this method to simulate the following continuous random variables.

(a) A normal distribution $N(\mu, \sigma^2)$.

(b) A student-t distribution $t_\nu(\mu, \sigma^2)$ with the density function

$$f(z; \mu, \sigma) = \frac{\Gamma\left(\frac{\nu+1}{2}\right)}{\Gamma\left(\frac{\nu}{2}\right)\sqrt{\nu\pi}\sigma}\left(1 + \frac{(z-\mu)^2}{\nu\sigma^2}\right)^{-\frac{\nu+1}{2}}.$$

(c) A Cauchy distribution $\text{Cauchy}(\mu, \sigma)$ with the density function

$$f(z; \mu, \sigma) = \frac{1}{\pi}\frac{\sigma}{(z-\mu)^2 + \sigma^2}.$$

(d) A Freéchet distribution $\text{Freéchet}(\alpha, \mu, \sigma)$ with the density function

$$f(z; \alpha, \mu, \sigma) = \frac{\alpha}{\sigma}\left(\frac{z-\mu}{\sigma}\right)^{-1-\alpha}\exp\left(-\left(\frac{z-\mu}{s}\right)^{-\alpha}\right).$$

(Hint: If a random variable $U \sim \mathcal{U}(0, 1)$, then $\mu + \sigma(-\log U)^{-1/\alpha} \sim$ Freéchet(α, μ, σ).)

(e) A Gumbel distribution $\text{Gumbel}(\mu, \sigma)$ distribution with the density function

$$f(z; \mu, \sigma) = \frac{1}{\sigma}\exp\left(\frac{z-\mu}{\sigma} + \exp\left(-\frac{z - mu}{\sigma}\right)\right).$$

(Hint: If X is an exponentially distributed variable with mean 1, then $-\log X \sim \text{Gumbel}(0, 1)$.)

(f) A Laplace distribution Laplace(μ, σ) with the density function

$$f(z; \mu, \sigma) = \frac{1}{2\sigma} \exp\left(-\frac{|z - \mu|}{\sigma}\right).$$

(Hint: If X, Y are exponentially distributed variables with mean 1, then $X - Y \sim$ Laplace$(0, 1)$.)

(g) A Logistic distribution Logistic(μ, σ) with the density function

$$f(z; \mu, \sigma) = \frac{\exp(-\frac{z-\mu}{\sigma})}{\sigma\left(1 + \exp(-\frac{z-\mu}{\sigma})\right)^2}.$$

(Hint: If $U \sim \mathcal{U}(0, 1)$, then $\log U - \log(1 - U) \sim$ Logistic$(0, 1)$.)

2. Use the reciprocal transformation method to simulate the following random variables.

(a) An inverse-gamma distribution InvGamma(α, λ), which has the density function

$$f_Z(z; \alpha, \lambda) = \frac{\lambda^\alpha z^{-\alpha-1} e^{-\lambda z^{-1}}}{\Gamma(\alpha)} = f_X(z^{-1})/z^2, \quad z \in \mathbb{R},$$

where $f_X(x)$ is the density function of Gamma(α, λ). (Hint: When α is an positive integer, say $\alpha = n$, then $X_1 + \cdots + X_n \sim$ Gamma(n, λ), where $X_1, \ldots, X_n$ are n i.i.d. random variables following an exponential distribution with rate parameter λ.)

(b) Suppose $\mathbf{Y}_1, \ldots, \mathbf{Y}_n$ are independent $N(\mathbf{0}, \mathbf{\Sigma})$ random vectors of dimension m. Then the random matrix $\mathbf{W} = \Sigma_{i=1}^n \mathbf{Y}_i \mathbf{Y}_i^T$ is said to have a Wishart distribution, denoted by $W_m(\mathbf{\Sigma}, n)$. The density function of the Wishart distribution $W_m(\mathbf{\Sigma}, n)$ is

$$f(\mathbf{W}) = \frac{\det(\mathbf{W})^{(n-m-1)/2} \exp\left\{-\frac{1}{2}\mathrm{tr}\left(\mathbf{\Sigma}^{-1}\mathbf{W}\right)\right\}}{[2^m \det(\mathbf{\Sigma})]^{n/2} \Gamma_m(n/2)}, \quad \mathbf{W} > \mathbf{0},$$

in which $\mathbf{W} > \mathbf{0}$ denotes that $\mathbf{W}$ is positive definite and $\Gamma_m(\cdot)$ denotes the multivariate gamma function

$$\Gamma_m(t) = \pi^{m(m-1)/4} \prod_{i=1}^m \Gamma\left(t - \frac{i-1}{2}\right).$$

If $\mathbf{W}$ has the Wishart distribution $W_m(\mathbf{\Phi}, k)$, then $\mathbf{V} = \mathbf{W}^{-1}$ is said to have the inverted Wishart distribution, which is denoted by $IW_m(\mathbf{\Psi}, k)$ with $\mathbf{\Psi} = \mathbf{\Phi}^{-1}$.

3. Let X be a continuous univariate random variable with density function $f(x)$. Then the truncated random variable X to an interval $[a, b]$, denoted by Z, has the following density and distribution functions

$$f_Z(z) = \frac{f(z)}{\int_a^b f(x)dx}, \quad F_Z(z) = \frac{F(z) - F(a^-)}{F(b) - F(a^-)}, \quad a \le z \le b,$$

where $F(a^-) = \lim_{x \uparrow a} F(x)$. If X can be generated by the inverse-transform method, then Z can be simulated by first generating $U \sim \mathcal{U}(0,1)$ and then letting $Z = F^{-1}(F(a^-) + U(F(b) - F(a^-)))$. Use this method to simulate the following truncated random variables.

   (a) An exponential random variables with mean 1 truncated to the interval $[0, 3]$.

   (b) A $N(\mu, \sigma^2)$ random variable truncated to the interval $[a, b]$.

   (c) (Sampling from the tail of a Normal distribution) A $N(0, 1)$ random variable truncated to the interval $[a, \infty)$. (Hint: The truncated normal Z can be obtained by $Z = \Phi^{-1}(\Phi(a) + U(1 - \Phi(a)))$, where $U \sim \mathcal{U}[0, 1]$.)

4. Consider the following two-dimensional SDE

$$dX_t = Y_t dt, \qquad dY_t = (X_t(\alpha - X_t^2) - Y_t)dt + \sigma X_t dW_t.$$

Simulate this process with parameters $\alpha = 1$ and $\sigma = 1$ for $t \in [0, 400]$ with a step size $\Delta t = 10^{-3}$ and initial value $(X_0, Y_0) = (-2, 0)$.

5. Consider the geometric Brownian motion $dX_t = \mu X_t dt + \sigma X_t dW_t$ for $t \in [0, T]$ and initial value X_0, where $\{W_t\}_{t \ge 0}$ is a Wiener process. The (strong) solution is given by

$$X_t = X_0 \exp\left((\mu - \frac{1}{2}\sigma^2)t + \sigma W_t\right).$$

For a given step size h, consider a sequence of i.i.d. $N(0, h)$ random variables $Z_1, Z_2, \dots$.

   (a) Show that, for $n = 1, 2, \dots$, the exact solution at time nh is written as $X_{nh} = X_0 \exp\left((\mu - \frac{\sigma^2}{2})nh + \sigma \sum_{k=1}^{n} V_k\right)$, the Euler's method gives the approximation $Y_n^{e,h} = Y_{n-1}^{e,h}(1 + \mu h + \sigma V_n)$, and the Milstein's method gives the approximation $Y_n^{m,h} = Y_{n-1}^{m,h}(1 + \mu h + \sigma V_n + \frac{1}{2}\sigma^2(V_n^2 - h))$.

   (b) Consider the case $\mu = 1$, $\sigma = 0.5$, and initial value $X_0 = 1$. Use the exact, Euler, and Milstein methods with time step size $h = 10^{-1}, 10^{-2}, 10^{-3}$ to simulate geometric Brownian motion processes. Compare the different approximations under different step sizes.

6. (**Gibbs sampler for a multivariate normal distribution**). Consider a trivariate normal distribution with mean $\boldsymbol{\mu} = (\mu_1, \mu_2, \mu_3)^T$ and the covariance

$$\boldsymbol{\Sigma}(\rho) = \begin{pmatrix} 1 & \rho & \rho^2 \\ \rho & 1 & \rho \\ \rho^2 & \rho & 1 \end{pmatrix}.$$

The three-step Gibbs sampler with the partition of $\mathbf{X} = (X_1, X_2, X_3)^T$ into X_1, X_2, and X_3 can be implemented as follows. Set a starting value $\mathbf{x}^{(0)} \in \mathbb{R}^3$, and iterate for $t = 1, 2, \ldots,$

(i) Generate $x_1^{(t)} \sim N(\mu_1 + \rho(x_2^{(t-1)} - \mu_2, 1 - \rho^2)$.

(ii) Generate $x_2^{(t)} \sim N\left(\mu_2 + \frac{\rho}{1+\rho^2}\left[x_1^{(t)} - \mu_1 + x_3^{(t-1)} - \mu_3\right], \frac{1-\rho^2}{1+\rho^2}\right)$.

(iii) Generate $x_3^{(t)} \sim N(\mu_3 + \rho(x_2^{(t)} - \mu_2), 1 - \rho^2)$.

Use the above Gibbs sampler to simulate trajectory of two Markov chains with $\boldsymbol{\mu} = (4, 5, 6)^T$ and $\rho = 0.3$ and 0.9, respectively.

7. (**Gibbs sampler for a mixture of normal distribution**). Consider a mixture of normal distribution

$$f(x) = \sum_{j=1}^{J} p_j \frac{e^{-(x-\mu_j)^2/(2\sigma_j^2)}}{\sqrt{2\pi}\sigma_j}.$$

(a) Show that the conjugate prior on (μ_j, σ_j) is

$$\mu_j | \sigma_j \sim N(\alpha_j, \sigma_j^2/\lambda_j), \qquad \sigma_j^2 \sim \text{IG}\left(\frac{\lambda_j + 3}{2}, \frac{\beta_j}{2}\right).$$

(b) Show that the two steps of the Gibbs sampler are as follows.

(i) Simulate for $i = 1, \ldots, n$,

$$Z_i \sim P(Z_i = j) \propto p_j \frac{e^{-(x-\mu_j)^2/(2\sigma_j^2)}}{\sqrt{2\pi}\sigma_j},$$

and let

$$n_j = \sum_{i=1}^{n} 1_{\{z_i = j\}}, \quad n_j \bar{x}_j = \sum_{i=1}^{n} 1_{\{z_i = j\}} x_i, \quad s_j^2 = \sum_{i=1}^{n} 1_{\{z_i = j\}} (x_i - \bar{x}_j)^2.$$

(ii) Generate

$$\mu_j | \sigma_j \sim N\left(\frac{\lambda_j \alpha_j + n_j \bar{x}_j}{\lambda_j + n_j}, \frac{\sigma_j^2}{\lambda_j + n_j}\right),$$

$$\sigma_j^2 \sim \text{IG}\left(\frac{\lambda_j + n_j + 3}{2}, \frac{\lambda_j \alpha_j^2 + \beta_j + s_j^2 - (\lambda_j + n_j)^{-1}(\lambda_j \alpha_j + n_j \bar{x}_j)^2}{2}\right),$$

$$p \sim \text{Dirichlet}_k(\gamma_1 + n_1, \ldots, \gamma_k + n_k).$$

(c) Use the Gibbs sampler in (b) to simulate trajectory of a Markov chain with $J = 2$, $p_1 = 1 - p_2 = 0.6$, $\alpha_1 = 0.5$, $\lambda_1 = 1$, $\beta_1 = 4$, $\alpha_2 = -0.5$, $\lambda_2 = 3$, $\beta_2 = 6$.

8. (**Metropolis-Hastings sampler for Student's** t). Consider the student's t_ν distribution with ν degrees of freedom

$$f(x|\nu) = \frac{\Gamma(\frac{\nu+1}{2})}{\Gamma(\frac{\nu}{2})} \frac{1}{\sqrt{\nu\pi}} \left(1 + \frac{x^2}{\nu}\right)^{-\frac{\nu+1}{2}}.$$

Compute the mean of a t_4 distribution using Algorithm 7.8 with candidate densities (a) a $N(0,1)$ distribution and (b) a t_2 distribution.

9. (**Bounded relative error of tail probability estimation for exponential distribution**). Let X be an exponentially distributed random variable with mean $1/\lambda$ and $p = P(X \geq \gamma)$. Note the exact value of p is $p = \exp(-\gamma/\lambda)$. We wish to estimate p via importance sampling method with a Cauchy importance sampling density $g(x; \mu, \sigma) = \sigma(\sigma^2 + (x - \mu)^2)^{-1}/\pi$, $\sigma > 0$. Let $Z = f(X)1_{\{X>\gamma\}}/g(X; \mu, \sigma)$ and consider the importance sampling estimator $\widehat{p} = n^{-1}\sum_{i=1}^{n} Z_i$.

   (a) Show that $(\mu^*, \sigma^*) = (\sigma + \lambda/2, \lambda/2)$ minimizes $E_g(Z^2)$ and the minimized value is $(\pi/2 - 1)p^2$.

   (b) Show that the relative error of the estimator $\widehat{p}$ is $\sqrt{\mathrm{Var}(\widehat{p})}/E(\widehat{p}) = \sqrt{(\pi - 2)/(2n)}$.

10. (**Stopping time probabilities in random walk**). Let $X_1, X_2, \ldots$ are i.i.d. $N(\mu, 1)$ ($\mu < 0$) random variables. Note that their moment generating function is $M_X(\theta) = \exp(\mu\theta + \theta^2/2)$ and the solution of $M_X(\theta) = 1$ is $\theta^* = -2\mu$. Let $\mu = -0.9$, $\gamma = 12$. Use Algorithm 7.10 to estimate the stopping time probability $p = P(\tau < \infty)$, where $\tau = \inf\{n : S_n \geq \gamma\}$, and find its relative error.

11. (**Estimating insurers' default risk**). Consider an insurance company that has a core capital of $\gamma = 10^3$ and processes on average $\lambda = 300$ insurance claims per year. Let N be the number of claims per year and assume $N \sim \mathrm{Poisson}(\lambda)$. Let $X_1, X_2, \ldots$ (measured in millions of dollars) be the size of the claims in an insurance company, and assume they are i.i.d. outcomes from the Gamma$(2, 1)$ distribution. To use Algorithm 7.12 to estimate the probability that the size of all claims in a given year exceeds the core capital (that is, $P(X_1 + \cdots + X_N \geq \gamma)$).

   (a) Show that $\theta^* = 1 - (2\lambda/\gamma)^{1/3}$ is the root of the equation $\lambda M'_X(\theta) = \gamma$, hence $M_{\theta^*}(t) = (1 - \theta^*)^2/(1 - \theta^* - t)^2$, which is the moment generating function of the Gamma$(2, 1 - \theta^*)$ distribution.

(b) The importance sampling distribution is given as follows. The number of claims should be simulated from Poisson(λ_{θ^*}) distribution with $\lambda_{\theta^*} = (\gamma^2\lambda/4)^{1/3}$, and the size of each claim X is simulated from the Gamma($2, (2\lambda/\gamma)^{1/3}$) distribution.

Use the above setting with $N = 10^3$ to calculate $P(X_1 + \cdots + X_N \geq \gamma)$ and estimate its relative error.

12. (**Brownian bridge**). Consider the standard Brownian bridge process X_t (7.16) and its strong solution (7.17). Show that X_t is a Gaussian process with mean 0 and covariance function $\mathrm{Cov}(X_t, X_s) = \min\{s, t\} - st$.

13. (**Control variates for option pricing**). Let S_t be the price of a risky asset at time t. Assume that the interest rate is a constant r. Then under the risk-neutral measure, $e^{-rt}S_t$ is a martingale and $E(e^{-rt}S_t) = S_0$. Suppose we want to price an option on S_t with discounted payoff X_0, which is a function of $\{S_t; 0 \leq t \leq T\}$. Under the risk-neutral measure, $e^{-rt}X_t$ is also a martingale and $E(e^{-rT}X_T) = X_0$. To use control variate methods to improve the efficiency of estimating X_0 via simulation, we can use the control variate estimator which is based on independent replications $\{S_t^{(i)}; 0 \leq t \leq T\}$, $i = 1, \ldots, n$,

$$\hat{\mu}_m(b) = \frac{1}{n}\sum_{i=1}^{n}\left(X_0^{(i)} - b[e^{-rT}S_T^{(i)} - S_0]\right).$$

When X is a standard European call option with maturity T and strike K so that $X_0 = E[e^{-rT}(S_T - K)_+]$, the effectiveness of the control variate depends on the correlation of S_T and $(S_T - K)_+$, or equivalently, the strike K. Suppose that S_t follows a geometric Brownian motion $dS_t = rS_t dt + \sigma S_t dW_t$, in which W_t is a Wiener process. Let $r = 3\%$, $\sigma = 25\%$, $S_0 = 30$, and $T = 0.5$.

(a) Estimate the correlation ρ_{XS} of S_T and $(S_T - K)_+$ for $K = 20, 25, 30, 35, 40, 45, 50$.

(b) Compute the optimal b^* which minimizes the variance of $\hat{\mu}_m(b)$ for each K.

(c) Compute ρ_{XS}^2, the fraction of variance in the call option payoff eliminated by using the underlying asset as a control variate, for each K.

Chapter 8

Surveillance and predictive analytics

In recent decades, as financial instruments have become more complex and market instability has increased, there has been a growing focus among researchers and practitioners in both academia and the financial industry on monitoring and predicting structural changes in financial products, portfolios, and even submarkets. To address these challenges, the utilization of sequential analysis, which offers statistical estimation or decision-making in real-time as data is gathered, is typically required. This chapter introduces several sequential analysis methods, surveillance techniques, and prediction models for detecting structural changes in financial variables.

Section 8.1 introduces two types of risk management and surveillance processes. The first is active risk management, which involves actively attempting to mitigate the potential for loss in an investment portfolio. The second is systematic risk surveillance, which entails monitoring the stability of financial markets. Sections 8.2–8.4 introduce statistical methods for sequential analysis and change-point detection. In particular, Section 8.2 introduces several sequential hypothesis testing methods, and Sections 8.3 and 8.4 present frequentist and Bayesian methods for quickest change-point detection, respectively. These methods are widely used in quality control in manufacturing and fault detection-diagnosis in automated control systems. Since much of the methodology was originally developed in the quality control context, for simplicity we still use the terminology from that field, such as "quality characteristic" in lieu of monitoring statistic or data summary. For financial risk management that involves human investors and employees rather than machines and robots, immediate corrective action is often not feasible and multiple change-points can occur during the period of vigilant surveillance. Methods for these problems are introduced in Section 8.4. In addition, Section 8.5 introduces two applications of multiple change-point methods in financial modeling. The first deals with surveillance methods for structural breaks in credit rating transition dynamics, and the second uses a multiple change-point time series model to make inference from asset prices on their volatility dynamics and contemporaneous jumps in autoregression coefficient and in volatilities. Section 8.6 extends the surveillance and estimation methods for multiple change-points to prediction and describes a predictive model that is based on hidden Markov filtering approach introduced in Section 8.4.3.

8.1 Financial risk surveillance

This section presents two kinds of risk management processes involving sequential surveillance. One is active risk management at the corporate level, and the other is the surveillance of the instability or structural change risk in complex financial systems for regulatory purposes. Sequential surveillance is also found in other financial applications. For further information, we direct interested readers to the survey by Frisén (2008, 2009).

8.1.1 Active risk management

The term *active risk management* seems somewhat new in finance, where *active portfolio management* is a much more familiar concept. To illustrate the idea, we start with active portfolio management and then highlight some of its principles that can be transformed to an active approach to risk management.

There are two main styles of portfolio management, *passive* and *active*. In passive portfolio management, the goal is to construct portfolios that mimic the return of a given index. Advocates of passive portfolio management argue that index tracking incurs low cost as it does not require much research and information gathering on individual stocks, and base their argument on market efficiency. The goal of active portfolio management, on the other hand, is to construct portfolios that aim at outperforming some index or benchmark; see Grinold and Kahn (1999). An active portfolio manager is rewarded for generating additional return, and penalized for subpar return or for adding excess volatility to the benchmark. For example, if an active portfolio manager merely replicates the benchmark portfolio and uses leverage to increase return, there will be a corresponding increase in the volatility of the return and hence the risk of the portfolio. The key to active risk management is to identify, through superior information gathering and efficient processing of the information, sources of the additional return that the active portfolio can generate relative to the benchmark. Transaction costs, which consist of commissions, bid-ask spread and market impact, have to be considered in active portfolio because they contribute to the portfolio's overall "implementation shortfall," which compares the actual portfolio return with that of an "ideal" portfolio without transaction costs; see Guo et al. (2016, Chapter 3).

As pointed out earlier, there are both active and passive styles of risk management in banking and bank regulation. The active style is manifested in asset and liability management and supervisory bank ratings. However, at the corporate level, active risk management is proposed to be the core of enterprise risk management, as discussed below.

Active risk management as core of enterprise risk management

Chapter 7 of National Research Council (2005), which was originally prepared to enhance the U.S. Department of Energy's risk management of its large projects and to provide a basic understanding of effective risk management, is entitled "active risk management." It says:

> Some projects appear to have a passive and ad hoc approach to the management of risk, without the benefits of either tracking the root causes of identified risks or making proactive decisions and actions to mitigate the risks. ... risks may be identified but they are largely ignored in the planning and execution process until undesired events occur, at which time solutions are sought. ... The previously discussed risk identification, analysis, and mitigation planning are important, but they are not sufficient. Active risk management includes the assignment of mitigation responsibilities to appropriate project participants and the oversight of follow-through regarding every risk factor.

The chapter describes two types of methods for active risk management. The first is an integrated risk management plan that "ties together all the components of risk management — i.e., risk identification, analysis, and mitigation — into a functional whole." The second is a risk tracking system, called a "risk register," that keeps data current to support management decisions and actions and to avoid delay. Delay is a serious problem in risk management, and can arise due to procrastination, over-optimism, or simple ignorance of the need for action.

Enterprise risk management (ERM) in business refers to methods and processes used by an organization for risk management. ERM typically involves identifying risks and benefits, assessing the likelihood and impact of events, determining response strategies, and monitoring progress. In 2003, the Casualty Actuarial Society introduced a framework for ERM applicable to any industry, to assess, control, exploit, finance, and monitor risks from all sources. In 2004, the COSO (Committee of Sponsoring Organization of the Treadway Commission) defined an integrated framework for ERM as a "process, effected by an entity's board of directors, management, and other personnel, applied in strategy setting and across the enterprise, designed to identify potential events that may affect the entity, and manage risk to be within its risk appetite, to provide reasonable assurance regarding the achievement of entity objectives."

In 2008, the US Air Force (USAF) awarded a contract valued at over $1 million to Active Risk for deploying its ERM software package *Active Risk Manager* (ARM) and assisting USAF in implementing active risk management to enable on-time and on-budget delivery of projects. ARM is an award-winning ERM package used by some of the world's best known organization, including London Underground, Lockhead Martin, Bechtel, Saudi Aramco, US Homeland Security besides USAF. In 2013, the Sword Group, an

NYSE/Euronext trade holding company for IT services, software and technology, acquired Active Risk and integrated it into the company's software decision.

Hierarchical approach to risk analytics and management

We now describe the basic framework and ideas underlying the hierarchical approach to active risk management. The first important ingredient of the hierarchical approach is data aggregation. For early warning signals, which is arguably the only feasible approach to extract information from the massive and ever-growing data collected by a financial institution for its trading and risk management, ad hoc methods are widely adopted in practice. Sequential methods for processing information that leads to quick fault detection and diagnosis, subject to a prescribed false positive error rate, are usually used here. These methods can be integrated into a multi-layer hierarchy, where each layer represents a component of the financial institution, its business, or a model that has encountered a flaw or fault affecting the enterprise. Timely intervention is crucial to alter the course or prevent further deterioration. In essence, we use a divide and conquer strategy to handle the "big data," with each fault-identified division further divided in this hierarchical approach.

Another important ingredient of the hierarchical approach is using statistical methodology for principled signal extraction in connection with change-point detection and diagnosis. This is followed by implementing corrective actions or intensifying surveillance before executing policy responses, considering cost and organizational constraints. The third key ingredient is the implementation of policy responses to the warning signals generated, typically involving adjustments or reallocations of portfolios based on risk constraints.

8.1.2 Monitoring systematic risk

In addition to active risk management focused on individual portfolios, a significant aspect of risk surveillance involves monitoring systematic risk within a well-structured system, which can be used for regulatory purposes. This process typically encompasses the following steps.

The first step involves defining a well-structured system and selecting a proxy for monitoring systematic risk. This typically includes identifying system components, understanding their dynamics and interactions over time. For instance, to monitor real-time changes in the stock market, all individual stocks are considered as system components, each with its price movements under certain assumptions. Since modeling the entire system's dynamics is often challenging, proxies are chosen for surveillance. In the case of monitoring stock market stability, a common approach is to use the stock market index as a proxy, where the index's volatility reflects the market's stability level. In complex systems, finding a single proxy is usually impractical. Therefore, it is necessary to identify a collection of proxies to infer information about system instability.

The second step involves applying appropriate surveillance methods to monitor the risk of structural changes in the chosen proxy. For instance, using the stock market index as an example, frequent abrupt changes in the long-run volatility level indicate market instability. In such cases, sequential change-point detection methods can be used to identify abrupt volatility changes in the market index. Detecting or estimating structural changes in the dynamics of a complex system is more intricate and typically requires a model for the proxy's dynamics.

In addition to the proxy itself, explanatory variables that indicate structural changes within the system can be integrated into the surveillance process. Modeling these variables' mechanisms may require separate consideration, and their causal relationship with the proxy must be carefully modeled for effective surveillance. When dealing with a large number of potential explanatory variables, regularization or other machine learning techniques may be employed in the surveillance process.

Sometimes, instead of solely monitoring the risk of structural changes within a system, there may be a need for early warnings or predictions regarding such changes. In addition to addressing the challenges encountered in the surveillance process, predictive models for structural changes within the system must be developed for early warning purposes. This task is typically more demanding than simply monitoring the risk of systematic changes, as it requires the development of predictive frameworks capable of anticipating structural shifts within the system.

8.2 Sequential hypothesis testing

Sequential hypothesis testing is considered the fundamental probelm in sequential analysis, often referred to as the "birth" of this field by Ghosh (1991). The problem has evolved from Wald (1945)'s seminal paper on testing a simple null hypothesis versus a simple alternative hypothesis to a relatively complete theory of sequential testing of composite hypotheses. In this section, we introduce the sequential probability ratio test, multiple simple hypothesis test, and some simple composite hypothesis problems.

8.2.1 Sequential probability ratio test

We first recall the *Neyman-Pearman Lemma* for testing a simple hypothesis against a simple alternative. Let X denote a random variable with probability density function f. To test $H_0 : f = f_0$ against $H_1 : f = f_1$, define the likelihood ratio $L(X) = f_1(X)/f_0(X)$, choose a constant $\eta > 0$, and then "reject H_0 if $L(X) \geq \eta$, and accept H_0 if $L(X) \leq \eta$." Denote by P_i the probability under the hypothesis H_i, $i = 0, 1$. Such test is optimal in the sense that any test of H_0 against H_1 that has significance level no larger than $\alpha = P_0(L(X) \geq \eta)$ must have power no larger than $P_1(L(X) \geq \eta)$.

In contrast to the hypothesis testing problem of fixed size samples, the *sequential probability ratio test* (SPRT), introduced by Wald (1945, 1947), has three possibilities, rejecting H_0 for large $L(X)$, accepting for small $L(X)$, and collecting more data for intermediate values of $L(X)$. In particular, let $X_1, X_2, \ldots$ be a sequence of random variables with joint density function $f(x_1, \ldots, x_n)$ $(n = 1, 2, \ldots)$. Consider testing the simple hypothesis

$$H_0 : f = f_0 \text{ for all } n \quad \text{against} \quad H_1 : f = f_1 \text{ for all } n.$$

Denote by $\mathbf{X}_{1:n}$ the sample $(X_1, \ldots, X_n)$. Let

$$L_n = L_n(\mathbf{X}_{1:n}) = \frac{f_1(X_1, \ldots, X_n)}{f_0(X_1, \ldots, X_n)} \tag{8.1}$$

be the *likelihood ratio* (LR) between the hypothesis H_1 and H_0 for the sample $\mathbf{X}_{1:n}$. The goal of the SPRT is to decide which hypothesis is correct as soon as possible (i.e., for the smallest value of n). Choose constants $0 < A \leq 1 \leq B$. After the nth observation X_n is obtained, the SPRT decides to either (i) stop and accept H_1 if $L_n \geq B$, or (ii) stop and accept H_0 if $L_n \leq A$, or (iii) continue sampling if $A < L_n < B$.

This is equivalent to sampling $X_1, X_2, \ldots$ until the random time

$$T = \begin{cases} \text{first } n \geq 1 & \text{such that } L_n \notin (A, B), \\ \infty & \text{if } L_n \in (A, B) \text{ for all } n \geq 1, \end{cases} \tag{8.2}$$

and stopping sampling at time T. If $T < \infty$, reject H_0 if $L_T \geq B$ or accept H_0 if $L_T \leq A$.

Example 8.1 (SPRT for Gaussian random variables) *Let* $X_1, X_2, \ldots$ *be independent and normally distributed random variables with mean* μ *and unit variance. For testing* $H_0 : \mu = \mu_0$ *and* $\mu = \mu_1$ *(assume* $\mu_0 < \mu_1$*), the likelihood ratio is*

$$L_n = \prod_{k=1}^{n} \frac{\phi(x_k - \mu_1)}{\phi(x_k - \mu_0)} = \exp\left[(\mu_1 - \mu_0)S_n - \frac{1}{2}n(\mu_1^2 - \mu_0^2)\right],$$

where $\phi(x) = (2\pi)^{-1/2}\exp(-x^2/2)$ *and* $S_n = \sum_{k=1}^{n} X_k$*. Hence, the stopping rule (8.2) can be written as*

$$T = \inf\left\{n \geq 1 \ : \ S_n - \frac{n}{2}(\mu_1 + \mu_0) \notin \left(\frac{\log A}{\mu_1 - \mu_0}, \frac{\log B}{\mu_1 - \mu_0}\right)\right\}.$$

Note that $T = \infty$ *if the constraint is not satisfied for all* $n \geq 1$*. The SPRT rejects* H_0 *if and only if*

$$S_T \geq \frac{\log B}{\mu_1 - \mu_0} + \frac{T}{2}(\mu_1 + \mu_0).$$

□

Size and power of the test

Assume that the SPRT procedure indeed terminates for a finite sample size T, i.e., $P_i(T < \infty) = 1$ for $i = 0, 1$. We want to find the size $\alpha = P_0(L_T \geq B)$ and the power $\beta = P_1(L_T \leq A\}$ of the test. Denote B_n the subset of n-dimensional space where $A < L_k(x_1, \ldots, x_k) < B$ for $k = 1, 2, \ldots, n-1$ and $L_n(x_1, \ldots, x_n) \geq B$. Then $\{T = n, L_n \geq B\} = \{(X_1, \ldots, X_n) \in B_n\}$ and

$$
\begin{aligned}
\alpha &= \sum_{n=1}^{\infty} P_0(T = n, L_n \geq B) = \sum_{n=1}^{\infty} \int_{B_n} f_0(x_1, \ldots, x_n) dx_1 \ldots dx_n \\
&= \sum_{n=1}^{\infty} E_1\left(L_n^{-1} 1_{\{T=n, L_n \geq B\}}\right) = E_1\left(L_T^{-1} 1_{\{L_T \geq B\}}\right) \\
&\leq B^{-1} P_1(L_T \geq B) = (1 - \beta)/B.
\end{aligned}
\tag{8.3}
$$

Interchange the role of A and B and use a similar argument, we obtain

$$
\beta = P_1(L_T \leq A) \leq A P_0(L_T \leq A) = A(1 - \alpha).
\tag{8.4}
$$

Equality in (8.3) and (8.4) cannot hold exactly due to the overshoot issue, that is, L_n does not have to hit the boundaries exactly when it first leaves (A, B). Wald suggested to ignore the overshoot and replace the inequality with approximate equalities, that is,

$$
\alpha \approx B^{-1}(1 - \beta), \qquad \beta \approx A(1 - \alpha).
\tag{8.5}
$$

Solving equations (8.5) for α and β yields the approximation to α and β,

$$
\alpha \approx \frac{1 - A}{B - A}, \qquad \beta \approx \frac{A(B - 1)}{B - A}.
\tag{8.6}
$$

Note that the above approximation to α and β holds not only in the iid case but also in the general non-iid case as long as the SPRT terminates with probability 1.

Expected value of sample size

To study the expected value of T we make the additional assumption that $X_1, X_2, \ldots$ are independent and identically distributed. The following are two propositions that help us find the expected value of T.

Proposition 8.1 (Wald's identity) *Let $Y_1, Y_2, \ldots$ be independent and identically distributed with mean value $\mu = E(Y_1)$. Let M be an integer valued random variable such that $\{M = n\}$ is an event determined by conditions on $Y_1, \ldots, Y_n$ for all $n = 1, 2, \ldots$. Assume that $E(M) < \infty$. Then $E(Y_1 + \cdots + Y_M) = \mu E(M)$.*

Proof. First assume that $Y_k > 0$. Note that $\sum_{k=1}^{M} Y_k = \sum_{k=1}^{\infty} 1_{\{M \geq k\}} Y_k$ and $\{M \geq k\} = (\cup_{j=1}^{k-1} \{M = j\})^c$ is independent of $Y_k, Y_{k+1}, \ldots$. Then by monotone convergence,

$$E\left(\sum_{k=1}^{M} Y_k\right) = \sum_{k=1}^{\infty} E(Y_k 1_{\{M \geq k\}}) = \mu \sum_{k=1}^{\infty} P(M \geq k) = \mu E(M).$$

For the general case, note that $Y_k = Y_k^+ - Y_k^-$ for all k, where $x^+ = \max(x, 0)$ and $x^- = -\min(x, 0)$. Applying the above argument to Y_k^+ and Y_k^- separately yields the result. $\qquad \square$

Proposition 8.2 (Stein's lemma) *Let $Y_1, Y_2, \ldots$ be independent and identically distributed with $P(Y_1 = 0) < 1$. Let $-\infty < a < b < \infty$, $S_m = \sum_{k=1}^{m} Y_m$ and define*

$$M = \begin{cases} \text{first } m \geq 1 & \text{such that } S_m \notin (a, b), \\ \infty & \text{if } S_m \in (a, b) \text{ for all } n \geq 1. \end{cases}$$

Then there exist constants $C > 0$ and $0 < \rho < 1$ such that $P(M > n) \leq C\rho^n$ $(n = 1, 2, \ldots)$. In particular, $E(M^k) < \infty$ for all $k = 1, 2, \ldots$ and $Ee^{\lambda M} < \infty$ for $\lambda < \log \rho^{-1}$.

Proof. If $P(Y_1 = 0) \neq 1$, then there is a $y > 0$ such that $P(|Y_1| \geq y) > 0$. Without loss of generality, assume that $P(Y_1 \geq y) = \delta > 0$. Let m be an integer for which $my > b - a$. Then

$$P(S_m \geq b - a) \leq P(S_m \geq b - a) \leq P(Y_1 \geq y, \ldots, Y_m \geq y) = \delta^m,$$

and hence for all $k \geq 1$

$$P(M > mk) = P(b < S_n < b, n = 1, \ldots, mk) \leq (1 - \delta^m)^k.$$

For any m, let k be such that $mk < n \leq (k+1)m$. Then

$$P(M > n) \leq P(M > km) \leq (1-\delta^m)^k \leq (1-\delta^m)^{\frac{n}{m}-1} = \frac{1}{1-\delta^m}(1-\delta^m)^{n/m} = C\rho^n,$$

where $C = 1/(1 - \delta^m)$ and $\rho = (1 - \delta^m)^{1/m}$, and the lemma follows. $\qquad \square$

We now approximate the expected value of sample size. Consider the likelihood $L_n = f_1(X_1, \ldots, X_n)/f_0(X_1, \ldots, X_n)$. Since $X_1, \ldots, X_n$ are assumed independent, $\log L_n = \sum_{k=1}^{n} \log\{f_1(X_k)/f_0(X_k)\}$ is a sum of independent, identically distributed random variables. Moreover, the stopping time (8.2) can be written as

$$T = \inf\{n \geq 1 \ : \ \log L_n \notin (\log A, \log B)\}.$$

By Proposition 8.1,

$$E_i[\log L_T] = \mu_i E_i(T), \qquad \mu_i := E_i[\log(f_1(X_1)/f_0(X_1))], \quad i = 0, 1. \quad (8.7)$$

The approximation (8.6) suggests that $\log L_T$ can be considered as a two-valued random variable taking on the values $\log A$ and $\log B$, so

$$E_i[\log L_T] \approx P_i(L_T \leq A) \log A + P_i(L_T \geq B) \log B. \qquad (8.8)$$

Plugging (8.3) and (8.7) into (8.8) yields the approximation

$$E_1 T \quad \approx \quad \frac{A(B-1)\log A + B(1-A)\log B}{(B-A)\mu_1}, \qquad (8.9a)$$

$$E_0 T \quad \approx \quad \frac{(B-1)\log A + (1-A)\log B}{(B-A)\mu_0}. \qquad (8.9b)$$

The above approximations for the expected sample sizes are valid only when $X_1, X_2, \ldots$ are independent and identically distributed or slightly more generally when the likelihood ratio process L_n is a random walk.

Optimality of the SPRT

For testing a simple hypothesis against a simple hypothesis with independent, identically distributed observations, a SPRT is optimal in the sense of minimizing the expected sample size both under H_0 and under H_1 among all tests having no larger error probabilities. To see this, we first consider a lower bound for $E_1(T)$ for any stopping time T such that $P_0(T < \infty) < 1$.

Proposition 8.3 *For any stopping time T with $P_0(T < \infty) < 1$,*

$$E_1(T) \geq -\log P_0(T < \infty)/E_1[\log(f_1(X)/f_0(X))].$$

Proof. Assume $E_1(T) < \infty$ otherwise the result is trivially true. Note that for any random variable Y with mean μ, since $e^x \geq 1 + x$, $E(e^{Y-\mu}) \geq 1 + E(Y - \mu) = 1$ and hence $E(e^Y) > e^{E(Y)}$. Then use Wald's identities,

$$P_0(T < \infty) = E_1 \exp\left(-\sum_{k=1}^{T} \log \frac{f_1(X_k)}{f_0(X_k)}\right)$$

$$\geq \exp\left(-\left[\sum_{k=1}^{T} \log \frac{f_1(X_k)}{f_0(X_k)}\right]\right) = \exp\left[-E_1(T)E_1\left(\log \frac{f_1(X_k)}{f_0(X_k)}\right)\right].$$

This implies the result since $E_1[\log(f_1(X_1)/f_0(X_1))] > 0$. $\square$

Consider a sequential test of $H_0 : f = f_0$ against $H_1 : f = f_1$ with error probabilities $\alpha = P_0(\text{Reject } H_0)$ and $\beta = P_1(\text{Accept } H_0)$. The following result extends Proposition 8.3 and asserts that the expected sample sizes in the SPRT are approximately minimal.

Proposition 8.4 *Let T be the stopping time of any test of $H_0 : f = f_0$ against $H_1 : f = f_1$ with error probabilities $\alpha \in (0,1)$, $\beta \in (0,1)$. Assume $E_i(T) < \infty$ $(i = 0,1)$. Let $\mu_i = E_i[\log(f_1(X_1)/f_0(X_1))]$. Then*

$$E_1(T) \geq \mu_1^{-1}\left[(1-\beta)\log\left(\frac{1-\beta}{\alpha}\right) + \beta\log\left(\frac{\beta}{1-\alpha}\right)\right],$$

$$E_0(T) \geq \mu_0^{-1}\left[\alpha\log\left(\frac{1-\beta}{\alpha}\right) + (1-\alpha)\log\left(\frac{\beta}{1-\alpha}\right)\right].$$

Proof. Let $R = \{\text{Reject } H_0\}$, $R^c = \{\text{Accept } H_0\}$. Use Wald's identity

$$\alpha = P_0(R) = E_1\left(\prod_{k=1}^{T}\frac{f_0(X_k)}{f_1(X_k)}\cdot 1_R\right) = P_1(R)E_1\left(e^{-\log T}|R\right)$$

$$\geq (1-\beta)\exp[-E_1(\log L_T|R)] = (1-\beta)\exp\left[-\frac{1}{1-\beta}E_1(\log L_T \cdot 1_R)\right].$$

Taking logarithms yields

$$(1-\beta)\log\left(\frac{\alpha}{1-\beta}\right) \geq -E_1[\log L_T \cdot 1_R].$$

A similar calculation gives

$$\beta\log\left(\frac{1-\alpha}{\beta}\right) \geq -E_1[\log L_T \cdot 1_{R^c}].$$

Adding the above two inequalities together yields the first assertion of the proposition, since $\mu_1 > 0$. The second assertion can be proved similarly. $\square$

The above optimality properties of the SPRT is for the discrete-time iid case. It can be extended to the continuous-time Brownian motion problems and general discrete-time or continuous-time stochastic models with almost no assumptions on the observation distribution; see discussion in Tartakovsky et al. (2014, Chapter 3).

8.2.2 Multiple simple hypothesis

Suppose there are $m > 1$ alternative hypotheses to test against the null. We generalize the SPRT for two simple hypothesis to the case of testing $m+1$ hypotheses $H_i : f = f_i$, $i = 0,1,\ldots,m$, where f_i are density functions of random variable X. Denote by $\mathbf{X}_{1:n}$ the sample $(X_1,\ldots,X_n)$. Let

$$L_{ij} = L_{ij}(\mathbf{X}_{1:n}) = \frac{f_j(X_1,\ldots,X_n)}{f_i(X_1,\ldots,X_n)}$$

be the likelihood ratio between the hypothesis H_j and H_i for the sample $\mathbf{X}_{1:n}$.

Denote by P_i and E_i the probability and the expectation under the hypothesis H_i $(i = 0,1,\ldots,m)$, respectively. A multi-hypothesis sequential test is a pair $\delta = (d, T)$, where T is a stopping time with respect to the filtration (or information set) generated by $\mathbf{X}_{1:n}$ and $d = d(\mathbf{X}_{1:n})$ is an decision function with values in $\{0, 1, \ldots, m\}$. Hence, $d = i$ means that the hypothesis H_i is accepted, that is, $\{d = i\} = \{T < \infty, \delta \text{ accepts } H_i\}$.

Let $\alpha_{ij}(\delta) = P_i(d = j)$, $i \neq j$, $i, j = 0, 1, \ldots, N$, be the error probabilities of the test δ, i.e., the probabilities of accepting a specific H_j when H_i is true. Note that the probabilities of rejecting the hypothesis H_i when it is true is given by

$$\alpha_i(\delta) = P_i(d \neq i) = \sum_{j \neq i} \alpha_{ij}(\delta), \qquad i = 0, 1, \ldots, m.$$

For a threshold or boundary matrix $\mathbf{A} = (A_{ij})$ in which $A_{ij} > 0$ and $A_{ii} = 0$, define the *matrix SPRT* (MSPRT) $\delta_m^* = (T_m^*, d_m^*)$ as follows:

Stop at the fist $n \geq 1$ such that, for some i, $L_{ij} \geq A_{ji}$ for all $j \neq i$, (8.10)

and accept the unique i that satisfies these inequalities. The MSPRT is built on $(N+1)N/2$ one-sided SPRTs between the hypothesis H_i and H_j. For the case $m = 1$, this test degenerates to the Wald's SPRT. If A_{ji} do not depend on j, then δ_N^* stops at the first time such that

$$\frac{L_i(t)}{\max_{j \neq i} L_j(t)} \geq A_i$$

for some i, this is referred to as the *generalized likelihood ratio test* (GLRT).

The structure of the optimal multi-hypothesis sequential test and its asymptotic optimality in the iid case can be analyzed analogously as the two hypothesis sequential test, though it is technically more complicated. Interested readers can find more details on this in Tartakovsky et al. (2014, Chapter 4).

8.2.3 Composite hypothesis

The sequential hypothesis tests in Sections 8.2.1 and 8.2.2 address the problems of testing simple hypotheses, this section considers problems involving composite hypotheses. To illustrate the idea, suppose that all observations are independent and identically distributed. Consider the problem of testing a simple null hypothesis $H_0 : \theta = \theta_0$ against a composite alternative $H_1 : \theta \in \Theta_1$. Wald (1947) suggested the following two approaches of modifying the SPRT for this problem.

The first approach is to replace the likelihood ratio (8.1) by a weighted likelihood ratio

$$\bar{L}_n = \int_{\Theta_1} w(\theta) L_n^\theta d\theta, \tag{8.11}$$

in which $w(\theta)$ is a weight function on $\theta \in \Theta_1$ such that $\int_{\Theta_1} w(\theta) d\theta = 1$. This implies the *weighted SPRT* with the stopping time

$$\bar{T} = \inf\{n \geq 1 \; ; \; \bar{L}_n \notin (A, B)\}, \qquad 0 < A < 1, B > 1. \tag{8.12}$$

The upper bound of the weighted SPRT on the type I error probability can be obtained analogously as that of the SPRT. In particular, following the argument in (8.3), we have

$$\alpha = P_0(\overline{L}_{\widetilde{T}} \geq B) = \int_{\Theta_1} P_\theta(\overline{L}_{\widetilde{T}} \geq B)w(\theta)d\theta = E_1(\overline{L}_{\widetilde{T}}^{-1} 1_{\{L_{\widetilde{T}} \geq B\}})$$

$$\leq B^{-1}P_1(\overline{L}_{\widetilde{T}} \geq B) \leq 1/B.$$

The upper bound on the weighted type II error probability is given by

$$1 - \beta = \int_{\Theta_1} P_\theta(\overline{L}_{\widetilde{T}} \leq A)w(\theta)d\theta = E_0\left[\overline{L}_{\widetilde{T}} 1_{\{\overline{L}_{\widetilde{T}} \leq A\}}\right] \leq A[1 - P_0(\overline{L}_{\widetilde{T}} \leq A)] \leq A.$$

The second approach is to consider the generalized likelihood ratio statistic

$$\widetilde{L} = \sup_{\theta \in \Theta_1} L_n^\theta,$$

which leads to the *generalized sequential likelihood ratio test* with the stopping time

$$\widetilde{T} = \inf\{n \geq 1 \; ; \; \widetilde{L}_n \notin (A, B)\}.$$

Different from the weighted SPRT, it is not easy to find the upper bounds of the generalized sequential likelihood ratio test on the type I and II error probabilities.

If the null hypothesis is also composite, $H_0 : \theta \in \Theta_0$, Wald (1947) suggested using the stopping rule (8.12) with the following weighted likelihood ratio,

$$\overline{L}_n = \frac{\int_{\Theta_1} w_1(\theta) \prod_{i=1}^n p_\theta(X_i)d\theta}{\int_{\Theta_0} w_0(\theta)) \prod_{i=1}^n p_\theta(X_i)d\theta}. \tag{8.13}$$

Under this rule, one can apply Wald's likelihood ratio identity and show that the upper bounds on the type I and II error probabilities are $\alpha \leq 1/B$ and $1 - \beta \leq A$; see Exercise 8.2.

In contrast to sequential tests of simple hypotheses that have remarkable optimality property, sequential tests with composite hypotheses lose such property due to two issues. One is the open continuation region that might lead to large sample sizes, and the other is the difficulty of estimating parameters in the SPRT tests; see discussion in Siegmund (1985, Section 2.5).

8.3 Frequentist methods for quickest change detection

Sequential change-point detection, or quickest change detection, or quickest disorder detection, is concerned with the design and analysis of techniques for online quickest detection of a change in the state of a process, subject to some tolerance on the risk of false detection. Specifically, a time process may undergo an abrupt or gradual change-of-state from normal to abnormal. If the state of the process changes and becomes abnormal, one is interested in detecting such a change as soon as possible after its occurrence, so that an appropriate response can be provided in a timely manner. Thus, with the arrival of every new observation, one needs to make the decision whether to let the process continue, or to stop and raise an alarm. The time instance at which the process undergoes an abrupt state change is called the *change-point*, and the location of change-point is usually unknown in practice.

In particular, the change-point problem assumes that one obtains a series of observations $X_1, X_2, \ldots$ such that, for some value $\nu \geq 0$, $X_1, X_2, \ldots, X_\nu$ have one distribution and $X_{\nu+1}, X_{\nu+2}, \ldots$ have another distribution. A sequential detection rule is defined by a stopping time T with respects to observations $X_1, \ldots, X_T$ with which a change is declared. Generally, such detection rule can be constructed by using Bayesian and non-Bayesian methods. This section describes some non-Bayesian methods for change-point estimation and surveillance and present their optimality and asymptotic optimality properties, and Bayesian methods for change-point detection will be introduced in the next section.

8.3.1   Control charts

The subject of statistical quality control concerns the monitoring and evaluation of product quality in continuous mass production processes. Shewhart (1931) introduced the fundamental concept of a "state of statistical control," where the behavior of a selected quality characteristic of products at time t follows a specified probability distribution that depends on some parameter θ_t. This parameter remains at a target level of within specified limits. To detect significant deviations from this state, Shewhart also introduced the process inspection scheme involving sampling of fixed sizes at regular intervals. From each sample at time t, a statistic Z_t is computed. These sequential statistics Z_t are graphically presented on a control chart. When a sample point falls outside the control limit(s) marked on the chart, the monitoring scheme signals that the production process has deviated from the state of statistical control. Shewhart's control charts are sometimes called *single-sample* schemes because each decision is based soly on the current sample without incorporating data from previous samples.

We now describe several simple and well-known change detection algorithms, derive some analytical properties of these algorithms, and discuss numerical approximations. Most of the algorithms presented here deal with samples of data with either a *fixed size* window or a *randomly sized* sliding window. In quality control, these elementary algorithms are usually called *Shewhart control charts, geometric moving average control charts* or *exponentially weighted moving average charts*, and *finite moving average control charts*.

Consider a parametric setting of quickest detection problem. Suppose that observations $\{X_n\}_{n\geq 1}$ are independent and $X_1,\ldots,X_\nu$ are each distributed according to a common distribution $F_{\theta_0}(x)$ (density $f_{\theta_0}(x)$), while $X_{\nu+1}, X_{\nu+2},\ldots$ each follow a common distribution $F_{\theta_1}(x)$ (density $f_{\theta_1}(x)$). Denote $\mathbf{X}_{i:j} = (X_i,\ldots,X_j)$. Let

$$L_{i:j} = L(\mathbf{X}_{i:j}) = \prod_{k=i}^{j} \frac{f_{\theta_1}(X_k)}{f_{\theta_0}(X_k)}, \qquad \lambda_{i:j} = \log L_{i:j}$$

be the likelihood ratio and log-likelihood ratio of $\mathbf{X}_{i:j}$ with respect to distributions f_{θ_1} and f_{θ_0}, respectively. One way to characterize the performance of change-point detection algorithms is to use the *average run length* (ARL) function.

Definition 8.1 *Let T be the time of alarm or a stopping time of a sequential change detection algorithm, i.e., the time at which the change is detected. The ARL function is defined as*

$$ARL(\theta) = ARL(\theta; z) = E_\theta(T), \tag{8.14}$$

where z is the starting value of the decision statistic.

Shewhart control charts

One of the simplest fixed sample size change detection algorithm is a repeated Neyman-Pearson test that is called *Shewhart control chart* in quality control. The stopping rule of this algorithm is

$$T_{NP} = m \cdot \inf\{K \geq 1 : L_{(K-1)m+1:Km} \geq \eta\}. \tag{8.15}$$

Denote by $\alpha_0(\theta)$ the size of a test or the level of significance of the test, and $\beta(\theta) = 1 - \alpha_1(\theta)$ the power of the test, where $\alpha_1(\theta)$ is the type II error probability. To compute the ARL function of this chart, we let $K^* = \inf\{K \geq 1 : L_{(K-1)m+1:Km} \geq \eta\}$ and note that the number of samples K^* has a geometric distribution, that is,

$$P_\theta(K^* = k) = [1 - \beta(\theta)]^{k-1}\beta(\theta).$$

Therefore, $E_\theta(K^*) = 1/\beta(\theta)$ and the ARL function of the Shewhart control chart at θ is given by

$$ARL(\theta) = E_\theta(T_{NP}) = \frac{m}{\beta(\theta)}. \tag{8.16}$$

In particular,

$$ARL(\theta_0) = E_{\theta_0}(T_{NP}) = \frac{m}{\alpha_0}, \qquad ARL(\theta_1) = E_{\theta_1}(T_{NP}) = \frac{m}{1-\alpha_1}, \tag{8.17}$$

where α_0 and α_1 are the probabilities of type I and type II errors, respectively. When the hypothesis before and after the change are composite, the formula (8.16) for the ARL function is still valid.

Example 8.2 (Change in normal mean) *Consider the case of a change in the mean of an independent Gaussian sequence $N(\theta, \sigma^2)$ from θ_0 to $\theta_1(> \theta_0)$ with known variance σ^2. The Neyman–Pearson test of samples $\mathbf{X}_{(K-1)m+1:Km}$ for hypothesis $H_0 : \theta = \theta_0$ against $H_1 : \theta = \theta_1$ can be written as*

$$L(\mathbf{X}_{(K-1)m+1:Km}) = \prod_{i=(K-1)m+1}^{Km} \frac{\phi((X_i - \theta_1)/\sigma)}{\phi((X_i - \theta_0)/\sigma)} \underset{H_0}{\overset{H_1}{\gtrless}} A(\alpha),$$

where $\phi(x)$ is the density function of $N(0, 1)$ distribution. Let $\overline{X}_{(K-1)m+1:Km} := m^{-1} \sum_{i=(K-1)m+1}^{Km} X_i$ The above test is equivalent to

$$\sqrt{m}(\overline{X}_{(K-1)m+1:Km} - \theta_0)/\sigma \underset{H_0}{\overset{H_1}{\gtrless}} \kappa, \quad \kappa := \frac{\sigma \log A(\alpha)}{\sqrt{m}(\theta_1 - \theta_0)} + \frac{\sqrt{m}(\theta_1 - \theta_0)}{2\sigma}.$$

$$(8.18)$$

Hence, the alarm is raised at the first time at which $\overline{X}_{(K-1)m+1:Km} - \theta_0 \geq \kappa\sigma/\sqrt{m}$, the power function is given by

$$\beta(\theta) = P_\theta\left(\overline{X}_{(K-1)m+1:Km} - \theta_0 \geq \kappa\frac{\sigma}{\sqrt{m}}\right) = 1 - \Phi\left(\kappa - \frac{\sqrt{m}}{\sigma}(\theta - \theta_0)\right).$$

where $\Phi(x) = (2\pi)^{-1} \int_{-\infty}^{x} e^{-u^2/2} du$, and the ARL function is given by

$$ARL(\theta) = \frac{m}{1 - \Phi\left(\kappa - \frac{\sqrt{m}}{\sigma}(\theta - \theta_0)\right)}. \qquad \square$$

Exponentially weighted moving average control charts

The *exponentially weighted moving average* (EWMA) control chart, also known as the *geometric moving average* (GMA) control chart, is defined recursively as follows

$$g_0 = 0, \qquad g_n = (1 - \alpha)g_{n-1} + \alpha Z_n, \qquad (8.19)$$

where $\alpha \in (0, 1]$ is the tuning parameter and $Z_n = \log[f_{\theta_1}(X_n)/f_{\theta_0}(X_n)]$. The stopping rule for a one-sided change-detection problem is given by

$$T_{\text{EWMA}} = \inf\{n \geq 1 : g_n \geq \eta\}, \qquad (8.20)$$

where η is a threshold. Note that the stopping rule (8.19) can be rewritten as

$$g_n = \alpha Z_n + (1 - \alpha)\alpha Z_{n-1} + \cdots + (1 - \alpha)^{n-1}\alpha Z_1.$$

Hence, g_k represents the weighted log-likelihood ratio with exponentially decreasing weights over time. For a two-sided change-detection problem, one can use the two-sided EWMA stopping rule which is given by

$$T_{\text{EWMA}} = \inf\{n \geq 1 : g_n \notin (\eta_1, \eta_2)\}. \qquad (8.21)$$

The ARL function of the EWMA control chart can be approximated numerically; see Crowder (1987).

Finite moving average charts

The *finite moving average* (FMA) chart is based on the following statistic

$$g_k = \sum_{i=0}^{n-1} c_i Z_{k-i},$$

where c_i is chosen so that $c_0 > 0, \ldots, c_{n-1} > 0$ and $c_k = 0$ when $k \geq n$. The stopping rule is

$$T_{\mathrm{FMA}} = \inf\{k \geq n : g_k \geq \eta\}, \tag{8.22}$$

where η is a threshold that controls false alarm rate. Assume the log-likelihood ratios $\{Z_k\}_{k\geq 1}$ are independent and identically distributed with mean $E(Z_1) = \mu$ and finite variance $\mathrm{Var}(Z_1) = \sigma^2 < \infty$. Consider the statistic

$$g_k = \sum_{i=1}^{k} c_{k-i}(Z_i - \mu_0), \quad k \geq n.$$

Then the sequence $\{g_k\}_{k\geq n}$ is stationary with mean $E_\mu(g_k) = \sum_{i=0}^{n-1} c_i(\mu - \mu_0)$ and covariance

$$\mathrm{Cov}(g_k, g_{k+l}) = \begin{cases} \sigma^2 \sum_{i=0}^{n-l-1} c_i c_{i+l} & \text{if } l = 0, \ldots, n-1, \\ 0 & \text{if } l \geq n. \end{cases}$$

Let p_l be the probabilities $p_0(h) = 1$ and

$$p_l(h) = P_\mu(g_n < h, \ldots, g_{n+l} < h) = P_\mu(T_{\mathrm{FMA}} > l + n)$$

for $l \geq 1$. Then the ARL is given by

$$\mathrm{ARL}(\mu) = n + \sum_{i=1}^{\infty} p_i(h).$$

As no analytical expression of the ARL function is available, upper and lower bounds for the above ARL can be obtained; see Lai (1974) and Bohm and Hackl (1990).

8.3.2 CUSUM algorithm

To improve the sensitivity of the Shewhart chart by incorporating the results of previous inspections, Page (1954) proposed the following class of *cumulative sum* (CUSUM) schemes. Let $Z_i = \log[f_1(X_i)/f_0(X_i)]$ be the log-likelihood ratio of X_i with respect to distributions $f_1(X)$ and $f_0(X)$. The typical behavior of the log-likelihood ratio $L_n = \sum_{i=1}^{n} Z_i$ shows a negative drift before change and a positive drift after change. This indicates that the decision rule on the change-point should be based on the difference between the value of L_n and its current (running) minimum value, which is given by

$$g_n = L_n - \min_{0 \le j \le n} L_j = \max_{1 \le j \le n+1} \sum_{i=k}^{n} Z_i = \max\left\{0, \max_{1 \le j \le n+1} \sum_{i=k}^{n} Z_i\right\}. \quad (8.23)$$

The CUSUM stopping rule is

$$T_{CS} = \inf\{n \ge 1 : g_n \ge \eta\}. \quad (8.24)$$

The statistic $L_n - \min_{0 \le i \le n} L_i$ is called *drawup* in technical trading rules. Similarly, $\max_{0 \le i \le n} L_i - L_n$ is called *drawdown* in technical analysis, in which the "filter rule" gives a buy signal as soon as the drawdown exceeds some prespecified threshold and gives a sell signal when the drawup exceeds the threshold; see more discussion on this in Lai and Xing (2008a, Section 11.1).

Page (1954) also suggested to consider the CUSUM procedure as a sequence of SPRTs (8.2) between two simple hypotheses $H_0 : \theta = \theta_0$ and $H_1 : \theta = \theta_1$ with thresholds 0 and η. In particular, the CUSUM stopping time can be defined as the first time at which λ_n reaches the threshold η. Consider the zero value of the lower threshold in the repeated SPRT, the resulting detection procedure can be rewritten in a recursive form as $g_0 = 0$ and

$$g_n = (g_{n-1} + Z_n)^+, \quad n = 1, 2, \ldots, \qquad T_{CS} = \inf\{n \ge 1 : g_n \ge \eta\}.$$

For the two-sided problem where the post-change θ is either $\theta_1^u = \theta_0 + \delta$ or $\theta_1^l = \theta - \delta$, with δ known, one can use the two parallel CUSUM algorithms which detect the increase and the decrease of θ, respectively. The stopping rule of such two-sided CUSUM is

$$T_{2CS} = \inf\{n \ge 1 : g_n^u \ge \eta \text{ or } g_n^l \ge \eta\}, \quad (8.25)$$

in which

$$g_n^u = \left[g_{n-1}^u + \log \frac{f_{\theta_1^u}(X_n)}{f_{\theta_0}(X_n)}\right]^+, \quad g_n^u = \left[g_{n-1}^u - \log \frac{f_{\theta_1^l}(X_n)}{f_{\theta_0}(X_n)}\right]^+.$$

The CUSUM algorithm can be extended for non-iid case. In particular, assume that the observations $\{X_n\}_{n \ge 1}$ are distributed according to conditional densities $f_{0,n}(X_n | \mathbf{X}_{1:(n-1)})$ for $n = 1, \ldots, \nu$ and according to conditional densities $f_{1,n}(X_n | \mathbf{X}_{1:(n-1)})$ for $n > \nu$. Then the CUSUM statistic in this case still has the form of (8.23) and can be computed recursively as follows

$$g_0 = 0, \quad g_n = \left[g_{n-1} + \log \frac{f_{1,n}(X_n | \mathbf{X}_{1:(n-1)})}{f_{0,n}(X_n | \mathbf{X}_{1:(n-1)})}\right]^+.$$

The CUSUM stopping rule is still defined by (8.24) with the newly defined g_n.

8.3.3 Lorden's lemma and minimax formulation

The ARL is defined by $E_\theta T$ assuming $\theta_t \equiv \theta$. In quality control, when the quality characteristic remains at a satisfactory level $\theta \in \Theta_0$, we want $E_\theta T$ to be large. When the quality characteristic stays at an unsatisfactory level $\theta' \notin \Theta_0$, we want $E_{\theta'} T$ to be small. Moustakides (1986) and Ritov (1990) showed that the likelihood ratio of CUSUM scheme (8.24) is optimal in the following sense. Let c_γ be so chosen that the ARL of the CUSUM scheme (8.24) is γ when $\nu = \infty$, and let $\mathcal{F}_\gamma$ be the class of all inspection schemes with ARL$\geq \gamma$ when $\nu = \infty$. When ν is finite, we are concerned with the conditional expected delay $E[(T - \nu + 1)^+ | X_1, \ldots, X_{\nu-1}]$, whose supremum over $(\nu, X_1, \ldots, X_{\nu-1})$ represents the worse case delay, which is minimized by the CUSUM scheme (8.24). More precisely, Moustakides (1986) and Ritov (1990) have shown that (8.24) minimizes the worse-case detection delay

$$\overline{E}_1(T) = \sup_{\nu \geq 1} \text{ess sup} \left\{ E^{(\nu)}[(T - \nu + 1)^+ | X_1, \ldots, X_{\nu-1}] \right\} \qquad (8.26)$$

over all rules T that belong to $\mathcal{F}_\gamma$, where $P^{(m)}$ denotes the probability measure under which $X_m, X_{m+1}, \ldots$, are i.i.d. with common density f_1 and $X_1, \ldots, X_{m-1}$ are i.i.d. with common density f_0. Earlier Lorden (1971) showed that his optimality property holds asymptotically as $\gamma \to \infty$ and that

$$\overline{E}_1(T) \sim \frac{\log \gamma}{I(f_1, f_0)},$$

where $I(f_1, f_0) = E_{f_1}[\log (f_1(X_1)/f_0(X_1))]$ denotes the Kullback-Leibler information number.

As pointed out by Lorden (1971), Page's likelihood ratio CUSUM scheme (8.24) corresponds to stopping when a one-sided sequential probability ratio test (SPRT) based on $X_k, X_{k+1}, \ldots$ shows significant evidence against the null hypothesis $H_0 : f = f_0$. The maximum over $1 \leq k \leq n$ in (8.24) corresponds to the maximum likelihood estimation, at stage n, of the time ν for which $X_1, \ldots, X_{\nu-1}$ are i.i.d. with density function f_0 and $X_\nu, X_{\nu+1}, \ldots$ are i.i.d. with density function f_1. Note that (8.24) can be expressed as $N = \min_{k \geq 1}(N_k + k - 1)$, where N_k is the stopping time of the one-sided SPRT applied to $X_k, X_{k+1}, \ldots$. Instead of the one-sided SPRT, we can use other sequential tests and apply their stopping rules to $X_k, X_{k+1}, \ldots$ to define N_k. Lorden (1971) has proved the following result that converts a hypothesis test into a change-point detection rule.

Proposition 8.5 (Lorden's lemma) *Let $X_1, X_2, \ldots$ be i.i.d. random vectors and let τ be a stopping time with respect to $X_1, X_2, \ldots$ such that $P(\tau < \infty) \geq \alpha$. For $k = 1, 2, \ldots$, let T_k denote the stopping time obtained by applying τ to $X_k, X_{k+1}, \ldots$. Then $T = \min_{k \geq 1}(T_k + k - 1)$ is a stopping time and $ET \geq 1/\alpha$.*

Consider the one-sided SPRT which stops sampling as soon as $\sum_{i=1}^{t} \log\left(f_1(X_i)/f_0(X_i)\right) \geq \log\gamma$ and rejects $H_0 : f = f_0$ on stopping. This test minimizes the expected sample size under the simple alternative $H_1 : f = f_1$ among all tests, sequential or otherwise, that have the same or smaller error probability of falsely rejection H_0. Consider the stopping rule

$$\tau^* = \begin{cases} m_\gamma & \text{if } \sum_{i=1}^{m_\gamma} \log\left(\dfrac{f_1(X_i)}{f_0(X_i)}\right) \geq \log\gamma, \\ \infty & \text{otherwise,} \end{cases} \tag{8.27}$$

which corresponds to the Neyman-Pearson test, with fixed sample size m_γ, of $H_0 : f = f_0$ versus $H_1 : f = f_1$. Since

$$P_0(\tau^* < \infty) = P_0\left\{ \sum_{i=1}^{m_\gamma} \log\left(\dfrac{f_1(X_i)}{f_0(X_i)}\right) \leq \log\gamma \right\} \leq \exp(-\log\gamma) = \gamma^{-1},$$

it follows from Lorden's lemma that $E_0(T^*) \geq \gamma$, where P_0 and E_0 denote the probability measure and expectation under which $X_1, X_2, \ldots$ are i.i.d. with common density function f_0 and

$$T^* = \inf\left\{ n : \sum_{i=n-m_\gamma+1}^{n} \log\left(\dfrac{f_1(X_i)}{f_0(X_i)}\right) \geq \log\gamma \right\} \tag{8.28}$$

is the moving average detection rule. For the CUSUM rule (8.24), we have

$$\overline{E}_1(T) = E^{(1)}T = E^{(1)}\tau \sim \frac{\log\gamma}{I(f_1, f_0)} \qquad \text{as } \gamma \to \infty. \tag{8.29}$$

For the moving average rule T^* defined by (8.28), Lai (1995) has shown that (8.29) also holds with T replaced by T^* if the window size m_γ satisfies

$$m_\gamma \sim \frac{\log\gamma}{I(f_1, f_0)} \quad \text{and} \quad \left(m_\gamma - \frac{\log\gamma}{I(f_1, f_0)}\right)\bigg/\sqrt{\log\gamma} \to \infty. \tag{8.30}$$

Hence, the moving average rule is asymptotically as efficient as the CUSUM rule if the window size m_γ is chosen appropriately.

8.3.4   GLR and window-limited GLR rules

Assume that the pre- and post-change density belong to a multivariate exponential family

$$f_{\boldsymbol{\theta}}(\mathbf{x}) = \exp\{\boldsymbol{\theta}'\mathbf{x} - \psi(\boldsymbol{\theta})\} \tag{8.31}$$

with respect to some measure ω on $\mathbb{R}^d$. We now consider the *generalized likelihood ratio* (GLR) and window-limited GLR rules for the composite hypotheses.

Known pre-change but unknown post-change distributions

We first assume that the pre-change distribution is known, but the post-change distribution is unknown. The GLR statistic for testing the null hypothesis of no change-point, based on $\mathbf{X}_1, \ldots, \mathbf{X}_n$, versus the alternative hypothesis of a single change-point prior to n but not before n_0 is

$$\max_{n_0 \leq k < n} \left\{ \sup_{\widetilde{\boldsymbol{\theta}}} \sum_{i=k+1}^{n} \log f_{\widetilde{\boldsymbol{\theta}}}(\mathbf{X}_i) - \sup_{\boldsymbol{\lambda}} \sum_{i=1}^{n} \log f_{\boldsymbol{\lambda}}(\mathbf{X}_i) \right\}. \tag{8.32}$$

For the special case of the $N(\theta, 1)$ family, Siegmund and Venkatraman (1995) considered this GLR statistics, for which the supremum is attained at $\theta = (X_k + \cdots + X_n)/(n - k + 1)$, giving the GLR rule

$$T_{\mathrm{SV}} = \inf\{n : \max_{1 \leq k \leq n} \{(X_k + \cdots + X_n)^2/[2(n - k + 1)]\} \geq c\}. \tag{8.33}$$

They show that, under P_0, $\{\sqrt{c} \exp(-c) T_{\mathrm{SV}}, c \geq 1\}$ is uniformly integrable and $K\sqrt{c} \exp(-c) T_{\mathrm{SV}}$ converges in distribution as $c \to \infty$ to the exponential distribution with mean 1 for some positive constant K, i.e.,

$$E_0(T_{\mathrm{SV}}) \sim K^{-1} c^{-1/2} \exp(c).$$

Hence, choosing $c = \log \gamma + \frac{1}{2} \log(\log \gamma) + \log K + o(1)$ yields $E_0(T_{\mathrm{SV}}) = \gamma\{1 + o(1)\}$.

Unknown pre- and post-change distributions

We now consider the case in which the pre- and post-change density functions belong to (8.31) but are not known in advance. The GLR statistic for testing the null hypothesis of no change-point, based on $\mathbf{X}_1, \ldots, \mathbf{X}_n$, versus the alternative hypothesis of a single change-point prior to n but not before n_0 is

$$\max_{n_0 \leq k < n} \left\{ \sup_{\boldsymbol{\theta}} \sum_{i=1}^{k} \log f_{\boldsymbol{\theta}}(\mathbf{X}_i) + \sup_{\widetilde{\boldsymbol{\theta}}} \sum_{i=k+1}^{n} \log f_{\widetilde{\boldsymbol{\theta}}}(\mathbf{X}_i) - \sup_{\boldsymbol{\lambda}} \sum_{i=1}^{n} \log f_{\boldsymbol{\lambda}}(\mathbf{X}_i) \right\}$$

$$= \max_{n_0 \leq k < n} \{k I(\bar{\mathbf{X}}_{1,k}) + (n - k) I(\bar{\mathbf{X}}_{k+1,n}) - n I(\bar{\mathbf{X}}_{1,n})\},$$

$$\tag{8.34}$$

where

$$\bar{\mathbf{X}}_{m,n} = \sum_{i=m}^{n} \mathbf{X}_i/(n - m + 1), \qquad I(\boldsymbol{\mu}) = \sup_{\boldsymbol{\theta}} \{\boldsymbol{\theta}^T \boldsymbol{\mu} - \psi(\boldsymbol{\theta})\} = \boldsymbol{\theta}_{\boldsymbol{\mu}}^T \boldsymbol{\mu} - \psi(\boldsymbol{\theta}_{\boldsymbol{\mu}}),$$

and $\boldsymbol{\theta}_{\boldsymbol{\mu}} = (\nabla \psi)^{-1}(\boldsymbol{\mu})$, noting that $\sup_{\boldsymbol{\lambda}}$ is related to maximizing the likelihood under the null hypothesis and that $\sup_{\boldsymbol{\theta}}$ and $\sup_{\widetilde{\boldsymbol{\theta}}}$ arise from maximizing the likelihood under the hypothesis that a single change-point occurs at $k+1$. Let

$$g(\alpha, x, y) = \alpha I(x) + (1 - \alpha) I(y) - I(\alpha x + (1 - \alpha)y).$$

Replacing the likelihood ratio statistics in the CUSUM rule (8.24) by the GLR statistics (8.34) yields the GLR rule

$$T_{\text{GLR}} = \inf\{n > n_0 : \max_{n_0 \le k < n} ng(k/n, \bar{\mathbf{X}}_{1,k}, \bar{\mathbf{X}}_{k+1,n}) \ge c\} \qquad (8.35)$$

for detecting a change in $\boldsymbol{\theta}$ when the pre- and post-parameters are unknown.

For the special case of the $N(\theta, 1)$ family, $ng(k/n, \bar{X}_{1,k}, \bar{X}_{k+1,n})$ in (8.35) can be expressed as

$$k(n-k)(\bar{X}_{k+1,n} - \bar{X}_{1,k})^2/n = nk(\bar{X}_{1,n} - \bar{X}_{1,k})^2/(n-k). \qquad (8.36)$$

Therefore, in this normal case, T_{GLR} is the same as the detection rule considered by Siegmund and Venkatraman (1995), which is related to an earlier rule proposed by Pollak and Siegmund (1991) under the assumption of known $\delta := \theta_1 - \theta_0$ and having the form

$$T_{GLR,\delta} = \inf\{n > n_0 : \sum_{k=n_0}^{n} \exp[\delta k(\bar{X}_{1,n} - \bar{X}_{1,k}) - \delta^2 k(1 - k/n)/2] \ge \gamma\},$$

$$(8.37)$$

with $\log \gamma \sim c$. In (8.35) or (8.37), it is assumed that the actual change-time, defined as ∞ if no change ever occurs, is larger than the initial sample size n_0. Note that if we replace $\sum_{k=k_0}^{n}$ by $\max_{n_0 \le k \le n}$ and maximize the exponential term in (8.37) over δ, we obtain (8.35) in this normal case in view of (8.36).

Window-limited GLR

The GLR scheme (8.33) and its generalizations to other parametric families do not have convenient recursive forms and require computation at every stage n the full data set $\{X_1, \ldots, X_n\}$. A natural modification to reduce the memory requirements is the window-limited GLR scheme, which replaces $\max_{1 \le k \le n}$ in (8.33) by maximizing k over the moving window $n - M \le k \le n - \widetilde{M}$ with $0 \le \widetilde{M} < M$. Lai (1995) and Chan and Lai (2007) consider how M and $\widetilde{M}$ should be chosen to yield asymptotically optimal detection procedures in general parametric models for Markov chains, which they have developed the more general GLR window-limited detection rule

$$T_W = \inf \left\{ n : \max_{k:n-k+1 \in \mathcal{N}} \sup_{\theta} \left[\sum_{i=k}^{n} \log \left(\frac{f_\theta(X_i|X_{i-1})}{f_0(X_i|X_{i-1})} \right) \right] \ge c_\gamma \right\}. \qquad (8.38)$$

in which $b > 1$ and $n_j = [b^j M]$, $1 \le j \le J = \min\{i : n_i \ge \gamma\}$, $M \sim a \log \gamma$, $J \log b = \log \gamma - \log \log \gamma - \log a + o(1)$, and $\mathcal{N} = \{1, \ldots, M+1\} \cup \{n_j : 1 \le j \le J\}$. They show how $c_\gamma \sim \log \gamma$ can be chosen such that $E_{\theta_0} T_W \ge \gamma$. Moreover, $E_\theta T_W \sim (\log \gamma)/I(f_\theta, f_{\theta_0})$ for $I(f_\theta, f_{\theta_0}) > 1/a$.

The window-limited scheme (8.38) involves $O(\log \gamma)$ log-likelihood ratios or score statistics $L_{n,n-j}$ which are moving averages with j varying within a fixed set that contains $O(\log \gamma)$ elements. These moving averages $L_{n,n-j}$ can be conveniently computed by using $O(\log \gamma)$ parallel algorithms, one for each j. The threshold c_γ for the window-limited GLR scheme to satisfy $E_{\theta_0} T \approx \gamma$ can be computed by using Monte Carlo simulations to evaluate $E_{\theta_0} T(c_\gamma)$ and the method of successive secant approximations supplemented by bisection search, as shown in Lai (1995, Section 3.3) that also proposes alternative baseline performance measures which are much less demanding than $E_{\theta_0} T$.

Lai (1995) begins by considering the classical Shewhart chart based on independent observations, for which the distribution of the waiting time T to false alarm under P_{θ_0} is geometric. In this case, $P_{\theta_0}(T = 1) = 1/E_{\theta_0} T$ and

$$E_{\theta_0} T \sim m / P_{\theta_0}(T \leq m) \tag{8.39}$$

uniformly in $m = o(E_{\theta_0} T)$, as $E_{\theta_0} T \to \infty$. Moreover, (8.39) still holds as $E_{\theta_0} T \sim \gamma \to \infty$ and $m/\log \gamma \to \infty$ but $\log m = o(\log \gamma)$, for the window-limited GLR scheme.

Instead of the commonly used ARL constraint $E_{\theta_0} T \approx \gamma$ on the time to false alarm of a detection rule, Lai (1995) and Lai and Shan (1999) intro-duce a more tractable constraint of the form $P_{\theta_0}(T \leq m) \approx m/\gamma$, where m is such that m/γ is small but $m/\log \gamma$ is large. They use importance sampling techniques to evaluate $P_{\theta_0}(T \leq m)$ and computed the threshold c_γ such that $P_{\theta_0}(T(c_\gamma) \leq m) \approx m/\gamma$. The ARL constraint stipulates a long expected du-ration to false alarm. However, a large mean of T does not necessarily imply that the probability of having a false alarm before some specified time m is small. In fact, it is easy to construct positive integer-valued random variables T with a large mean γ and also having a high probability that $T = 1$. This high probability of false alarm at the initial stage is clearly unacceptable, and the mean may be too crude as a summary of the desired features of T under P_{θ_0}. Note also that the constraint $E_{\theta_0}(T) \geq \gamma$ only yields the asymptotic lower bound

$$E^{(\nu)}(T - \nu | T \geq \nu) \geq (1/I(f_1, f_0) + o(1)) \log \gamma \qquad \text{as } \gamma \to \infty$$

for the expected detection delay under $P_{\theta_1}^{(\nu)}$ for some unspecified ν, where $P_{\theta_1}^{(\nu)}$ denotes the probability measure corresponding to a change of parameters from θ_0 to θ_1 at time ν. Subject to the constraint

$$P_{\theta_0}(\nu \leq T < \nu + m) \leq m/\gamma$$

with $m/\log \gamma \to \infty$ but $\log m = o(\log \gamma)$, Lai (1998) develops asymptotic lower bounds for $E_{\theta_1}^{(\nu)}(T - \nu + 1)^+$. He also shows that under certain regularity conditions these asymptotic lower bounds are attained by the moving average scheme and the likelihood ratio CUSUM scheme assuming θ_1 to be known, and by the composite of moving average schemes without assuming θ_1 to be known.

8.4   Bayesian sequential change-point detection

This section presents the Bayesian approach to change-point detection for iid and general non-iid models. The key feature of the Bayesian sequential change point detection is that the change point is a random variable or process following a prior distribution. In this section, we first introduce the detection rule for a single change-point which follows a prior distribution and then the rule for a sequence of change-points that follows a geometric renewal process.

8.4.1   Shiryaev procedure and its optimality

Let T be a detection rule (i.e., a stopping time) for a single change-point. Denote by $\{\pi_k\}_{k\geq 1}$ the prior distribution of the change-point ν, $\pi_k = P(\nu = k)$. From the Bayesian point of view, it is reasonable to measure the false alarm risk with the *probability of false alarm* (PFA), which is defined as

$$\mathrm{PFA}^\pi(T) := P^\pi(T \leq \nu) = \sum_{k=1}^{\infty} \pi_k P(T \leq k | \nu = k). \tag{8.40}$$

The detection delay of the stopping rule can be measured by the *expected delay to detection* (EDD), which is defined as

$$\mathrm{EDD}^\pi(T) = E^\pi(T - \nu | T > \nu) = \frac{E^\pi((T - \nu)^+)}{P^\pi(T > \nu)}, \tag{8.41}$$

where $x^+ = \max(0, x)$ and E^π denotes the expectation with respect to P^π.

Let $C_\alpha = \{T : \mathrm{PFA}^\pi(T) \leq \alpha\}$ be the class of stopping times for which the PFA does not exceed a preset level $\alpha \in (0, 1)$. Then under the Bayesian approach, the goal is to find an optimal stopping time T_{opt} such that

$$\mathrm{EDD}^\pi(T_{\mathrm{opt}}) = \inf_{T \in C_\alpha} \mathrm{EDD}^\pi(T) \qquad \text{for every } \alpha \in (0, 1).$$

Suppose that the observations $\{X_t\}_{t \geq 1}$ are independent and such that $X_1, \ldots, X_\nu$ are each distributed according to a common density $f_0(x)$, while $X_{\nu+1}, X_{\nu+2}, \ldots$ each follows a common density $f_1(x) \neq f_0(x)$. Moreover, assume that the change point ν has a geometric prior distribution with success probability p, that is,

$$P(\nu = k) = p(1 - p)^k \quad \text{for } k = 0, 1, 2, \ldots. \tag{8.42}$$

Assuming a loss of c for each observation taken after ν and a loss of 1 for a false alarm before ν, Shiryaev (1963, 1978) used optimal stopping theory to show that the optimal detection procedure is based on comparing the posterior probability of a change with a certain detection threshold sequentially over time. Such procedure is called the *Shiryaev procedure*. Specifically, let

$$R_{n,p} = \sum_{k=1}^{n} \prod_{i=k}^{n} \frac{f_1(X_i)}{(1-p)f_0(X_i)}. \tag{8.43}$$

Then by Bayes rule, it can be shown that

$$P(\nu < n | X_1, \ldots, X_n) = \frac{R_{n,p}}{R_{n,p} + 1/p}. \tag{8.44}$$

Hence, thresholding the posterior probability $P(\nu < n | X_1, \ldots, X_n)$ is equivalent to thresholding the process $\{R_{n,p}\}_{n \geq 1}$. Therefore, the Shiryaev detection procedure has the form

$$T_S = \inf\{n \geq 1 : R_{n,p} \geq \eta\}, \tag{8.45}$$

where η is chosen in such a way that the PFA is exactly equal to α, i.e., $\text{PFA}^\pi(T_S(\eta_\alpha)) = \alpha$. The following proposition shows first-order approximations for all positive moments of the detection delay; see its proof in Tartakovsky et al. (2014, Theorem 7.1.4).

Proposition 8.6 (First-order approximation to moments of the EDD) *Condition on $\nu = k$, let $X_1, \ldots, X_k$ be i.i.d. with the pdf $f_0(x)$ and $X_{k+1}, X_{k+2}, \ldots$ be iid with the pdf $f_1(x)$. Let the prior distribution of the change point ν be geometric distribution (8.42). Suppose that the Kullback-Leibler information $I = I(f_0, f_1) = E_0[\log(f_0/f_1)]$ is positive and finite. Then for any $m > 0$,*

$$E^\pi\left[(T_S(\eta) - \nu)^m | T_S(\eta) > \nu\right] \sim \left(\frac{\log \eta}{I + |\log(1-p)|}\right)^m \quad \text{as } \eta \to \infty.$$

If η_α is chosen so that $\text{PFA}(T_S(\eta_\alpha)) \leq \alpha$ and $\log \eta_\alpha \sim \log(1/\alpha)$, in particular, $\eta_\alpha = (1-\alpha)/\alpha p$, then for all $m > 0$,

$$\inf_{T \in \mathcal{C}_\alpha} E^\pi[(T - \nu)^m | T > \nu] \sim E^\pi\left[(T_S(\eta_\alpha) - \nu)^m | T_S(\eta_\alpha) > \nu\right]$$

$$\sim \left(\frac{|\log \alpha|}{I + |\log(1-p)|}\right)^m \quad \text{as } \alpha \to 0.$$

The Shiryaev procedure can be extended to more general cases. For instance, when $X_1, \ldots, X_n$ are not independent, $R_{n,p}$ can be modified as

$$R_{n,p} = \sum_{k=1}^{n} \prod_{i=k}^{n} \frac{f_{1,i}(X_i | X_1, \ldots, X_{i-1})}{(1-p)f_{0,i}(X_i | X_1, \ldots, X_{i-1})}. \tag{8.46}$$

Then the posterior probability (8.44) and the Shiryaev's rule (8.46) stay the same form.

As the parameter of the geometric prior distribution vanishes, i.e., $p \to 0$, the Shiryaev detection statistic (8.43) or (8.46) converges to the so-called *Shiryaev-Roberts detection statistic*

$$\tilde{R}_{n,p} = \sum_{k=1}^{n} \prod_{i=k}^{n} \frac{f_1(X_i)}{f_0(X_i)} \quad \text{or} \quad \tilde{R}_{n,p} = \sum_{k=1}^{n} \prod_{i=k}^{n} \frac{f_{1,i}(X_i|X_1,\ldots,X_{i-1})}{f_{0,i}(X_i|X_1,\ldots,X_{i-1})},$$

and the *Shiryaev-Roberts* rule can be obtained as

$$T_{\mathrm{SR}} = \inf\{n \geq 1 : \tilde{R}_{n,p} \geq \eta\},$$

Pollak (1985) showed that the Shiryaev-Roberts rule is asymptotically Bayes risk efficient as $p \to 0$.

8.4.2 Detection procedures for composite hypotheses

The preceding section shows the Bayesian change point detection procedure for two simple hypotheses, we now consider the case in which the pre- and post-change density functions are not known in advance, but belong to a multivariate exponential family (8.31). Let π be a prior density function (with respect to Lebesgue measure) on $\boldsymbol{\Theta} := \{\boldsymbol{\theta} : \int e^{\boldsymbol{\theta}' \mathbf{X}} d\omega(\mathbf{X}) < \infty\}$ given by

$$\pi(\boldsymbol{\theta}; a_0, \boldsymbol{\mu}_0) = c(a_0, \boldsymbol{\mu}_0) \exp\left\{ a_0 \boldsymbol{\mu}_0' \boldsymbol{\theta} - a_0 \psi(\boldsymbol{\theta}) \right\}, \qquad \boldsymbol{\theta} \in \boldsymbol{\Theta}, \tag{8.47}$$

where

$$1/c(a_0, \boldsymbol{\mu}_0) = \int_{\boldsymbol{\Theta}} \exp\left\{ a_0 \boldsymbol{\mu}_0' \boldsymbol{\theta} - a_0 \psi(\boldsymbol{\theta}) \right\} d\boldsymbol{\theta}, \qquad \boldsymbol{\mu}_0 \in (\boldsymbol{\nabla}\psi)(\boldsymbol{\Theta}),$$

in which $\boldsymbol{\nabla}$ denotes the gradient vector of partial derivatives. The posterior density of $\boldsymbol{\theta}$ given the observations $\mathbf{X}_1, \ldots, \mathbf{X}_m$ drawn from $f_{\boldsymbol{\theta}}$ is

$$\pi\left(\boldsymbol{\theta}; a_0 + m, \frac{a_0 \boldsymbol{\mu}_0 + \sum_{i=1}^{m} \mathbf{X}_i}{a_0 + m}\right). \tag{8.48}$$

Therefore, (8.47) is a conjugate family of priors and (8.48) shows that a_0 can be interpreted as an additional sample size associated with the prior and $\boldsymbol{\mu}_0$ is the prior mean of the $\mathbf{y}_i$.

$$\int_{\boldsymbol{\Theta}} f_{\boldsymbol{\theta}}(\mathbf{X})\pi(\boldsymbol{\theta}; a, \boldsymbol{\mu}) d\boldsymbol{\theta} = \frac{c(a, \boldsymbol{\mu})}{c(a+1, (a\boldsymbol{\mu} + \mathbf{X})/(a+1))}.$$

Suppose that the parameter $\boldsymbol{\theta}$ takes the value $\boldsymbol{\theta}_0$ for $t < \nu$ and another value $\boldsymbol{\theta}_1$ for $t \geq \nu$, and that the change-time ν and the pre- and post-change values $\boldsymbol{\theta}_0$ and $\boldsymbol{\theta}_1$ are unknown. Following Shiryaev (1963, 1978), one can use the Bayesian approach that assumes ν to be geometric with parameter p but constrained to be larger than n_0 and that $\boldsymbol{\theta}_0, \boldsymbol{\theta}_1$ are independent, have the same density function (8.47) and are also independent of ν. Let $\pi_n = P\{\nu \leq$

$n|\mathbf{X}_1,\ldots,\mathbf{X}_n\}$. Whereas π_n is a Markov chain in the case of known $\boldsymbol{\theta}_0$ and $\boldsymbol{\theta}_1$, it is no longer Markovian in the present setting of unknown pre- and post-change parameters and Shiryaev's rule that triggers an alarm when π_n exceeds some threshold is no longer optimal; see Zacks (1991, pp.540–541) who suggests using dynamic programming to find the optimal stopping rule but also notes that "it is generally difficult to determine the optimal stopping boundaries."

Because of the complexity of Bayes rule, Zacks and Barzily (1981) introduced a more tractable myopic (two-step-ahead) policy in the univariate Bernoulli case ($X_i = 0$ or 1). Here, we introduce a modification of Shiryaev's rule, which was shown by Lai and Xing (2010) to be asymptotically Bayes as $p \to 0$. Denote by $\mathcal{F}_t$ the σ-field generated by $\mathbf{X}_1,\ldots,\mathbf{X}_t$. Let

$$\pi_{0,0} = c(a_0,\boldsymbol{\mu}_0), \quad \pi_{i,j} = c\left(a_0 + j - i + 1, \frac{a_0\boldsymbol{\mu}_0 + \sum_{t=i}^{j} \mathbf{X}_t}{a_0 + j - i + 1}\right).$$

Note that for $n_0 < i \le n$,

$$P\{\nu = i|\mathcal{F}_n\} \propto p(1-p)^{i-1}\pi_{0,0}^2/\pi_{1,i-1}\pi_{i,n}, \quad P\{\nu > n|\mathcal{F}_n\} \propto (1-p)^n\pi_{0,0}/\pi_{1,n}. \tag{8.49}$$

The normalizing constant is determined by the fact that all the probabilities in (8.49) sum to 1. Let $p_{i,n} = P\{\nu = i|\mathcal{F}_n\}$ be the posterior probability given the observed samples up to time n, we then have

$$P(n_0 < \nu \le n|\mathcal{F}_n) = \sum_{i=n_0+1}^{n} p_{i,n} = \frac{\sum_{i=n_0+1}^{n} P(\nu = i|\mathcal{F}_n)}{\sum_{i=n_0+1}^{n} P(\nu = i|\mathcal{F}_n) + P(\nu > n|\mathcal{F}_n)},$$

in which $P\{\nu = i|\mathcal{F}_n\}$ and $P\{\nu > n|\mathcal{F}_n\}$ are given by (8.49). Therefore, Shiryaev's stopping rule in the present setting of unknown pre- and post-change parameters can again be written in the form of (8.45) with

$$R_{p,n} = \sum_{i=n_0+1}^{n} \frac{\pi_{0,0}\pi_{1,n}}{(1-p)^{n-i}\pi_{1,i}\pi_{i,n}}$$

in place of (8.43). Although this stopping rule is no longer optimal, as noted earlier, it can be modified to give

$$T_{\mathrm{exShi}} = \inf\{n > n_p : P(\nu \le n|\nu \ge n - k_p, \mathcal{F}_n) \ge \eta_p\}, \tag{8.50}$$

which is shown by Lai and Xing (2010) to be asymptotically optimal as $p \to 0$, for suitably chosen k_p, η_p and $n_p \ge n_0$. Since

$$P(\nu \le n|\nu \ge n - k_p, \mathcal{F}_n) = \frac{\sum_{i=n-k_p}^{n} P(\nu = i|\mathcal{F}_n)}{\sum_{i=n-k_p}^{n} P(\nu = i|\mathcal{F}_n) + P(\nu > n|\mathcal{F}_n)},$$

we can use (8.49) to rewrite (8.50) in the form

$$T_{\mathrm{exShi}} = \inf\left\{n > n_p : \sum_{i=n-k_p}^{n} \frac{\pi_{0,0}\pi_{1,n}}{(1-p)^{n-i+1}\pi_{1,i-1}\pi_{i,n}} \ge \gamma_p\right\}. \tag{8.51}$$

This has essentially the same form as Shiryaev's rule (8.45) with the obvious changes to accommodate the unknown $f_{\boldsymbol{\theta}_0}$ and $f_{\boldsymbol{\theta}_1}$, and with the important sliding window modification $\sum_{i=n-k_p}^{n}$ of Shiryaev's sum $\sum_{i=n_0+1}^{n}$, which has too many summands to trigger false alarms when $\boldsymbol{\theta}_0$ and $\boldsymbol{\theta}_1$ are estimated sequentially from the observations.

Lai et al. (2009) have modified the above Bayesian model and detection rule to handle the case where multiple change-points with unknown pre- and post-change parameters can occur, as in sequential surveillance applications. The key idea is still to use sliding windows $\sum_{i=n-k_p}^{n}$ as in (8.51) but with the summands modified to be the posterior probabilities p_{in} that the most recent change-point up to time n occurs at i. Then the detection rule based on the assumption of *multiple change-points* (MCP) can be expressed as

$$T_{\text{MCP}} = \inf \Big\{ n > n_p : \sum_{i=n-k_p}^{n} p_{in} \geq \gamma_p \Big\}. \tag{8.52}$$

8.4.3 Hidden Markov filtering with multiple change-points

The above Bayesian methods address the issue of detecting a change-point which follows a prior distribution. The setting of detecting a single change-point is standard in engineering applications, such as fault detection and isolation in complex dynamical systems and industrial processes, integrity monitoring of navigation systems, and radar and sonar signal processing. However, in many financial and biological applications, the system often involves many unknown change-points and complex dynamics. For these problems, Lai and Xing (2008b, 2011) proposed a general hidden Markov filtering approach which allow recursive and tractable estimators of multiple change-points and is computationally very efficient. This section introduces this approach for multiple change-points problems, and the extension of the approach to estimate dynamical systems with unknown multiple change-points is presented in Section 8.5.2.

Suppose that all observations $\mathbf{y}_1, \ldots, \mathbf{y}_t, \ldots$ follow the multiparameter exponential family (8.31), i.e.,

$$f_{\boldsymbol{\theta}}(\mathbf{y}) = \exp\{\boldsymbol{\theta}'\mathbf{y} - \psi(\boldsymbol{\theta})\} \tag{8.53}$$

with respect to some measure ν on $\mathbb{R}^d$, and the prior density π (with respect to Lebesgue measure) is given by (8.47), or

$$\pi(\boldsymbol{\theta}; a_0, \boldsymbol{\mu}_0) = c(a_0, \boldsymbol{\mu}_0) \exp\{a_0 \boldsymbol{\mu}_0' \boldsymbol{\theta} - a_0 \psi(\boldsymbol{\theta})\}, \quad \boldsymbol{\theta} \in \boldsymbol{\Theta}, \tag{8.54}$$

where $c(a_0, \boldsymbol{\mu}_0)$ are defined in (8.47). Suppose that, instead of being time-invariant, the parameter vector $\boldsymbol{\theta}_t$ may undergo occasional changes such that for $t > 1$, the indicator variables $I_t := 1_{\{\boldsymbol{\theta}_t \neq \boldsymbol{\theta}_{t-1}\}}$ are independent Bernoulli random variables with $P(I_t = 1) = p$. When there is a parameter change at time t (i.e., $I_t = 1$), the changed parameter $\boldsymbol{\theta}_t$ is assumed to be sampled from π. The simplicity of the conjugate family plays an important role in the explicit formulas for the sequential (filtering) estimates $E(\boldsymbol{\mu}_t|\mathcal{Y}_t)$ and for the fixed-sample (smoothing) estimates $E(\boldsymbol{\mu}_t|\mathcal{Y}_n)$, where $\boldsymbol{\mu}_t = \nabla\psi(\boldsymbol{\theta}_t)$ and $\mathcal{Y}_t$ denotes $(\mathbf{y}_1, \ldots, \mathbf{y}_t)$. We also use $\mathcal{Y}_{i,j}$ to denote $(\mathbf{y}_i, \ldots, \mathbf{y}_j)$ for $i \le j$.

Forward filter $\boldsymbol{\theta}_t|\mathcal{Y}_t$

An important ingredient in deriving the forward filter is the most recent change-time K_t up to t, i.e., $K_t = \max\{s \le t : I_s = 1\}$. Denote by $p_{it} = P(K_t = i|\mathcal{Y}_t)$ the probability that the most recent change-time up to time t is K_t and $f(\cdot|\cdot)$ the conditional density. Since K_t can take values from 1 to t, we obtain that

$$f(\boldsymbol{\theta}_t|\mathcal{Y}_t) = \sum_{i=1}^{t} P(K_t = i|\mathcal{Y}_t)f(\boldsymbol{\theta}_t|\mathcal{Y}_{i,t}, K_t = i) = \sum_{i=1}^{t} p_{it} f(\boldsymbol{\theta}_t|\mathcal{Y}_{i,t}, K_t = i).$$

In the above equality, the conditional density $f(\boldsymbol{\theta}_t|\mathcal{Y}_{i,t}, K_t = i)$ is the posterior density function of $\boldsymbol{\theta}_t$ given $\mathcal{Y}_{i,j}$ and $K_t = i$. Then it follows from (8.48) that

$$f(\boldsymbol{\theta}_t|\mathcal{Y}_{i,t}, K_t = i) = \pi(\boldsymbol{\theta}_t; a_0 + t - i + 1, \bar{\mathbf{Y}}_{i,t}),$$

where $\bar{\mathbf{Y}}_{i,j} = (a_0\boldsymbol{\mu}_0 + \sum_{k=i}^{j} \mathbf{y}_k)/(a_0 + j - i + 1)$, $j \ge i$, is the posterior mean. Combining the above two equations yields that

$$f(\boldsymbol{\theta}_t|\mathcal{Y}_t) = \sum_{i=1}^{t} p_{it} \pi(\boldsymbol{\theta}_t; a_0 + t - i + 1, \bar{\mathbf{Y}}_{i,t}). \tag{8.55}$$

This shows that the posterior distribution of $\boldsymbol{\theta}_t$ given $\mathcal{Y}_t$ is a mixture of distributions. To compute the mixture weights p_{it}, Lai and Xing (2011) showed that p_{it} can be represented recursively by noting that $\sum_{i=1}^{t} p_{it} = 1$ and

$$p_{it} = \frac{p_{it}^*}{\sum_{j=1}^{t} p_{jt}^*}, \qquad p_{it}^* := \begin{cases} pf(\mathbf{y}_t|I_t = 1) & \text{if } i = t, \\ (1-p)p_{i,t-1}f(\mathbf{y}_t|\mathcal{Y}_{i,t-1}, K_t = i) & \text{if } i \le t-1. \end{cases}$$

Making use of

$$f(\mathbf{y}_t|\mathcal{Y}_{i,t-1}, K_t = i) = \int f_{\boldsymbol{\theta}_t}(\mathbf{y}_t)f(\boldsymbol{\theta}_t|\mathcal{Y}_{i,t}, K_t = i)d\boldsymbol{\theta}_t$$

and (8.31), one can represent p_{it}^* as follows,

$$p_{it}^* = \begin{cases} p\pi_{0,0}/\pi_{t,t} & \text{if } i = t, \\ (1-p)p_{i,t-1}\pi_{i,t-1}/\pi_{i,t} & \text{if } i < t, \end{cases} \tag{8.56}$$

where $\pi_{0,0} = c(a_0, \boldsymbol{\mu}_0)$ and $\pi_{i,j} = c(a_0 + j - i + 1, \bar{\mathbf{Y}}_{i,j})$; see Exercise 8.5. Then the change-point probability and the posterior mean at t are given by

$$P(I_t = 1|\mathcal{Y}_t) = p_{tt}, \qquad E(\boldsymbol{\mu}_t|\mathcal{Y}_t) = \sum_{i=1}^{t} p_{it} \bar{\mathbf{Y}}_{i,j}.$$

Moreover, the probability that the most recent change-point occurs at one of the times $\{s, s+1, \ldots, t\}$ is

$$P(K_t \in \{s, s+1, \ldots, t\}) = \sum_{i=s}^{t} p_{it}.$$

The above probability will be used in Section 8.5.1 to construct a surveillance rule for structural breaks in credit rating transition dynamics that may have multiple unknown structural breaks.

The forward filter (8.55) with mixture weights (8.56) has a very neat structure. Its computational complexity linearly increases with time t. The following example illustrates how to apply it to multiple change-point problems with Gaussian distributions.

Example 8.3 (Forward filter in the Normal mean shift model) *Suppose that all observations $y_1, y_2, \ldots$ follow a Gaussian distribution $N(\theta_t, \sigma^2)$, in which σ^2 is known and θ_t are piecewise constant means. Assume that $I_1 = 1$ and $I_t = 1_{\{\theta_t \neq \theta_{t-1}\}}$ are independent Bernoulli random variables with $P(I_t = 1) = p$ for $t \geq 2$. Given $\{I_t\}$, the mean θ_t satisfies the following dynamics*

$$\theta_t = (1 - I_t)\theta_{t-1} + I_t z_t,$$

where z_t are independent and identically distributed normal random variables with mean μ and variance ν^2. Given $\mathcal{Y}_t$ and the most recent change time up to time t is i, i.e., $K_t = i$, the posterior distribution of θ_t is a normal distribution $N(\mu_{it}, \nu_{it}^2)$ in which

$$\mu_{it} = \frac{1}{\nu_{it}^2}\left(\frac{\mu}{\nu^2} + \frac{\sum_{s=i}^{t} y_s}{\sigma^2}\right), \qquad \nu_{it}^2 = \left(\frac{1}{\nu^2} + \frac{t-i+1}{\sigma^2}\right)^{-1}.$$

The weights or probabilities $p_{it} = P(K_t = i | \mathcal{Y}_t)$ are given by $p_{it} = p_{it}^* / \left(\sum_{j=1}^{t} p_{jt}^*\right)$ and p_{it}^* are given by (8.56) with

$$\pi_{00} = (2\pi\nu^2)^{-1/2} e^{-\mu^2/(2\nu^2)}, \qquad \pi_{it} = (2\pi\nu_{it}^2)^{-1/2} e^{-\mu_{it}^2/(2\nu_{it}^2)}.$$

Then the forward filter for $\theta_t | \mathcal{Y}_t$ is a mixture of $N(\mu_{it}, \nu_{it}^2)$ with weights p_{it} $(i = 1, \ldots, t)$. The posterior mean of θ_t given $\mathcal{Y}_t$ is

$$E(\theta_t | \mathcal{Y}_t) = \sum_{i=1}^{t} p_{it} \mu_{it}.$$

Besides, the filtering estimate of the change-point probability at time t given $\mathcal{Y}_t$ is p_{tt}. $\square$

Smoother $\boldsymbol{\theta}_t|\mathcal{Y}_n$

The forward filter $\boldsymbol{\theta}_t|\mathcal{Y}_t$ provides an online estimate of $\boldsymbol{\theta}_t$ at each time t. Suppose that one is interested in estimating $\boldsymbol{\theta}_t$ at each time $t = 1, \ldots, n$ for a sequence of observations $\mathbf{y}_1, \ldots, \mathbf{y}_n$. This is referred to as the smoothing estimate of $\boldsymbol{\theta}_t$ given $\mathcal{Y}_t$. Since the number of change-points at the n time points is unknown, estimating $\boldsymbol{\theta}_t$ directly is very time consuming and challenging. However, by making use of the forward filter and its backward analog, one can compute the smoother in a much easier way.

In particular, Lai and Xing (2011) showed how to derive the posterior distribution of $\boldsymbol{\theta}_t|\mathcal{Y}_n$ by using Bayes' theorem to combine the forward filter $\boldsymbol{\theta}_t|\mathcal{Y}_t$ and the backward filter $\boldsymbol{\theta}_t|\mathcal{Y}_{t+1,n}$. The backward filter is obtained by reversing time, noting that the $\tilde{I}_t = 1_{\{\boldsymbol{\theta}_t \neq \boldsymbol{\theta}_{t+1}\}}$ are still independent Bernoulli. Using the time-reversed counterpart

$$\widetilde{K}_t = \min\{s > t : \tilde{I}_s = 1\}$$

of K_t and

$$P(\boldsymbol{\theta}_t \in A|\mathcal{Y}_{t+1,n}) = \int P(\boldsymbol{\theta}_t \in A|\boldsymbol{\theta}_{t+1})dP(\boldsymbol{\theta}_{t+1}|\mathcal{Y}_{t+1,n}),$$

the backward (time-reversed) filter can be expressed as

$$f(\boldsymbol{\theta}_t|\mathcal{Y}_{t+1,n}) = p\pi(\boldsymbol{\theta}_t; a_0, \boldsymbol{\mu}_0) + (1-p)\sum_{j=t+1}^{n} q_{j,t+1}\pi(\boldsymbol{\theta}_t; a_0 + j - t, \bar{\mathbf{Y}}_{t+1,j}),$$

$$(8.57)$$

where $q_{jt} = q_{jt}^*/\left(\sum_{l=t}^{n} q_{lt}^*\right)$ and

$$q_{j,t}^* = \begin{cases} p\pi_{0,0}/\pi_{t,t} & \text{if } j = t, \\ (1-p)q_{j,t+1}\pi_{t+1,j}/\pi_{t,j} & \text{if } j > t. \end{cases} \qquad (8.58)$$

By Bayes' theorem, the smoother is proportional to the product of forward and backward filters divided by the prior distribution, that is,

$$f(\boldsymbol{\theta}_t|\mathcal{Y}_n) \propto f(\boldsymbol{\theta}_t|\mathcal{Y}_t)f(\boldsymbol{\theta}_t|\mathcal{Y}_{t+1,n})/\pi(\boldsymbol{\theta}; a_0, \boldsymbol{\mu}_0). \qquad (8.59)$$

Combining (8.57) with (8.55), and noting that

$$\pi(\boldsymbol{\theta}; a_0 + t - i + 1, \bar{\mathbf{Y}}_{i,t})\frac{\pi(\boldsymbol{\theta}; a_0 + j - t, \bar{\mathbf{Y}}_{t+1,j})}{\pi(\boldsymbol{\theta}; a_0, \boldsymbol{\mu}_0)} = \frac{\pi_{it}\pi_{t+1,j}}{\pi_{ij}\pi_{00}}\pi(\boldsymbol{\theta}; a_0 + j - i + 1, \bar{\mathbf{Y}}_{ij}),$$

one can use (8.59) to obtain the smoother $\boldsymbol{\theta}_t|\mathcal{Y}_n$, which is expressed as

$$f(\boldsymbol{\theta}_t|\mathcal{Y}_n) = \sum_{1 \leq i \leq t \leq j \leq n} \beta_{ijt}\pi(\boldsymbol{\theta}_t; a_0 + j - i + 1, \bar{\mathbf{Y}}_{i,j}), \qquad (8.60)$$

where $\beta_{ijt} = \beta^*_{ijt}/P^*_t$, $P^*_t = p + \sum_{1 \le i \le t < j \le n} \beta^*_{ijt}$, and

$$\beta^*_{ijt} = \begin{cases} pp_{it} & \text{if } i \le t = j, \\ (1-p)p_{it}q_{j,t+1}\pi_{it}\pi_{t+1,j}/\pi_{ij}\pi_{00} & \text{if } i \le t < j. \end{cases} \tag{8.61}$$

From the above, it follows that the change-point probability and posterior mean at time t are expressed as

$$P(I_{t+1} = 1|\mathcal{Y}_n) = p/P^*_t, \qquad E(\boldsymbol{\mu}_t|\mathcal{Y}_n) = \sum_{1 \le i \le t \le j \le n} \beta_{ijt}\bar{\mathbf{Y}}_{i,j}.$$

Example 8.4 (Smoother in the Normal mean shift model) *Consider the Normal mean shift model in Example 8.3. Define μ_{it}, v^2_{it}, π_{00}, and π_{it} in the same way as those in Example 8.3. We now derive the smoothing estimate of $\theta_t|\mathcal{Y}_t$. Since π_{00} and π_{it} $(i \le t)$ are defined, the forward weights p_{it} and backward weights $q_{j,t+1}$ are given by (8.56) and (8.58), respectively. Computing the smoothing weights β_{ijt} by (8.61), the posterior distribution of $\theta_t|\mathcal{Y}_n$ is a mixture of $N(\mu_{ij}, v^2_{ij})$ with weight β_{ijt}. The change-point probability at time t and posterior mean at time t are expressed as p/P_t and $E(\theta_t|\mathcal{Y}_n) = \sum_{1 \le i \le t \le j \le n} \beta_{ijt}\mu_{ij}$.* □

Hyperparameter estimation and efficient approximation

For problems of multiple change-point detection and estimation, the above filters and smoothers successfully reduce the computational complexity to linear and quadratic complexity. However for multiple change-point problems of large sample size, say $n \approx 10^6$, there are still two issues to be addressed. One is hyperparameter estimation and the other is computational efficiency. Lai and Xing (2011) proposed the following procedures for these two concerns and analyzed their asymptotic properties.

First, consider the issue that the forward filter $\theta_t|\mathcal{Y}_t$ and the smoother $\theta_t|\mathcal{Y}_n$ involve the hyperparameters p, a_0, and $\boldsymbol{\mu}_0$. These hyperparameters of course can be specified by users, but they can also be replaced by their estimates in the empirical Bayes approach. From the definition of p^*_{it}, it follows that the likelihood function of p, a_0, and $\boldsymbol{\mu}_0$ is

$$\prod_{t=1}^n f(\mathbf{y}_t|\mathcal{Y}_{t-1}) = \prod_{t=1}^n \left(\sum_{i=1}^t p^*_{it} \right), \tag{8.62}$$

in which p^*_{it} is a function of p, a_0 and $\boldsymbol{\mu}_0$ given by (8.56). Since the $\mathbf{y}_t$ are exchangeable random vectors with mean $\boldsymbol{\mu}_0$ in the Bayesian model, we can estimate $\boldsymbol{\mu}_0$ by the sample mean $\widehat{\boldsymbol{\mu}} = n^{-1}\sum_{t=1}^n \mathbf{y}_t$. The hyperparameter a_0 is used to weight the sample mean $\widehat{\boldsymbol{\mu}}$ with the sample data between change-points in (8.55). A simple choice is $a_0 = 1$, which can be interpreted as having an additional observation at $\widehat{\boldsymbol{\mu}}$ at a change-time when there is little information on the changed parameter. The important hyperparameter in the change-point model is the relative frequency p of change-points. Putting the above simple choice of the hyperparameters a_0 and μ_0 in (8.62), one can estimate p by maximizing the log-likelihood function $l(p) = \sum_{t=1}^{n} \log(\sum_{i=1}^{t} p_{it}^*)$, which can be conveniently computed by grid search, over a grid of the form $\{2^j/n : j_0 \le j \le j_1\}$.

Second, we want to reduce the linear computational complexity in (8.56) to a bounded level. To do that, we consider consider a *bounded complexity mixture* (*BCMIX*) approximation. The basic idea of the BCMIX approximation to the filter is to keep the most recent m weights, remove the smallest weight from the rest, and then reweight the M weights in each time step. Specifically, consider posterior density (8.55) with weights $p_{j,t}$. Let $\mathcal{K}_{t-1}$ be the set of indices i for which $p_{i,t-1}$ is kept at stage $t-1$; thus, $\mathcal{K}_{t-1} \supset \{t-1, , \cdots, t-m\}$. At stage t, define $p_{i,t}^*$ as in (8.56) for $i \in \{t\} \cup \mathcal{K}_{t-1}$, and let i_t be the index not belonging to $\{t, \cdots, t-m+1\}$ such that

$$p_{i_t,t}^* = \min\{p_{j,t}^* : j \in \mathcal{K}_{t-1} \quad \text{and} \quad j \le t - m\}.$$

We choose i_t to be the minimizer farthest from t if the above set has two or more minimizers. Define $\mathcal{K}_t = \{t\} \cup (\mathcal{K}_{t-1} - \{i_t\})$, and let

$$p_{i,t} = \left(p_{i,t}^* \Big/ \sum_{j \in \mathcal{K}_t} p_{j,t}^*\right), \quad i \in \mathcal{K}_t.$$

Note that if $|\mathcal{K}_{t-1}| \le M$, we still have $|\mathcal{K}_t| \le M$. Figure 8.1 shows a schematic plot of updating forward filters using BCMIX.

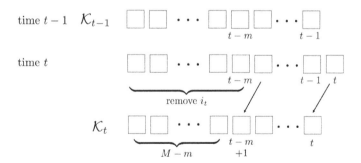

FIGURE 8.1: A schematic plot of updating forward filters using BCMIX.

The BCMIX idea can be applied to the smoother (8.60). First, for backward filters $\theta_t|\mathcal{Y}_{t+1,n}$. Let $\widetilde{\mathcal{K}}_{t+1}$ denote the set of indices j for which $q_{j,t+1}$ in (8.57) is kept at stage $t+1$; thus, $\widetilde{\mathcal{K}}_{t+1} \supset \{t+1, , \cdots, t+m\}$. At stage t, define $q_{j,t}^*$ as in (8.58) for $j \in \{t\} \cup \widetilde{\mathcal{K}}_{t+1}$, and let j_t be the index not belonging to $\{t, \cdots, t+m-1\}$ such that

$$q^*_{j_t,t} = \min\{q^*_{j,t} : j \in \widetilde{\mathcal{K}}_{t+1} \quad \text{and} \quad j \geq t+m\},$$

choosing j_t to be the minimizer farthest from t if the above set has two or more minimizers. Define $\widetilde{\mathcal{K}}_t = \{t\} \cup (\widetilde{\mathcal{K}}_t - \{j_t\})$ and let

$$q_{j,t} = \left(q^*_{j,t} \Big/ \sum_{j \in \widetilde{\mathcal{K}}_t} q^*_{j,t}\right), \quad j \in \widetilde{\mathcal{K}}_t.$$

It yields a BCMIX approximation to the backward filter $\boldsymbol{\theta}_t | \mathcal{Y}_{t+1,n}$. Then for each $t = 1, \ldots, n-1$, given the BCMIX approximation with selected weights indices $\mathcal{K}_t$ and $\widetilde{\mathcal{K}}_{t+1}$, the BCMIX approximation to the smoother (8.60) can be obtained by combining the forward and backward BCMIX filters via Bayes' theorem:

$$f(\boldsymbol{\theta}_t | \mathcal{Y}_n) \approx \sum_{i \in \mathcal{K}_t, \, j \in \widetilde{\mathcal{K}}_{t+1}} \widetilde{\beta}_{ijt} \pi(\boldsymbol{\theta}_t; a_0 + j - i + 1, \bar{\mathbf{Y}}_{i,j}),$$

in which

$$\widetilde{\beta}_{ijt} = \beta^*_{ijt} / \widetilde{P}^*_t, \qquad \widetilde{P}^*_t = p + \sum_{1 \leq t \leq n, i \in \mathcal{K}_t, j \in \widetilde{\mathcal{K}}_{t+1}} \beta^*_{ijt},$$

and β^*_{ijt} given by (8.61) for $i \in \mathcal{K}_t$ and $j \in \widetilde{\mathcal{K}}_{t+1}$. The BCMIX approximation to the change-point probability and posterior mean at time t are therefore

$$\widehat{P}(I_{t+1} = 1 | \mathcal{Y}_n) = \frac{p}{\widetilde{P}^*_t}, \qquad \widehat{E}(\boldsymbol{\mu}_t | \mathcal{Y}_n) = \sum_{i \in \mathcal{K}_t, \, j \in \widetilde{\mathcal{K}}_{t+1}} \widetilde{\beta}_{ijt} \bar{\mathbf{Y}}_{i,j}.$$

The above BCMIX approximation greatly reduces the computational complexity for the filtering and smoothing estimate of $\boldsymbol{\theta}_t$. For multiple change-points problems with large sample size n (say $n \sim 10^6$), the estimate of $\{\boldsymbol{\theta}_t\}_{1 \leq t \leq n}$ has only $O(n)$ complexity. This suggests that the Markov filtering approach and the BCMIX approximation can be embedded into more complex systems to estimate change-points; see Xing et al. (2012a,b) for two such applications in finance and biology.

8.5 Multiple change-points in financial modeling

In this section, two applications of multiple change-points and surveillance methods in financial modeling are introduced. The first study applies the detection and surveillance rules introduced earlier to monitor structural breaks in credit rating transition dynamics, and the second study introduces an change-point autoregressive GARCH models that can be used to make inference from asset prices on discrete-time volatility dynamics and contemporaneous jumps in autoregression coefficients and in volatilities.

8.5.1 Surveillance of credit transition dynamics

Several market structural breaks have occurred in financial markets during the last few decades, including the stock market crash of 1987, the credit market turmoil of 1998, the dot-com bubble burst and corporate scandals of 2001–2002, the global financial crisis of 2007–2008 stemming from the U.S. subprime mortgage crisis, the European debt crisis since the end of 2009, and the economic downturn triggered by the pandemic during 2020–2023. These market structural changes raise important questions for economists, financial practitioners, and government regulators, including whether quantitative tools can be used to effectively monitor, estimate, or predict the occurrence of structural breaks in financial markets or their sub-markets.

To address this concern, we need to first identify dynamic models capable of extracting and aggregating information from the credit market to represent its overall movement over time. Xing et al. (2020) proposed using firms' rating transition dynamics as a proxy of the credit market movement for several reasons. First, credit ratings, provided by *credit rating agencies* (CRAs), assess entities' relative ability to meet financial commitments, serving as an information good that reduces information asymmetry in the credit market, thereby enhancing capital market efficiency and transparency (Langohr and Langohr, 2010). Second, firms' credit rating transitions reveal various types of information, including conflicts of interest between investors and CRAs (Opp et al., 2013; Skreta and Veldkamp, 2009), the impact of a CRA's reputational concern on rating quality (Mariano, 2012; Mathis et al., 2009), the interaction between the business cycle and firms' incentives (Povel et al., 2007). Moreover, studies have shown that firms' credit rating transitions contain the information about structural breaks in the credit market and can be extracted (Xing et al., 2012a; Xing and Chen, 2018). Third, credit rating models are crucial in modern credit risk management. They are used in bond and credit derivatives pricing, structural models, and reduced-form models to assess firms' credit risk and credit ratings (Merton, 1974b; Leland, 1994; Duffie et al., 2007, 2009). Furthermore, in bank regulation, banks are required to construct credit rating models to stress test their portfolios and evaluate evidence of rating transitions in external ratings.

The next step is to characterize firms' credit rating transition dynamics. Section 3.6 presents several ways to model the dynamics of credit rating transitions based on transition probability matrices or their generators. Here, we consider the homogeneous and piecewise homogeneous continuous-time models of rating transition dynamics in Section 3.6. Suppose that firms' rating migrations follow a K-state continuous-time Markov chain whose probability transition matrices are characterized by their generator matrices $\mathbf{\Lambda}(t)$ for $t \geq 0$. Denote by $\mathcal{Y}_{s,t}$ the observed ratings during the period (s,t); then its sufficient statistic is given by $\{K_{s,t}^{(i,j)}, S_{s,t}^{(i)} \mid 1 \leq i,j \leq K\}$, where $K_{s,t}^{(i,j)}$ is the number of transitions from rating category i to rating category j, $S_{s,t}^{(i)}$ are the amount of time that the entity spent in category i. Then given the generator $\mathbf{\Lambda}(t) \equiv \mathbf{\Lambda} = (\lambda^{(i,j)})$ and observation $\mathcal{Y}_{0,t}$ during $[0,t)$, the log-likelihood function is expressed as

$$\log L(\mathbf{\Lambda}|\mathcal{Y}_{0,t}) = \sum_{i=1}^{K} \Big[\sum_{j\neq i} K_{0,t}^{(i,j)} \log \lambda^{(i,j)} - \Big(\sum_{j\neq i} \lambda^{(i,j)} + 1 - K \Big) S_{0,t}^{(i)} \Big]; \quad (8.63)$$

see Section 3.6.1.

Suppose that there is a structural break in $\mathbf{\Lambda}(t)$ at time ν and $\mathbf{\Lambda}(t) = \mathbf{\Lambda}_0$ for $t < \nu$ and $\mathbf{\Lambda}(t) = \mathbf{\Lambda}_1$ for $t \geq \nu$. If both $\mathbf{\Lambda}_0$ and $\mathbf{\Lambda}_1$ are known, one may consider the CUSUM rule to detect the change-point ν,

$$N_{\mathrm{CUSUM}} = \inf\{n : \max_{0 \leq k < n} \log[L(\mathbf{\Lambda}_1; \mathcal{Y}_{k+1,n})/f(\mathbf{\Lambda}_0; \mathcal{Y}_{k+1,n})] \geq c\}$$

with c satisfying $E_0(N_{\mathrm{CUSUM}}) = \gamma$. Since the pre- and post-changes of generator matrices are unknown in reality, we cannot use the CUSUM rule to detect change-points (or structural breaks) in rating transition dynamics. Hence, we next assume that either $\mathbf{\Lambda}_1$ is unknown or both $\mathbf{\Lambda}_0$ and $\mathbf{\Lambda}_1$ are unknown. To better present the idea, we ignore the assumption that state K is an absorbing state (or default state) in this section.

LR and GLR rules

Consider first a hypothesis testing problem for observed ratings transitions $\mathcal{Y}_{t-1,t}$ at the period $(t-1,t]$, $H_0 : \mathbf{\Lambda} = \mathbf{\Lambda}_0$ versus $H_1 : \mathbf{\Lambda} \neq \mathbf{\Lambda}_0$, where the generator matrix $\mathbf{\Lambda} = (\lambda^{(i,j)})$ and $\mathbf{\Lambda}_0 = (\lambda_0^{(i,j)})_{1 \leq i,j \leq K}$. Assume that $\mathbf{\Lambda}_0$ is known. Since the log-likelihood function of the data for generator matrix $\mathbf{\Lambda}$ is given by (8.63), the logarithm of the likelihood ratio statistic for this hypothesis testing problem is

$$Z_{\mathrm{LR}}(\mathbf{\Lambda}_0; \mathcal{Y}_{t-1,t}) := 2\Big(\max_{\mathbf{\Lambda} \neq \mathbf{\Lambda}_0} \log L(\mathbf{\Lambda}|\mathcal{Y}_{t-1,t}) - \log L(\mathbf{\Lambda}_0|\mathcal{Y}_{t-1,t}) \Big)$$

$$= 2\sum_{i=1}^{K} \Big[\sum_{j\neq i} K_{t-1,t}^{(i,j)} \log \frac{\widehat{\lambda}^{(i,j)}}{\lambda_0^{(i,j)}} - \sum_{j\neq i} \big(\widehat{\lambda}^{(i,j)} - \lambda_0^{(i,j)}\big) S_{t-1,t}^{(i)} \Big],$$

where $\widehat{\lambda}^{(i,j)} = K_{t-1,t}^{(i,j)}/S_{t-1,t}^{(i)}$ is the maximum likelihood estimate (MLE) of $\lambda^{(i,j)}$ ($1 \leq i \neq j \leq K$).

Besides using the MLE of $\mathbf{\Lambda}$, one may specify gamma prior distributions for elements of $\mathbf{\Lambda}$ and obtain a *mixture likelihood ratio* (MLR). Suppose the element $\lambda^{(i,j)}$ of the generator matrix $\mathbf{\Lambda}$ follow the gamma distribution

Gamma(α_{ij}, β_i) with density function (3.78). The MLR of the data is expressed as

$$Z_{\mathrm{MLR}}(\mathbf{\Lambda}_0; \mathcal{Y}_{t-1,t}) := 2\log\left(\int L(\mathbf{\Lambda}|\mathcal{Y}_{t-1,t})g(\mathbf{\Lambda})d\mathbf{\Lambda}\right) - 2\log L(\mathbf{\Lambda}_0|\mathcal{Y}_{t-1,t})$$

$$= 2\sum_{i=1}^{K}\left\{\sum_{j\neq i}\left[\alpha_{ij}\log\beta_i - \log\Gamma(\alpha_{ij}) - (K_{t-1,t}^{(i,j)} + \alpha_{ij})\log(S_{t-1,t}^{(i)} + \beta_i)\right.\right.$$

$$\left.\left. + \log\Gamma(K_{t-1,t}^{(i,j)} + \alpha_{ij}) - K_{t-1,t}^{(i,j)}\log\lambda_0^{(i,j)}\right] + \left(\sum_{j\neq i}\lambda_0^{(i,j)} + 1 - K\right)S_{t-1,t}^{(i)}\right\}.$$

Based on the LR and MLR statistics, we obtain the following rules that detect whether a change-point occurs at period $(t, t+1]$ in the generator of rating migration matrices,

$$T_{\mathrm{LR}} = \inf\{t \geq t_0 : Z_{\mathrm{LR}}(\mathbf{\Lambda}_0; \mathcal{Y}_{t-1,t}) \geq c\},$$

and

$$T_{\mathrm{MLR}} = \inf\{t \geq t_0 : Z_{\mathrm{MLR}}(\mathbf{\Lambda}_0; \mathcal{Y}_{t-1,t}) \geq c\}.$$

The above two rules assume the pre-change distribution of firms' rating transitions is known. Since $\mathbf{\Lambda}_0$ is unknown in reality, one can replace it by its MLE using historical observations. In particular, given the observed rating transitions $\mathcal{Y}_{0,t-1}$ during the period $(0, t-1]$, let $\widehat{\mathbf{\Lambda}}_0(\mathcal{Y}_{0,t-1})$ be the MLE of $\mathbf{\Lambda}_0$ during the period $(0, t-1]$; see the MLE (3.77) in Section 3.6.1 for details. Then the LR rules T_{LR} and T_{MLR} can be modified as follows:

$$\widehat{T}_{\mathrm{LR}} = \inf\{t \geq t_0 : Z_{\mathrm{LR}}(\widehat{\mathbf{\Lambda}}_0(\mathcal{Y}_{0,t-1}); \mathcal{Y}_{t-1,t}) \geq c\},$$
$$\widehat{T}_{\mathrm{MLR}} = \inf\{t \geq t_0 : Z_{\mathrm{MLR}}(\widehat{\mathbf{\Lambda}}_0(\mathcal{Y}_{0,t-1}); \mathcal{Y}_{t-1,t}) \geq c\}.$$

Besides constructing detection rules directly, the LR and MLR statistics $Z_{\mathrm{LR}}(\mathbf{\Lambda}_0; \mathcal{Y}_{t-1,t})$ and $Z_{\mathrm{MLR}}(\mathbf{\Lambda}_0; \mathcal{Y}_{t-1,t})$ can also be used to construct the following EWMA control charts for generator matrices of rating transition matrices:

$$W_{t,LR} = (1-\alpha)W_{t,LR} + \alpha Z_{\mathrm{LR}}(\widehat{\mathbf{\Lambda}}_0(\mathcal{Y}_{0,t-1}); \mathcal{Y}_{t-1,t}), \qquad t \geq 1,$$

$$W_{t,MLR} = (1-\alpha)W_{t,LR} + \alpha Z_{\mathrm{MLR}}(\widehat{\mathbf{\Lambda}}_0(\mathcal{Y}_{0,t-1}); \mathcal{Y}_{t-1,t}), \qquad t \geq 1.$$

As it is usually difficult to find the distributions of the above two EWMA statistics, one may use bootstrap methods to estimate the distribution of the EWMA statistics for real data implementation.

Since both $\mathbf{\Lambda}_0$ and $\mathbf{\Lambda}_1$ are unknown in practice, one may consider the GLR statistic for testing the null hypothesis of no change-point based on $\mathcal{Y}_{0,1}, \ldots, \mathcal{Y}_{t-1,t}$, versus the alternative hypothesis of a single change-point prior to t but not before t_0:

$$Z_{\mathrm{GLR}}(\mathcal{Y}_{0,t}) := \max_{n_0 \leq k \leq n}\left\{\sup_{\mathbf{\Lambda}_1}\log L(\mathbf{\Lambda}_1; \mathcal{Y}_{0,k}) + \sup_{\mathbf{\Lambda}_2}\log L(\mathbf{\Lambda}_2; \mathcal{Y}_{k,t}) - \sup_{\mathbf{\Lambda}_0}\log L(\mathbf{\Lambda}_0; \mathcal{Y}_{0,t})\right\}$$

$$= \max_{n_0 \leq k \leq n}\sum_{1 \leq i \neq j \leq K}\left\{K_{0,k}^{(i,j)}\log\frac{K_{0,k}^{(i,j)}}{S_{0,k}^{(i)}} + K_{k,t}^{(i,j)}\log\frac{K_{k,t}^{(i,j)}}{S_{k,t}^{(i)}} - K_{0,t}^{(i,j)}\log\frac{K_{0,t}^{(i,j)}}{S_{0,t}^{(i)}}\right\}$$

where $\sup_{\mathbf{\Lambda}_0}$ is the maximum likelihood under the null hypothesis, and $\sup_{\mathbf{\Lambda}_1}$ and $\sup_{\mathbf{\Lambda}_2}$ are obtained by maximizing the likelihood under the hypothesis of a single change-point occurring at $k + 1$. Then the GLR rule with unknown pre- and post-change parameters is

$$T_{\text{GLR}} = \inf\left\{t > t_0 : Z_{\text{GLR}}(\mathcal{Y}_{0,t}) \geq c\right\}. \tag{8.64}$$

Bayes-type rules

Besides the frequentist rules based on the likelihood ratio, we can also construct surveillance rules using Bayesian methods. Following the Shiryaev procedure in Section 8.4, we find the probability that a structural break occurs prior to or equal to time n is

$$P\{v \leq n | \mathcal{Y}_{0,n}\} = R_{p,n}/(R_{p,n} + p^{-1}),$$

where p is the parameter of the geometric distribution $P\{v = n\} = p(1-p)^{n-1}$ and

$$R_{p,n} = \sum_{k=1}^{n} \frac{1}{(1-p)^{n-k+1}} \frac{L(\mathbf{\Lambda}_1|, \mathcal{Y}_{k,n})}{L(\mathbf{\Lambda}_0|, \mathcal{Y}_{k,n})}.$$

Then the Shiryaev rule declares a change at time

$$T_{\text{S}} = \inf\{n \geq 1 : R_{p,n} \geq c\}.$$

Letting $p = 0$, we obtain the Shiryaev-Roberts rule for detecting a change-point in firms' rating transition dynamics:

$$T_{\text{SR}} = \inf\left\{n \geq 1 : \sum_{k=1}^{n} \frac{L(\mathcal{Y}_{k,n}|\mathbf{\Lambda}_1)}{L(\mathcal{Y}_{k,n}|\mathbf{\Lambda}_0)} \geq c\right\}.$$

Since both the pre-change and post-change generator matrices are unknown in practice, we can consider the extended Shiryaev's rule (8.50) by assuming that v follows a geometric prior distribution with parameter p but is constrained to be lager than n_0, and $\mathbf{\Lambda}_0, \mathbf{\Lambda}_1$ are generator matrices in which the ijth elements are independent and identically distributed and follow the prior distribution Gamma(α_{ij}, β_i) with density function (3.78). In such a case, the extended Shiryaev's rule (8.50) can be reduced to (8.51) with π_{00} and π_{ml} given by

$$\pi_{00} = \prod_{i,j \in \mathcal{K}} \frac{\beta_i^{\alpha_{i,j}}}{\Gamma(\alpha_{ij})} \qquad \pi_{ml} = \prod_{i,j \in \mathcal{K}} \frac{(S_{m,l}^{(i)} + \beta_i)^{(K_{m,l}^{(i,j)} + \alpha_{ij})}}{\Gamma(K_{m,l}^{(i,j)} + \alpha_{ij})}.$$

The above Bayes-type rules assume that there is only one change-point during the sample period. In the surveillance of firms' rating transition dynamics, one cannot expect that structural breaks will always be detected.

Therefore, surveillance rules that allow misdetection of change-points have the advantage of not disregarding samples that contain no change-points or change-points of small sizes. To design surveillance rules that allow misdetection of structural breaks in firms' rating transition dynamics, we use the piecewise time-homogeneous continuous-time Markov chains model for rating transition dynamics in Section 3.6.2. The inference procedure there provides the probability that the most recent change-point occurs in the period $[m, t]$; see equation (3.83) in Section 3.6.2. Then applying the detection rule (8.52) here yields the following surveillance rule

$$N_{\mathrm{MCP}} = \inf \left\{ n > n_p : \sum_{h=m}^{l} p_{m,l} \geq c \right\}. \qquad (8.65)$$

Using the above surveillance rules, Xing et al. (2020) analyzed Standard and Poor's monthly credit rating of firms from January 1986 to February 2017. They found that the LR and GLR rules suggest more structural breaks than other rules.

8.5.2 Change-point autoregressive GARCH models

As introduced in Section 2.3.3, since the seminal works of Engle (1982) and Bollerslev (1986), ARCH type volatility models have been widely used to model and forecast volatilities of financial time series. In many empirical studies of stock returns and exchange rates, estimation of the parameters ν, a, and b in the GARCH(1,1) model

$$y_t = \sigma_t \epsilon_t, \quad \sigma_t^2 = (1 - a - b)\nu^2 + a y_{t-1}^2 + b \sigma_{t-1}^2 \qquad (8.66)$$

reveals high volatility persistence, with the maximum likelihood estimate of $a + b$ close to 1. To model such persistence, the "integrated" GARCH (Engle and Bollerslev, 1986) and the "fractionally integrated" GARCH (Baillie et al., 1996) have been introduced to quantify the long memory of volatilities. However, it has been pointed out that if the model parameters undergo occasional changes, then the fitted models that assume constant parameters tend to exhibit long memory; see Diebold (1986), Lamoureux and Lastrapes (1990), Mikosch and Starica (2004), and Hillebrand (2005). Although many models have been proposed to incorporate (multiple) change-points into regression type models, it is not easy to do so for GARCH dynamics. This is because the combination of multiple change-points and nonlinear time series dynamics generates tremendously complicated time series trajectories and hence the inference procedure becomes computationally infeasible. For example, Cai (1994) noted the "tremendous complication" of the normal equations of the EM algorithm, even for regime-switching ARCH models with a finite number of regimes, making it "extremely difficult" to implement for sample sizes exceeding 50.

To overcome this difficulty and allow a continuum number of regimes, Lai and Xing (2013) introduced a new class of change-point ARX-GARCH models. These models contain multiple change-points in the regression coefficients and the unconditional variance of the random disturbances in an *autoregression with exogenous covariates* (ARX), while allowing the conditional variances to follow a GARCH model. Since their models decouple the long-run and short-term volatilities in the GARCH model and incorporate change-points in the long-run volatility, their inference procedure is able to combine the hidden Markov filtering approach in Section 8.4.3 and the likelihood method for GARCH and hence becomes computationally tractable. Specifically, a change-point ARX-GARCH model is expressed as

$$y_t = \boldsymbol{\beta}_t' \mathbf{x}_t + \nu_t \sqrt{h_t} \epsilon_t, \tag{8.67}$$

in which the parameter vector $\boldsymbol{\beta}_t$ and the unconditional variance ν_t^2 are piecewise constant, with jumps at change times, the vector $\mathbf{x}_t$ consists of exogenous variables and the past observations $y_{t-1}, y_{t-2}, \ldots, y_{t-\kappa}$, and we use h_t to represent short-term proportional fluctuations in variance generated by the GARCH model

$$h_t = \left(1 - \sum_{i=1}^{k} a_i - \sum_{l=1}^{k'} b_l\right) + \sum_{i=1}^{k} a_i w_{t-i}^2 + \sum_{l=1}^{k'} b_l h_{t-l}, \quad \text{with } w_s = \sqrt{h_s} \epsilon_s. \tag{8.68}$$

The ϵ_t are assumed to be i.i.d. standard normal random variables such that ϵ_t are independent of $\mathbf{x}_t$, and the time-invariant GARCH parameters $a_1, \ldots, a_k, b_1, \ldots, b_{k'}$ are assumed to satisfy $a_i \geq 0$, $b_l \geq 0$ and $\sum_{i=1}^{k} a_i + \sum_{l=1}^{k'} b_l \leq 1$. Let $\tau_t = 1/(2\nu_t^2)$. Then $\boldsymbol{\theta}_t = (\boldsymbol{\beta}_t', \tau_t)'$ are assumed to be piecewise constant and satisfy the following conditions.

(i) For $t > t_0 = \max(k, k')$, the change-times of $\boldsymbol{\theta}_t$ form a renewal process with i.i.d. inter-arrival times that are geometrically distributed with parameter p or, equivalently,

$$I_t := 1_{\{\boldsymbol{\theta}_t \neq \boldsymbol{\theta}_{t-1}\}} \text{ are i.i.d. Bernoulli}(p) \text{ with } P(I_t = 1) = p,$$

$I_{t_0} = 1$, and there is no change-point prior to t_0.

(ii) $\boldsymbol{\theta}_t = (1 - I_t)\boldsymbol{\theta}_{t-1} + I_t(\mathbf{z}_t', \gamma_t)'$, where $(\mathbf{z}_1', \gamma_1)', (\mathbf{z}_2', \gamma_2)', \ldots$ are i.i.d. random vectors, $\mathbf{z}_t | \gamma_t \sim \text{Normal}(\mathbf{z}, \mathbf{V}/(2\gamma_t))$, $\gamma_t \sim \chi_d^2/\rho$, and χ_d^2 is the chi-square distribution with d degrees of freedom.

(iii) The processes $\{I_t\}$, $\{(\mathbf{z}_t', \gamma_t)\}$, and $\{(\mathbf{x}_t, \epsilon_t)\}$ are independent.

Inference procedure

Conditions (i)–(iii) specify a Markov chain with unobserved states $(I_t, \boldsymbol{\theta}_t)$. The observations $(\mathbf{x}_t, y_t)$ are such that $(y_t - \boldsymbol{\beta}_t' \mathbf{x}_t)/\nu_t$ forms a

GARCH process. This hidden Markov model has hyperparameters $\Phi = \{p, \mathbf{z}, \mathbf{V}, \rho, d, a_1, \ldots, a_k, b_1, \ldots, b_{k'}\}$. Suppose that Φ is known for the time being, we can extend the hidden Markov filtering approach in Section 8.4.3 here; see also the special case of $\mathbf{x}_t = (y_{t-1}, \ldots, y_{t-k})'$ and $h_t \equiv 1$ in Lai et al. (2005).

Define $\mathcal{Y}_t = (\mathbf{x}_1, y_1, \ldots, \mathbf{x}_t, y_t)$ and $\mathcal{Y}_{i,t} = (\mathbf{x}_i, y_i, \ldots, \mathbf{x}_t, y_t)$. We first consider the estimation of $\boldsymbol{\theta}_t$. Let $K_t = \max\{s \leq t : I_s = 1\}$ and note that $t - K_t \geq k$ by assumption (i). By the argument in Section 8.4.3, the distribution of $(\boldsymbol{\beta}, \nu^2)$ given $\mathcal{Y}_t$ are mixtures of distributions of $(\boldsymbol{\beta}, \nu^2)$ based on $\mathcal{Y}_{i,t}$, with weights p_{it} to be specified. Let $(\widehat{\boldsymbol{\beta}}_t, \widehat{\nu}_t^2)$ be the estimates of $(\boldsymbol{\beta}, \nu^2)$ given $\mathcal{Y}_t$, and $\widehat{\boldsymbol{\beta}}_{i,t}$ and $\widehat{\nu}_{i,t}^2$ the estimates of $(\boldsymbol{\beta}, \nu^2)$ given $\mathcal{Y}_{i,t}$ and no-change-points constraint between i and t. Then for fixed i and increasing t, $\widehat{\boldsymbol{\beta}}_{i,t}$ and $\widehat{\nu}_{i,t}^2$ can be computed recursively as follows. Initializing at $t = i - 1$ with $\widehat{\boldsymbol{\beta}}_{i,t} = \mathbf{z}$, $\widehat{\mathbf{V}}_{i,t} = \mathbf{V}$, and $\widehat{\nu}_{i,t}^2 = \rho/(2d)$, define for $t \geq i$,

$$\widehat{h}_{i,t} = \left(1 - \sum_{j=1}^{k} a_j - \sum_{l=1}^{k'} b_l\right) + \sum_{l=1}^{k'} b_l \widehat{h}_{i,t-l} + \sum_{l=1}^{k} a_i \frac{(y_{t-l} - \widehat{\boldsymbol{\beta}}_{i,t-l}' \mathbf{x}_{t-l})^2}{\widehat{\nu}_{i,t-l}^2},$$

$$\mathbf{V}_{i,t} = \mathbf{V}_{i,t-1} - \frac{\mathbf{V}_{i,t-1} \mathbf{x}_t \mathbf{x}_t' \mathbf{V}_{i,t-1}}{\widehat{h}_{i,t} + \mathbf{x}_t' \mathbf{V}_{i,t-1} \mathbf{x}_t},$$

$$\widehat{\boldsymbol{\beta}}_{i,t} = \widehat{\boldsymbol{\beta}}_{i,t-1} + \frac{\mathbf{V}_{i,t-1} \mathbf{x}_t \left(y_t - \widehat{\boldsymbol{\beta}}_{i,t-1}' \mathbf{x}_t\right)}{\widehat{h}_{i,t} + \mathbf{x}_t' \mathbf{V}_{i,t-1} \mathbf{x}_t},$$

$$\widehat{\nu}_{i,t}^2 = \frac{d+t-i-2}{d+t-i-1} \widehat{\nu}_{i,t-1}^2 + \frac{1}{d+t-i-1} \cdot \frac{\left(y_t - \widehat{\boldsymbol{\beta}}_{i,t-1}' \mathbf{x}_t\right)^2}{\widehat{h}_{i,t} + \mathbf{x}_t' \mathbf{V}_{i,t-1} \mathbf{x}_t}.$$

For the weights p_{it}, we note that the argument in Section 8.4.3 shows that $p_{it} = P(K_t = i | \mathcal{Y}_n)$ are given recursively by $p_{it} = p_{it}^* / \sum_{l=1}^{t} p_{lt}^*$ and p_{it}^* are given by (8.56), in which

$$\pi_{00} = |\mathbf{V}|^{-1/2} \frac{[\rho/2]^{d/2}}{\Gamma(d/2)}, \qquad \pi_{it} = |\mathbf{V}_{i,t}|^{-1/2} \frac{\rho_{i,t}^{(d+t-i+1)/2}}{\Gamma((d+t-i+1)/2)},$$

and

$$\rho_{i,t} = \frac{1}{2}\rho + \mathbf{z}'\mathbf{V}^{-1}\mathbf{z} - \mathbf{z}_{i,t}'\mathbf{V}_{i,t}^{-1}\mathbf{z}_{i,t} + \sum_{s=i}^{t} \frac{y_s^2}{\widehat{h}_{i,t}}, \qquad \mathbf{z}_{i,t} = \mathbf{V}_{i,t}\left(\mathbf{V}^{-1}\mathbf{z} + \sum_{s=i}^{t} \frac{y_s}{\widehat{h}_{i,s}}\mathbf{x}_s\right).$$

Now that $\boldsymbol{\theta}_t = (\boldsymbol{\beta}_t', \tau_t)'$ are estimated, one can obtain the estimated "short-term" noises $\widehat{\epsilon}_t = (y_t - \mathbf{x}_t'\widehat{\boldsymbol{\beta}}_t)/\nu_t$. Since the hyperparameters Φ are known, given initial values of $h_1, \ldots, h_{t_0}$, the "short-term" volatility h_t can be computed recursively by (8.66).

The above inference procedure deals with the case of known hyperparameters. When the hyperparameters Φ are unknown, Lai and Xing (2013) proposed to use the method of moments to estimate $\mathbf{z}, \mathbf{V}, \rho, d$ and the likelihood method in Section 8.4.3 estimate the rest of hyperparameters. Besides, using the BCMIX approximation in Section 8.4.3 can simplify the inference procedure and reduce the computational complexity.

Implications in financial econometrics

One implication of the change-point ARX-GARCH model in financial econometrics is its interpretation of spurious long-range dependence via parameter jumps. Lai and Xing (2013) used the change-point ARX-GARCH model to analyze the weekly returns of the SP500 index and made inference on model hyperparameters, the long-run volatility level with change-points, the short-term volatility and the dynamics of the return series. They showed that incorporating change-points in parameters can remove the spurious long memory in volatility exhibited by fitting the AR-GARCH model to the entire time series without incorporating possible parameter changes during a long period that undergoes several structural changes. In other words, the apparent long memory in volatility arises from parameter changes in the long time-scale.

Another interesting implication of the change-point ARX-GARCH model is that it helps us understand components of asset return dynamics. In option pricing, in order to estimate the magnitude and assess the significance of volatility dynamics and jump risk premia, contemporaneous jumps in prices and in volatility have been incorporated into dynamic models of asset prices in the finance literature; see Broadie et al. (2007) for a review and discussion in support of contemporaneous jumps in both price and volatility. Among these discussion, Duffie et al. (2000) introduced a continuous-time *stochastic volatility* (SV) model that incorporates *contemporaneous jumps* (CJ) in returns and volatility, and developed analytic methods for pricing under the model, which has since become very popular in the finance literature. However, parameter estimation and empirical analysis of the SVCJ model has been a challenging problem. Eraker et al. (2003) developed a simulation-based Bayes estimator, using MCMC methods to estimate both the hidden states and the model parameters after discretizing the continuous-time bivariate process of returns and their volatilities into an HMM. In contrast, the stochastic change-point AR-GARCH model proposed by Lai and Xing (2013) is much simpler to implement, and offers a promising alternative to the SVCJ model. Moreover, it can easily incorporate exogenous covariates, as we have shown in the more general ARX setting.

8.6   Prediction of change-point probabilities

The discussion so far has concentrated on surveillance methods for change-points or abrupt parameter changes of the system. If such changes are caused by certain factors that are lagged or exogenous variables, one would be curious to known if the probabilities of change-points could be predicted. Intuitively, one can use the logistic regression model to connect the change-point event with lagged or exogenous variables. The main difficulty here is to embed change-points in the system dynamics appropriately to reduce the computational complexity. This section introduces a generic predictive model for change-points by using the logistic modeling and the hidden Markov filtering approach in Section 8.4.3.

The predictive model consists of two parts. The first part of the model is similar to the one presented in Section 8.4.3. We assume that all observations $\mathbf{y}_1, \ldots, \mathbf{y}_t, \ldots$ follow the multiparameter exponential family (8.53) and the prior density π is given by (8.54). Suppose that the parameter vector $\boldsymbol{\theta}_t$ may undergo occasional changes such that for $t > 1$, the indicator variables $I_t :=$ $1_{\{\boldsymbol{\theta}_t \neq \boldsymbol{\theta}_{t-1}\}}$ are independent Bernoulli random variables with $P(I_t = 1) = p_t$. When there is a parameter change at time t (i.e., $I_t = 1$), the changed parameter $\boldsymbol{\theta}_t$ is assumed to be sampled from π. The second part of the model is to connect the change-point probability p_t with a set of covariates via the logistic regression, i.e.,

$$\text{logit}(p_t) = \log \frac{p_t}{1 - p_t} = \mathbf{b}'\mathbf{x}_t + \mathbf{c}'\mathbf{u}_t, \quad \text{or} \quad p_t = \frac{e^{\mathbf{b}'\mathbf{x}_t + \mathbf{c}'\mathbf{u}_t}}{1 + e^{\mathbf{b}'\mathbf{x}_t + \mathbf{c}'\mathbf{u}_t}}, \quad (8.70)$$

in which $\mathbf{x}_t \in \mathbb{R}^d$ are d-dimensional observable lagged or exogenous variables, $\mathbf{u}_t$ are h-dimensional unobservable random variables, and $\mathbf{b}$ and $\mathbf{c}$ are the corresponding regression coefficients. Denote by $\Phi = \{\mathbf{b}, \mathbf{c}, a_0, \boldsymbol{\mu}_0\}$ the model parameters.

To make the inference on Φ, we consider a simple case that the logistic regression (8.70) doesn't contain any unobserved variables $\mathbf{u}_t$. There are two ways to make inference on model parameters. The first way is the maximum likelihood approach. Repeating the argument in Section 8.4.3, one can show that the forward filter $\boldsymbol{\theta}_t | \mathcal{Y}_t$ is still given by the mixture distribution (8.55) but with the following modified weights p_{it},

$$p_{it} = \frac{p_{it}^*}{\sum_{j=1}^t p_{jt}^*}, \quad p_{it}^* = \begin{cases} p_t \pi_{0,0}/\pi_{t,t} & \text{if } i = t, \\ (1 - p_t)p_{i,t-1}\pi_{i,t-1}/\pi_{i,t} & \text{if } i < t, \end{cases} \quad (8.71)$$

in which $\pi_{0,0}$ and $\pi_{i,t}$ are defined same as the ones in (8.56). Then the likelihood of the model is still given by the likelihood function (8.62) can be maximized over the parameter space of Φ. The second way of estimating model parameters is the *expectation-maximization* (EM) approach which treats $\boldsymbol{\theta}_t$ as the missing data. In particular, the log-likelihood of the complete data $\{\mathbf{y}_t, \boldsymbol{\theta}_t\}$ in such case consists of two parts,

$$l_c(\Phi) = l_1(\mathbf{b}|\mathcal{Y}_n) + l_2(a_0, \boldsymbol{\mu}_0|\mathcal{Y}_n),$$

in which

$$l_1(\mathbf{b}|\mathcal{Y}_n) = \sum_{t=1}^{n} \left\{ \log(1-p_t) \cdot 1_{\{\boldsymbol{\theta}_t \neq \boldsymbol{\theta}_{t-1}\}} + \log(p_t) \cdot 1_{\{\boldsymbol{\theta}_t = \boldsymbol{\theta}_{t-1}\}} \right\}$$

$$= \sum_{t=1}^{n} \left\{ (\mathbf{x}_t'\mathbf{b}) 1_{\{\boldsymbol{\theta}_t \neq \boldsymbol{\theta}_{t-1}\}} - \log\left[1 + e^{\mathbf{x}_t'\mathbf{b}}\right] \right\}$$

and

$$l_2(a_0, \boldsymbol{\mu}_0|\mathcal{Y}_n) = \sum_{t=1}^{n} \left\{ \boldsymbol{\theta}_t'\mathbf{y}_t - \psi(\boldsymbol{\theta}_t) + \left[\log c(a_0, \boldsymbol{\mu}_0) + a_0\boldsymbol{\mu}_0'\boldsymbol{\theta}_t - a_0\psi(\boldsymbol{\theta}_t)\right] \cdot 1_{\{\boldsymbol{\theta}_t \neq \boldsymbol{\theta}_{t-1}\}} \right\}.$$

In the expectation step of an EM algorithm, one needs to take expectation for l_c given observed data $\mathbf{Y}_n$ so that the "missing data" $\{\boldsymbol{\theta}_t\}$ can be integrated out. This requires to replace $1_{\{\boldsymbol{\theta}_t \neq \boldsymbol{\theta}_{t-1}\}}$, $\boldsymbol{\theta}_t$, $\psi(\boldsymbol{\theta}_t)$, $\boldsymbol{\theta}_t 1_{\{\boldsymbol{\theta}_t \neq \boldsymbol{\theta}_{t-1}\}}$, and $\psi(\boldsymbol{\theta}_t)1_{\{\boldsymbol{\theta}_t \neq \boldsymbol{\theta}_{t-1}\}}$ by their expectations respectively. In particular, $1_{\{\boldsymbol{\theta}_t \neq \boldsymbol{\theta}_{t-1}\}}$ replaced by the change-point probability $\widetilde{y}_t = P(I_t = 1|\mathcal{Y}_n)$, and hence

$$E(l_1(\mathbf{b}|\mathcal{Y}_n)) = \sum_{t=1}^{n} \left\{ (\mathbf{x}_t'\mathbf{b})\widetilde{y}_t - \log\left(1 + e^{\mathbf{x}_t'\mathbf{b}}\right) \right\}.$$

Then the estimated regression coefficient $\mathbf{b}$ can be updated in the M-step by solving the first order condition of the above log-likelihood. For other parameters, since $E(\boldsymbol{\theta}_t 1_{\{\boldsymbol{\theta}_t \neq \boldsymbol{\theta}_{t-1}\}}|\mathcal{Y}_n)$ and $E(\psi(\boldsymbol{\theta}_t)1_{\{\boldsymbol{\theta}_t \neq \boldsymbol{\theta}_{t-1}\}}|\mathcal{Y}_n)$ can also be computed explicitly based on the smoother $\boldsymbol{\theta}_t|\mathcal{Y}_t$, a_0 and $\boldsymbol{\mu}_0$ can also be updated in the M-step. Then all parameters of Φ can be estimated by the EM algorithm.

A more complicated case is that there are unobserved factors $\mathbf{u}_t$ in the logistic regression (8.70). Note that we may assume the factors $\{\mathbf{u}_t\}_{t=1}^{n}$ are independent identically distributed random variables or follow some dynamic models. Suppose that the factor series $\mathbf{U} := \{\mathbf{u}_t\}_{t=1}^{n}$ has a joint density function $g(\mathbf{U})$. Then the likelihood function of Φ is expressed as

$$L(\Phi|\mathcal{Y}_n) = \int f(\mathcal{Y}_n|\mathbf{U})g(\mathbf{U})d\mathbf{U},$$

in which the likelihood function $f(\mathcal{Y}_n|\mathbf{U})$ has the same form as (8.62). Assume that $\widehat{\Phi}$ maximizes the likelihood function $L(\Phi|\mathcal{Y}_n)$. Let $h_\Phi(\mathcal{Y}_n, \mathbf{U}) = f(\mathcal{Y}_n|\mathbf{U})g(\mathbf{U})$. Then $\widehat{\Phi}$ solves the first-order condition

$$\int \frac{\partial h_\Phi(\mathcal{Y}_n, \mathbf{U})}{\partial \Phi} d\mathbf{U} = 0. \tag{8.72}$$

Since it is difficult to find an analytical solution for the above equation, one can use stochastic approximation algorithms to solve (8.72).

Xing and Chen (2018) applied the above predictive model to credit rating dynamics and investigated whether structural break in credit rating dynamics can be predicted using market fundamentals and macroeconomic variables. They showed that certain economic variables significantly influence the probabilities of structural breaks (i.e., change-points), whether dynamic latent factors are included in the predictive model or not.

In general, this generic predictive model can be extended to address change-point prediction in other applications, provided that a hidden Markov model for change-points is appropriately specified. However, several challenges may complicate the prediction process. For instance, dealing with a large number of candidate factors related to change-points often requires variable selection in the inference procedure, and modeling the dynamics of unobserved variables and incorporating it into the predictive model may increase the difficulty of inference.

Supplements and problems

Our introduction on methods of sequential analysis is very selective. For example, methods of multivariate change-point detection are not presented here, interested readers can find an excellent overview on that in Tartakovsky et al. (2014).

In terms of overviews, surveys, or introductions on sequential analysis, we name a few references here. The first introduction on concepts and procedures of sequential analysis is given by Wald (1947). Since then, there are extensive references in statistics and engineering on the subject of sequential hypothesis testing and quick detection. Siegmund (1985) provide a up-to-date presentation on methods for sequential analysis. The handbook edited by Ghosh and Sen (1991) present studies of important topics in sequential analysis. Reviews on the research development of sequential analysis can be found Lai (1995) and Lai (2001). The most recent systematic and up-to-date presentation on sequential hypothesis testing and change-point detection is given by Tartakovsky et al. (2014).

There are also discussions and surveys on specific topics in sequential analysis. Woodroofe (1982) introduce nonlinear renewal theory and its application to sequential analysis. Bartroff et al. (2013) present methods of sequential experimentation and their applications in clinical trials. Frisén (2008) discuss some application of surveillance methods in finance.

1. (**SPRT for Bernoulli distribution**). Suppose $X_1, X_2, \ldots$ are independent and identically distributed with $P_p(X_k = 1) = p = 1 - P_p(X_k = -1)$. Let $S_n = \sum_{k=1}^{n} X_k$. Consider the test $H_0 : p = p_0$ against $H_1 : p_1$ $(p_0 < p_1)$.

(a) Show that the likelihood ratio of samples $X_1, \ldots, X_n$ is expressed as

$$\Lambda_n = \left[\frac{p_1(1-p_1)}{p_0(1-p_0)} \right]^{n/2} \left[\frac{p_1(1-p_0)}{p_0(1-p_1)} \right]^{S_n/2}.$$

(b) Assume that $p_0 + p_1 = 1$ and $B = A^{-1}$. Denote $b = -(\log B)/\log[p_0(1-p_0)]$. Show the stopping rule (8.2) becomes

$$N = \begin{cases} \text{first } n \geq 1 & \text{such that } |S_n| \geq b \\ \infty & \text{if } |S_n| < b \text{ for all } n \geq 1. \end{cases}$$

2. (**Upper bounds on error probabilities of SPRT with composite hypothesis**). Suppose that $X_1, X_2, \ldots$ are independent and identically distributed with the density function $p_\theta(x)$. Consider the problem of testing a simple null hypothesis $H_0 : \theta \in \Theta_0$ against a composite alternative $H_1 : \theta \in \Theta_1$. Define the weighted likelihood ratio by (8.13) and the stopping rule by (8.12). Denote by α and β the type I error probability and the power of the test. Show that $\alpha \leq 1/B$ and $1 - \beta \leq A$.

3. (**ARL of Shewhart control chart for change in the Gaussian mean**). Consider a Shewhart control chart for a two-sided change in the mean of an independent Gaussian sequence $N(\theta, \sigma^2)$ from θ_0 to either $\theta_1^+ = \theta_0 + \delta$ or $\theta_1^- = \theta_0 - \delta$. The alarm is raised when

$$\overline{X}_{(K-1)m+1:Km} - \theta_0 \geq \kappa \frac{\sigma}{\sqrt{m}}, \qquad \overline{X}_{(K-1)m+1:Km} = \frac{1}{m} \sum_{i=(K-1)m+1}^{Km} X_i.$$

Show that the ARL function at θ is

$$\mathrm{ARL}(\theta) = \frac{m}{1 - \Phi\left(\kappa - \frac{\sqrt{m}}{\sigma}(\theta - \theta_0) \right) + \Phi\left(-\kappa - \frac{\sqrt{m}}{\sigma}(\theta - \theta_0) \right)}.$$

In particular, $\mathrm{ARL}(\theta_0) = 2^{-1}m/(1 - \Phi(\kappa))$.

4. (**Two-sided CUSUM for change in the Gaussian mean**). Consider a CUSUM algorithm for a two-sided change in the mean of an independent Gaussian sequence $N(\theta, \sigma^2)$ from θ_0 to either $\theta_1^+ = \theta_0 + \delta$ or $\theta_1^- = \theta_0 - \delta$. Given the stopping rule (8.25), show that

$$g_n^u = \left[g_{n-1}^u + X_n - \theta_0 - \frac{\delta}{2} \right]^+, \qquad g_n^l = \left[g_{n-1}^u - X_n + \theta_0 - \frac{\delta}{2} \right]^+,$$

5. (**Hidden Markov filtering with multiple change-points**). Consider the multiple change-points model in Section 8.4.3. Show that the the forward filter for θ_t given $\mathcal{Y}_t$ is given by the mixture distribution (8.55) with weights (8.56).

6. (**A Poisson mean shift model**). Suppose that all observations $y_1, y_2, \ldots, y_n$ follow a Poisson distribution with mean λ_t, in which λ_t are piecewise constant means. Assume that $I_t = 1$ and $I_t = 1_{\{\lambda_{t-1} \neq \lambda_t\}}$ are independent Bernoulli random variables with $P(I_t = 1) = p$ for $t \geq 2$. Given $\{I_t\}$, the mean λ_t satisfies the dynamics $\lambda_t = (1 - I_t)\lambda_{t-1} + I_t z_t$, where z_t are independent and identically distributed Gamma random variables Gamma(α, β) with the density function $f(\lambda) = \beta^\alpha \lambda^{\alpha-1} e^{-\beta\lambda}/\Gamma(\alpha)$. Given $\{\lambda_t\}$, observations $y_1, y_2, \ldots, y_n$ are mutually independent. Let $\mathcal{Y}_{i,j} = (y_i, \ldots, y_j)$.

(a) Denote by $K_t = i$ the most recent change-point up to time t. Show that the posterior distribution of λ_t given $K_t = i$ and $\mathcal{Y}_{1,t}$ is Gamma$(\alpha_{it}, \beta_{it})$, in which $\alpha_{it} = \alpha + \sum_{s=i}^t y_s$ and $\beta_{it} = (\beta^{-1} + t - i + 1)^{-1}$.

(b) Denote by $\pi_{00} = \beta^{-\alpha}/\Gamma(\alpha)$ and $\pi_{it} = \beta_{it}^{-\alpha_{it}}/\Gamma(\alpha_{it})$. Show that the posterior distribution of λ_t given $\mathcal{Y}_{1,t}$ is a mixture of Gamma distribution

$$\lambda_t | \mathcal{Y}_{1,t} \sim \sum_{i=1}^t p_{it} \text{Gamma}(\alpha_{it}, \beta_{it}),$$

in which the weight $p_{it} = p_{it}^*/\sum_{s=1}^t p_{st}^*$ and p_{it}^* is defined by (8.56).

(c) Let $q_{jt} = q_{jt}^*/\sum_{s=t}^n q_{st}^*$ and q_{st}^* be defined by (8.58). Show that the posterior distribution of λ_t given $\mathcal{Y}_{1,n}$ is a mixture of Gamma distribution

$$\lambda_t | \mathcal{Y}_{1,n} \sim \sum_{1 \leq i \leq j \leq n} \beta_{ijt} \text{Gamma}(\alpha_{it}, \beta_{it}), \qquad 1 \leq t < n,$$

in which $\beta_{ijt} = \beta_{ijt}^*/P_t^*$, $P_t^* = p + \sum_{1 \leq i \leq t < j \leq n} \beta_{ijt}^*$ and β_{ijt}^* are defined by (8.61). Besides, given $\mathcal{Y}_{1,n}$, the probability of having a change-point between t and $t + 1$ is $P(I_{t+1}|\mathcal{Y}_{1,n}) = p/P_t^*$.

Part III

Data and Risk Analytics in FinTech

Chapter 9

FinTech ABCD and analytics

FinTech or *Financial Technology* is the term used to describe innovative technologies and software applications that aim to improve and automate the delivery and usage of financial services. FinTech integrates various new technologies into financial services, including asset and wealth management, digital currencies, usage-based insurance, compliance management, and others. It has significantly transformed the traditional way how financial services, transactions, and interactions are conducted. The *ABCDs* of FinTech is an acronym representing four key trends and concepts within the FinTech industry. It refers to:

- *Artificial Intelligence* (AI). It utilizes AI technologies to enhance decision-making processes, automate tasks, and provide personalized financial services.

- *Blockchain*. It implements distributed ledger technology for secure and transparent transactions, particularly in areas such as digital currencies and smart contracts.

- *Cloud computing*. It leverages cloud-based infrastructure to enable scalability, flexibility, and accessibility of financial services and data.

- *Data (big data and data analytics)*. It harnesses large volumes of data and advanced analytics techniques to gain insights, improve risk management, and enhance customer experiences within financial services.

Together, these trends encompass the innovative advancements driving the evolution of the FinTech landscape.

This chapter provides a non-technical description on basic concepts of AI, blockchain and cloud computing, and their applications in FinTech. Section 9.1 presents a brief history of AI and their applications in finance. Section 9.2 describes the evoluation of blockchain technology, its major components and applications to finance. Section 9.3 explains service delivery and deployment models of cloud computing, and introduces the application of could computing in data analytics and finance. Section 9.4 explains the main areas of FinTech services and describes the current and future applications of data and risk analytics in FinTech.

9.1   Artificial intelligence in FinTech

In recent decades, AI has undergone significant development and changed the world of finance in unprecedented ways. AI, characterized by the utilization of algorithms and expert systems to execute tasks traditionally reliant on human intelligence, is driving a paradigm shift in the operations of financial institutions and regulatory bodies. This section offers a succinct overview of the concept and historical evolution of AI, discusses the interconnection between AI and data science, and explores the diverse applications of AI within the realm of finance.

9.1.1   Emergence and development of AI

As early as in 1950, the renowned mathematician Alan Turing posed a simple question: can machines think? To deal with this question, Turing (1950) proposed the so-called "Turing test," which is also known as the "imitation game." Turing test is a simple behavioral test for the presence of mind or intelligence in entities considered to be minded. In the test, a human judge uses a natural language to communicate with a human and a machine, each of which is hidden from the judge. If the judge cannot distinguish the human from the machine, then the machine is said to have passed the test. The Turing test provides a practical way to determine whether the entity is intelligent or not. Although it is not easy for a machine to pass the Turing test, there have been several criticisms on the Turing test as a measure of machine intelligence. For example, the test is not a definitive measure of true intelligence, it focuses only on the ability of mimicking human-like behavior and communication, rather than on the cognitive capability or consciousness of the machine. While the Turing test has limitations and critiques, it remains an important historical milestone in the development of AI.

Birth and early development of AI

Besides Turing's development of the Turing test, initial efforts at AI also involved modeling the neurons in the brain. McCulloch and Pitts (1943) first proposed the notion of artificial neuron, and Hebb (1949) developed *Hebbian learning* for neural networks. In 1951, Marvin Minsky and Dean Edmonds, supported by John von Neumann, built the first neural network computer, the *Stochastic Neural Analog Reinforcement Calculator* or the SNARC, which uses 3000 vacuum tubes to simulate a network of 40 neurons and tries to learn from experience and improve its performance through a process of trial and error. Following these accomplishment, John McCarthy organized the famous *Dartmouth workshop* in 1956 and coined the term *artificial intelligence* at the workshop. The attendees included Marvin Minsky, Nathaniel Rochester, Claude Shannon, and many other researchers who were interested in simulating human intelligence using machines. Since then, AI has begun to emerge as a new discipline that aims to create intelligent systems that can perform tasks, solve problems, and make decisions in a complex, changing environment.

The early development of AI focused on solving mathematical problems and performing symbolic reasoning. In 1955–1956, a program called the *Logic Theorist* was developed by Allen Newell and Herbert Simon to mimic the problem-solving skills of a human being, this is considered as the first AI program. In 1961, Newell and Simon created a computer program called the *General Problem Solver* (GPS) intended to work as a universal problem solver machine. The GPS works with *means-end analysis* and is the first computer program that separated its knowledge of problems from its strategy of solving problems. GPS was able to solve simple problems such as the *Towers of Hanoi*, but it could not solve any real-world problems due to the combinatorial explosion in the search space. In 1959, Gelernter (1959) created the *Geometry Theorem Prover* that was able to prove theorems in elementary Euclidean plane geometry. Different from the above programs, McCarthy (1958) proposed a hypothetical program called the *Advice Taker* "whose performance could improve over time as a result of receiving advice, rather than by being reprogrammed." McCarthy (1958) was probably the first paper on logical AI, i.e., AI in which logic is the method of representing information in computer memory and not just the subject matter of the program, and it may also be the first paper to propose common sense reasoning ability as the key to AI.

In addition to the effort on using machines to solve mathematical problems and perform symbolic reasoning, some precursor ideas that contributed to the evolution of machine learning were also developed in 1950s–1960s. Rosenblatt (1958) developed the perceptron, the first modern neural network, which was a type of supervised learning algorithm that could learn to make binary classifications; see Section 5.5. This laid the foundation for machine learning and artificial neural neworks, which have now become essential components of AI. Moreover, some early machine translation systems were developed during 1950s–1960s, where computers were used to translate text from one language to another. Some statistical techniques were incorporated into these systems, which laid the groundwork for later development in natural language processing.

Development of optimal control and adaptive control

Parallel to the development of AI, the fields of optimal control and adaptive control have experienced significant progress during the 1950s and 1960s. In the field of optimal control, two powerful mathematical frameworks were developed to solve optimal control problems, Pontryagin's maximum principle (Pontryagin et al., 1962) and the dynamic programming principle (Bellman, 1954). The maximum principle was formulated by the Russian mathematician Lev Pontryagin and his students in 1956, and it states that any optimal control requires solving a so-called Hamiltonian system. The dynamic programming principle is both a mathematical opimization method and an algorithmic paradigm, and it solves complex control problems by breaking them down into simpler subproblems. Both principles have found numerous applications in fields ranging from science and engineering to economics and finance today; see Chapter 6 for methods of solving Markov decision processes using these principles. Different from optimal control that aims to optimize the system's performance based on a pre-defined cost or reward funciton, adaptive control focuses on adjusting the control law in real-time based on uncertainties and variations of the system's dynamics. During 1950s and 1960s, methods and algorithms for system identification and adaption were developed. The approaches developed in the fields of optimal control and adaptive control made significant progress in solving their respective problems. Importantly, they also laid theoretical foundations for reinforcement learning and approximate dynamic programming.

Expert or knowledge-based systems

The AI efforts during 1950s and 1960s were primarily focused on developing general purpose intelligent programs. While these programs could solve relatively simple problems within limited domains, they failed to scale up to handling complex problems. This limitation motivated many researchers to shift their focus in the 1970s and 1980s towards developing systems capable of solving difficult problems in specialized domains. These systems are based on domain-specific knowledge and are called *knowledge-based systems* or *expert systems*. Initially, these expert systems made exact inference and provided categorical decisions. A notable example is the MYCIN system (Buchanan and Shortliffe, 1984), a medical expert system that utilized *certainty factors* for diagnosing bacterial infections and recommending treatments.

However, categorical decisions and certain conclusions are not sufficient for many applications. To incorporate uncertainty in the rules in the expert systems, probabilistic approaches were developed to perform uncertain inference in AI. These systems are now called *Bayesian networks*; see a summary on them in Neapolitan (1989) and Pearl (1988).

Machine learning, big data, and deep learning

The development of AI during the 1970s and 1980s was relatively slow due to reduced funding and public attention in AI research. However, with advancements of machine learning techniques such as support vector machines, decision trees, and reinforcement learning, there has been a resurgence in the AI research. Moreover, with the emergence of big data and improvements in computing power, deep learning has gradually become the most widely used computational approach in machine learning. Deep neural networks such as convolutional neural networks, recurrent neural networks, and generative adversarial networks have achieved outstanding results on several complex cognitive tasks, matching or even outperforming the human performance.

One example of an AI program is *AlphaGo*, which was developed by a team at DeepMind (Silver et al., 2016) and is the first computer program to defeat a professional human Go player and a Go world champion. *AlphaGo* combines deep artificial neural networks, supervised learning, Monte Carlo tree search, and reinforcement learning, and relies on supervised learning from a large database of expert human moves. Later on, the team at DeepMind developed a successor program *AlphaGo Zero* (Silver et al., 2017). Different from *AlphaGo* which needs a large database of expert human moves, *AlphaGo Zero* uses only reinforcement learning and no human data or guidance beyond the basic rules of the game, but is stronger than any previous version (Silver et al., 2017).

9.1.2 Application of AI in finance

Artificial intelligence (AI) has had a significant impact on the finance industry in recent years. It has transformed various aspects of financial services, including trading, risk management, fraud detection, and asset and portfolio management. Here are some examples of key applications of AI in finance:

- *Algorithmic trading.* AI can be used in algorithmic trading systems to analyze large amounts of financial data and execute high-frequency trades. Machine learning algorithms can identify patterns and trends in market data, helping traders make more informed decisions.

- *Risk management.* AI can be used for risk assessment and management. Machine learning models, combined with data and risk analytics, can analyze credit risk, market risk, and operational risk to help financial institutions make better lending decisions and manage their portfolios more effectively.

- *Fraud and anomaly detection.* AI can detect fraudulent transactions by analyzing historical data and identifying unusual patterns or anomalies. AI algorithms, such as machine learning and neural networks, can be trained on historical transaction data to establish a baseline of normal behavior. Any deviations from this baseline can be treated as potential anomalies that might indicate fraud.

- *Credit scoring.* Credit scoring is an important component of the financial industry, as it assesses an individual's creditworthiness, determining their eligibility for loans, credit cards, mortgages, and other financial products. AI can enhance credit scoring models by incorporating a wide range of alternative data sources beyond traditional credit reports, such as social media activity, online behavior, utility payments, and so on, to provide a more comprehensive view of an individual's financial behavior and creditworthiness more accurately.

- *Portfolio management.* AI-driven robo-advisors help individual investors create and manage diversified investment portfolios. AI models can process and analyze vast amounts of financial data, news, social media sentiment, satellite imagery, and supply chain data to gain insights into investment opportunities and risks that may not be evident from traditional financial data.

- *Compliance and regulatory reporting.* Compliance and regulatory reporting in financial industries refers to processes that are designed to ensure that financial institutions adhere to laws, regulations, and industry standards, and accurately report their activities to regulatory authorities. AI-driven systems can monitor financial transactions in real-time to identify potentially suspicious activities.

- *Asset management.* AI is used to analyze investment portfolios, recommend changes, and optimize asset allocation. The use of AI in asset management has the potential to enhance the accuracy and efficiency of investment decisions, optimize portfolio performance, and improve risk management.

The applications of AI in finance are continually evolving and expanding. As AI technologies improve and become more accessible, the finance industry is likely to see even more significant advancements and innovations in the future.

9.2   Blockchain technology

Blockchain technology is a revolutionary digital ledger system that has transformed the way we record and verify transactions. Unlike traditional centralized databases, a blockchain is a decentralized, transparent, and highly secure ledger that is maintained by a distributed network of computers. Blockchain was originally developed as the technology underpinning cryptocurrencies like Bitcoin. Since then, it has rapidly expanded its scope, finding diverse applications across numerous industries.

9.2.1   Origin and evolution

The origin of blockchain technology can be traced back to the late 20th century. In 1991, Stuart Haber and W. Scott Stornetta proposed a cryptographically secure chain of blocks to timestamp digital documents, ensuring their immutability. Although this early concept didn't gain immediate widespread attention, it laid the foundation for future developments.

The true birth of blockchain technology is in 2008 when an anonymous individual or group using the pseudonym Satoshi Nakamoto published a whitepaper titled "Bitcoin: A Peer-to-Peer Electronic Cash System." This paper outlined the principles of a decentralized, digital cryptocurrency called Bitcoin, powered by a blockchain to enable secure, peer-to-peer transactions without the need for intermediaries. On January 3, 2009, the first Bitcoin block, known as the "genesis block" or "Block 0," was mined by Nakamoto, marking the beginning of the Bitcoin blockchain. This event is a historical moment in the development of blockchain technology. Since then, bitcoin's popularity has begun to grow, and the concept of blockchain technology has begun to get attention. Developers and enthusiasts explored its potential for various applications beyond cryptocurrency, such as secure data sharing, identity management, and more.

In late 2013, a programmer and blockchain enthusiast named Vitalik Buterin proposed the concept of *Ethereum*. Buterin published the Ethereum whitepaper in November 2013, outlining the idea of a decentralized platform with its cryptocurrency and a Turing-complete programming language to create *smart contracts*. The Ethereum network went live on July 30, 2015, with the creation of its Genesis Block. This marked the official launch of the Ethereum blockchain, and the cryptocurrency associated with it was named "Ether." In the 2020s, Ethereum gained widespread attention for its role in the rapid growth of *decentralized finance* (DeFi) applications and *non-fungible tokens* (NFTs). DeFi platforms on Ethereum enabled lending, borrowing, and trading without traditional intermediaries, while NFTs allowed for unique digital assets like collectibles and digital art.

Besides the above two important developments, other applications of blockchain technology have also emerged. For example, the surge of bitcoin's value has attracted more investors and spurred the creation of numerous cryptocurrencies. *Initial coin offerings* (ICOs) have also emerged as a new way to fund blockchain-based projects. Blockchain technology continues to evolve, and it is expected to shape the future of finance, technology, and many other industries.

9.2.2 Major components

Blockchain technology consists of several key components that work together to create a secure, decentralized, and transparent ledger for recording transactions. The following lists the main components of blockchain technology.

Distributed ledger

The core component of a blockchain is a distributed ledger. The distributed ledger is a digital database that is stored and maintained across a network of multiple computers or nodes. Unlike a traditional centralized ledger, which is controlled by a single entity (e.g., a central bank, corporation, or government), a distributed ledger is decentralized and maintained collectively by the participants in the network. Each participant, or node, has a copy of the entire ledger, and changes to the ledger are agreed upon through a consensus mechanism.

Blocks

Another fundamental components of blockchain technology are blocks, which play a crucial role in organizing and recording transactions on the blockchain. In a blockchain, data is grouped into blocks, and each block is linked to the previous block, creating a chain of blocks. This chaining ensures the chronological order of transactions and the immutability of data.

Cryptographic Hash

A cryptographic hash, often simply referred to as a "hash," is a fixed-size string of characters generated by applying a specific mathematical algorithm to an input data of arbitrary size. The primary purpose of cryptographic hashing is to take data and produce a fixed-length output that appears random, making it extremely difficult to reverse-engineer the original input data from the hash. Cryptographic hashes are widely used in computer science, cryptography, and cybersecurity for various purposes. Cryptographic hashing is a fundamental tool in computer security, ensuring data integrity, privacy, and authentication in a wide range of applications.

Consensus Mechanism

A *consensus mechanism* is a fundamental component of blockchain technology and distributed ledger systems. It is a protocol or algorithm that enables nodes in a network to agree on the contents of a blockchain or the validity of transactions. Consensus mechanisms are critical for maintaining the integrity, security, and trustworthiness of a decentralized and distributed system. There are several consensus mechanisms, each with its own set of rules and characteristics. For example, *proof of work* (PoW) is the consensus mechanism used in Bitcoin and many other cryptocurrencies. Miners compete to solve complex mathematical puzzles, and the first one to solve it gets the right to add a new block to the blockchain. PoW is known for its security but is computationally intensive and energy-consuming. Besides PoW, there are other consensus mechanisms for blockchains. The choice of consensus mechanism depends on the specific goals and requirements of a blockchain network.

Transactions

In a blockchain, transactions are fundamental units of data that represent various activities or changes of state on the network. These activities can involve the transfer of assets (cryptocurrency tokens or other digital assets), the execution of smart contracts, or the recording of data. Transactions are central to the functioning of blockchain systems, as they are grouped into blocks and added to the blockchain, creating an immutable and transparent ledger of activity.

Public and private keys

Public and private keys are pairs of cryptographic keys used in various encryption and authentication processes, including in blockchain and cryptocurrency systems. These keys play a crucial role in securing digital communications, enabling secure transactions, and providing user identity verification.

The public key is a long, randomly generated string of characters. It is publicly known and can be freely shared with anyone. The private key is a secret, securely stored piece of information that must be kept confidential. It is used to decrypt data encrypted with the corresponding public key and to sign digital messages or transactions. Public and private keys work together in public key cryptography, enabling secure communication and authentication without the need to share a secret key between parties.

In blockchain technology and cryptocurrencies, public keys are used as addresses to receive funds, while private keys are used to access and control the funds associated with those addresses. It's crucial to keep your private key secret and secure, as anyone with access to it can control your assets. Hardware wallets and secure software wallets are commonly used to manage and protect private keys in the world of cryptocurrencies.

Peer-to-Peer Network

A peer-to-peer (P2P) network in the context of blockchain technology refers to the decentralized and distributed nature of the network where nodes (computers) connect directly to each other to form a network. In a blockchain-based P2P network, there is no central authority, server, or intermediary controlling the network. Instead, every participating node in the network has equal status and functionality, which fosters decentralization, transparency, and resilience. Peer-to-peer networks play a crucial role in the success of blockchain technology. They enable the secure and decentralized operation of blockchain networks, facilitating trustless transactions and data management. This decentralized architecture has applications beyond cryptocurrencies and includes various blockchain use cases like supply chain tracking, smart contracts, and decentralized applications.

Smart contracts

Smart contracts are self-executing contracts with the terms and conditions of the agreement between parties directly written into code. These contracts automatically execute and enforce themselves when predefined conditions or trigger events are met, without the need for intermediaries, such as lawyers or notaries. Smart contracts are a fundamental component of blockchain technology, particularly on platforms like Ethereum, which introduced the concept.

Smart contracts are a powerful and innovative technology with the potential to automate and streamline various business processes and agreements. They are a critical building block for decentralized applications and are central to the growth of blockchain ecosystems.

9.2.3   Applications of blockchain

Blockchain technology has been found applications in many industries including finance and insurance. The following lists some applications in finance and data analytics.

Cryptocurrency

Cryptocurrencies are digital or virtual currencies that use cryptography for security. They are decentralized and typically operate on a technology called blockchain. Cryptocurrencies have gained significant attention and popularity due to their potential to offer secure, borderless, and peer-to-peer transactions. Popular cryptocurrencies include Bitcoin (BTC), Ethereum (ETH), Ripple (XRP), Litecoin (LTC), and many others, with new cryptocurrencies continually emerging. While cryptocurrencies offer numerous advantages, they also come with risks, including regulatory uncertainty, security concerns, and market volatility.

Data Analytics

Data analytics is the process of finding actionable information from large data sets. Blockchain technology has several potential applications in the field of data analytics. It can enhance the security, transparency, and integrity of data, making it a valuable tool for data analytics and insights. In terms of data integrity and verification, blockchain's immutability and transparency make it ideal for ensuring the integrity of data. Data records can be stored on a blockchain, and once recorded, they cannot be altered without consensus from the network. This ensures that data remains tamper-proof, providing a reliable source for analytics. In terms of data provenance, blockchain can track the provenance of data, recording its origin and every step of its journey. This is crucial for data analytics, as it allows analysts to trace the source and history of data, providing context and increasing trust in the data's accuracy. Besides the above two aspects, there are other ways that blockchain can be applied in data analytics. However, the application of blockchain to data analytics also has challenges, such as scalability, energy consumption, regulatory compliance, and the limit of technology infrastructure.

Finance and insurance

Blockchain technology has found significant applications in the FinTech and insurance industries, revolutionizing traditional practices. In finance,

besides cryptocurrencies, blockchain offers the potential to streamline transactions, enhance security, and create new financial instruments. For examples,
in payments and remittances, blockchain enables fast and cost-effective cross-
border payments and remittances, reducing the need for intermediaries and
minimizing transaction fees. In decentralized finance, the platforms are built
on blockchain, offering financial services such as lending, borrowing, and trading without traditional banks or intermediaries. Users have direct control of
their assets. In smart contracts, FinTech companies use smart contracts to
automate financial agreements, such as loans, insurance policies, and derivatives. These self-executing contracts can streamline processes and reduce the
risk of disputes.

In insurance, blockchain is transforming claims processing, reducing fraud,
and improving transparency. Specifically, in claims processing, blockchain can
simplify and accelerate claims processing by providing a secure and transparent platform for verifying and settling claims. The decentralized nature of
blockchain reduces fraud and automates the claims process. In reinsurance,
blockchain can improve transparency in the reinsurance process, streamlining
communication and settlements among primary insurers, reinsurers, and other
stakeholders. In microinsurance, blockchain can make it cost-effective to offer
microinsurance to underserved populations, providing affordable coverage for
small-scale risks.

In both finance and insurance, blockchain delivers benefits like cost reduction, transparency enhancement, security improvement, and innovative opportunities. Though these advantages are substantial, it's also important to
consider the regulatory and technical challenges that need to be addressed.

9.3   Cloud computing

In cloud computing, a *cloud* is defined as a space over network infrastructure where computing resources, including computer hardware, storage,
databases, networks, operating systems, and even entire software applications, are available instantly, on-demand. Cloud computing is a technology
that enables the delivery of computing services over the internet. These services are provided by cloud service providers, and users can access and utilize
them on-demand without the need for on-premises hardware or infrastructure.
Compared with service oriented architectures, grid computing, utility computing, and cluster computing, cloud computing represents a new way of managing information technology. Cloud computing offers scalability, flexibility,
and cost-efficiency, making it a valuable asset in various industries, including
finance.

9.3.1   Service delivery and deployment models

A cloud delivery model represents a specific, pre-packaged combination of IT resources offered by a cloud provider. Three common cloud delivery models—Infrastructure as a Service (IaaS), Platform as a Service (PaaS), and Software as a Service (SaaS)—have become widely established and formalized, and are explained as follows:

- *Infrastructure as a Service* (IaaS). The capability provided to the consumer allows for provisioning processing, storage, networks, and other essential computing resources. For example, one can run CPU/memory-intensive applications using Amazon's IaaS Cloud.

- *Platform as a Service* (PaaS). The capability provided to the consumer allows for deploying consumer-created or acquired applications onto the cloud infrastructure using supported programming languages, libraries, services, and tools provided by the cloud provider. For example, this includes building and deploying applications using the Google Cloud Platform.

- *Software as a Service* (SaaS). The capability provided to the consumer is to run their applications on a cloud infrastructure. For example, users can open Word or PDF files using Google Apps without needing to install MS Office or Adobe Reader software on their local system.

Cloud hosting deployment models categorize cloud environments based on ownership, scale, and accessibility, defining their purpose and nature. Understanding these models is essential for matching workload requirements. The following lists the four deployment models.

- *Public cloud.* Public clouds are hosted and managed by third-party cloud providers, serving multiple organizations and users. They offer cost-effective and scalable solutions. Examples of public clouds include Amazon Web Services (AWS), Microsoft Azure, and Google Cloud Platform (GCP).

- *Private cloud.* Private clouds are exclusively dedicated to a single organization, providing enhanced control over resources, security, and compliance. They can be hosted either on-premises or by a third-party provider.

- *Hybrid cloud.* Hybrid clouds integrate both public and private cloud resources, allowing seamless movement of data and applications between them. This approach enhances flexibility and facilitates efficient data integration.

- *Community cloud.* Community clouds are shared by a specific group of organizations with similar interests, such as regulatory requirements or industry standards.

9.3.2 Characteristics of cloud computing

Cloud computing is characterized by several key attributes that distinguish it from traditional computing paradigms. These characteristics include the following.

- *On-demand self-service.* Cloud services can be provisioned and managed by users as needed, without requiring human intervention from the service provider. This enables rapid access to computing resources.

- *Broad network access.* Cloud services are accessible over the internet from various devices, providing flexibility and remote access for users and applications.

- *Resource pooling.* Cloud providers pool computing resources, such as servers and storage, and share them among multiple users or tenants. This multi-tenant model optimizes resource utilization and cost-efficiency.

- *Rapid elasticity.* Cloud resources can be quickly scaled up or down to meet varying workloads or demands. This elasticity ensures that users have access to the resources they need without overprovisioning.

- *Measured service.* Cloud usage is metered and billed based on actual resource consumption. Users are charged for the specific amount of computing, storage, or other resources they use, aligning expenses with consumption.

9.3.3 Cloud computing in data analytics and finance

Cloud computing has gained significant traction in the finance industry due to its potential to enhance agility, reduce costs, and improve data security. The following lists some applications of cloud computing in data analytics and finance:

- *Data storage and management.* Cloud storage solutions enable financial institutions to securely store and manage vast amounts of data, from customer records and transaction histories to compliance documents and market data.

- *Data analytics.* Cloud platforms provide the computing power and storage needed for data analytics and modeling. Financial firms can analyze large datasets to gain insights for risk assessment, investment strategies, and customer profiling.

- *Risk assessment.* Cloud-based risk assessment tools can process complex financial data, assess market and credit risks, and provide real-time risk management, facilitating firms make informed decisions.

- *Algorithmic trading.* High-frequency trading and algorithmic trading strategies rely on cloud-based solutions to process large volumes of market data and execute orders with low latency.

- *Blockchain and digital assets.* Cloud platforms support the deployment and management of blockchain nodes, which are crucial for participating in blockchain networks or managing digital assets and cryptocurrencies.

- *Mobile banking and payments.* Cloud platforms facilitate the delivery of mobile banking and payment services, ensuring that customers can access their accounts, make transactions, and manage their finances from smartphones and other devices.

Besides the above mentioned applications, cloud computing can also be applied to other areas of data analytics and finance. We shall note that, in spite of numerous advantages that cloud computing has, it also has some challenges and limitations, such as security concerns, limited controls, data transfer costs, data portability, and so on. Hence, financial institutions need to evaluate the pros and cons when adopting and managing cloud services.

9.4 Big data and data analytics

Financial institutions deal with a large amount of data. Hence, the ability to store more data and process information more efficiently is critically important for financial companies. The AI, blockchain, cloud computing and other innovative technologies have revolutionized many financial processes and data analytics procedures. This section describes the main areas of FinTech services and data analytics in FinTech.

9.4.1 FinTech services

FinTech encompasses a wide range of services and innovations that leverage technology to improve and streamline financial processes. Besides blockchain and cryptocurrency (see Section 9.2), other main services in FinTech can be categorized into several areas:

- *Payments and transfers.* FinTech has revolutionized how financial transactions are initiated, processed, and completed. This includes mobile payments, digital wallets, peer-to-peer transfers, online banking and transfers, contactless payments, and so on. These services are marked by their convenience, speed, and accessibility.

- *Lending and borrowing.* This includes peer-to-peer lending, small business loans, online maket place lending, crowdfunding, and so on. FinTech companies can provide online lending services to their clients, with quick response and disbursement of loans.

- *Robo-advisors and wealth management* (WealTech). FinTech companies can provide automated investment platforms and digital tools for investment portfolio optimization, wealth management, and financial planning.

- *Insurance technology* (InsurTech). This usually includes digitial insurance platforms, AI-powered underwriting, and claims processing. FinTech companies could offer online platforms for purchasing and managing insurance policies, assessing risk, and automating the claims process for faster payouts.

- *Regulatory technology* (RegTech). FinTech provides advanced software and data analytics to help businesses and financial institutions comply with regulations efficiently and at a lower cost.

- *Personal finance and budgeting.* FinTech companies can provide apps and tools for budgeting, expense tracking, setting financial goals, and offering resources to help customers improve their financial literacy.

- *Real estate and property technology* (PropTech). FinTech allows people to invest in real estate projects through crowdfunding and provides tools to help property owners and managers streamline operations.

- *Alternative banking.* This is also referred to as "neobanking" or "challenger banking." FinTech companies can offer more convenient, cost-effective, and user-friendly financial services outside of traditional banking institutions.

- *Credit scoring and reporting.* FinTech companies can use data alternative to traditional sources to assess creditworthiness and provide more accurate credit scores.

9.4.2   Data and risk analytics in FinTech

Data analytics plays a crucial role in the FinTech industry. It helps businesses derive insights from vast volumes of financial data. The following lists some main areas of data and risk analytics in fintech.

- *Risk management.* FinTech companies use risk analytics to assess manage market risk, credit risk, operational risk, liquidity risk, and other financial risks. With advanced analytics models and procedures, large datasets can be processed and analyzed efficiently and more accurate predicitons and risk assessment can be obtained.

- *Fraud detection and prevention.* Advanced machine learning and other data analytics methods can be used to analyze vast historical data to detect and/or identify unusual patterns, transactions, and potential fraud in real-time, hence providing protection for FinTech platforms and customers.

- *Algorithmic trading.* FinTech companies use advanced data analytics to develop complex trading strategies and algorithms that make data-driven decisions in real-time.

- *Digital banking and payments.* Data analytics tools can monitor transactions for anomalies and enhance the efficiency and speed of payment processing.

- *Wealth management.* FinTech companies use data analytics to create and manage diversified investment portfolios based on customers' risk preference and financial goals.

- *Regulatory compliance.* Financial institutions can use data analytics tools to monitor transactions and report suspicious activities to comply with regulatory rules.

- *Insurance underwriting and claims processing.* Data analytics models and algorithms can assess insurance risks and calculate premiums based on customer data and behavioral patterns, and expedite claims processing with improved efficiency.

- *Peer-to-peer lending.* Risk analytics can be used to assess credit risk and compute credit scores with incorporation of alternative data sources like social media.

- *Market sentiment and predictive analytics.* Data analytics are used to analyze market and social sentiment and to predict market trends and customer behaviors.

In addition to the aforementioned domains, there are many other areas in FinTech that data and risk analytics can be applied to enhance risk management, customer experience, operational efficiency, and compliance. Although data and risk analytics are widely used in FinTech industry and provides many advantages, one should also realize their limitations. For example, some advanced analytics models, such as deep learning and neural networks, can be difficult to interpret. Moreover, effective data analytics depends on the accuracy and quality of data, but maintaining data quality can be challenging, especially when dealing with large amounts of data.

Supplements

This chapter presents the ABCDs of FinTech and how artificial intelligence, blockchain technology, cloud computing are connected with big data and data analytics in finance and insurance. The presentation here is non-technical and only means to be a starting point for interested readers. The following lists some textbook references on each subject.

There are many overviews and surveys on AI with different focuses during the past decades, so we only name a few reference here. Russell and Norvig (2021) provide the most comprehensive, up-to-date introduction to the theory and applications of AI. The book is highly regarded for its depth and breadth of coverage and is a standard reference for students, researchers, and practitioners in the AI field. For introduction on specific topics in AI, Goodfellow et al. (2016) and Sutton and Barto (2018) provide thorough introductions to deep learning and reinforcement learning, respectively. Neapolitan and Jiang (2018) provide an accessible and selective introduction to models and methods in AI, including logic-based methods, probability-based methods, emergent intelligence, evolutionary computational methods, neural network, deep learning, and techniques for natural language understanding. Ertel (2018) introduces concrete algorithms and applications to some selective topics in AI, such as logic, search, reasoning with uncertainty, machine learning, neural networks, and deep learning. Joshi (2020) presents an overview of some machine learning methods and AI with applications. Xiong (2022) presents the recent development of relationships between AI and causal inference.

For reference on blockchain technology, Narayanan et al. (2016) provide a comprehensive and accessible introduction to cryptocurrencies, focusing on Bitcoin and other blockchain-based digital currencies. Antonopoulos (2014) gives a complete guide to Bitcoin and cryptocurrencies. Antonopoulos and Wood (2020) explain systematically how to understand and work with Ethereum, one of the leading blockchain platforms for building smart contracts and decentralized applications. For blockchain technology and its applications to finance, Borovykh (2018) provides a concise but authoritative introduction on the subject.

For reference on cloud computing, Erl et al. (2013) give a complete overview of cloud computing concepts, technologies, and architectural patterns. Marinescu (2023) provides a comprehensive and in-depth analysis of cloud computing for students and IT professionals. Faynberg et al. (2016) first introduce basic concepts of cloud computing and then analyze the business and technology trends in cloud computing.

Bibliography

O. O. Aalen. A model for non-parametric regression analysis of counting processes. In W. Klonechi, A. Kozek, and J. Rosinski, editors, *Lecture Notes in Statistics-2: Mathematical Statistics and Probability Theory*, pages 1–25. Springer-Verlag, New York, 1980.

O. O. Aalen, O. Borgan, and H. K. Gjessing. *Survival and Event History Analysis: A Process Point of View*. Springer, New York, USA, 2008.

D. H. Ackley, G. E. Hinton, and T. J. Sejnowski. A learning algorithm for boltzmann machines. *Cognitive Science*, 9:147–169, 1985.

C. C. Aggarwal. *Neural Networks and Deep Learning*. Springer, New York, USA, 2nd edition, 2023.

F. AitSahlia and T. L. Lai. Exercise boundaries and efficient approximations to american option prices and hedge parameters. *Journal of Computational Finance*, 4:85–103, 2001.

E. Altman. Measuring corporate bond mortality and performance. *Journal of Finance*, 44:909–922, 1989.

S. Amari and S. C. Douglas. Why natural gradient? In *Proceedings of the 1998 IEEE International Conference on Acoustics, Speech and Signal Processing*, pages 1213–1216, Seattle, USA., 1998.

L. Andersen and R. Brotherton-Ratcliffe. The equity option volatility smile: An implicit finite-difference approach. *Journal of Computational Finance*, 1:5–37, 1997.

L. B. G. Andersen and V. V. Piterbarg. *Interest Rate Modeling*. Atlantic Financial Press, New York, USA, 2010.

P. K. Andersen, O. Borgan, R. D. Grill, and N. Keiding. *Statistical Models Based on Counting Processes*. Springer, New York, USA, 1995.

K. Antonio and J. Beirlant. Actuarial statistics with generalized linear mixed models. *Insurance: Mathematics and Economics*, 40:58–76, 2007.

A. M. Antonopoulos. *Mastering Bitcoin: Unlocking Digital Cryptocurrencies*. O'Reilly Media, Sebastopol, CA, USA, 2014.

A. M. Antonopoulos and G. Wood. *Mastering Ethereum: Building Smart Contracts and DApps*. O'Reilly Media, Sebastopol, CA, USA, 2020.

P. Artzner, F. Delaen, J.-M. Eber, and D. Heath. Thinking coherently. *Risk*, 10 (November):68–71, 1997.

P. Artzner, F. Delaen, J.-M. Eber, and D. Heath. Coherent measures of risk. *Mathematical Finance*, 9:203–228, 1999.

S. Asmussen and H. Albrecher. *Ruin Probabilities*. Springer, Singapore, 2nd edition, 2010.

S. Asmussen and P. W. Glynn. *Stochastic Simulation: Algorithms and Analysis*. Springer-Verlag, New York, USA, 2007.

S. Asmussen and M. Steffensen. *Risk and Insurance*. World Scientific, New York, USA, 2010.

R. T. Baillie, T. Bollerslev, and H. O. Mikkelsen. Fractionally integrated generalized autoregressive conditional heteroskedasticity. *Journal of Econometrics*, 74:3–30, 1996.

M. S. Bartlett. The statistical conception of mental factors. *The British Journal of Psychology*, 28:97–104, 1937.

M. S. Bartlett. A note on multiplying factors for various chi-squared approximations. *Journal of the Royal Statistical Society, Series B*, 16:296–298, 1954.

J. Bartroff, T. L. Lai, and M.-C. Shih. *Sequential Experimentation in Clinical Trials: Design and Analysis*. Springer, New York, USA, 2013.

N. Bäuerle and U. Rieder. *Markov Decision Processes with Applications to Finance*. Springer, New York, USA, 2011.

J. Beirlant, Y. Goegebeur, J. Teugels, and J. Segers. *Statistics of Extremes: Theory and Applications*. John Wiley & Sons, Ltd., Hoboken, New Jersey, USA, 2004.

R. Bellman. The theory of dynamic programming. *Bulletin of the American Mathematical Society*, 60:503–515, 1954.

R. Bellman. A problem in the sequential design of experiments. *Sankhya*, 16: 221–229, 1956.

R. E. Bellman. *Dynamic Programming*. Princeton University Press, New Jersey, USA, 1957.

L. Berlyand and P.-E. Jabin. *Mathematics of Deep Learning: An Introduction*. De Gruyter, Berlin, Germany, 2023.

D. P. Bertsekas. *Dynamic Programming and Optimal Control.* Athena Scientific, Belmont, Massachusetts, USA, 4th edition, 2012.

D. P. Bertsekas. *Reinforcement Learning and Optimal Control.* Athena Scientific, Belmont, Massachusetts, USA, 2019.

D. P. Bertsekas and S. E. Shreve. *Stochastic Optimal Control: The Discrete-Time Case.* Athena Scientific, Belmont, Massachusetts, USA, 1978.

J. Besag. Spatial interaction and the statistical analysis of lattice systems (with discussion). *Journal of the Royal Statistical Society, Series B*, 36: 192–326, 1974.

J. Beyersmann, A. Allignol, and M. Schumacher. *Competing Risks and Multistate Models with R.* Springer, New York, USA, 2012.

S. Bhatnagar, R. S. Sutton, M. Ghavamzadeh, and M. Lee. Natural actor-critic algorithms. *Automatica*, 45:2471–2482, 2009.

T. R. Bielecki and M. Rutkowski. *Credit Risk: Modeling, Valuation, and Hedging.* Springer, New York, USA, 2004.

M. Bilodeau and D. Brenner. *Theory of Multivariate Statistics.* Springer, New York, USA, 1999.

F. Black. The pricing of commodity contracts. *Journal of Financial Economics*, 3:167–179, 1976.

F. Black and J. Cox. Valuing corporate securities: Some effects of bond indenture provisions. *Journal of Finance*, 31:351–367, 1976.

W. Bohm and P. Hackl. Improved bounds for the average run length of control charts based on finite weighted sums. *Annals of Statistics*, 18:1895–1899, 1990.

T. Bollerslev. Generalized autoregressive conditional heteroscedasticity. *Journal of Econometrics*, 31:307–327, 1986.

P. Borovykh. *Blockchain Applications in Finance.* BlockchainDriven, 2018 edition, 2018.

A. Brace, D. Gatarek, and M. Musiela. The market model of interest rate dynamics. *Mathematical Finance*, 7:127–154, 1997.

L. Breiman. Bagging predictors. *Machine Learning*, 24:123–140, 1996.

N. E. Breslow and D. G. Clayton. Approximate inference in generalized linear mixed models. *Journal of the American Statistical Association*, 88:9–25, 1993.

D. Brigo and F. Mercurio. *Interest Rate Models: Theory and Practice.* Springer, Berlin, Germany, 2nd edition, 2006.

M. Broadie, M. Chernov, and M. Johannes. Model specification and risk premia: Evidence from futures options. *Journal of Finance*, 62:1453–1490, 2007.

D. S. Broomhead and D. Lowe. Multivariate functional interpolation and adaptive networks. *Complex Systems*, 2:321–355, 1988.

S. Bubeck and N. Cesa-Bianchi. Regret analysis of stochastic and nonstochastic multi-armed bandit problems. *Foundations and Trends in Machine Learning*, 5:1–122, 2012.

B. G. Buchanan and E. H. Shortliffe. *Rule-Based Expert System: The MYCIN Experiments of the Stanford Heuristic Programming Project.* Addison-Wesley, Massachusetts, USA, 1984.

J. A. Bucklew, P. Ney, and J. S. Sadowsky. Monte Carlo simulation and large deviations theory for uniformly recurrent markov chains. *Journal of Applied Probability*, 27:44–59, 1990.

H. Bühlmann and A. Gisler. *A Course in Credibility Theory and its Applications.* Springer-Verlag, Berlin, Germany, 2005.

H. Bühlmann and Straub. Glaubwürdigkeit für schadensätze. *Bulletin of Swiss Association of Actuaries*, 1:111–133, 1970.

J. Cai. A markov model of switching-regime arch. *Journal of Business and Economic Statistics*, 12:309–316, 1994.

A. J. G. Cairns. *Interest Rate Models: An Introduction.* Princeton University Press, New Jersey, USA, 2004.

R. A. Carmona and M. R. Tehranchi. *Interest Rate Models: an Infinite Dimensional Stochastic Analysis Perspective.* Springer, New York, USA, 2006.

P. Carr, R. Jarrow, and R. Myneni. Alternative characterizations of american put options. *Mathematical Finance*, 2:87–106, 1992.

H. P. Chan and T. L. Lai. Saddlepoint approximations and nonlinear boundary crossing probabilities of markov random walks. *The Annals of Applied Probability*, 13:395–429, 2007.

G. Chaplin. *Credit Derivatives: Trading, Investing and Risk Management.* John Wiley & Sons, Ltd., Chichester, England, 2nd edition, 2010.

C. Chatfield and H. Xing. *The Analysis of Time Series: An Introduction with R.* Chapman & Hall/CRC, New York, USA, 7th edition, 2019.

U. Cherubini, E. Luciano, and W. Vecchiato. *Copula Methods in Finance.* John Wiley & Sons, Ltd., Chichester, England, 2004.

U. Cherubini, F. Gobbi, S. Mulinacci, and S. Romagnoli. *Dynamic Copula Methods in Finance.* John Wiley & Sons, Ltd., Chichester, England, 2011.

S. Chib and E. Greenberg. Markov chain Monte Carlo simulation methods in econometrics. *Econometric Theory*, 12:409–431, 1996.

S. Chib, F. Nardari, and N. Shephard. Analysis of high dimensional multivariate stochastic volatility models. *Journal of Econometrics*, 134:341–371, 2006.

N. Chopin and O. Papaspiliopoulos. *An Introduction to Sequential Monte Carlo.* Springer, Cham, Switzerland, 2020.

R. Cole and J. W. Gunther. Separating the likelihood and timing of bank failure. *Journal of Banking and Finance*, 19:1073–1089, 1995.

S. Coles. *An Introduction to Statistical Modeling of Extreme Values.* Springer, New York, USA, 2001.

R. Cont and P. Tankov. *Financial Modelling with Jump Processes.* Chapman & Hall/CRC, New York, USA, 2003.

R. J. Cook and J. F. Lawless. *The Statistical Analysis of Recurrent Events.* Springer, New York, USA, 2007.

National Research Council. *The Owner's Role in Project Risk Management.* The National Academies Press, Washington, DC, USA, 2005.

D. R. Cox. Regression models and life tables (with discussion). *Journal of the Royal Statistics Society, Series B*, 34:187–220, 1972.

D. R. Cox. Partial likelihood. *Biometrika*, 62:269–276, 1975.

J. C. Cox and S. A. Ross. The valuation of options for alternative stochastic processes. *Journal of Financial Economics*, 3:145–166, 1976.

J. C. Cox, S. A. Ross, and M. Rubinstein. Option pricing: A simplified approach. *Journal of Financial Economics*, 7:229–263, 1975.

J. C. Cox, J. E. Ingersoll, and S. A. Ross. A theory of the term structure of interest rates. *Econometrica*, 53:385–407, 1985.

D. Crisan and A. Doucet. A survey of convergence results on particle filtering for practitioners. *IEEE Transactions on Signal Processing*, 50:736–746, 2002.

M. J. Crowder. *Multivariate Survival Analysis and Competing Risks.* Chapman & Hall/CRC, Boca Raton, Florida, USA, 2012.

S. V. Crowder. A simple method for studying run-length distributions of exponentially weighted moving average charts. *Technometrics*, 29:401–407, 1987.

J. S. Dagpunar. *Simulation and Monte Carlo: With Applications in Finance and MCMC.* John Wiley & Sons, Ltd., Chichester, England, 2007.

P. Del Moral. *Feynman-Kac Formulae: Genealogical and Interacting Aarticle Systems with Applications.* Springer-Verlag, New York, USA, 2004.

E. Demidenko. *Mixed Models: Theory and Applications with R.* John Wiley & Sons, Ltd., New York, USA, 2nd edition, 2013.

S. Deng, K. Giesecke, and T. L. Lai. Sequential importance sampling and resampling for dynamic portfolio credit risk. *Operations Research*, 60:78–91, 2012.

E. Derman and I. Kani. Riding on a smile. *Risk*, 7:32–39, 1994.

D. C. M. Dickson. *Insurance Risk and Ruin.* Cambridge University Press, Cambridge, UK, 2nd edition, 2016.

F. X. Diebold. Comment on modeling the persistence of conditional variance. *Econometric Reviews*, 5:51–56, 1986.

Z. Ding, W. J. Grander, and R. F. Engle. A long memory property of stock market returns and a new model. *Journal of Empirical Finance*, 1:83–106, 1993.

M. F. Dixon, I. Halperin, and P. Bilokon. *Machine Learning in Finance: From Theory to Practice.* Springer, New York, USA, 2020.

R. Douc, E. Moulines, and D. Stoffer. *Nonlinear Time Series: Theory, Methods and Applications with R Examples.* Chapman & Hall/CRC, New York, USA, 2014.

A. Doucet and A. Johansen. A tutorial on particle filtering and smoothing: Fifteen years later. In D. Crisan and B. Rozovskii, editors, *The Oxford Handbook of Nonlinear Filtering*, pages 656–704. Oxford University Press, New York, 2011.

A. Doucet, N. de Freitas, N. Gordon, and A. Smith. *Sequential Monte Carlo Methods in Practice.* Springer, New York, USA, 2001.

L. Duchateau and P. Janssen. *The Frailty Model.* Springer, New York, USA, 2008.

D. Duffie and K. J. Singleton. *Credit Risk: Pricing, Measurement, and Management.* Princeton University Press, Princeton, New Jersey, USA, 2003.

D. Duffie, J. Pan, and K. Singleton. Transform analysis and asset pricing for affine jump-diffusions. *Econometrica*, 68:1343–1376, 2000.

D. Duffie, L. Saita, and K. Wang. Multi-period corporate default prediction with stochastic covariates. *Journal of Financial Economics*, 83:635–665, 2007.

D. Duffie, A. Eckner, G. Horel, and L. Saita. Frailty correlated default. *Journal of Finance*, 64:2089–2123, 2009.

W. Dunsmuir. A central limit theorem for parameter estimation in stationary vector time series and its applications to models for a signal observed with noise. *The Annals of Statistics*, 7:490–506, 1979.

B. Dupire. Pricing with a smile. *Risk*, 7:18–20, 1979.

P. Embrechts, C. Kluppelberg, and T. Mikosch. *Modelling Extremal Events for Insurance and Finance*. Springer, New York, USA, 1997.

P. Embrechts, A. J. McNeil, and D. Straumann. Correlation and dependency in risk management: Properties and piffalls. In M. Dempster, editor, *Risk Management: Value at Risk and Beyond*, pages 176–223. Cambridge University Press, 2002.

R. F. Engle. Autoregressive conditional heteroscedasticity with estimates of the variance of united kingdom inflation. *Econometrica*, 50:987–1007, 1982.

R. F. Engle and T. Bollerslev. Modeling the persistence of conditional variances. *Econometric Reviews*, 5:1–50, 1986.

R. F. Engle and K. F. Kroner. Multivariate simultaneous generalized arch. *Econometric Theory*, 11:122–150, 1995.

B. Eraker, M. S. Johannes, and N. G. Polson. The impact of jumps in returns and volatility. *Journal of Finance*, 53:1269–1300, 2003.

T. Erl, R. Puttini, and Z. Mahmood. *Cloud Computing: Concepts, Technology & Architecture*. Prentice Hall, Upper Saddle River, NJ, USA, 2013.

W. Ertel. *Introduction to Artificial Intelligence*. Springer, Weingarten, Germany, 2nd edition, 2018.

J. Fan and Q. Yao. *Nonlinear Time Series: Nonparametric and Parametric Methods*. Springer, New York, USA, 2003.

I. Faynberg, H.-L. Lu, and D. Skuler. *Cloud Computing: Business Trends and Technologies*. John Wiley & Sons, Hoboken, New Jersey, USA, 2016.

C. Francq and J.-M. Zakoian. *GARCH Models: Structure, Statistical Inference and Financial Applications*. John Wiley & Sons Ltd., Chichester, UK, 2019.

E. W. Frees, V. R. Young, and Y. Lou. A longitudinal data analysis interpretation of credibility models. *Insurance: Mathematics and Economics*, 24: 229–247, 1999.

Y. Freund and R. E. Schapire. A decision-theoretic generalization of on-line learning and an application to boosting. *Journal of Computer and System Sciences*, 55:119–139, 1997.

J. Friedman, T. Hastie, and R. Tibshirani. Additive logistic regression: a statistical view of boosting (with discussion). *Annals of Statistics*, 28:337–407, 2000.

M. Frisén. *Financial Surveillance*. Edited volume. Wiley, Chichester, UK, 2008.

M. Frisén. Optimal sequential surveillance for finance, public health, and other areas. *Sequential Analsyis*, 28:310–337, 2009.

D. Gamerman and H. F. Lopes. *Markov Chain Monte Carlo: Stochastic Simulation for Bayesian Inference*. Chapman & Hall/CRC, Boca Raton, Florida, USA, 2nd edition, 2006.

J. Gatheral. *The Volatility Surface: A Practitioner's Guide*. John Wiley & Sons, Ltd., New York, USA, 2006.

H. Gelernter. Realisation of a geometry-proving machine. In *Proceedings of the International Conference on Information Processing*, pages 273–282, Paris, June 15-20, 1959.

A. Gelman and T. Speed. Characterizing a joint probability distri- bution by conditionals. *Journal of the Royal Statistical Society, Series B*, 55:185–188, 1993.

S. Geman and D. Geman. Stochastic relaxation, gibbs distributions and the bayesian restoration of images. *IEEE Transactions on Pattern Analysis and Machine Intelligence*, 6:721–741, 1984.

J. Geweke. Modeling the persistence of conditional variances: A comment. *Econometric Review*, 5:57–61, 1986.

B. K. Ghosh. A brief history of sequential analysis. In B. K. Ghosh and P. K. Sen, editors, *Handbook of Sequential Analysis*, pages 2672–2680. Dekker, New York, 1991.

B. K. Ghosh and P. K. Sen. *Handbook of Sequential Analysis*. CRC, New York, USA, 1991.

K. Giesecke, H. Kakavand, M. Mousavi, and H. Takada. Exact and efficient simulation of correlated defaults. *SIAM Journal on Financial Mathematics*, 1:868–896, 2010.

F. Girosi, M. Jones, and T. Poggio. Regularization theory and neural networks architectures. *Neural Computation*, 7:219–269, 1995.

J. C. Gittins. Bandit processes and dynamic allocation indices. *Journal of the Royal Statistical Society*, 2:148–177, 1979.

J. C. Gittins and D. M. Jones. A dynamic allocation index for the sequential design of experiments. In J. Gani, editor, *Progress in Statistics*, pages 241–266. North Holland, Amsterdam, 1974.

J. C. Gittins, K. D. Glazebrook, and R. R. Weber. *Multi-Armed Bandit Allocation Indices*. John Wiley & Sons, Ltd., Chichester, UK, 2nd edition, 2011.

P. Glasserman. *Monte Carlo Methods in Financial Engineering*. Springer, New York, USA, 2003.

P. Glasserman and J. Li. Importance sampling for portfolio credit risk. *Management Science*, 51:1643–1656, 2005.

L. R. Glosten, R. Jagannathan, and D. Runkle. On the relation between the expected value and the volatility of the nominal excess return on stocks. *Journal of Finance*, 48:1779–1801, 1993.

P. W. Glynn. Likelihood ratio gradient estimation: An overview. In *Proceedings of Winter Simulation Conference*, pages 366–375, Atlanta, GA, 1987. ACM Press.

P. W. Glynn and D. L. Iglehart. Importance sampling for stochastic simulations. *Management Science*, 35:1367–1392, 1989.

I. Goodfellow, J. Pouget-Abadie, M. Mirza, B. Xu, D. Warde-Farley, S. Ozair, A. Courville, and Y. Bengio. Generative adversarial nets. In *Advances in Neural Information Processing Systems*, pages 2672–2680, 2014.

I. Goodfellow, Y. Bengio, and A. Courville. *Deep Learning*. The MIT Press, Massachusetts, USA, 2016.

C. Graham and D. Talay. *Stochastic Simulation and Monte Carlo Methods: Mathematical Foundations of Stochastic Simulation*. Springer, New York, USA, 2013.

P. J. Green. Reversible jump Markov chain Monte Carlo computation and Bayesian model determination. *Biometrika*, 82:711–732, 1995.

M. Greenwood. The natural duration of cancer. *Reports on Public Health and Medical Subjects*, 33:1–26, 1926.

R. Grinold and R. Kahn. *Active Portfolio Management: A Quantitative Approach for Producing Superior Returns and Controlling Risk*. McGraw Hill, New York, USA, 2nd edition, 1999.

I. Grondman, L Busoniu, G. A. D. Lopes, and R. Babuska. A survey of actor-critic reinforcement learning: Standard and natural policy gradients. *IEEE Transactions on Systems, Man, and Cybernetics. Part C: Applications and Reviews*, 42:1291–1306, 2012.

X. Guo, T. L. Lai, H. Shek, and S. P.-S. Wong. *Quantitative Trading Algorithms: Analytics, Data, Models, Optimization*. Chapman and Hall/CRC, Boca Raton, Florida, USA, 2016.

L. Haan and A. Ferreira. *Extreme Value Theory: An Introduction*. Springer, New York, USA, 2006.

C. A. Hachemeister. Credibility for regression models with applications to trend. In P. M. Kahn, editor, *Credibility: Theory and Applications*, pages 129–163. Academic Press, 1975.

J. Hammersley and M. Clifford. Markov fields on finite graphs and lattices. Unpublished manuscript, 1970.

D. D. Hanagal. *Modeling Survival Data Using Frailty Models*. Springer, New York, USA, 2nd edition, 2019.

J. Hasbrouck. *Empirical Market Microstructure: The Institutions, Economics, and Econometrics of Securities Trading*. Oxford University Press, New York, USA, 2007.

T. Hastie, R. Tibshirani, and J. Friedman. *The Elements of Statistical Learning: Data Mining, Inference, and Prediction*. Springer, New York, USA, 2nd edition, 2016.

W. K. Hastings. Monte Carlo sampling methods using markov chains and their applications. *Biometrika*, 57:92–109, 1970.

D. Heath, R. Jarrow, and A. Morton. Bond pricing and the term structure of interest rates: A new methodology for contigent claims valuation. *Econometrica*, 60:77–105, 1992.

D. O. Hebb. *The Organization of Behavior: A Neuropsychological Theory*. Wiley, New York, USA, 1949.

S. L. Heston. A closed-form solution for options with stochastic volatility with applications to bond and currency options. *Review of Financial Studies*, 6: 327–343, 1993.

M. L. Higgins and A. K. Bera. A class of nonlinear arch models. *International Economic Review*, 33:137–158, 1992.

E. Hillebrand. Neglecting parameter changes in garch models. *Journal of Econometrics*, 129:121–138, 2005.

G. E. Hinton. A practical guide to training restricted boltzmann machines. In G. Montavon, G. B. Orr, and K.-R. Müller, editors, *Neural Networks: Tricks of the Trade*, pages 599–619. Springer Berlin Heidelberg, Berlin, Heidelberg, 2nd edition, 2012.

K. Hornik, M. Stinchcombe, and H. White. Multilayer feedforward networks are universal approximators. *Neural Network*, 2:359–366, 1989.

P. Hougaard. *Analysis of Multivariate Survival Data*. Springer, New York, USA, 2000.

C. Hsiao, C. Kim, and G. Taylor. A statistical perspective on insurance rate-making. *Journal of Econometrics*, 44:5–24, 1990.

J. Hull and A. White. The pricing of options on assets with stochastic volatilities. *Journal of Finance*, 42:281–300, 1987.

J. Hull and A. White. Pricing interest rate derivative securities. *Review of Financial Studies*, 3:573–592, 1990.

J. C. Hull. *Options, Futures, and Other Derivatives*. Pearson, Hoboken, New Jersey, USA, 11th edition, 2021.

J. C. Hull. *Risk Management and Financial Institutions*. Wiley, Hoboken, New Jersey, USA, 6th edition, 2023.

R. Israel, J. Rosenthal, and J. Wei. Finding generators for Markov chain via empirical transition matrices. *Mathematical Finance*, 11:245–265, 2001.

S. D. Jacka. Optimal stopping and the american put. *Mathematical Finance*, 1:1–14, 1991.

E. Jacquier, N. G. Polson, and P. E. Rossi. Bayesian analysis of stochastic volatility models. *Journal of Business and Economic Statistics*, 20:69–87, 1994.

E. Jacquier, N. G. Polson, and P. E. Rossi. Bayesian analysis of stochastic volatility models with fat-tails and correlated errors. *Journal of Econometrics*, 122:185–212, 1997.

P. James. *Option Theory*. John Wiley & Sons Ltd., West Sussex, England, 2003.

W. James and C. Stein. Estimation with quadratic loss. *Proceedings of the Fourth Berkeley Symposium on Mathematical Statistics and Probability*, 1: 361–379, 1961.

F. Jamshidian. An exact bond pricing formula. *Jounal of Finance*, 44:205–209, 1989.

F. Jamshidian. Libor and swap market models and measures. *Finance and Stochastics*, 1:293–330, 1997.

C. M. Jarque and A. K. Bera. A test for normality of observations and regression residuals. *International Statistical Review*, 55:163–172, 1987.

R. A. Jarrow. *Continuous-Time Asset Pricing Theory*. Springer, New York, USA, 2nd edition, 2021.

J. L. Jensen. *Saddlepoint Approximations*. Clarendon Press, Oxford, UK, 1995.

T. Jo. *Machine Learning Foundations: Supervised, Unsupervised, and Advanced Learning*. Springer, New York, USA, 2021.

N. L. Johnson and S. Kotz. *Distributions in Statistics: Discrete Distributions*. Houghton Mifflin, Boston, MA, USA, 1969.

P. D. Jong and G. Z. Heller. *Generalized Linear Models for Insurance Data*. Cambridge University Press, Cambridge, UK, 2008.

P. Jorion. *Value at Risk: The New Benchmark for Managing Financial Risk*. McGraw-Hill, New York, USA, 3rd edition, 2006.

A. V. Joshi. *Machine Learning and Artificial Intelligence*. Springer, New York, USA, 2020.

N. Ju. Pricing an american option by approximating its early exercise boundary as a multipiece exponential function. *The Review of Financial Studies*, 11:627–646, 1998.

R. Kaas, M. Goovaerts, J. Dhaene, and M. Denuit. *Modern Actuarial Risk Theory: Using R*. Springer, New York, USA, 2nd edition, 2008.

J. D. Kalbfleisch and R. L. Prentice. *The Statistical Analysis of Failure Time Data*. John Wiley & Sons, Inc., Hoboken, New Jersey, USA, 2nd edition, 2002.

R. E. Kalman. A new approach to linear filtering and prediction problems. *Transactions of the ASME–Journal of Basic Engineering*, 82(Series D):35–45, 1960.

E. L. Kaplan and P. Meier. Nonparametric estimation from incomplete observations. *Journal of American Statistical Association*, 53:457–481, 1958.

I. Karatzas and S. E. Shreve. *Brownian Motion and Stochastic Calculus*. Springer-Verlag, New York, USA, 2nd edition, 1991.

Kari Karhunen. Über lineare methoden in der wahrscheinlichkeitsrechnung. *Annales Academiæ Scientiarum Fennicæ Mathematica. Series A, I. Mathematica-physica*, 37:3–79, 1947.

M. G. Kendall. External risk measures and basel accords. *Biometrika*, 30: 81–93, 1938.

P. E. Kloeden and E. Platen. *Numerical Solution of Stochastic Differential Equations*. Springer-Verlag, Berlin, Germany, 1999.

V. R. Konda and J. N. Tsitsiklis. On actor-critic algorithms. *SIAM Journal of Control and Optimization*, 42:1143–1166, 2003.

R. Korn, E. Korn, and G. Kroisandt. *Monte Carlo Methods and Models in Finance and Insurance*. Chapman & Hall/CRC, Boca Raton, Florida, USA, 2010.

S. G. Kou. A jump diffusion model for option pricing. *Management Science*, 48:1086–1101, 2002.

S. G. Kou, X. Peng, and C. C. Heyde. External risk measures and basel accords. *Mathematics of Operations Research*, 38:393–417, 2013.

V. Krishnamurthy. *Partially Observed Markov Decision Processes: From Filtering to Controlled Sensing*. Cambridge University Press, Cambridge, UK, 2016.

D. P. Kroese, T. Taimre, and Z. I. Botev. *Handbook of Monte Carlo Methods*. John Wiley & Sons Ltd., New York, USA, 2003.

U. Küchler and M. Sørensen. *Exponential Families of Stochastic Processes*. Springer-Verlag, New York, USA, 1997.

T. L. Lai. Control charts based on weighted sums. *Annals of Statistics*, 2: 134–147, 1974.

T. L. Lai. Sequential changepoint detection in quality control and dynamical systems. *Journal of the Royal Statistical Society. Series B*, 57:613–658, 1995.

T. L. Lai. Information bounds and quick detection of parameter changes in stochastic systems. *IEEE Transactions on Information Theory*, 44:2917–2929, 1998.

T. L. Lai. Sequential analysis: Some classical problems and new challenges. *Statistica Sinica*, 11:303–408, 2001.

T. L. Lai and H. Robbins. Asymptotically efficient adaptive allocation rules. *Advances in Applied Mathematics*, 6:4–22, 1985.

T. L. Lai and J. Shan. Efficient recursive algorithms for detection of abrupt changes in signals and control systems. *IEEE Transactions on Automatic Control*, 44:952–966, 1999.

T. L. Lai and M.-C. Shih. A hybrid estimator in nonlinear and generalized linear mixed effects models. *Biometrika*, 90:859–879, 2003.

T. L. Lai and H. Xing. *Statistical Models and Methods for Financial Markets*. Springer, New York, USA, 2008a.

T. L. Lai and H. Xing. A hidden markov filtering approach to multiple change-point models. In *2008 47th IEEE Conference on Decision and Control*, pages 1914–1919, Cancun, Mexico, 2008b. doi: 10.1109/CDC.2008.4739184.

T. L. Lai and H. Xing. Sequential change-point detection when the pre- and post-change parameters are unknown. *Sequential Analysis*, 29:162–175, 2010.

T. L. Lai and H. Xing. A simple bayesian approach to multiple change-points. *Statistica Sinica*, 21:539–569, 2011.

T. L. Lai and H. Xing. Stochastic change-point arx-garch models and their applications to econometric time series. *Statistica Sinica*, 23:1573–1594, 2013.

T. L. Lai, H. Liu, and H. Xing. Autoregressive models with piecewise constant volatility and regression parameters. *Statistica Sinica*, 15:279–301, 2005.

T. L. Lai, M.-C. Shih, and S. P. Wong. A new approach to modeling covariate effects and individualization in population pharmacokinetics-pharmacodynamics. *Journal of Pharmacokinetics and Pharmacodynamics*, 33:49–74, 2006.

T. L. Lai, T. Liu, and H. Xing. A bayesian approach to sequential surveillance in exponential families. *Communications in Statistics, Theory and Methods*, 38:2958–2968, 2009.

C. G. Lamoureux and W. D. Lastrapes. Persistence in variance, structural change and the garch model. *Journal of Business and Economic Statistics*, 8:225–234, 1990.

D. Lando. *Credit Risk Modeling: Theory and Applications*. Princeton Unversity Press, New Jersey, USA, 2004.

H. M. Langohr and P. T. Langohr. *The Rating Agencies and their Credit Ratings: What They Are, How They Work and Why They Are Relevant*. John Wiley & Sons Ltd., Chichester, UK, 2010.

T. Lattimore and C. Szepesvári. *Bandit Algorithms*. Cambridge University Press, Cambridge, UK, 2020.

H. E. Leland. Corporate debt value, bond covenants, and optimal capital structure. *Journal of Finance*, 49:1213–1252, 1994.

F. Liang, C. Liu, and R. J. Carroll. *Advanced Markov Chain Monte Carlo Methods: Learning from Past Samples.* John Wiley & Sons, Ltd., Chichester, UK, 2010.

D. Y. Lin and Z. Ying. Semiparametric analysis of the additive risk model. *Biometrika*, 81:61–71, 1994.

D. Y. Lin and Z. Ying. Semiparametric analysis of general additive-multiplicative hazard models for counting processes. *The Annals of Statistics*, 23:1712–1734, 1995.

R. S. Liptser and A. N. Shiryaev. *Statistics of Random Processes I. General Theory.* Springer, New York, USA, 2001.

M. Loéve. Fonctions aléatoires du second ordre. *Revue Science*, 84:195–206, 1946.

F. A. Longstaff and E. S. Schwartz. Valuing american options by simulation: A simple least-squares approach. *The Review of Financial Studies*, 14:113–147, 2001.

G. Lorden. Procedures for reacting to a change in distribution. *The Annals of Mathematical Statistics*, 41:520–527, 1971.

D. B. Madan and E. Seneta. The variance gamma (v.g.) model for share market returns. *Journal of Business*, 63:511–524, 1990.

K. V. Mardia. Measures of multivariate skewness and kurtosis with applications. *Biometrika*, 57:519–530, 1970.

K. V. Mardia. Tests of univariate and multivariate normality. In P. R. Krishnaiah, editor, *Handbook of Statistics*, volume 1, pages 279–320. North Holland, 1980.

B. Mariano. Market power and reputational concerns in the ratings industry. *Journal of Monetary Economics*, 36:1616–1626, 2012.

D. C. Marinescu. *Cloud Computing: Theory and Practice.* Morgan Kaufmann, Burlington, Massachusetts, USA, 3rd edition, 2023.

A. W. Marshall and I. Olkin. A generalized bivariate exponential distribution. *Journal of Applied Probability*, 4:291–302, 1967a.

A. W. Marshall and I. Olkin. A multivariate exponential distribution. *Journal of the American Statistical Association*, 62:30–44, 1967b.

D. Martin. Early warning of bank failure: A logit regression approach. *Journal of Banking and Finance*, 1:249–276, 1977.

T. Martinussen and T. H. Scheike. A flexible additive multiplicative hazard model. *Biometrika*, 89:283–298, 2002.

T. Martinussen and T. H. Scheike. *Dynamic Regression Models for Survival Data*. Springer, New York, USA, 2006.

S. P. Mason and S. Bhattacharya. Risky debt, jump process, and safety covenants. *Journal of Financial Economics*, 9:281–307, 1981.

J. Mathis, J. McAndrews, and J. C. Rochet. Rating the raters: Are reputation concerns powerful enough to discipline rating agencies? *Journal of Monetary Economics*, 56:657–674, 2009.

J. McCarthy. Programs with common sense. In *Proceedings of the Teddington Conference on the Mechanization of Thought Processes*, pages 75–91, Teddington, England, 1958. National Physical Laboratory.

W. McCulloch and W. Pitts. A logical calculus of the ideas immanent in nervous activity. *Bulletin of Mathematical Biophysics*, 5:115–133, 1943.

A. J. McNeil, R. Frey, and P. Embrechts. *Quantitative Risk Management: Concepts, Techniques and Tools*. Princeton University Press, Princeton, New Jersey, USA, 2nd edition, 2015.

A. Melnikov. *Risk Analysis in Finance and Insurance*. Chapman & Hall/CRC, Boca Raton, Florida, USA, 2011.

R. C. Merton. Option pricing when underlying stock returns are discontinuous. *Journal of Financial Economics*, 3:125–144, 1976.

R. D. Merton. Theory of rational option pricing. *The Bell Journal of Economics and Management Science*, 4:141–183, 1974a.

R. D. Merton. On the pricing of corporate debt: The risk structure of interest rates. *Journal of Finance*, 29:449–470, 1974b.

N. Metropolis, A. W. Rosenbluth, M. N. Rosenbluth, A. H. Teller, and E. Teller. Equations of state calculations by fast computing machines. *Journal of Chemical Physics*, 21:1087–1092, 1953.

T. Mikosch. *Non-Life Insurance Mathematics: An Introduction with the Poisson Processes*. Springer, Princeton, New Jersey, USA, 2nd edition, 2009.

T. V. Mikosch and C. Starica. Non-stationarities in financial time series, the long-rangedependence and the igarch effects. *Review of Economics and Statistics*, 86:378–390, 2004.

G. V. Moustakides. Optimal stopping times for detecting changes in distributions. *The Annals of Statistics*, 14:1379–1387, 1986.

A. Narayanan, J. Bonneau, E. Felten, A. Miller, and S. Goldfeder. *Bitcoin and Cryptocurrency Technologies: A Comprehensive Introduction*. Princeton University Press, New Jersey, USA, 2016.

R. E. Neapolitan. *Probabilistic Reasoning in Expert Systems*. John Wiley & Sons, Ltd., New York, NY, USA, 1989.

R. E. Neapolitan and X. Jiang. *Artificial Intelligence: With an Introduction to Machine Learning*. Chapman & Hall/CRC, New York, NY, USA, 2nd edition, 2018.

R. B. Nelsen. Copulas and association. In G. Dall'Aglio, S. Kotz, and G. G. Salinetti, editors, *Advances in Probability Distributions with Given Marginals*, pages 51–74. Kluwer Academic Publishers, Dordrecht, 1991.

R. B. Nelsen. *An Introduction to Copulas*. Springer, New York, NY, USA, 2nd edition, 2006.

D. B. Nelson. Conditional heteroskedasticity in asset returns: A new approach. *Econometrica*, 59:347–370, 1991.

M. O'Hara. *Market Microstructure Theory*. Blackwell Inc., Cambridge, Massachusetts, USA, 1998.

C. Opp, M. Opp, and M. Harris. Rating agencies in the face of regulation. *Journal of Financial Economics*, 108:46–61, 2013.

E. S. Page. Continuous inspection schemes. *Biometrika*, 41:100–114, 1954.

S. Pantula. Modelling the persistence of conditional variance: A comment. *Econometric Reviews*, 5:71–73, 1986.

J. Park and I. W. Sangberg. Universal approximation using radial-basis-function networks. *Neural Computation*, 3:246–257, 1991.

J. Pearl. *Probabilistic Reasoning in Intelligent Systems*. Morgan Kaufmann, San Mateo, California, USA, 1988.

G. Peskir and A. N. Shiryaev. *Optimal Stopping and Free-Boundary Problems*. Birkhäuser Verlag, Basel, Switzerland, 2006.

J. Peters and S. Schaal. Natural actor-critic. *Neurocomputing*, 71:1180–1190, 2008.

V. V. Petrov. On the probabilities of large deviations for sums of independent random variables. *Theory of Probability and Its Applications*, 19:394–416, 1965.

John C. Platt. Sequential minimal optimization: A fast algorithm for training support vector machines. Technical report, Advances in Kernel Methods - Support Vector Learning, 1998.

M. Pollak. Optimal detection of a change in distribution. *The Annals of Statistics*, 18:1464–1469, 1985.

M. Pollak and D. Siegmund. Sequential detection of a change in a normal mean when the initial value is unknown. *The Annals of Statistics*, 10:287–298, 1991.

L. S. Pontryagin, V. G. Boltyanskii, R. V. Gamkrelidze, and E. F. Mishchenko. *The Mathematical Theory of Optimal Processes*. John Wiley & Sons, Inc., New York, USA, 1962.

P. E. Potter. *Stochastic Integration and Differential Equations*. Springer, New Jersey, USA, 2005.

P. Povel, R. Singh, and A. Winton. Booms, busts, and fraud. *The Review of Financial Studies*, 20:1219–1254, 2007.

W. B. Powell. *Approximate Dynamic Programming: Solving the Curses of Dimensionality*. John Wiley & Sons, Inc., New Jersey, USA, 2nd edition, 2011.

R. L. Prentice and S. Zhao. *The Statistical Analysis of Multivariate Failure Time Data: A Marginal Modeling Approach*. Chapman & Hall/CRC, Boca Raton, Florida, USA, 2019.

R. Rebonato. *Modern Pricing of Interest Rate Derivatives*. Princeton University Press, New Jersey, USA, 2002.

R. Rebonato. Interest-rate term-structure pricing models: A review. *Proceedings of the Royal Society of London, Series A*, 460:667–728, 2004.

G. E. Rejda and M. J. McNamara. *Principles of Risk Management and Insurance*. Pearson, New Jersey, USA, 14th edition, 2020.

A. C. Rencher and W. F. Christensen. *Methods of Multivariate Analysis*. John Wiley & Sons, Inc., Hoboken, 2012.

Y. Ritov. Decision theoretic optimality of the cusum procedure. *The Annals of Statistics*, 18:1464–1469, 1990.

H. Robbins. Asymptotically subminimax solutions of compound statistical decision problems. *Proceedings of the Second Berkeley Symposium on Mathematical Statistics and Probability*, 2:131–149, 1951.

H. Robbins. Some aspects of the sequential design of experiments. *Bulletin of the Americal Mathematical Society*, 58:527–535, 1952.

H. Robbins. An empirical bayes approach to statistics. *Proceedings of the Third Berkeley Symposium on Mathematical Statistics and Probability*, 1:157–163, 1956.

H. Robbins. Some thoughts on empirical bayes estimation. *The Annals of Statistics*, 11:713–723, 1983.

C. P. Robert and G. Casella. *Monte Carlo Statistical Methods*. Springer-Verlag, New York, USA, 2nd edition, 2004.

C. P. Robert and G. Casella. *Introducing Monte Carlo Methods with R*. Springer-Verlag, New York, USA, 2010.

R. T. Rockafellar and S. Uryasev. Optimization of conditional value-at-risk. *Journal of Risk*, 2:21–41, 2000.

R. T. Rockafellar and S. Uryasev. Conditional value-at-risk for general loss distributions. *Journal of Banking and Finance*, 26:1443–1471, 2002.

R. Roll and S. Ross. An empirical investigation of the arbitrage pricing theory. *Journal of Finance*, 5:1073–1103, 1980.

T. Roncalli. *Handbook of Financial Risk Management*. Financial Mathematics Series. CRC Press, 2020.

F. Rosenblatt. The perceptron: A probabilistic model for information storage and organization in the brain. *Psychological Review*, 65:386–408, 1958.

V. I. Rotar. *Actuarial Models: The Mathematics of Insurance*. Chapman & Hall/CRC, Boca Raton, Florida, USA, 2nd edition, 2014.

M. Rubinstein. Implied binomial trees. *Journal of Finance*, 49:771–818, 1994.

S. J. Russell and P. Norvig. *Artificial Intelligence: A Modern Approach*. Pearson, New York, USA, 4th edition, 2021.

K.-I. Sato. *Lévy Processes and Infinitely Divisible Distributions*. Cambridge University Press, Cambridge, UK, 1999.

B. Schölkopf, A. J. Smola, R. C. Williamson, and P. L. Bartlett. New support vector algorithms. *Neural Computation*, 12:1207–1245, 2000.

G. Schwarz. Estimating the dimension of a model. *The Annals of Statistics*, 6:461–464, 1978.

G. W. Schwert. Why does stock market volatility change over time? *Journal of Finance*, 44:1115–1153, 1989.

M. Sewak. *Deep Reinforcement Learning: Frontier of Artificial Intelligence*. Springer, New York, USA, 2019.

W. A. Shewhart. *Economic Control of Quality of Manufactured Products*. D. Van Nostrand Co., New York, USA, 1931.

A. N. Shiryaev. On optimum methods in quickest detection problems. *Theory of Probability and Its Application*, 8:22–46, 1963.

A. N. Shiryaev. *Optimal Stopping Rules*. Springer-Verlag, New York, USA, 1978.

S. E. Shreve. *Stochastic Calculus for Finance I: The Binomial Asset Pricing Model.* Springer, New York, USA, 2004a.

S. E. Shreve. *Stochastic Calculus for Finance II: Continuous-Time Models.* Springer, New York, USA, 2004b.

R. H. Shumway and D. S. Stoffer. *Time Series Analysis and Its Applications With R Examples.* Springer, New York, USA, 4th edition, 2017.

D. Siegmund. Importance sampling in the Monte Carlo study of sequential tests. *The Annals of Statistics*, 4:673–684, 1976.

D. Siegmund. *Sequential Analysis: Tests and Confidence Intervals.* Springer-Verlag, New York, USA, 1985.

D. Siegmund and E. S. Venkatraman. Using the generalized likelihood ratio statistic for sequential detection of a change-point. *The Annals of Statistics*, 23:255–271, 1995.

D. Silver, A. Huang, C. J. Maddison, A. Guez, L. Sifre, G. van den Driessche, J. Schrittwieser, I. Antonoglou, V. Panneershelvam, M. Lanctot, S. Dieleman, D. Grewe, J. Nham, N. Kalchbrenner, I. Sutskever, T. Lillicrap, M. Leach, K. Kavukcuoglu, T. Graepel, and D. Hassabis. Mastering the game of go with deep neural networks and tree search. *Nature*, 529:484–489, 2016.

D. Silver, J. Schrittwieser, K. Simonyan, I. Antonoglou, A. Huang, A. Guez, T. Hubert, L. Baker, M. Lai, A. Bolton, Y. Chen, L. Lillicrap, F. Hui, L. Sifre, G. van den Driessche, T. Graepel, and D. Hassibis. Mastering the game of go without human knowledge. *Nature*, 550:354–359, 2017.

S. P. Singh and R. S. Sutton. Reinforcement learning with replacing eligibility traces. *Machine Learning*, 22:123–158, 1996.

A. Sklar. Fonctions de réepartition à n dimensions et leurs marges. *Publications de l'Institut de Statistique de l'Universit é de Paris*, 8:229–231, 1959.

V. Skreta and L. Veldkamp. Rating shopping and asset complexity: A theory of rating inflation. *Journal of Monetary Economics*, 56:678–695, 2009.

A. Slivkins. Introduction to multi-armed bandits. *Foundations and Trends in Machine Learning*, 12:1–286, 2019.

C. Stanfill and D. Waltz. Toward memory-based reasoning. *Communications of the ACM*, 29:1213–1228, 1986.

C. Stein. Inadmissibility of the usual estimator for the mean of a multivariate normal distribution. *Proceedings of the Third Berkeley Symposium on Mathematical Statistics and Probability*, 1:197–206, 1956.

D. W. Stroock and S. R. S. Varadhan. *Multidimensional Diffusion Processes.* Springer-Verlag, New York, USA, 2005.

W. W. Stroup. *Generalized Linear Mixed Models: Modern Concepts, Methods and Applications.* Chapman & Hall/CRC, Boca Raton, Florida, USA, 2012.

B. Sundt and W. S. Jewell. Further results on recursive evaluation of compound distributions. *ASTIN Bulletin*, 12:27–39, 1981.

R. S. Sutton. Learning to predict by the methods of temporal differences. *Machine Learning*, 3:9–44, 1988.

R. S. Sutton and A. G. Barto. *Reinforcement Learning: An Introduction.* The MIT Press, Massachusetts, USA, 2nd edition, 2018.

R. S. Sutton, D. McAllester, S. Singh, and Y. Mansour. Policy gradient methods for reinforcement learning with function approximation. *Advances in Neural Information Processing Systems*, 12:1057–1063, 1999.

P. Tankov and R. Cont. *Financial Modelling with Jump Processes.* Chapman and Hall/CRC, New York, USA, 2003.

A. Tartakovsky, I. Nikiforov, and M. Basseville. *Sequential Analysis: Hypothesis Testing and Changepoint Detection.* Chapman & Hall/CRC, New York, USA, 2014.

S. J. Taylor. *Modeling Financial Time Series.* John Wiley and Sons, Chichester, UK, 1986.

W. R. Thompson. On the likelihood that one unknown probability exceeds another in view of the evidence of two samples. *Biometrika*, 25:275–294, 1933.

R. Tibshirani. Regression shrinkage and selection via the lasso. *Journal of the Royal Statistical Society: Series B*, 58:267–288, 1996.

A. N. Tikhonov and V. Y. Arsenin. *Solutions of Ill-posed Problems.* Winston, Washington DC, USA, 1977.

N. H. Timm. *Applied Multivariate Analysis.* Springer, New York, USA, 2002.

R. S. Tsay. *Analysis of Financial Time Series.* Wiley, New Jersey, USA, 3rd edition, 2010.

R. S. Tsay. *Multivariate Time Series Analysis: With R and Financial Applications.* Wiley, New Jersey, USA, 2013.

R. S. Tsay and R. Chen. *Nonlinear Time Series Analysis.* John Wiley & Sons, Inc., New Jersey, USA, 2019.

A. Turing. Computing machinery and intelligence. *Mind*, 59:433–460, 1950.

R. van der Weide. Go-garch: A multivariate generalized orthogonal garch model. *Journal of Applied Econometrics*, 17:549–564, 2002.

O. Vasicek. An equilibrium characterization of the term structure. *Journal of Financial Economics*, 5:177–188, 1977.

E. J. Vaughan and T. M. Vaughan. *Fundamentals of Risk and Insurance*. Wiley, New York, USA, 11th edition, 2014.

W. Vogel. An asymptotic minimax theorem for the two armed bandit problems. *The Annals of Mathematical Statistics*, 31:444–451, 1960.

A. Wald. Sequential tests of statistical hypotheses. *The Annals of Mathematical Statistics*, 16:117–186, 1945.

A. Wald. *Sequential Analysis*. John Wiley & Sons, Inc., New York, USA, 1947.

A. Wienke. *Frailty Models in Survival Analysis*. Chapman & Hall/CRC, Boca Raton, Florida, USA, 2010.

M. Woodroofe. *Nonlinear Renewal Theory in Sequential Analysis*. Society for Industrial and Applied Mathematics, Pennsylvania, USA, 1982.

E. P. Xing, A. Y. Ng, M. Jordan, and S. Russell. Distance metric learning with application to clustering with side-information. In *Advances in Neural Information Processing Systems*, pages 505–512, 2003.

H. Xing and Y. Chen. Dependence of structural breaks in rating transition dynamics on economic and market variations. *Review of Economics and Finance*, 11:1–18, 2018.

H. Xing, Y. Mo, W. Liao, and M. Zhang. Genomewide localization of protein-dna binding and histone modification by bcp with chip-seq data. *PLoS Computational Biology*, 8(7):e1002613, 2012a. doi: 10.1371/journal.pcbi.1002613.

H. Xing, N. Sun, and Y. Chen. Credit rating dynamics in the presence of unknown structural breaks. *Journal of Banking and Finance*, 36:78–89, 2012b.

H. Xing, K. Wang, Z. Li, and Y. Chen. Statistical surveillance of structural breaks in credit rating dynamics. *Entropy*, 22:1072, 2020. doi: 10.3390/e22101072.

M. Xiong. *Artificial Intelligence and Causal Inference*. Chapman & Hall/CRC, Boca Raton, Florida, USA, 2022.

S. Yao, H. Zou, and H. Xing. l_1 regularization for high-dimensional multivariate garch models. *Risks*, 12, 2024. doi: 10.3390/risks12020034.

M. Yuan and Y. Lin. Model selection and estimation in regression with grouped variables. *Journal of the Royal Statistical Society, Series B*, 68: 49–67, 2007.

S. Zacks. Detection and change-point problems. In B. K. Ghosh and P. K. San, editors, *Handbook of Sequential Analysis*, pages 531–562. Dekker, New York, 1991.

S. Zacks and Z. Barzily. Bayes procedures for detecting a shift in the probability of success in a series of bernoulli trials. *Journal of Statistical Planning and Inference*, 5:107–119, 1981.

J.-M. Zakoian. Threshold heteroskedastic models. *Journal of Economic Dynamics and Control*, 18:931–955, 1994.

Q. Zhao. *Multi-Armed Bandits: Theory and Applications to Online Learning in Networks*. Morgan & Claypool Publishers, 2019.

H. Zou and T. Hastie. Regularization and variable selection via the elastic net. *Journal of the Royal Statistical Society, Series B*, 67:301–320, 2005.

Index